A.F. Thilmann D.J. Burke
W.Z. Rymer (Eds.)

Spasticity

Mechanisms and Management

With 125 Figures and 12 Tables

Springer-Verlag
Berlin Heidelberg New York
London Paris Tokyo
Hong Kong Barcelona
Budapest

Priv.-Doz. Dr. A.F. THILMANN
Neurologische Klinik mit klinischer Neurophysiologie
Alfried-Krupp-Krankenhaus
Alfried-Krupp-Straße 21
D-45131 Essen, Germany

Professor DAVID JAMES BURKE, M.D., D.Sc.
Department of Neurology
The Prince Henry and Prince of Wales Hospitals and the
Prince of Wales Medical Research Institute, Randwick
Sydney, NSW 2031, Australia

Professor WILLIAM ZEV RYMER, M.D., Ph.D.
Rehabilitation Institute of Chicago
345 East Superior Street
Chicago, Illinois 60611-4496, USA

ISBN-13: 978-3-642-78369-2 e-ISBN-13: 978-3-642-78367-8
DOI: 10.1007/978-3-642-78367-8

Library of Congress Cataloging-in-Publication Data. Spasticity: mechanisms and management/A.F. Thilmann, D.J. Burke, W.Z. Rymer, editors. p. cm. "The editors organized a workshop on spasticity, held in Essen Germany, in November, 1992"—Pref. Includes bibliographical references and index. ISBN 3-540-56981-2:—ISBN 0-387-56981-2 1. Spasticity—Congresses. I. Thilmann, A.F. (Alfred F.), 1950– .II. Burke, D.J. (David James), 1944– . III. Rymer, W.Z. (William Zev), 1939– . RC935.S64S63 1993 616.8′3—dc20 93-34850

Softcover reprint of the hardcover 1st edition 1993

Typesetting by Best-set Typesetter Ltd., Hong Kong
25/3130/SPS-5 4 3 2 1 0 – Printed on acid-free paper

Preface

Over the past 10–20 years there has been an explosion of work on the disturbance of motor control that is manifested as "spasticity." This work has been based on new insights from animal experiments into the basis of normal motor control and the development of experimental techniques that could be used in patients with neurological disturbances. For this reason, on the initiative of Alfred Thilmann, the editors of this volume organized a workshop on spasticity, held in Essen, Germany, in November, 1992. The intention of the meeting was to bring together leading scientists in an atmosphere that would generate discussion and hopefully consensus and ideas for future work. The attendance was restricted and deliberately selective, with a mix of basic scientists, clinical scientists and practicing clinicians. The product of their presentations and deliberations is this volume. The book is organized in the same sequence as the workshop, so that the discussion sessions could be summarized by the appropriate chairman.

A notable feature of the workshop, reflected in the chapters in this book, is the emphasis on the movement disorder accompanying spasticity. For too long, too many clinicians have treated spasticity as if the patient's reflex abnormality were the cardinal mechanism underlying the motor disturbance. This workshop and this volume place the emphasis on the true disability, an inability to generate or to control appropriate movements, which may be causally independent of the abnormalities of muscle tone. This deficit is clearly multifactorial, as will become clear to the reader as he or she peruses the chapters and discussions. Ultimately, a clinically acceptable therapy will modify abnormal movement *including* disturbances of muscle tone, and help restore a better control of residual function.

The workshop was generously supported by Sandoz AG/ Wander Pharma, Nürnberg, Germany, who also undertook the organization and underwrote the costs of the production of this monograph. The editors would like to express their thanks, first, to the individual contributors, without each of whom a volume of lesser scientific value would have been produced, and secondly, to Sandoz AG/Wander Pharma for their wholehearted sponsorship.

This volume is dedicated to Peter Rack, an outstanding neuroscientist whose studies into physiological mechanisms, both normal and disturbed, have provided inspiration for a generation of young enquiring minds and a stimulus for a number of the chapters in this monograph.

Essen, Germany — Alfred F. Thilmann
Sydney, Australia — David J. Burke
Chicago, IL, USA — William Zev Rymer

Contents

Part II Disturbance of Voluntary Movement

Part III Spinal Pathophysiology: Animal Models

List of Contributors

Addresses are given at the beginning of the respective contribution.

Spasticity: History, Definitions, and Usage of the Term

A.F. Thilmann

History of the Term "Spasticity"

The term "spasticity" is derivated from the Greek *spasticos* (δπαδτικσς) and *spaon* (δπασν, to draw out, stretch). While the term spasm (Greek *spasmos*, δπαδμσς) was already in use by Hippocrates to describe an epileptic fit (Skinner 1970), the term spasticity has been used only in the more recent literature. It was first used in English by Good in 1829. He describes "spastic wryneck: from excess of muscular action on the contracted side." Thus, by this time usage was similar to that of today.

Although spasticity was recognized and treated by orthopedic surgeons (Stromeyer 1838, cited according to Dralle et al. 1988b), who performed subcutaneous tenotomy of contracted extremities, the first description of spasticity was given by Little in 1843. William John Little himself suffered from spasticity and after being treated by Stromeyer he dedicated his dissertation "Symbolae ad talipedum varum cognoscendum" to the surgeon (Little 1837, cited according to Dralle et al. 1988). Six years later, Little gave the first description – still valid today – of a cerebral palsy with spastic diplegia, now called Little's disease (Little 1843; see Fig. 1).

In 1868 Charcot described the symptoms of multiple sclerosis and introduced the term spastic paraplexie (Charcot 1868). Earlier descriptions of spasticity did not use this term, and descriptions of strokes and spinal cord diseases (Canstatt 1843) referred only to spasms or rigidity. Sigmund Freud, for example, who started his career as a neurologist, did not use the term spasticity in his descriptions on infantile cerebral palsy (Freud 1897). Even Sherrington (1898) used exclusively the term "rigidity" to describe the increased tone in decerebrate animals, although this preparation was later temporarily used as a model of spasticity. The differentiation between spasticity and rigidity remained difficult for a long time, and Little employed the term "spastic rigidity of muscles" in his description of the cerebral palsy. In the papers of Charcot (1884–1887, cited according to Lakke 1989) the

Alfried Krupp von Bohlen und Halbach Krankenhaus, Neurologische Klinik mit Klinischer Neurophysiologie, Alfried Krupp-Straße 21, D-45131 Essen, Deutschland

Thilmann et al. (Eds.)
Spasticity: Mechanisms and Management

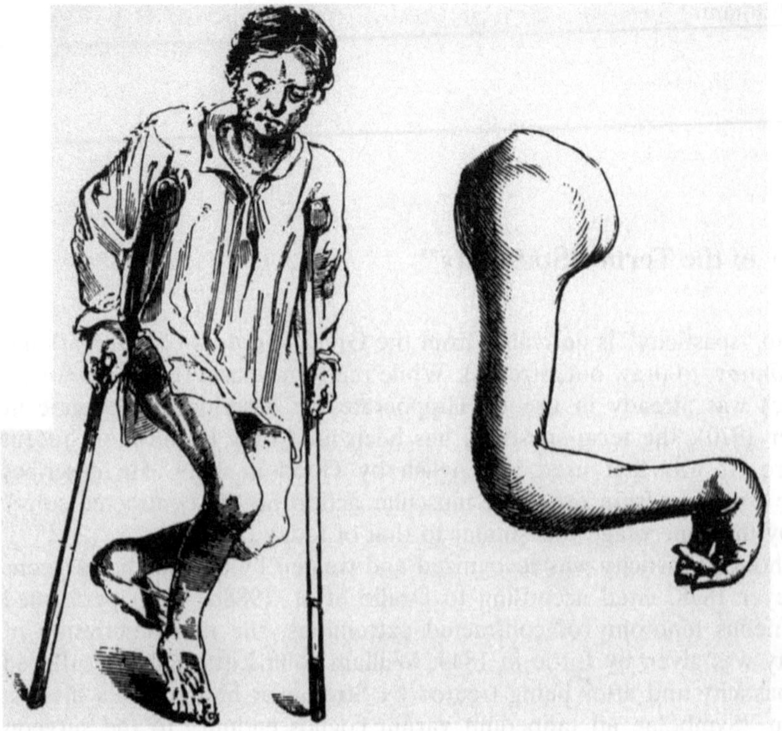

Fig. 1. Reproduction of the original drawings from Little (1843)

terms spasticity and rigidity appeared without systematic differences, but a distinction between "elastic" and "plastic" rigidity was made. Today "elastic" rigidity is usually termed spasticity while "plastic" rigidity corresponds to the modern concept of rigidity. More detailed descriptions of the history of spasticity and its therapy are given in the reviews of Dralle (1988) and Denny-Brown (1980).

Use of the Term "Spasticity"

Nowadays, the expression "spasticity" is found so often in medical literature and is such an elementary neurological term that no one really expects a definition. This is the reason for a very confusing usage of this term. In nonmedical literature it is very often treated synonymously with Little's disease. In the *Concise Oxford English Dictionary* (1980), for example, a spastic is described as "a person suffering from cerebral palsy with tonic spasms of muscles."

In the neurological literature there are varieties of meaning, and Landau (1974) has identified the following six meanings of spasticity:

1. Proprioceptive reflex release: quiescent unstimulated muscle with increased proprioceptive reflexes, including, in various degrees, increased phasic tendon jerk reflexes, tonic stretch reflexes, and the tonic clasp-knife reaction.
2. Generalized reflex release: reflex release including not only proprioceptive but also and especially polysynaptic flexion reflexes, e.g., the flexor spasms of chronic paraplegia.
3. Upper motor neuron syndrome: the entire upper motor neuron complex syndrome, including motor performance disability as well as reflex release, e.g., spastic paraparesis.
4. Dystonic-rigid state: a spectrum of ill-defined dystonic and rigid states of many origins and pathologic features, all characterized by some degree of involuntary continuous muscle contractions, e.g., athetoid cerebral palsy.
5. Mixed: combinations of the above definitions, especially 3 and 4, usually associated with hemispheral or brainstem lesions.
6. Undefined: no basis for determining narrower contextual meaning.

In recent publications, the term spasticity is used for two different sets of symptoms, creating some degree of conflict and confusion. First, spasticity refers to a velocity-dependent increase in muscle tone during externally imposed passive movements, and, second, spasticity is used as a synonym for the upper motoneuron syndrome.

Not all papers describe their exact usage or give enough details of patients to allow such an allocation.

Epidemiology

Worldwide, no real data for morbidity, incidence, or numbers of spastic patients are available. The reason is that spasticity is not a disease but a syndrome, appearing in association with various diseases. Spasticity may appear following almost all lesions of the central nervous system, with causes such as trauma, multiple sclerosis, ischemia, hemorrhage, or degeneration.

The costs for treatment of spasticity are very high as most patients suffer from their disorder for a very long time and receive expensive treatment, for example, physiotherapy over many years or even decades. If one considers that the long-term costs for treatment of strokes in the United States are estimated to be 7.5–11.2 billion dollars (Feigenson 1979), it is likely that spasticity is one of the most expensive disorders. Nevertheless, no exact figures on the costs are available, and until today the etiology remains unclear.

Definition of Spasticity

The most commonly used definition of spasticity is given by Lance (1980): "*Spasticity is a motor disorder characterized by velocity-dependent increase in tonic stretch reflexes ("muscle tone") with exaggerated tendon jerks, resulting from hyperexcitability of the stretch reflex as one component of the upper motor neuron syndrome.*" While this definition is far from perfect, it remains the best basis for agreement. At the present workshop substantial discussion was devoted to formulating a better definition, but without success. Either the description was too simplistic, so that it was unhelpful, or, alternatively, the participants could not agree on the detailed features of the quantitative behavior of the stretch reflexes in spastic hypertonia. Even the editors of this volume disagreed on the quantitative characteristics of stretch reflex behavior, especially with regard to the relative impact of changes in stretch reflex threshold versus changes in stretch reflex gain (or stiffness).

Nevertheless, this discussion of definitions gave rise to a description of the upper motoneuron system which many participants found acceptable. Specifically, it was proposed to rename the upper motoneuron syndrome the "corticofugal syndrome" to reflect the fact that many components of the descending pathways are involved, not simply corticospinal projections. The alternative definition proposed, of the corticofugal syndrome is "*a disorder of motor control due in part to reduction in cortical influences on the spinal cord, characterized by weakness, impaired coordination, spasticity, increased tendon jerks, and release of cutaneomuscular reflexes such as the Babinski response.*"

What Is Confusing About Spasticity?

As shown above, spasticity was introduced into the literature by clinicians, and today neurologists worldwide have few difficulties in agreeing on the diagnosis of spasticity in most of their patients. The reason is that they use the term in an empirical way and only for the description of a phenomen upon which they agree, even though many variations are seen among patients. The problems arise when quantitative and theoretical considerations are added. As soon as rigorous anatomical, physiological, biochemical, mechanical, and pharmacological features are considered, each research group looks at only one part of the spastic syndrome, because today it is impossible to explain all symptoms of the patients with one theoretical model. This results in a confusing situation in that various researchers may draw conclusions from their results which cannot be extended to other patients in whom the spasticity results from different lesions of the central nervous system. The reason for the confusion may be that in part our simplified view of the spastic syndrome is limiting and incorrect, and it may be that spasticity is not a single entity but in fact a variety of disorders, which express their disturbances over a common peripheral effector, namely the spinal motoneuron.

Certainly the differentiation into "plus" and "minus" syndromes as introduced by Jackson (1958) should be discontinued, as a "minus" symptom may be due to the fact that a "plus" symptom is appearing in the antagonistic muscle.

The statement of Denny-Brown (1980) that "Spasticity is a complex disability not identifiable with any single reflex or synaptic transmitter" is still valid. Nevertheless, the editors of this volume considered it worthwhile to bring interested research groups in the field of spasticity together to discuss the problems. This book tries to give an overview, however incomplete, of the basic mechanisms, new ideas, new experiments, and new treatment methods for spasticity. It is still far from solving the problem of spasticity, but may help us to see the different trials and errors on the way to a more complete knowledge of this important motor disorder.

References

Canstatt C (1843) Handbuch der medicinischen Klinik. Ferdinand Enke, Erlangen
Charcot JM (1868) Histologie de la sclérose en plaques. Gaz Hop (Paris) 41:554–555
Concise Oxford Dictionary of Current English, 6 edn (1980) Sykes JB (ed) Clarendon, Oxford
Denny-Brown D (1980) Historical aspects of the relation of spasticity to movement. In: Feldmann RG, Young RR, Koella WP (eds) Spasticity: disordered motor control. Year Book Medical Publishers, Chicago, pp 1–16
Dralle D, Müller H, Zierski J (1988) A short historical review of spasticity and its therapy. In: Müller H, Zierski J, Penn RD (eds) Local-spinal therapy of spasticity. Springer, Berlin Heidelberg New York
Feigenson JS (1979) Stroke rehabilitation: effectiveness and cost: some practical considerations (Editorial). Stroke 10:1–4
Freud S (1887) Die infantile Cerebrallähmung. In: Nothnagel A (ed) Spezielle Pathologie und Therapie, vol 92. Hölder, Vienna
Good JM (1829) The study of medicine, vol 4. Baldwin, Cradock and Joy, London
Jackson JH (1958) On certain relations of the cerebrum and cerebellum (on rigidity of hemiplegia and on paralysis agitans). In: Taylor S (ed) Selected writings of JH Jackson, vol 2. Basic, New York, pp 452–458 (Original 1877)
Lakke JPWF (1989) The clinical spectrum of spastic syndromes. In: Benecke R, Emre M, Davidoff RA (eds) The origin and treatment of spasticity. Parthenon, Carnforth
Lance JW (1980) Symposium synopsis: In: Feldmann RG, Young RR, Koella WP (eds) Spasticity: disordered motor control. Year Book Medical Publishers, Chicago
Landau WM (1974) Spasticity: the fable of a neurological demon and the emperor's new therapy. Arch Neurol 31:217–219
Little WJ (1843) Course of lectures on the deformities of the human frame. Lecture IX. Lancet I:350–354
Skinner HA (1970) The origin of medical terms. Hafner, New York

Part I
Descending Control of Movement

Organisation of Cortico-motoneuronal Projections from the Primary Motor Cortex: Evidence for Task-Related Function in Monkey and in Man

A. Turton, C. Fraser, D. Flament, W. Werner, K.M.B. Bennett, and R.N. Lemon

Organisation of Corticospinal Projections from the Primary Motor Cortex

In primates the fibres of the corticospinal tract are derived from extensive regions of the cerebral cortex including the motor, premotor, supplementary motor and cingulate areas in the frontal lobe and also from the postcentral gyrus of the parietal lobe. Dum and Strick (1991) have recently shown that in the macaque monkey only about half of the projections derived from the frontal lobe originate from the primary motor cortex. The corticospinal tract arises from large pyramidal neurones in the deepest part of layer V (Catsman-Berrevoets and Kuypers 1976; Jones and Wise 1977). The principal targets of corticospinal projections from the primary motor cortex appear to be interneurones and propriospinal neurones located in laminae VII and VIII (contralaterally) and lamina VIII (bilaterally; Schoen 1964; see Armand 1982). In addition, there are the direct cortico-motoneuronal (CM) projections. In the macaque monkey most of these direct CM projections influence motoneurones supplying muscles concerned with movements of the hand and fingers, but in man there are also strong projections to the more medial motoneuronal cell groups concerned with axial and truncal movements. Convincing evidence for the existence of these different projections has been obtained using magnetic stimulation in man (see Rothwell et al. 1991).

Although the corticospinal tract is derived from a large number of different cortical territories, the CM system is, as far as we know, derived almost exclusively from the caudal part of the primary motor cortex. The suggestion that this system might be in some way concerned with skilled hand movements was first made by Bernhard et al. (1953), who demonstrated a monosynaptic connection between the motor cortex and spinal motoneurones. A large amount of evidence has now been acquired implicating the CM system in the capacity to perform relatively independent finger movements (see Lemon 1993 for a review).

Anatomy Dept., Cambridge University, Cambridge CB2 3DY, England, UK

Thilmann et al. (Eds.)
Spasticity: Mechanisms and Management

In all primates studied it has been shown that the pyramidal tract comprises a very wide range of axon diameters; in the macaque monkey these range from below 1 μm, with conduction velocities of a few metres per second, to around 12 μm, conducting at around 60–70 $m s^{-1}$. Since both the larger, faster axons and the direct CM connections are, in evolutionary terms, recent developments, it might be argued that only the faster axons make such connections. However, by using the spike-triggered averaging technique to look for facilitation of electromyographic (EMG) activity from motor cortex neurones with slowly conducting axons, it has been demonstrated that a proportion of the slower fibres do make functional CM connections (Lemon et al. 1991, 1993).

The Pattern of Intraspinal Branching of Cortico-motoneuronal Axons

The axons of the corticospinal tract arborise extensively within the grey matter, and this branching has been elegantly demonstrated by intra-axonal labelling of single corticospinal axons. Shinoda et al. (1981) first showed that, in the monkey, the terminals of a single axon could be distributed among a number of different motor nuclei. Lawrence et al. (1985) also showed that a characteristic feature of CM terminal distribution is its elongation in the rostro-caudal plane, along the length of the spinal cord, with terminals being mostly restricted to lamina IX. This pattern is in keeping with CM innervation of the long motoneuronal columns running up and down the spinal cord, and Lawrence et al. provided light microscopic evidence for CM terminals making synaptic contacts on identified motoneurones. As far as is known, all direct CM connections are excitatory in nature; corticospinal inhibition of motoneuronal activity appears to be mediated by disynaptic connections (Jankowska et al. 1976).

Spike-triggered averaging studies have made it possible to look for the facilitatory influence of single motor cortical neurons upon the EMG of contralateral hand and forearm muscles (Fetz and Cheney 1980; Lemon et al. 1986; Cheney et al. 1991; Porter and Lemon 1993, pp. 155–186). An important conclusion from these studies is that single CM cells usually facilitate EMG activity of several muscles and only rarely facilitate a single muscle. This is probably a reflection of the intraspinal branching pattern described above. The term "muscle field" (Fetz and Cheney 1980) applies to the set of muscles whose motoneurones are facilitated by a CM cell. Lemon et al. (1991) summarized the results from 80 identified CM cells recorded from the motor cortex of macaque monkeys during performance of a precision grip between thumb and index finger. The muscle fields of these CM cells were restricted; the majority facilitated just over a quarter of the muscles sampled (typically 8–10 hand and forearm muscles). Most of these effects were found in the intrinsic hand muscles. Of the 80 CM cells, 19 facilitated only one muscle in the sample. Six of these cases involved an intrinsic thumb muscle; this may reflect the focused nature of cortical control for this digit.

Thus the fundamental principle underlying the organisation of the CM output appears to be one of multiple muscle control. Skilled movements of the hand are characterised by a fractionated pattern of activity, and the focused excitatory influence of the CM system would appear to be essential in producing this type of fractionated activity amongst the many different muscles acting on the hand and fingers.

Buys et al. (1986) analysed the combinations of muscles facilitated by different CM cells. Some combinations reflected the anatomical relationship between the target muscles: for example, some CM cells facilitated both the extensor digitorum communis (EDC) and the first dorsal interosseous (1DI). These muscles are both involved in extension of the interphalangeal joints of the index finger. Other CM cells had muscle fields which reflected task-related synergies, such as the facilitation of both the 1DI and the thumb muscle, adductor pollicis (AdP). These muscles are two of the "prime movers" within the intrinsic group for the production of precision grip. This analysis revealed a rather specific pattern of facilitation exerted by each CM cell.

Because the amplitude of post-spike facilitation in the different muscles within a CM cell's muscle field shows considerable variation, it is possible that the CM input influences the balance of fractionated activity across the muscles during independent finger movements. During the movement phase of the task, most CM cells are particulary active when one of their target muscles is active independently of the other, and this correlation appears to be reflected in the relative strength of post-spike facilitation exerted upon the different muscles. It has also been demonstrated that the strength of this facilitation varies according to the pattern and level of EMG activity. Thus the recruitment of a particular CM cell, combined with the increased influence of its impulses, enhances the fractionation of intrinsic hand muscle activity during movement (Bennett and Lemon 1991; Bennett 1992).

Representation of Cortico-motoneuronal Outputs Within the Motor Cortex

It has been known since the classic studies of Sherrington that a large portion of the primary motor cortex is devoted to the hand, and this is clearly demonstrated in the famous homunculus of Penfield. Recent studies of the monkey motor cortex have shown that the detailed output map of the motor cortex does not take the form of a somatotopic mapping of individual muscles or of specific movements but rather consists of a complex of overlapping output zones, each controlling a different part of the limb (Sato and

Tanji 1989; Huntley and Jones 1991; Donoghue et al. 1992; see Lemon 1988, 1990, for reviews). A particular muscle or movement is not represented once but many times, and the representations are discontinuous. This multiple representation is particularly marked for the muscles moving the digits with the total area occupied by the representation of a single hand muscle being as large as several square millimetres (Andersen et al. 1975; Lemon 1990).

But why is it necessary for a single hand muscle to be represented repeatedly within the primary motor cortex? Jackson (1932) thought of it in terms of the many different functions to be controlled: "The small muscles of the hand will be represented by much more grey matter in the highest centres than will be the large muscles of the upper arm, because they serve in more numerous different movements" (p. 262). Given the specific pattern of output connections of single CM cells described above, it is possible to think of the "more numerous different movements" as being encoded in the muscle field of a particular CM cell. Landgren et al. (1962) used the term "colony" to define all those CM cells whose axons converge onto a single motoneurone. The extensive colony of CM cells that might project to a 1DI motoneurone is shown in Fig. 1A. Figure 1B indicates some of the connections made at the spinal level by these CM cells. This suggests that the colony of CM neurones is in fact made up of different subpopulations, which are defined by the connections made with motoneurones of other muscles, some of which may act as functional synergists of 1DI. We know very little yet about the cortical distribution of these different subpopulations, but it is possible that this distribution may influence the responses evoked in different muscles when the cortex is stimulated non-invasively (see below).

Evidence for Task-Related Activity in the Monkey Motor Cortex

Although CM cells make direct connections with the motoneurones of their target muscles, there is considerable plasticity in their functional relationship: a CM cell and its target muscle are not always recruited together. Cheney and Fetz (1980) demonstrated that some CM cells recruited during a controlled wrist movement fall silent during a ballistic movement. They saw similar differences when monkeys performed a power grip (Fetz and Cheney 1987). Muir and Lemon (1983) found that CM neurones facilitating hand muscles are particularly active during precision grip, while they showed much less discharge during a power grip task. Although the target muscles of the CM neurones were active during both tasks, the precision grip was associated with a fractionated pattern of muscle activity, in contrast to the cocontracted pattern of power grip activity.

During the "movement period" of the precision grip task, most CM cells are very active, and exhibit a strong correlation with activity in at least

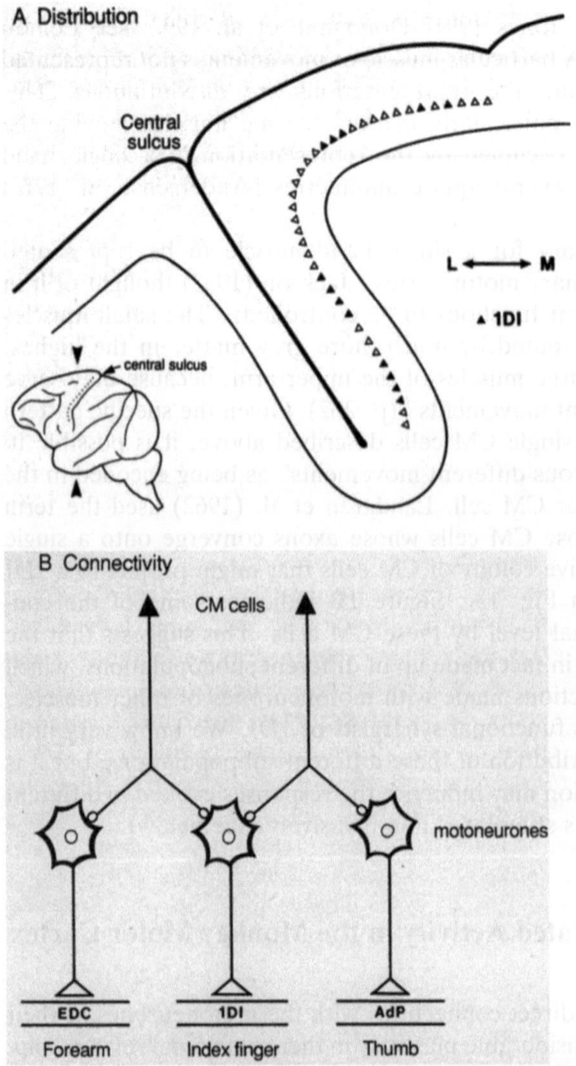

Fig. 1A,B. Schematic diagram illustrating the organisation that might exist for the colonies of CM cells projecting to the motoneurone pool of the 1DI. **A** CM cells in lamina V in the primary motor cortex. Discontinuous, multiple representation of CM cells projecting to 1DI (*filled triangles*). CM cells projecting to the hand are particularly numerous deep within the anterior bank of the central sulcus. **B** Possible organisation of output from two CM cells projecting to 1DI (*filled triangles*). One cell has a muscle field including a forearm muscle, the digit extensor EDC; the other projects to the thumb adductor AdP. These cells may be recruited differently according to the combination of muscles required for a given task one of their target muscles. However, during the "hold period" of the task, when the monkey exerts a steady gripping force, while some CM cells show activity that is highly correlated with that of their target muscles, the correlation is weak or absent for others (Bennett and Lemon 1991; Maier et al. 1993). CM cell activity during the movement is usually much greater than during the hold period. This difference in activity during the different phases of one task may underlie the striking responses in hand muscles caused by stimulating the human CM system during the early stages of a precision lifting, compared to much smaller effects observed during the steady hold period (Johansson et al. 1993).

Humphrey and Reed (1983) have shown that for the wrist muscle, extensor carpi radialis, some parts of its cortical representation are concerned with the action of this muscle as a "prime mover", while others are concerned with its secondary role as a stabiliser of the wrist during movement at other joints. Thus different cortical outputs to the same muscle may be involved when a muscle is used to produce a movement (reciprocal pattern of activity in agonists and antagonists) than when it is used to stiffen the limb (cocontraction pattern).

All of this evidence supports the notion that subpopulations of the CM cell group projecting to a single muscle are recruited preferentially according to the task in which the muscle is employed. It may be hypothesised that the particular movement to which a CM cell contributes might depend upon the other muscles in that cell's muscle field; this proposal is set out diagrammatically in Fig. 1B. This shows a CM cell, belonging to the colony of 1DI, branching to innervate motoneurones of 1DI and those of EDC. This CM cell could contribute to the synergistic action of EDC and 1DI in producing extension of the index finger at the interphalangeal joints, when the finger is flexed at the metacarpophalangeal joint. Another member of the 1DI colony is shown to innervate motoneurones supplying the thumb adductor (AdP). This type of CM cell would be active preferentially in tasks requiring synergy of these muscles, such as occurs during precision grip between the tips of the index finger and thumb.

Since many CM cells show some degree of activity during a variety of movements, it is probable that there is considerable overlap between the subpopulations of CM cells within a given colony, and that none is recruited in an exclusive fashion. Bennett, Flament and Lemon (unpublished observations) examined the activity of CM cells during two different types of independent finger movement task: a precision grip, in which the monkey used its index finger and thumb to move two levers towards each other, and a second task, in which the monkey rotated a lever in an arc at approximately 90° to the index finger movement. EMG recordings confirmed that a similar group of muscles was used in each task, but in a different pattern. Although some of the CM cells showed clear differences in activity between the two tasks, none was silent or unmodulated for one task while being active for the other.

Task-Related Behaviour of Hand Muscle Responses to Magnetic Stimulation in Man

We now move on to consider whether that which we have learnt from animal models of motor cortex control can be used to understand how the human hand is moved, and how its control is disturbed by brain damage. Our knowledge of corticospinal function in man has undergone a rapid expansion since the introduction of non-invasive methods of cortical stimulation (Merton and Morton 1980; Barker et al. 1985). There is a substantial body of evidence to suggest that magnetic stimulation of the cortex in man can excite CM cells and produce powerful, short-latency responses in limb muscles (see Rothwell et al. 1991). The strongest effects are seen in the intrinsic hand muscles (Hess et al. 1987; Palmer and Ashby 1992a).

The susceptibility of a CM neurone to excitation by transcranial magnetic stimulation (TMS) is determined by a number of different factors, including its size and its orientation and position with respect to the currents induced by TMS. CM cells can probably be activated both directly (at the initial segment of the neurone) and indirectly, or trans-synaptically (Edgley et al. 1990; Berardelli et al. 1990; Burke et al., this volume). Both types of activation depend upon the excitability of the neurone, that is, the amount of excitatory synaptic drive to the neurone prevailing at the moment when TMS is delivered. Changes in the excitability of, say, the CM colony projecting to a single motoneurone could influence the number of cells in the colony discharged by TMS, and this would change the size and composition of the descending volley impinging on the target motoneurone. Several investigations have demonstrated variation in the amplitude of EMG responses to TMS that could have resulted from changes in the level of cortical excitability. In some studies this level was increased at the time of TMS by delivering a peripheral input (muscle stretch) known to produce short-latency excitation of the motor cortex (Day et al. 1991; Palmer and Ashby 1992b). Datta et al. (1989) used a different approach. They argued that performance of some hand tasks was more likely to involve the motor cortex since cutaneomuscular reflex responses, of supposed cortical origin, were enhanced during this type of task. They found that responses in the 1DI muscle to TMS are enhanced during an independent finger movement compared to a power grip.

Flament et al. (1993) pursued this approach using a number of different tasks. They examined the amplitude of responses in 14 normal volunteers who performed the isometric gripping tasks illustrated in Fig. 2. These tasks consisted of (a) simple isometric abduction of the index finger, (b) isometric grasp of a 10 cm diameter petri dish involving all digits, (c) pincer grip between thumb and index finger in opposition, (d) isometric power grip of a 4 cm diameter brass cylinder involving all digits, and (e) isometric clockwise rotational grip of a bottle top (3 cm diameter) using thumb and index finger.

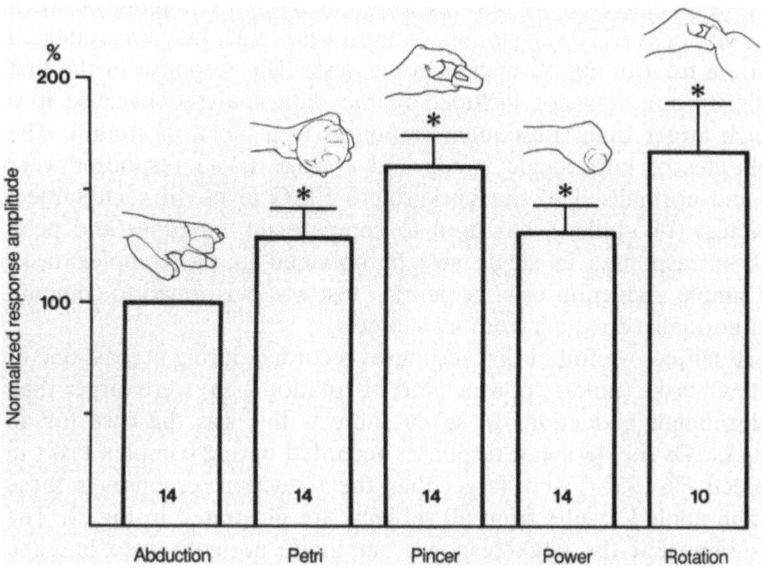

Fig. 2. Task dependence of responses in 1DI to magnetic stimulation of the brain in man. Responses were recorded during performance of the five different isometric tasks shown above each column. See text for details of tasks. Mean (±SE) 1DI peak response amplitudes are normalised to baseline EMG level for all subjects and are expressed as a percentage of the response in the index finger abduction task (100%), shown at the extreme left. *Asterisk*, statistically significant difference from the abduction value (paired t test, $p < 0.05$). *Numbers* (in each bar) denote the number of subjects in which the task was tested. (Reproduced with permission from Flament et al. 1993)

For the abduction task digits III, IV and V were taped together to further isolate and restrict movement to the index finger; EMG recordings from a number of selected muscles showed that, in this task, activity is largely confined to 1DI. The other tasks also involved activity in 1DI, but in association with different levels of activity among other intrinsic hand and forearm muscles.

The EMG recorded from 1DI with either surface or intramuscular electrodes was displayed on a meter, and subjects were instructed to maintain their contraction level at 5% MVC during performance of each task. TMS was delivered to the brain using a Novametrix 200 electromagnetic stimulator with a standard 9 cm diameter coil. Stimuli ranging from 35% to 80% of the maximum stimulator output (1.5 T) were used. The coil was placed over the vertex and orientated to optimally activate the tested (dominant) hand muscles in each subject. The position of the vertex was marked on a tightly fitting plastic cap placed on the subject's head, and coil position relative to the head was maintained constant by means of a rigid bite-bar arrangement and a clamp to hold the coil.

Stimuli were delivered pseudo-randomly at intervals ranging between 3.5 and 7 s, in blocks of 20–25 stimuli. In most cases each task was repeated twice, giving a total of 40–50 responses per task. The response to the first TMS administration was not included in the data analysed because it is usually much larger than subsequent responses in a block of stimuli. The peak amplitudes of both single sweep and average EMG responses were measured and normalised to the background EMG level. In each subject Student's *t* test (two-tailed) was used to compare the baselines and peak amplitudes of responses in single sweeps obtained in the complex tasks versus the simple abduction task. A paired *t* test was performed to compare the pooled normalised results from all subjects.

In every subject we found that responses recorded during at least one of the "complex" tasks (pincer, power, petri or rotation grip) were larger than during index finger abduction. In seven subjects this was the case for all complex tasks. Of the 48 mean responses recorded during complex tasks in the 14 subjects, 36 (75%) were larger than the abduction responses in these subjects. The pooled results from all subjects are illustrated in Fig. 2. The histograms represent the peak response amplitude normalised to baseline and expressed as a percentage of the abduction value. The mean response amplitudes were significantly greater in four of the complex tasks than in abduction (paired *t* test, $p < 0.05$). In general, the largest responses were obtained during the pincer and rotation grips. A possible explanation of this result is that, because more muscles were active during the complex tasks, there was contamination of the 1DI response to TMS by responses in neighbouring muscles. We attempted to exclude this possibility by making intramuscular recordings from 1DI in four subjects. The same pattern of response described for the surface recorded EMG was obtained.

We also tested, in six subjects, the effect of changing the intensity of TMS upon the responses. The first intensity was set to be just supra-threshold for evoking a response after each stimulus. The second intensity was stronger by 5%–8% of stimulator maximum output. It might be predicted that the most dramatic task-related differences in response amplitude would occur at juxta-threshold intensities, which might be expected to recruit many additional CM neurones, brought to discharge by TMS because of a raised level of excitability during performance of a particular task. But this was not the case: at the higher intensity the difference in response amplitude between abduction and the complex tasks was more likely to be statistically significant. At the low intensity an average of two complex tasks was associated with significantly larger responses (Student's *t* test, $p < 0.05$); at the higher it was 3.8 tasks. The EMG response amplitudes recorded from one subject at two intensities of TMS are shown in Fig. 3. In this subject 63% of maximum stimulator output was adequate to elicit a response of 1DI after each stimulus; the higher intensity was 70%. The response amplitudes normalised to baseline are illustrated for the five tasks. It can be seen that stimuli at the higher intensity (70%) produced significantly larger responses

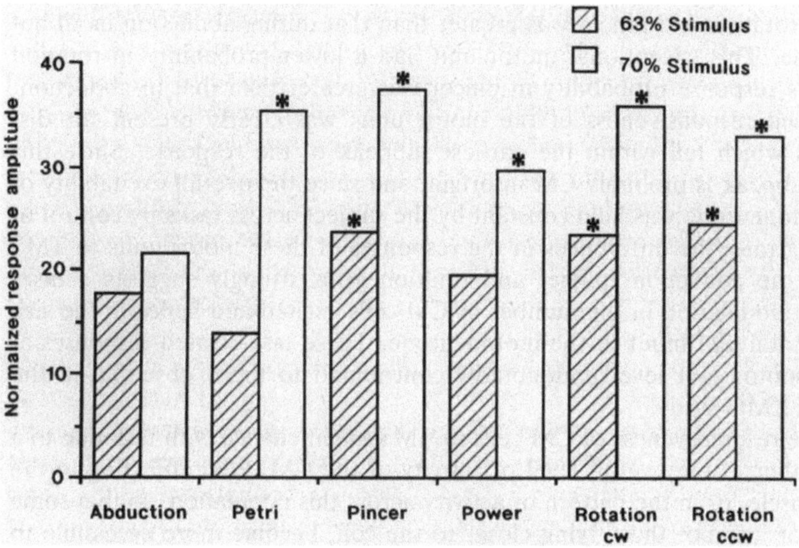

Fig. 3. Mean amplitude of the EMG response in 1DI to magnetic stimulation at two stimulus intensities (63% and 70% of stimulator output). Peak response amplitude is normalised to baseline EMG level. *Asterisks*, statistically significant differences from the abduction result at each stimulus strength (Student's t test, $p < 0.05$). The responses show a greater degree of task dependence with the stronger stimuli. (Reproduced with permission from Flament et al. 1993)

during the complex tasks than during abduction (Student's t test, $p < 0.05$) while at 63% the response in one of the tasks (petri) was slightly less than abduction, and only three of the complex task responses were significantly larger.

A change in the size of the descending corticospinal volley is not the only mechanism that could have caused the changes illustrated in Fig. 2. There were probably large changes in peripheral afferent input across the tasks; for example, the 1DI muscle was active at different lengths, and there were different patterns of joint and cutaneous inputs associated with the different grips. It could be argued that the differences in 1DI response amplitude might have resulted from a change in the responsiveness of the 1DI motoneurone pool to a constant corticospinal input. We tried to control for this by maintaining the activity of the 1DI muscle at 5% MVC throughout the experiment. Further control experiments were undertaken in which the responses of single 1DI motor units to TMS were recorded across different tasks. In these experiments, carried out in six subjects, the firing rate of the motor unit was kept constant across tasks, and post-stimulus time histograms were compiled for 50–100 stimuli. Nine motor units were recorded during both index finger abduction and rotation tasks and six motor units during both abduction and pincer tasks. The probability of evoking a response during rotation and pincer was greater than that during abduction in all but one case. This exceptional motor unit had a lower probability in rotation only; its response probability in pincer was greater than that in abduction. Enhanced responsiveness of the motor units was clearly present for discharges which fell within the earliest subpeak of the response. Since this early subpeak is probably CM in origin, and since the overall excitability of the motoneurone was held constant by the subject across tasks by control of its firing rate, the differences in the responses of these motor units to TMS during the abduction, pincer and rotation grips strongly suggests a task-related modulation in the number of CM cells excited and hence in the size of the total CM input to the motoneurone. These task-related responses at single motor unit level undoubtedly contributed to those observed in the surface EMG study.

The responsiveness of CM cells to TMS might change with task due to a shift either in the overall level of activity of the CM cells projecting to the 1DI muscle, or in the pattern of activity across this population, so that some cells, for instance those lying closer to the coil, become more accessible to stimulation. A further possibility is that there are differences in either the size or strength of connectivity of the CM cells recruited across tasks, and that these differences are reflected in the responses to TMS (Edgley et al. 1992).

We have proposed that each task is characterised by a different pattern of activity within the CM cell population for 1DI, and that it is these differences which result in the changing size of the corticospinal volley. Apart from differences in the central drive to these CM cells for the different tasks, the changes in peripheral feedback from the hand described above also contribute to the patterns of cortical activity associated with a given task (Lemon 1981; Doemges and Rack 1992).

As shown in Fig. 1A, it is possible that some of the CM cells projecting to 1DI are located at a considerable distance from the surface of the brain and from the stimulating coil, deep within the anterior bank of the central sulcus. If this were the case, the more striking effects of task with stronger stimuli would result from the fact that weaker stimuli did not activate these deep-lying CM neurones.

Why were the responses during abduction consistently smaller than during the complex tasks? It was the only task that required activation of a single muscle, and because corticospinal projections to the upper limb motoneurone pools are divergent in nature (see above), it is probable that inhibitory mechanisms are required in order to achieve contraction of a single muscle. We have suggested that this inhibition reduces both the size of the corticospinal volley destined for the 1DI motoneurones and/or the responses of these motoneurones to this volley (Flament et al. 1993). It is also possible that the abduction manoeuvre, a rather unnatural task, may be more dependent on subcortical mechanisms less accessible to TMS.

Lack of Task-Dependence in Hand Muscle Responses to Magnetic Stimulation in Stroke Patients

One of the most persistent and devastating results of stroke affecting the motor cortex or internal capsule is the loss of skilled hand movement. In the majority of patients, both strength and skill are more greatly affected in the distal muscles than in the more proximal muscle groups (Walshe 1963, p. 21; Colebatch and Gandevia 1989). Even when patients make a substantial recovery after stroke, their hand movements on the affected side can still appear slow and clumsy (Carr and Shepherd 1987, p. 48; Brodal 1973). When magnetic stimulation is applied to the damaged hemisphere of stroke patients, the EMG responses in hand muscles are small and/or delayed. The presence of an EMG response and its latency following the magnetic stimulus has been found to correlate with functional recovery after stroke (Heald et al. 1990).

These observations suggest a loss in the number of CM cells projecting to the motoneurones of a given muscle. Given that the task-related changes in responses to TMS appear to reflect the number of CM cells active for a given task, it is interesting to discover whether stroke affects the degree of task-related activity exhibited by patients. To test this possibility, we carried out a similar investigation to that described above in five stroke patients, four men and one woman, aged 34–72 years. All five had acute onset of neurological deficit of presumed vascular origin between 8 months and 10 years before they were studied; in four patients (WH, BH, GJ, AP) computed tomography indicated striatocapsular lesions. Immediately after stroke, all five patients had little or no upper limb movement; at the time of the study they showed no significant level of spasticity.

To assess the patients' hand function at the time of the study, they performed two tests, a timed peg transfer task and a second timed task requiring nuts to be unscrewed from bolts (Turton and Fraser 1986). The proficiency of hand function in the selected patients varied considerably. WH was the most affected; he was unable to do the nuts and bolts test and was very slow on the peg test (he took 26 s longer to transfer 10 pegs with the affected compared to the unaffected hand). BH, GJ and RB were less affected (5, 6 and 4 s longer on the peg test, respectively) while AP, who was by far the best recovered patient, was delayed by only 2 s on the affected side. She had resumed her job as a teller in a local bank, and showed great proficiency in counting bank notes with her "affected" hand.

There was generally a good correspondence between these tests of hand function and the results of the study using TMS. Thresholds were higher by 5%–18% of stimulator output for responses in 1DI on the affected side in WH, BH and RB (tested during active contraction of the muscle), compared to responses in the unaffected hand. These same three patients showed relatively small responses, with long latencies (in WH response latency increased by up to 7 ms) and rather protracted durations (up from 8–10 ms on the unaffected side to as much as 23 ms on the affected side).

After a short practice period all the patients were able to perform the isometric gripping tasks and to maintain their EMG level at a relatively constant level during delivery of TMS. The approach chosen was to stimulate the damaged side first with stimuli that were higher than the threshold value by around 10% of the stimulator's maximum output (1.5 T). After responses had been obtained from the 1DI during the abduction, pincer, power and rotation tasks, the intact hemisphere was stimulated during task performance with the unaffected hand. The stimulus intensity was adjusted so that the peak amplitude of the response matched that obtained from the affected hand recorded in the previous block of trials. Once again, peak response amplitude was normalised as a multiple of background EMG level, and the results obtained during the three complex tasks were expressed as a percentage of the response recorded during index abduction.

Figure 4 illustrates the response size across tasks for the patients' unaffected and affected hands. Four of the five patients (WH, GJ, RB and BH) showed a similar pattern of results: on the unaffected side there was a clear task-related variation in response amplitude, similar to that described for the neurologically normal subjects discussed above. These changes were statistically significant for one complex task (rotation) in WH and GJ and for two complex tasks (pincer and rotation) in BH. In RB a similar trend was seen, but the differences were not statistically significant. In none of these four patients was there any significant differentiation of response amplitude across tasks on the affected side. There was a trend in RB for responses in *all* the complex tasks to be smaller than in abduction; this was never found in the control subjects.

The exception was the best-recovered patient (AP). She exhibited an essentially "normal" set of responses on her "affected" side. During performance of both pincer and rotation tasks, responses were significantly larger than abduction for her unaffected hand and on the affected side responses during pincer were also significantly larger than during abduction.

These results show that when patients with clear signs of damage to descending corticospinal fibres perform a variety of isometric grips with the affected hand, the responses evoked in the 1DI muscle lack the task-dependent variation in amplitude found on the unaffected side of the same patients. This finding can be interpreted as a sign of a much reduced number of corticospinal neurones: it is speculated that the most affected patients in our small sample were hampered by a greatly diminished number of functioning CM connections to the 1DI muscle. Unlike the case in the healthy cortex, this surviving population may be entirely non-redundant in that patients may use the *same* CM cells for performance of each of the different tasks, thereby leading to a loss of the task-related variation in response amplitude. Such a strategy would be useful for developing some

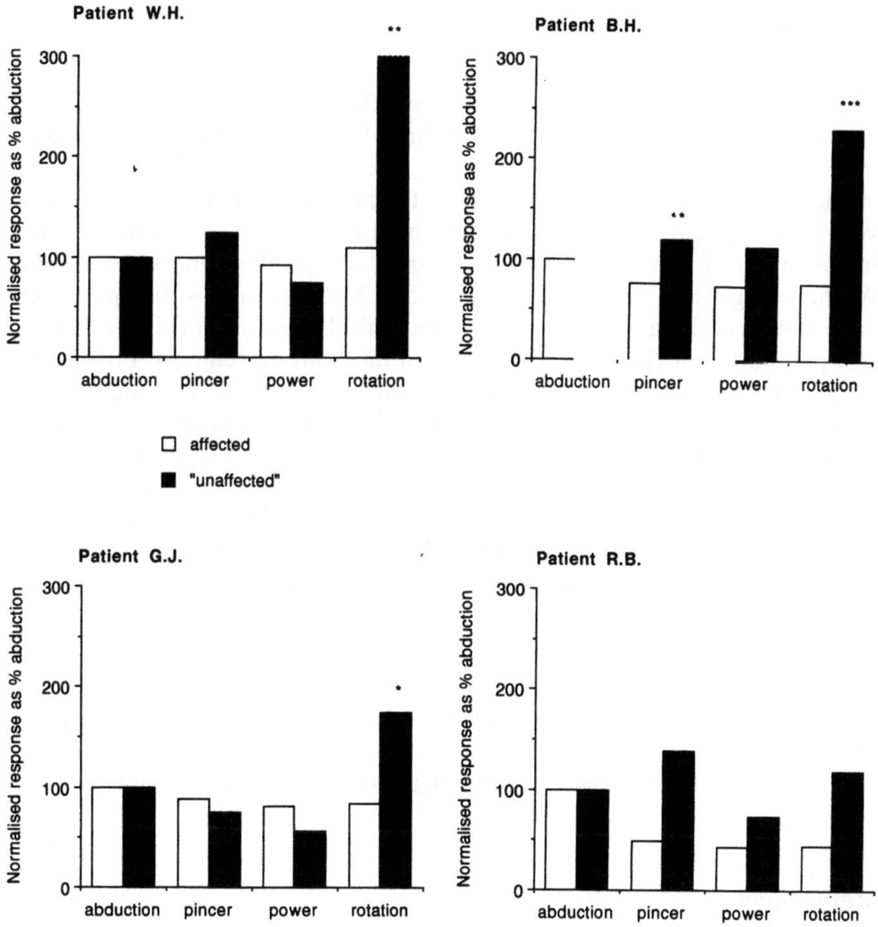

Fig. 4. Responses elicited by magnetic stimulation in the 1DI muscle of four stroke patients (WH, BH, GJ and RB) during performance of the tasks illustrated in Fig. 2. Bar charts show mean response amplitudes, normalised to background EMG level and expressed as percentage of the response amplitude in index finger abduction. All but one patient (RB) exhibited task-related changes in the responses of the unaffected hand (*hatched bars*). None of them showed significant changes on the affected side (*open bars*). *Asterisks*, tasks with responses significantly larger than abduction: *, $p < 0.05$; **, $p < 0.01$; ***, $p < 0.001$

contraction in the distal muscles but would not be appropriate for production of the specific patterns of muscular contraction demanded by the different tasks. If one considers the normal cortex as possessing a large number of possible muscle combinations, specified by the muscle field of different CM cells (Lemon 1993), then this "library" of possible synergies appears to be greatly reduced after stroke. The reduction in the number of possible strategies which can be applied to a given task might therefore explain the clumsiness of the patients on hand function tests.

Concluding Remarks

Much of our present understanding about the corticospinal system in man has stemmed from animal studies. In this short review we have described some of the characteristic features of the CM system in the macaque monkey, and in particular its pattern of branching within the spinal cord. We have speculated as to how such branching might contribute to the fractionation of hand muscle activity during independent finger movement. We have suggested that large numbers of CM cells, each with a particular pattern of branching, might form the anatomical substrate for the varied use of a particular muscle in a wide variety of different movements. Finally, we have suggested that this system, once damaged, gives rise to clumsiness of hand movements. This may be interpreted as the recruitment of a small population of surviving CM elements, in a rather stereotyped fashion not always appropriate for the task in hand. Further studies, examining the way in which this population is used and organised after stroke are now needed to help us understand the effects of stroke upon this important motor system and to provide new insights into the assessment and therapy of stroke sufferers.

Acknowledgements. This work was supported by the Medical Research Council, Action Research and the East Anglian Regional Health Authority. We would like to thank Paul Goldsmith and Camilla Buckley, who assisted with this project, and Rosalyn Cummings for expert technical support.

References

Andersen P, Hagan PJ, Phillips CG, Powell TPS (1975) Mapping by microstimulation of overlapping projections from area 4 to motor units of the baboon's hand. Proc R Soc B 188:31–60

Armand J (1982) The origin, course and terminations of corticospinal fibers in various mammals. Prog Brain Res 57:329–360

Barker AT, Jalinous R, Freeston IL (1985) Non invasive magnetic stimulation of the human motor cortex. Lancet 2:1106–1107

Bennett KMB (1992) Corticomotoneuronal control of precision grip tasks. PhD Thesis, Cambridge University

Bennett KMB, Lemon RN (1991) The activity of monkey corticomotoneuronal (CM) cells is related to their pattern of post-spike facilitation of intrinsic hand muscles. J Physiol (Lond) 435:53P

Berardelli A, Inghilleri G, Crucci G, Manfredi M (1990) Descending volley after electrical and magnetic transcranial stimulation in man. Neurosci Lett 112:54–58

Bernhard CG, Bohm e, Petersén I (1953) Investigations on the organization of the corticospinal system in monkeys (Macaca Mulatta). Acta Physiol Scand 29 [Suppl 106]:79–105

Brodal A (1973) Self observations and neuro-anatomical considerations after a stroke. Brain 96:675–694

Buys EJ, Lemon RN, Mantel GWH, Muir RB (1986) Selective facilitation of different hand muscles by single corticospinal neurones in the conscious monkey. J Physiol (Lond) 381: 529–549

Carr JH, Shepherd RB (1987) A motor relearning programme for stroke. Heinemann, London

Catsman-Berrevoets CE, Kuypers HGJM (1976) Cells of origin of cortical projections to dorsal column nuclei, spinal cord and bulbar medial reticular formation in the rhesus monkey. Neurosci Lett 3:245–252

Cheney PD, Fetz EE (1980) Functional classes of primate corticomotoneuronal cells and their relation to active force. J Neurophysiol 44:773–791

Cheney PD, Fetz EE, Mewes K (1991) Neural mechanisms underlying corticospinal and rubrospinal control of limb movements. Prog Brain Res 87:213–252

Colebatch JG, Gandevia SC (1989) The disribution of muscular weakness in upper motor neuron lesions affecting the arm. Brain 112:749–763

Datta AK, Harrison LM, Stephens JA (1989) Task-dependent changes in the size of response to magnetic brain stimulation in human first dorsal interosseous muscle. J Physiol (Lond) 418:13–23

Day BL, Riescher H, Struppler A, Rothwell JC, Marsden CD (1991) Changes in the response to magnetic and electrical stimulation of the motor cortex following muscle stretch in man. J Physiol (Lond) 433:41–57

Doemges F, Rack PMH (1992) Changes in the stretch reflex of the human first dorsal interosseous muscle during different tasks. J Physiol (Lond) 447:563–573

Donoghue JP, Leibovic S, Sanes JN (1992) Organization of the forelimb area in primate motor cortex: representation of individual digit, wrist, and elbow muscles. Exp Brain Res 89:1–19

Dum RP, Strick PL (1991) The origin of corticospinal projections from the premotor areas in the frontal lobe. J Neurosci 11:667–689

Edgley SA, Eyre JA, Lemon RN, Miller S (1990) Excitation of the corticospinal tract by electromagnetic and electrical stimulation of the scalp in the macaque monkey. J Physiol (Lond) 425:301–320

Edgley SA, Eyre JA, Lemon RN, Miller S (1992) Direct and indirect activation of corticospinal neurones by electrical and magnetic stimulation in the anaesthetized macaque monkey. J Physiol (Lond) 446:224P

Fetz EE, Cheney PD (1980) Postspike facilitation of forelimb muscle activity by primate corticomotoneuronal cells. J Neurophysiol 44:751–772

Fetz EE, Cheney PD (1987) Functional relations between primate motor cortex cells and muscles: fixed and flexible. Ciba Found Symp 132:98–117

Flament D, Goldsmith P, Buckley JC, Lemon RN (1993) Task-dependence of EMG responses in first dorsal interosseous muscle to magnetic brain stimulation in man. J Physiol (Lond) 464:361–378

Heald A, Bates D, Cartlidge NEF, Miller S (1990) Integrity of descending motor pathways in stroke patients: initial results of a longitudinal study. Eur J Neurosci Suppl 3:56

Hess CW, Mills KR, Murray NMF (1987) Responses in small hand muscles from magnetic stimulation of the human brain. J Physiol (Lond) 388:397–419

Humphrey DR, Reed DJ (1983) Separate cortical systems for control of joint movement and joint stiffness: reciprocal activation and coactivation of antagonist muscles. In: Desmedt J (ed) Motor control mechanisms in health and disease. Raven, New York, pp 347–372

Huntley GW, Jones EG (1991) Relationship of intrinsic connections to forelimb movement representations in monkey motor cortex: a correlative anatomic and physiological study. J Neurophysiol 66:390–413

Jankowska E, Padel Y, Tanaka R (1976) Disynaptic inhibition of spinal motoneurones from the motor cortex in the monkey. J Physiol (Lond) 258:467–487

Johansson RS, Lemon RN, Westling G (1993) Cortical influence over precision grip in man is strongly modulated during different phases of the task. J Physiol (Lond) 459:469P

Jones EG, Wise SP (1977) Size, laminar and columnar distribution of efferent cells in the sensory-motor cortex of monkeys. J Comp Neurol 175:391–438

Landgren S, Phillips CG, Porter R (1962) Cortical fields of origin of the monosynaptic pyramidal pathways to some alpha motoneurones of the baboon's hand and forearm. J Physiol (Lond) 161:112–125
Lawrence DG, Porter R, Redman SJ (1985) Corticomotoneuronal synapses in the monkey: light microscopic localization upon motoneurons of intrinsic muscles of the hand. J Comp Neurol 232:499–510
Lemon RN (1981) Functional properties of monkey motor cortex neurones receiving afferent input from the hand and fingers. J Physiol (Lond) 311:497–519
Lemon R (1988) The output map of the primate motor cortex. TINS 11:501–506
Lemon RN (1990) Mapping the output functions of the motor cortex. In: Edelman G, Gall E, Cowan WM (eds) Signal and sense: local and global order in perceptual maps. Wiley, Chichester, pp 315–356
Lemon RN (1993) Cortical control of the primate hand. Exp Physiol 78:263–301
Lemon RN, Mantel GWH, Muir RB (1986) Corticospinal facilitation of hand muscles during voluntary movement in the conscious monkey. J Physiol (Lond) 381:497–527
Lemon RN, Bennett KMB, Werner W (1991) The cortico-motor substrate for skilled movements of the primate hand. In: Requin J, Stelmach GE (eds) Tutorials in motor neuroscience, vol 62. Kluwer, Dordredrt pp 477–495
Lemon RN, Werner W, Bennett KMB, Flament DA (1993) The proportion of slow and fast pyramidal tract neurones producing post-spike facilitation of hand muscles in the conscious monkey. J Physiol (Lond) 459:166P
Maier M, Bennett KMB, Hepp-Reymond M-C, Lemon RN (1993) Contribution of the monkey cortico-motoneuronal system to the control of force in precision grip. J Neurophysiol 69:772–785
Merton PA, Morton HB (1980) Stimulation of the cerebral cortex in the intact human subject. Nature 285:227
Muir RB, Lemon RN (1983) Corticospinal neurons with a special role in precision grip. Brain Res 261:312–316
Palmer E, Ashby P (1992a) Corticospinal projections to upper limb motoneurones in humans. J Physiol (Lond) 448:397–412
Palmer E, Ashby P (1992b) Evidence that a long latency stretch reflex in humans is transcortical. J Physiol (Lond) 449:429–440
Phillips CG (1975) Laying the ghost of "muscles versus movements". Can J Neurol Sci 2:209–218
Porter R, Lemon RN (1993) Corticospinal function and voluntary movement. Oxford University Press, Oxford
Rothwell JC, Thompson PD, Day BL, Boyd S, Marsden CD (1991) Stimulation of the human motor cortex through the scalp. Exp Physiol 76:159–200
Sato KC, Tanji J (1989) Digit-muscle responses evoked from multiple intracortical foci in monkey precentral motor cortex. J Neurophysiol 62:959–970
Schoen JHR (1964) Comparative aspects of the descending fibre systems in the spinal cord. Prog Brain Res 11:203–222
Shinoda Y, Yokota J-I, Futami T (1981) Divergent projection of individual corticospinal axons to motoneurons of multiple muscles in the monkey. Neurosci Lett 23:7–12
Taylor J (ed) (1932) Selected writings of John Hughlings Jackson. Hodder Stoughton, London
Turton AJ, Fraser CM (1986) A test battery to measure the recovery of voluntary movement control following stroke. Int Rehabil Med 8:74–78
Walshe FMR (1963) Diseases of the nervous system, 10th edn. Churchill Livingstone, Edinburgh

Investigation of Descending Motor Pathways in Man

J.C. Rothwell, T.C. Britton, P. Brown, B.L. Day, J. Fong,
P.D. Thompson, and K. Werhahn

The cerebral cortex, reticular formation, and vestibular nuclei all send descending motor fibres to the spinal cord. However, with the exception of the corticospinal tract, there is remarkably little information on the function of these individual systems, especially in man. Recently, several new techniques have appeared which make it possible to study some of these pathways in intact conscious subjects.

Corticospinal Tract

The human motor cortex can now be stimulated directly through the scalp by two methods. Merton and Morton (1980) developed the electric method of stimulation using a specially designed stimulator capable of giving brief high voltage stimuli. Such stimuli applied to the scalp through conventional electrodes can activate brain structures beneath but are uncomfortable to use because only a small fraction of the administered current travels through the high-resistance pathway of the scalp and skull into the brain. Most current flows along the surface of the scalp, producing local pain and contraction of scalp muscles. Because of this, most clinical studies use the magnetic method of brain stimulation developed by Barker et al. (1985). With this technique, a large capacitance is discharged through an insulated coil of wire held on the scalp. Very high currents flow in the wire and generate a large magnetic field (about 2 T, the same as in a conventional magnetic resonance imaging scanner) perpendicular to the windings of the coil. This field passes easily through the scalp and skull into the brain. Because it changes rapidly over time, reaching a peak within about 100 μs, it induces electric current to flow in any conductive structure, such as the brain and cerebrospinal fluid, which may be nearby. Thus, in some respects, the magnetic field can be regarded as a carrier for electric current into the brain. Although we refer to the stimulation as magnetic, it is the conventional

MRC Human Movement & Balance Unit, Institute of Neurology, Queen Square, London WC1N 3BG, England, UK

Thilmann et al. (Eds.)
Spasticity: Mechanisms and Management

electric current in the brain which is responsible for stimulating nervous structures.

The advantage of magnetic stimulation is that it produces very little sensation on the scalp. Because the skull offers little impedance to the passage of magnetic field, the currents induced on the scalp are little different from those in the brain, and hence pain or muscle contraction is minimised. The disadvantage of the technique is that the locus of stimulation is not well defined. With the usual round stimulating coils the induced electric current flows in the brain in an annulus under the coil. More focal stimulation can be achieved using figure-of-eight coils in which the magnetic field under the junction region is twice that at the edge of each wing (e.g., Ueno et al. 1988). With such stimulators, it is possible to map the motor responses on the scalp with sufficient precision to define a face, hand, arm and leg representation in the lateral to medial direction over the motor areas.

It is thought that both electric and magnetic forms of cortical stimulation preferentially activate the large diameter component of the corticospinal tract (see Rothwell et al. 1991). However, the precise locus of excitation is not yet known. In the past, the debate has concentrated on the latency difference between responses obtained in hand muscles after either electric stimulation or after magnetic stimulation with a large round coil centred at the vertex. With these forms of stimulation in active muscle, electromyogram (EMG) latencies in hand muscles (both from surface EMG and in single motor unit studies) are 2–3 ms longer if magnetic stimulation is used than if electric stimulation is used. This effect disappears at high-stimulus intensities (Fig. 1). One explanation for the different latencies is that electric stimulation activates the axons of pyramidal tract neurones directly in the white matter (D activation), whereas magnetic stimulation activates the same neurones transynaptically through cortico-cortical connections (Day et al. 1989). Thus, the longer latency of magnetic responses is due to the time taken for synaptic transmission at the motor cortex. A second explanation is that electrical stimulation activates the pyramidal tract axons very deep in the white matter at the base of the brain, whilst magnetic stimulation activates the axons within the cortical grey matter at the initial segment region of the pyramidal tract neurones. The difference in peripheral motor latencies with the two forms of stimulation is therefore the result of distal versus proximal activation of the pyramidal pathway (Edgley et al. 1990).

Although there is good evidence that at high intensities the point of activation after electric stimulation may spread deep into the brain, the second hypothesis outlined above seems unlikely to hold at *threshold* intensities. First, even at threshold, the latency of EMG responses is still 2 ms or so shorter with electric than with magnetic stimulation. At such low intensities it seems unlikely that electric stimulation could activate axons 10–12 cm deep into the white matter. Second, in about 30% of subjects, if a figure-of-eight magnetic coil is used for stimulation, and positioned so that the junction region is oriented in the medio-lateral direction of the scalp, it

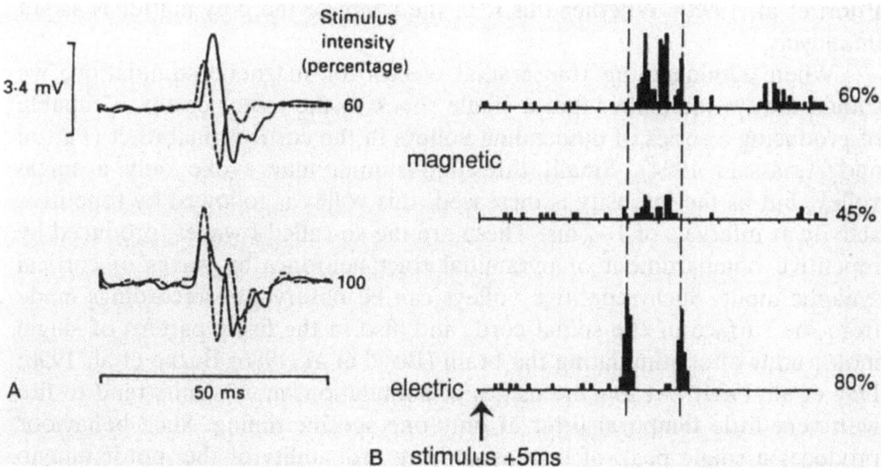

Fig. 1A,B. Comparison between EMG responses evoked in human muscle by electrical and magnetic transcranial stimulation. **A** Surface EMG responses in pre-activated first dorsal interosseous muscle after different intensities of stimulation. *Continuous lines*, the response to magnetic stimulation at the vertex with a round coil (anticlockwise current flow in coil); *dashed lines*, anodal electric stimulation. At an intensity of 60%, the latency of electrically evoked responses is about 2 ms shorter than those evoked by magnetic stimulation. At an intensity of 100%, the response is the same latency to both forms of stimulation (from Day et al. 1989). **B** Comparison of the effects of anodal electrical and anticlockwise magnetic stimulation on the probability of single motor unit firing in the first dorsal interosseous muscle. The same unit is illustrated in all traces; calibration intervals are at 2.5 ms. Electrical stimulation produced two peaks of increased firing probability separated by about 4 ms. Magnetic stimulation gave two or three peaks, but the earliest was 1–2 ms later than the earliest peak after electrical stimulation. (From Rothwell et al. 1991)

is possible to obtain latencies at threshold which are as short as those with electric stimulation (Amassian et al. 1990; Werhahn et al. 1993). If electric stimulation excited structures deep in the brain, then this result would imply that threshold magnetic stimulation with this orientation is also capable of penetrating to the level of the brainstem. However, experiments on monkeys have repeatedly confirmed that magnetic stimulation never acts so deep within the brain, and that in general its effect is limited to neurones within the grey matter (e.g., Edgley et al. 1990). Thus, we can conclude that electrical stimulation at threshold activates axons near the surface of the cortex. The corollary is that at least at threshold, with the usual round magnetic coil centred at the vertex, pyramidal tract activation occurs through indirect, synaptic mechanisms. These issues are discussed further in this volume by Burke et al.

Finally, we should note that these arguments do not hold for stimulation over the leg area of cortex. For the tibialis anterior muscle the latency of EMG responses after electric and those after magnetic stimulation of the brain are equal, implying a similar site of action for both methods (Iles 1990;

Priori et al. 1993). Whether this is in the white or the grey matter is as yet unknown.

When administering transcranial electric or magnetic stimulation, we should always remember that a single shock to the motor cortex is capable of producing a series of descending volleys in the corticospinal tract (Patton and Amassian 1954). Small, threshold stimuli may evoke only a single volley, but as the intensity is increased, this volley is followed by repetitive activity at intervals of 1–2 ms. These are the so-called I waves, produced by repetitive bombardment of pyramidal tract neurones by waves of cortical synaptic input. Such repetitive volleys can be observed in recordings made from the surface of the spinal cord, and also in the firing pattern of single motor units after stimulating the brain (Boyd et al. 1986; Burke et al. 1990; Day et al. 1989). At low intensities of stimulation, motor units tend to fire with very little temporal jitter at only one specific timing. Such behaviour produces a single peak of increased firing probability of the motor unit in post-stimulus time histograms of unit discharge after the shock. As the intensity of the stimulation to the brain is increased, several peaks appear in the post-stimulus time histogram, each separated by 1–2 ms or so. It is thought that these peaks of increased firing probability of the motor unit represent arrival at the parent spinal motoneurone of excitatory post-synaptic potentials released by each of the descending corticospinal volleys (e.g., Fetz and Gustafsson 1983).

In man, short latency EMG responses to cortical stimulation can be observed in virtually every muscle of the body including the external anal sphincter, the diaphragm and even the oesophagus (see Rothwell et al. 1991). The latency jitter of these responses is very small and suggests very strongly that they are produced by monosynaptic activation of the spinal motoneurones. If so, then in man, the cortico-motoneuronal component of the corticospinal tract must be much more widespread than in the monkey. For example, in the monkey direct electric stimulation of the pyramids yields monosynaptic excitatory post-synaptic potentials in motoneurones of the hand and arm, and muscles such as deltoid receive no monosynaptic input. Projections to oesophagus or sphincter muscles have never been sought.

Magnetic stimulation is only capable of activating the corticospinal tract at the level of the motor cortex. It is not possible to stimulate descending pathways in the spinal cord because the small volume of nervous tissue within the spinal column limits the amount of stimulating current which can be induced by the magnetic field. However, using high voltage electric stimulation, it is possible to stimulate the descending motor pathways in the cord by applying brief single stimuli over the spinal column (Marsden et al. 1982). This procedure, as with electrical stimulation of the brain, causes local muscle contraction and hence is rather uncomfortable.

It is also possible to use electrical stimulation to activate the descending motor pathways at a relatively high level approximating the cervicomedullary

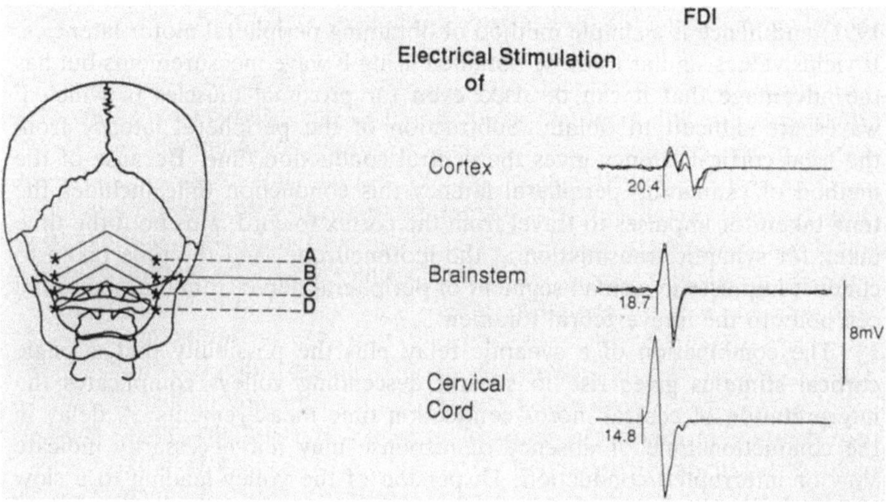

Fig. 2. *Left*, positions of electrodes for brainstem stimulation. The electrodes at position *B* were at the posterior edge of each mastoid process. Electrodes *A*, *C* and *D* are spaced vertically at 2-cm intervals. *Right*, three superimposed single EMG responses from the right first dorsal interosseous (*FDI*) muscle of a normal subject following motor cortical, brainstem (at level *B*), and cervical electrical stimulation, with the subject contracting the muscle to 5% of maximum. Stimulation occurs at the start of the sweep, and the onset latencies are marked next to the vertical lines in the figure. The cortico-brainstem conduction time is 1.7 ms, brainstem-cervical conduction time is 3.9 ms. (From Ugawa et al. 1991)

junction. Ugawa et al. (1991) showed that if stimulating electrodes are placed 4 cm either side of the inion, responses can be obtained in hand and arm muscles at a latency some 2 ms shorter than those seen after electric stimulation of the motor cortex. They termed this brainstem stimulation (Fig. 2). Using a collision technique, they showed that the same descending fibres are involved in producing the brainstem responses as mediate cortically evoked responses, and hence they proposed that the brainstem stimulus is activating corticospinal fibres. Finally, because the latency of EMG responses to brainstem stimulation was approximately constant even when the electrodes were moved several centimetres higher or lower than the optimal position, they proposed that a horizontally oriented structure is the likely point of stimulation. Taken together with the short latency of these responses, this would be consistent with stimulation of the pyramidal tract at the level of the decussation.

The most widespread use of transcranial stimulation of the motor cortex is to measure the central motor conduction time in corticospinal pathways. This is achieved by recording the latency of muscle action potentials to (a) magnetic stimulation over the motor cortex and (b) stimulation over the cervical or lumbar enlargements. The latter activates spinal nerve roots as they exit the intervertebral foramen (Mills and Murray 1986; Epstein et al.

1991) and hence is a simple method of obtaining peripheral motor latencies. It yields values similar to those obtained using F wave measurements but has the advantage that it can be used even for proximal muscles in which F waves are difficult to obtain. Subtraction of the peripheral latency from the total cortical latency gives the central conduction time. Because of the method of estimating peripheral latency this conduction time includes the time taken for impulses to travel from the cortex to cord, plus both the time taken for synaptic transmission at the motoneurone, and the time taken to conduct impulses in a small segment of peripheral nerve from motoneuronal cell body to the intevertebral foramen.

The combination of a synaptic relay plus the possibility that a single cortical stimulus gives rise to several descending volleys complicates the interpretation of central motor conduction time measurements. A delay in the conduction time or absence of response may not necessarily indicate slow or interrupted conduction. Dispersion of the volley leading to a slow build-up of excitation at the motoneurones or an abnormally high threshold for motoneuronal firing may also cause an increase in latency of EMG responses, which may not reflect any change in the maximum conduction velocity of corticospinal tract fibres (see Thompson et al. 1987). The responses to cortical stimulation in patients with multiple sclerosis illustrate these effects quite clearly. Demyelination of central motor pathways causes conduction delay, conduction block and failure to transmit high-frequency volleys faithfully (McDonald and Sears 1970). This delays and disperses the volleys arriving at the spinal motoneurone and decreases their size. Thus, in multiple sclerosis, the EMG responses are usually delayed, polyphasic, reduced in amplitude and dispersed (Cowan et al. 1984; Hess et al. 1987).

Central motor conduction time is increased in many diseases (e.g., motoneurone disease, cervical spondylosis) and hence cannot be used alone as a unique diagnostic indicator. However, the ease of administering magnetic stimulation makes this a particularly useful clinical method. The yield of the technique in terms of abnormal findings in neurological disease compares favourably with other methods of monitoring conduction in central pathways. Several authors (Hess et al. 1987; Berardelli et al. 1988; Caramia et al. 1988) have compared the percentage of abnormalities revealed using magnetic stimulation of the motor cortex with those seen in using conventional visual or sensory evoked potential methods in groups of patients with multiple sclerosis. The percentage of abnormal motor responses is as high as if not higher than the percentage abnormality seen with evoked potential testing. Thus, in this respect, the technique has quite a high yield. A related question is how often measurement of central motor conduction times reveals deficiencies unsuspected from clinical examination. In this respect, responses to cortical stimulation usually fare less well. In all studies, the percentage of sub-clinical motor abnormalities was less using motor cortical stimulation than the percentage of sub-clinical abnormalities detected in the sensory system by sensory or visual evoked potentials. The reason for this is probably related to the sensitivity of clinical testing methods for motor and sensory function (Murray 1991). The motor system is usually said to be more eloquent in displaying its abnormalities than the sensory system.

Reticulospinal Pathways

There are no direct electrical methods of stimulating the reticulo-spinal tracts. Nevertheless, it seems likely that the onset of the startle response is mediated by one of these systems.

The startle reflex in man consists of a generalised flexion response (Landis and Hunt 1939) which habituates rapidly over two to six trials. It may be generated by any form of stimulation but is most usually provoked by a short, loud auditory tone. Although the absolute latency of the EMG responses in the startle is rather variable, the pattern of activity between different muscles is stereotyped (e.g., Wilkins et al. 1986; Brown et al. 1991b). Figure 3 shows the typical pattern of activity in a single normal subject.

The response appears to start with activity in the orbicularis oculi muscle. However, both Colebatch et al. (1990) and Brown et al. (1991b) have argued that this initial activity is not part of the true startle pattern but represents simply a blink reflex to the auditory stimulus. The reason for this is two-fold. First, it is sometimes possible to see a clear separation in the orbicularis oculi response between an early, short period of activity and a later, long-lasting portion of activity. Second, the early component does not habituate as rapidly to repeated presentation of the auditory tone as the startle pattern. Thus, presentation of the auditory stimulus on the first occasion may give a full blown startle pattern on the first trial, but after two or three presentations, the only response remaining is the initial portion of the orbicularis oculi burst.

If we assume that the initial part of the orbicularis oculi response is a blink reflex, then the startle pattern proper begins with activity in the sternocleidomastoid muscle after a latency in normals of around 60 ms. It is followed by activity in the masseter muscle and then by activity in limb and trunk muscles. The pattern and onset latencies of responses in the startle are very different from those seen in the same muscles after electrical or magnetic stimulation of the motor cortex. Table 1 gives a comparison of the interval between the onset of EMG activity in the sternomastoid and masseter or rectus abdominis muscles, and the interval between the activity in the biceps and abductor pollicis brevis muscles in normal subjects after a startle stimulus or after stimulation of the motor cortex. Three things are evident. (a) In the startle response, activity in sternocleidomastoid occurs earlier than that in masseter, whereas after cortical stimulation masseter is innervated first. This is consistent with an origin of the startle response

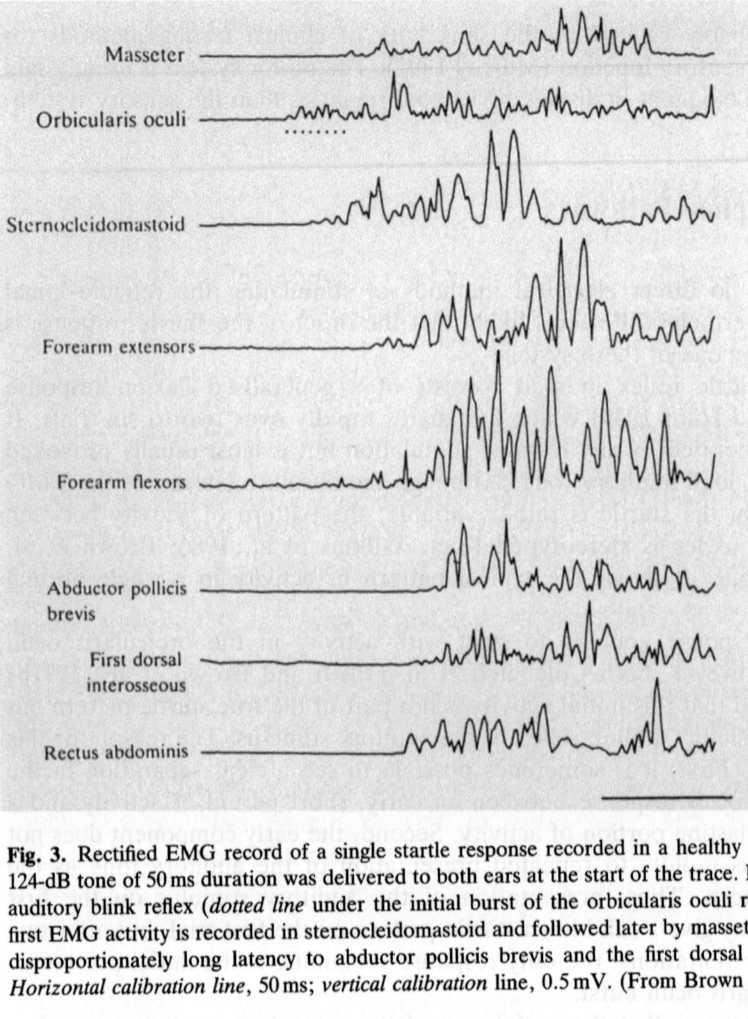

Fig. 3. Rectified EMG record of a single startle response recorded in a healthy individual. A 124-dB tone of 50 ms duration was delivered to both ears at the start of the trace. Excluding the auditory blink reflex (*dotted line* under the initial burst of the orbicularis oculi response), the first EMG activity is recorded in sternocleidomastoid and followed later by masseter. There is a disproportionately long latency to abductor pollicis brevis and the first dorsal interosseous. *Horizontal calibration line*, 50 ms; *vertical calibration* line, 0.5 mV. (From Brown et al. 1991b)

in the brainstem. Impulses are then conducted up and down from this centre innervating the XII nerve nucleus first then the VII and V (masseter) nuclei. (b) The excess latency of the rectus abdominis response over the sternocleidomastoid is some three times longer in the startle than it is after cortical stimulation. Assuming that the peripheral conduction pathways are the same in both cases, then the conduction velocity of the efferent pathway within the spinal cord must be much slower for the startle than it is for the pathway activated by cortical stimulation. (c) In the startle, the latency to onset of EMG responses in intrinsic hand muscles is disproportionately long even when allowance is made for the slow conduction in spinal efferent pathways. Again, this pattern is unlike the preferential accessibility of the

Table 1. A comparison of the efferent pathways subserving the normal auditory startle reflex with corticobulbar and pyramidal pathways in man

	Auditory startle reflex (ms)	Magnetic stimulation of motor cortex (ms)
Excess latency of masseter over SCM	0.7 ms	−2.9 ms (masseter precedes SCM)
Excess latency of RA over SCM	24.0 ms	8.0 ms
Excess latency of APB over biceps	22.4 ms	10.5 ms

The differences in the latencies to onset of EMG activity between individual muscles in the normal auditory startle reflex are calculated from the median latencies from 12 healthy seated subjects. The stimulus was a 1-kHz, 124-dB tone of 50 ms duration delivered to both ears at intervals of about 20 min. The differences in the latency to onset of EMG activity between masseter and sternocleidomastoid (SCM), between sternocleidomastoid and rectus abdominis (RA), and between biceps and abductor pollicis brevis (APB), following magnetic stimulation of the motor cortex, are calculated from the mean latencies reported by Cruccu et al. (1989) and Thompson (1991) in normal subjects. The pattern of recruitment of cranial nerves seen following magnetic stimulation of the motor cortex is reversed in the normal auditory startle reflex. The differences in latency to onset of EMG activity between sternocleidomastoid and rectus abdominis, and between biceps and abductor brevis, are longer in the auditory startle reflex than following magnetic stimulation of the motor cortex.

hand seen after cortical stimulation and indicates that quite different pathways are involved in the startle.

In summary, the electrophysiological evidence suggests that the physiological auditory startle reflex in man is mediated by an efferent system with its origin in the caudal brainstem. The spinal projections of this system are relatively slowly conducting and distributed predominantly to axial muscles. The pattern of muscle recruitment in the response is determined by the distance of each segmental level from the caudal brainstem, with two exceptions. First, an auditory blink reflex precedes the auditory startle reflex in the orbicularis oculi. Second, the latencies of intrinsic hand muscle responses are disproportionately delayed.

Studies in animals also suggest that the startle response originates in the caudal brainstem. Lesioning experiments in the rat have implicated the medial bulbo-pontine reticular formation (Szabo and Hazafi 1965), particularly the nucleus reticularis pontis caudalis (Hammond 1973; Leitner et al. 1980; Davis et al. 1982), as a primary centre subserving the acoustic startle response. The efferent limb may be the reticulo-spinal and bulbobulbar tracts (Scheibel and Scheibel 1958; Torvik and Brodal 1957) which originate in this area. If so, then the startle may be a method of analysing reticulospinal function in man.

Such a hypothesis for the site of origin of the startle response is also consistent with pathological studies in man. Brown et al. (1991c) studied several patients with hyperekplexia, a condition characterised by excessive sensitivity and lack of habituation to startle stimuli. These patients showed the usual pattern of EMG activity in the startle, although this was elicited much more readily than in normal subjects. Several of the patients had pathology affecting the brainstem. In a further study Vidailhet et al. (1992) showed that the startle reflex was virtually absent in patients with progressive supranuclear palsy. Although the pathology in this condition is widespread, one area of particularly high neuronal loss is in the lower pontine reticular formation.

Although the startle is often thought of as a relatively stereotyped, flexion response, we have recently shown that it may vary with posture (Brown et al. 1991a). Reflex activity in the tibialis anterior and soleus muscles is often absent or very rare in normal subjects when they are seated but can be recorded twice as frequently when standing, and at a latency 40–60 ms shorter. We suggest that the importance of the normal startle response may be in the rapid accomplishment of a defensive stance with maximum postural stability.

Vestibulo-spinal Tract

It has been known for many years that application of a small (1 mA) DC (galvanic) electric current between the mastoid processes can produce body sway in normal subjects standing with their eyes closed. Since these responses are absent in patients with either labyrinthine or VIII nerve lesions, it seems that the galvanic stimulus acts somewhere at a retrolabyrinthine location (Coats and Stoltz 1969; Watanabe et al. 1989). In the squirrel monkey, externally administered galvanic currents probably modulate the tonic firing rate of vestibular afferents by acting directly on the fibres close to their post-synaptic trigger site (Goldberg et al. 1982). Cathodal currents increase the firing rate, whereas anodal currents decrease it.

As shown by Lund and Broberg (1983), one remarkable phenomenon is that the motor response to the galvanic stimulus is organised to take into account the position of the head relative to the feet. The direction of body sway occurs primarily in the direction of the anode irrespective of the degree of head or trunk rotation. Motor centres which issue the response must receive and act upon inputs from various sources. Nevertheless, it is not clear through which central pathways the motor response is produced.

Nashner and Wolfson (1974) found that in unsupported standing subjects, galvanic vestibular stimulation produces a major EMG response in the soleus muscle with a latency of about 150 ms. This response was appropriate to contribute to the observed sway of the body (i.e., towards the ear behind which the stimulating anode was placed). We have recently reinvestigated this response in more detail and have described two new phenomena (Britton et al. 1993). First, the major EMG response seen by Nashner and Wolfson

(1974) is preceded by a smaller response in the opposite direction which appears at the considerably shorter latency of around 60 ms in leg muscles. Second, if subjects stand on an unstable base and use their arms to steady the movement of their body, then the same galvanic stimulus can also evoke short and long latency EMG responses in arm muscles. An example of responses from a typical subject is shown in Fig. 4. The combined data from ten subjects showed that the latency of the early response in the arm was much shorter than that in the leg (arm, 41 ± 2.6 ms; leg, 62 ± 2.4 ms). In contrast, the latency of the late response was longer in the arm (138 ± 4.3 ms) than in the leg (115 ± 5.2 ms).

If we assume that the early component of the EMG response in arm and leg results from synchronous discharge of a common supraspinal structure, then this allows us to calculate the approximate spinal conduction velocity of the descending pathway involved. In fact, the latency difference of about 20 ms for production of EMG responses in these two muscles is very similar to that seen after magnetic stimulation of the motor cortex. Thus, the descending pathway mediating the galvanic response must have a spinal descending conduction velocity in the range of 60–80 m/s. We have suggested that this may involve the vestibulospinal tract, although it is clear that components of the reticulospinal system may also have high-velocity axons.

In absolute terms the latency of the galvanic EMG responses was some 30 ms longer than the minimum latency after magnetic stimulation of the motor cortex. Presumably, most of this figure must represent the time taken for the galvanic stimulus to modulate the firing of vestibular afferents plus the additional central processing time needed to initiate activity in the descending pathway.

The assumption that the EMG responses in the arm and leg arise from a common source in the brain at a fixed latency is not valid when considering the late component of the galvanic vestibular response. This part of the response had a longer latency for muscles in the arm than in the leg. Thus, it is more likely that the late component is part of a patterned, functional response which is issued independently of the early component. It may be similar to the patterned reflex responses described by Nashner (1977) after disturbances of the support surface. Unfortunately, since the muscles were activated in a distal to proximal sequence, it is not possible to calculate any conduction velocity for the descending pathway mediating this effect. It is interesting to note that in the cat both vestibulospinal and reticulospinal tracts are known to receive input from the vestibular nuclei (Ladpli and Brodal 1968; Peterson and Abzug 1975). Since these pathways appear to have opposite effects on any given lumbosacral spinal motoneurones (Grillner et al. 1971), it is tempting to suggest that they may contribute to the oppositely directed early and late responses seen in the present experiments. Why there should be two systems exerting opposite influences on a motoneurone pool in response to a single stimulus is not clear. It may be

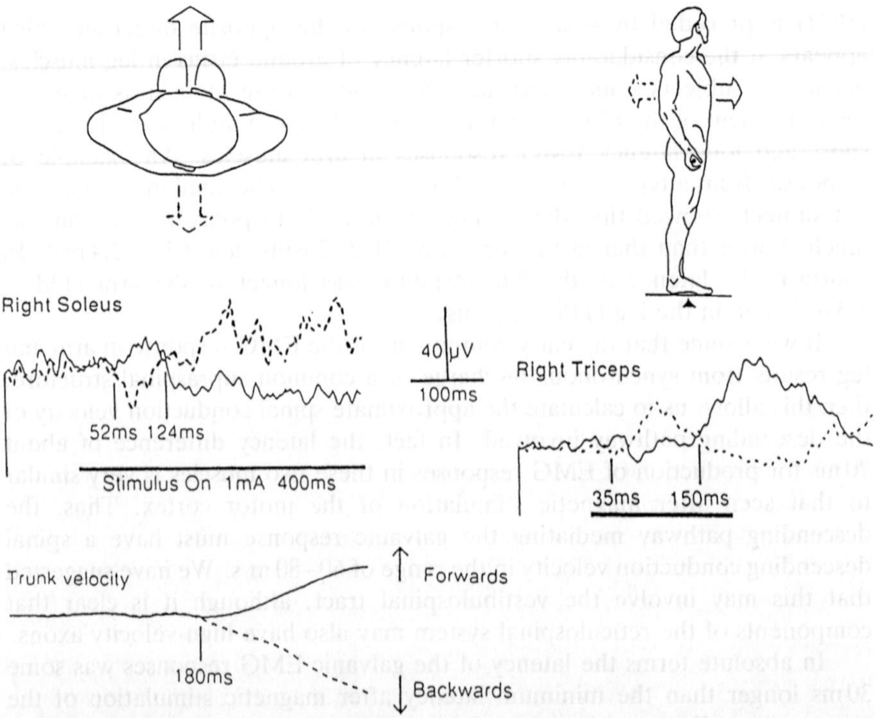

Fig. 4. *Left*, galvanic vestibular stimulation produces body sway towards the anodal ear and modulates EMG activity in soleus in a biphasic manner. Results are shown from a single subject standing erect with his eyes closed and head turned to the left. The subject was leaning slightly forwards to maintain a small tonic activity in soleus. The traces show averaged (n = 40) rectified EMG records from soleus; *lower traces*, integrated averaged signals from a linear accelerometer taped to the middle of the back. A 1-mA galvanic current was administered between the mastoid processes for 400 ms starting 100 ms after the beginning of the trace (*solid line*), the polarity of the current being varied randomly between left ear positive (*dotted traces*) and right ear positive (*solid traces*). Trunk movement, which begins at 180 ms, was towards the anodal ear irrespective of whether the head was turned left or right. The soleus EMG response consisted of two components. The second component at 124 ms was appropriate for producing the observed body sway (with forward sway there was a reduction in the activity, whereas with backward stray there was an increase). The first component was in the opposite direction to the second component and began at 52 ms. It failed to produce any measurable movement of the trunk. *Right*, galvanic vestibular stimulation modulates triceps brachii EMG activity when the arm is used to maintain balance. Average data from five subjects. Subject stood on a transversely pivoted platform and maintained balance in the antero-posterior plane by holding a fixed handle placed by their right hip with their right hand. The head was turned to the left. Galvanic stimuli (1 mA) of 150 ms duration were delivered 100 ms after the start of the sweep. The polarity of the current was varied randomly between left ear positive (*dotted traces*) and right ear positive (*solid traces*). Modulation of right triceps EMG activity was biphasic similar to that seen in the soleus. The late component begins at 150 ms and the early component at 35 ms. (From Britton et al. 1993)

that in man the early response represents the action of a vestigial system which in the course of evolution has been largely replaced by one which produces responses more appropriate to bipedal stance.

Conclusion

Techniques are now available to study the activity in three different descending motor pathways in man. Provisionally, we have designated these pathways the corticospinal, reticulospinal and vestibulospinal tracts. Precisely which structures are involved may become clear after further pathophysiological investigations in neurological disease.

Acknowledgement. We should like to thank Mr. R. Bedlington for his unfailing assistance in building and maintaining much of the equipment used in these experiments.

References

Amassian VE, Quirk GJ, Stewart M (1990) A comparision of corticospinal activation by magnetic coil and electric stimulation of monkey motor cortex. Electroencephalogr Clin Neurophysiol 77:390–401

Barker AT, Jalinous R, Freeston IL (1985) Non-invasive stimulation of human motor cortex. Lancet II:1106–1107

Berardelli A, Inghilleri M, Cruccu G, Fornanelli M, Accornero N, Manfredi M (1988) Stimulation of motor tracts in multiple sclerosis. J Neurol Neurosurg Psychiatry 51:677–683

Boyd SG, Rothwell JC, Cowan JMA, Webb PJ, Morley T, Asselman P, Marsden CD (1986) A method of monitoring function in cortical pathways during scoliosis surgery with a note of motor conduction velocities. J Neurol Neurosurg Psychiatry 49:251–257

Britton TC, Day BL, Brown P, Rothwell JC, Thompson PD, Marsden CD (1993) Postural electromyographic responses in the arm and leg following galvanic vestibular stimulation in man. Exp Brain Res (in press)

Brodal A (1981) Neurological anatomy in relation to clinical medicine, 3rd edn. Oxford University Press, New York

Brown P, Day BL, Rothwell JC, Thompson PD, Marsden CD (1991a) The effect of posture on the normal and pathological auditory startle reflex. J Neurol Neurosurg Psychiatry 54:892–897

Brown P, Rothwell JC, Thompson PD, Britton TC, Day BL, Marsden CD (1991b) New observations on the normal auditory startle reflex in man. Brain 114:1891–1902

Brown P, Rothwell JC, Thompson PD, Britton TC, Day BL, Marsden CD (1991c) The hyperekplexias and their relationship to the normal startle reflex. Brain 114:1903–1928

Burke D, Hicks RG, Stephen JPH (1990) Corticospinal volleys evoked by anodal and cathodal stimulation to the human motor cortex. J Physiol (Lond) 425:283–299

Caramia MD, Bernardi G, Zarda F, Rossini PM (1988) Neurophysiological evaluation of De central nervous impulse propagation in patients with sensorimotor disturbances. Electroencephalogr Clin Neurophysiol 70:16–25

Coats AC, Stoltz MS (1969) The recorded body sway response to galvanic stimulation of the labyrinth: a preliminary study. Laryngyoscope 79:85–103
Colebatch JG, Barrett G, Lees AJ (1990) Exaggerated startle reflexes in an elderly woman. Mov Disord 5:167–169
Cowan JMA, Rothwell JC, Dick JPR, Thompson PD, Day BL, Marsden CD (1984) Abnormalities in central motor pathway conduction in multiple sclerosis. Lancet II:304–307
Cruccu G, Berardelli A, Inghilleri M, Manfredi M (1989) Functional organisation of the trigeminal motor system in man: a neurophysiological study. Brain 112:1333–1350
Davis M, Gendelman DS, Tischler MD, Gendelman PM (1982) A primary acoustic startle circuit: lesion and stimulation studies. J Neurosci 2:791–805
Day BL, Dressler D, Maertens de Noordhout A, Marsden CD, Nakashima K, Rothwell JC, Thompson PD (1989) Electric and magnetic stimulation of human motor cortex: surface EMG and single motor unit responses. J Physiol (Lond) 412:449–473
Edgley SA, Eyre JA, Lemon RN, Miller S (1990) Excitation of the corticospinal tract by electromagnetic and electrical stimulation of the scalp in the Macaque monkey. J Physiol (Lond) 425:301–320
Epstein CM, Fernandez-Beer E, Weissman JD, Matsuura S (1991) Cervical magnetic stimulation: the role of the neural foramen. Neurology 41:677–680
Fetz EE, Gustafsson B (1983) Relation between shapes of post-synaptic potentials and changes in firing probability of cat motoneurones. J Physiol (Lond) 341:387–410
Goldberg JM, Fernandez C, Smith CE (1982) Responses of vestibular nerve afferents in the squirrel monkey to externally applied galvanic currents. Brain Res 252:156–160
Grillner S, Hongo T, Lund S (1971) Convergent effects of alpha-motoneurones from the vestibulospinal tract and a pathway descending in the medial longitudinal fasciculus. Exp Brain Res 12:457–479
Hammond GR (1973) Lesions of the pontine and medulary reticular formation and prestimulus inhibition of the acoustic startle reaction in rats. Physiol Behav 10:239–243
Hess CW, Mills KR, Murray NMF, Schriefer TN (1987) Magnetic brain stimulation: central motor conduction studies in multiple sclerosis. Ann Neurol 22:744–752
Iles JF (1990) Use of magnetic brain stimulation in the study of corticospinal action on spinal motor mechanisms in man. J Physiol (Lond) 429:39P
Ladpli R, Brodal A (1968) Experimental studies of commissural and reticular formation projections from the vestibular nuclei in the cat. Brain Res 8:65–96
Landis C, Hunt WA (1939) The startle pattern. Farrar and Rinehart, New York
Leitner DS, Powers AS, Hoffman HS (1980) The neural substrate of the startle response. Physiol Behav 25:291–297
Lund S, Broberg C (1983) Effects of different head positions on postural body sway in man induced by a reproducible vestibular error signal. Acta Physiol Scand 117:307–309
Marsden CD, Merton PA, Morton HB (1982) Percutaneous stimulation of spinal cord and brain: pyramidal tract conduction velocities in man. J Physiol (Lond) 328:6P
McDonald WI, Sears TA (1970) The effects of central demyelination on conduction in the central nervous system. Brain 93:583–598
Merton PA, Morton HB (1980) Stimulation of the cerebral cortex in the intact human subject. Nature 285:227
Mills KR, Murray NMF (1986) Electrical stimulation of the human vertebral column: which neural elements are excited? Electroencephalogr Clin Neurophysiol 63:582–589
Murray NMF (1991) Magnetic stimulation of cortex: clinical applications. J Clin Neurophysiol 8:66–76
Nashner LM (1977) Fixed patterns of rapid postural responses among leg muscles during stance. Exp Brain Res 30:13–24
Nashner LM, Wolfson P (1974) Influence of head position and proprioceptive cues on short latency postural reflexes evoked by galvanic stimulation of the human labyrinth. Brain Res 67:255–268

Patton HD, Amassian VE (1954) Single and multiple unit analysis of cortical state of pyramidal tract activation. J Neurophysiol 17:345–363
Peterson BW, Abzug G (1975) Properties of projections from vestibular nuclei to medial reticular formation in the cat. J Neurophysiol 38:1421–1435
Priori A, Bertolasi L, Dressler D, Rothwell JC, Day BL, Thompson PD, Marsden CD (1993) Transcranial electric and magnetic stimulation of the leg area of human motor cortex: single motor unit and surface EMG responses in the tibialis anterior muscle. Electroencephalogr Clin Neurophysiol (in press)
Rothwell JC, Thompson PD, Day BL, Boyd S, Marsden CD (1991) Stimulation of the human motor cortex through the scalp. Exp Physiol 76:159–200
Scheibel ME, Scheibel AB (1958) Structural substrates for integrative patterns in the brainstem reticular core. In: Jasper HH, Proctor LD, Knighton RS, Noshay WC, Costello RT (eds) Reticular formation of the brain. Little, Brown, Boston, pp 31–55
Szabo I, Hazafi K (1965) Elicitability of acoustic startle reaction after brainstem lesions. Acta Physiol Acad Sci Hung 27:155–165
Thompson PD (1991) Motor cortical stimulation in man. PhD thesis, University of London
Thompson PD, Day BL, Rothwell JC et al. (1987) The interpretation of electromyographic responses to electrical stimulation of the motor cortex in diseases of the upper motor neurone. J Neurol Sci 80:91–110
Torvik A, Brodal A (1957) The origin of reticulospinal fibres in the cat: an experimental study. Anat Rec 128:113–137
Ueno S, Tashiro T, Harada K (1988) Localised stimulation of neural tissues in the brain by means of a paired configuration of time-varying magnetic fields. J Appl Physiol 64:5862–5864
Ugawa Y, Rothwell JC, Day BL, Thompson PD, Marsden CD (1991) Percutaneous electrical stimulation of the corticospinal pathways at the level of the pyramidal decussation in humans. Ann Neurol 29:418–427
Vidailhet M, Rothwell JC, Thompson PD, Lees AJ, Marsden CD (1992) The auditory startle response in the Steele-Richardson-Olszewski syndrome and Parkinson's disease. Brain 115:1181–1192
Watanabe Y, Mizukoshi K, Ohi H, Yasuda K, Ohashi N, Kobayashi H (1989) Retrolabyrinthine disorders detected by galvanic body sway responses in routine equilibrium examinations. Acta Otolaryngol Suppl (Stockh) 468:343–348
Werhahn KJ, Fong JKY, Meyer B-U, Rothwell JC, Day BL, Thompson PD (1993) Effects of coil orientation on the latency of EMG responses produced by transcranial magnetic stimulation over the motor cortex in man. J Physiol (Lond) (in press)
Wilkins DE, Hallett M, Wess MM (1986) Audiogenic startle reflex of man and its relationship to startle syndromes: a review. Brain 109:561–573

Propriospinal Transmission of Voluntary Movement in Humans

E. Pierrot-Deseilligny and D. Mazevet

Introduction: The C3–4 Propriospinal System

Unmasking of previously depressed pathways may provide one of the mechanisms underlying motor recovery following upper motor neurone lesions (Bach-y-Rita 1981). It can be assumed that the assumption by one structure of a function lost by another is facilitated if the output from the two structures converge onto common neurones projecting onto motoneurones (MNs). In this connection the C3–4 propriospinal system, which receives extensive excitatory input from several descending tracts and primary afferents, could play an important role in the process of recovery from hemiplegia.

The C3–4 Propriospinal System in the Cat

The C3–4 propriospinal system, which has been investigated extensively in the cat (see Alstermark and Lundberg 1992), is the only spinal network with a clearly demonstrated function in motor control: transmission to forelimb MNs of the descending command for visually guided target-reaching (Alstermark et al. 1981b).

Propriospinal neurones (PNs) receive extensive monosynaptic excitation from cortico-, rubro-, reticulo-, and tectospinal fibers and also from low-threshold cutaneous and muscle afferents in the forelimb nerves (Illert et al. 1978). Because of this extensive convergence the cortical command can be reshaped and updated at this premotoneuronal level to take into account the ultimate changes in the external and internal environment. Moreover, descending systems evoke disynaptic inhibition of PNs, via inhibitory interneurones, which provides feedforward inhibition of PNs that should not be activated in a given movement (Alstermark et al. 1984a,b). PNs also receive disynaptic inhibition from forelimb afferents (Alstermark et al. 1984a,b), and this feedback inhibition has been shown to be indispensable for the

Neurophysiologie Clinique, Rééducation, Hôpital de la Salpêtrière, 47, boulevard de l'Hôpital, F-75651, Paris cedex 13, France

Thilmann et al. (Eds.)
Spasticity: Mechanisms and Management

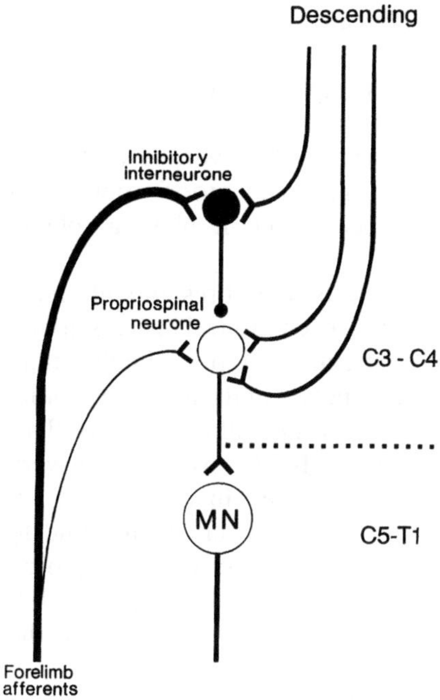

Fig. 1. Schematic diagram of some excitatory (*bars*) and inhibitory (*small filled circles*) connections to C3–4 propriospinal neurones in the cat. Monosynaptic projection of propriospinal neurone to a forelimb MN is represented. *Large filled circle*, inhibitory interneurone. Primary afferents evoke both excitation and inhibition of propriospinal neurones, but inhibitory effects are very prominent (*thick line*). (Modified from Lundberg 1979 and adapted from Malmgren and Pierrot-Deseilligny, 1988b)

smooth termination of reaching (Alstermark et al. 1986). Some of the main connections of the propriospinal system are sketched in Fig. 1.

The stem axon of the C3–4 PNs bifurcates into two branches. The descending branch projects to MNs (Illert et al. 1978) and to Ia inhibitory interneurones (Illert and Tanaka 1978) in the forelimb segments. Individual PNs project onto several motor nuclei, with many of the nuclei in the distribution of a cell's projection innervating muscles acting at different joints (Alstermark et al. 1990). This premotoneuronal linkage appears to be adequate for production of the coordinated muscle synergies which underlie reaching movements. Descending branches of C3–4 PNs also excite long PNs which project to the lumbar cord (Alstermark et al. 1987), and it is thought that this system might be responsible for postural adjustements which are required when the center of gravity is changed during reaching. The ascending collateral projects onto the lateral reticular nucleus (Alstermark et al. 1981a), providing a perfect example of an efference copy.

Evidence for a Propriospinal-like System in Humans

Given the considerable development of monosynaptic-cortico-motoneuronal projections in primates, the question arises of whether such a propriospinal system is still functioning or even exists in humans. However, several arguments support the existence of a similarly organized (propriospinal-like) system in man, and evidence is presented below that it plays a role in normal human movements.

The first step is to show that stimulation of low-threshold afferents evokes in upper limb MNs a nonmonosynaptic excitation. This was found while using two independent methods: construction of poststimulus time histograms (PSTHs) of voluntarily activated motor units and the spatial facilitation of the H reflex (Malmgren and Pierrot-Deseilligny 1988a). This nonmonosynaptic excitation has a long central delay (3–6 ms) and a very low threshold. A similar excitation has been observed in units of all explored muscles, and the central delay of this excitation has been found to be increased systematically with the rostrocaudal location of the explored motor nucleus: the more caudal, the longer the delay (Graciès et al. 1991). This strongly suggests a transmission through neurones located above the cervical enlargement.

Depression of propriospinal-like excitation is found constantly when the afferent input is increased. Evidence has been presented that this depression, evoked by both cutaneous and muscle afferents, reflects an inhibition of neurones mediating propriospinal-like excitation, namely a disfacilitation and not an inhibition exerted directly on MNs (Malmgren and Pierrot-Deseilligny 1988b; Nielsen and Pierrot-Deseilligny 1991). This depression fits feedback inhibition of PNs described in the cat.

Evidence for corticospinal excitation of propriospinal-like neurones has been provided using magnetic stimulation of the brain (Meunier and Pierrot-Deseilligny, unpublished). The spatial facilitation technique was used, the flexor carpi radialis (FCR) H reflex being conditioned by two volleys (ulnar and cortical) delivered separately and together. It was found that the reflex facilitation on combined stimulation was significantly larger than the sum of effects by separate stimuli, which suggests summation at the premotoneuronal level (see Malmgren and Pierrot-Deseilligny 1988a). The features (threshold, time course) of this additional facilitation are consistent with those of propriospinal-like excitation, thus indicating corticospinal excitation of propriospinal-like neurones. Moreover, it was found that the synaptic delay introduced by propriospinal-like neurones in the transmission of corticospinal excitation to MNs was compatible with a disynaptic linkage, which fits the connection described in the cat for PNs.

Thus, in humans, there is a neuronal system which is organized as the propriospinal system in the cat: neurones located above the cervical enlargement transmit disynaptic excitation from the corticospinal tract to upper limb MNs and receive both excitation and inhibition from upper limb afferents. Other similarities with the C3–4 propriospinal system in the cat are described below, and this system is referred to as propriospinal.

The investigations described here were undertaken to examine whether (and if so, to what extent) the propriospinal system is involved in normal human movements.

Evidence for Descending Excitation of PNs During Movement

The first step was to investigate whether the descending pathway evidenced in experiments using magnetic stimulation of the brain is activated during voluntary movement.

Facilitation of the Transmission in Propriospinal Pathways During Movement

Facilitation of transmission in the pathway of propriospinal excitation from ulnar afferents to FCR MNs was sought at the onset of voluntary wrist flexion (Baldissera and Pierrot-Deseilligny 1989; Burke et al. 1992a). For this, changes in the ulnar-induced excitation of the FCR H reflex were compared at rest and at the onset of movement.

Ulnar-Evoked Propriospinal Excitation at Rest. The time course of the variations in the FCR H reflex when preceded by a weak (0.7 × MT) stimulation of the ulnar nerve at rest is shown in Fig. 2A. A significant ($p < 0.01$) reflex facilitation was present when the conditioning volley reached spinal level 3.5–6 ms in advance of the test volley. The absence of facilitation at earlier intervals confirms the absence of monosynaptic projections from Ia afferents in the ulnar nerve to MNs of the forearm flexors innervated by the median nerve (Malmgren and Pierrot-Deseilligny 1988a). In Fig. 2B the test reflex size is plotted against the conditioning stimulus intensity (4.5 ms conditioning test interval) to show that the threshold for the facilitation was low (0.6 × MT). A similar ulnar nerve induced facilitation of the reflex (long latency, brief duration, low threshold, weak amplitude) could be observed in all explored subjects, but it was rarely (24 of 93 experiments) significant ($p < 0.05$), and the mean value of the effects observed in all experiments was a very small facilitation (6% of the control reflex size). The threshold of the ulnar-induced facilitation of the reflex was below that of the monosynaptic Ia excitation evoked by stimulation of this nerve in homonymous motoneurones (flexor carpi ulnaris). This indicates that the afferents responsible for the reflex facilitation do not have a smaller diameter and thus a slower conduction velocity than Ia afferents evoking monosynaptic Ia excitation. Under these conditions, a latency longer by 3–4 ms than the monosynaptic latency cannot be attributed to a peripheral

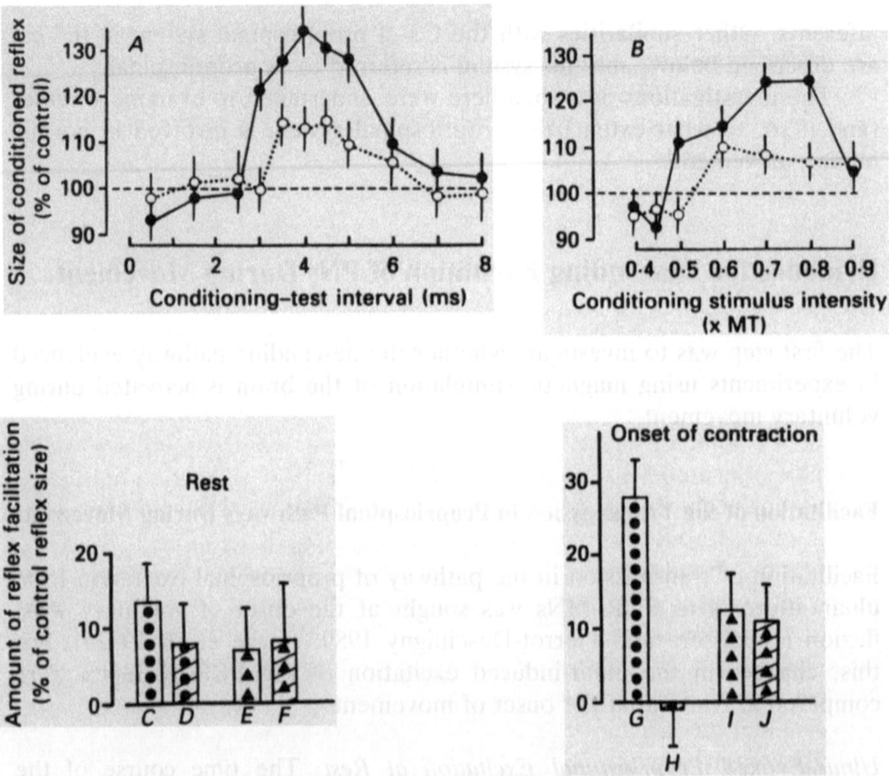

Fig. 2. **A, B** Changes in the amplitude of the FCR H reflex evoked by stimulation of the ulnar nerve are compared at rest (○) and at the onset of a voluntary wrist flexion (●). The amplitude of the reflex (expressed as a percentage of its control value) is plotted against the conditioning test interval (**A**, ulnar nerve stimulus intensity, 0.7 × MT) and the intensity of the conditioning stimulus (**B** 4.5 ms conditioning test interval). Each point represents the mean of 200 (**A**) or 40 (**B**) measurements for a single subject. *Vertical bars*, one standard error of the mean. **C–J** comparison between the effects of radial nerve stimulation on ulnar- and cutaneous-induced facilitations of the FCR H reflex, at rest (**C–F**) and at the onset of FCR contraction (**G–J**). Each column represents the amount of reflex facilitation evoked by ulnar (*dotted columns*) and cutaneous (*columns with triangles*) stimulation when not preceded (**C, E, G, I**) and preceded (*hatched columns*, **D, F, G, J**) by radial nerve stimulation. Each column represents the mean of 100 measurements for a single subject. *Vertical bars*, one standard error of the mean. (**A, B** Adapted from Burke et al. 1992a; **C–J**, adapted from Burke et al. 1992b)

conduction time longer than that of Ia afferents evoking the reflex but reflects the central delay of the facilitation. The characteristics of the reflex facilitation are thus consistent with those of propriospinal-like excitation described in PSTH and reflex experiments (Malmgren and Pierrot-Deseilligny 1988a), namely low threshold, 3–4 ms central delay, abrupt onset, and short duration (Fig. 2A).

Facilitation of Ulnar-Evoked Propriospinal Excitation at the Onset of Voluntary Wrist Flexion. Experiments were performed at the onset of a phasic stereotyped anisometric contraction of relatively weak strength. In these experiments the first stimulus to be delivered was triggered by the onset of EMG activity, which was recorded with a higher amplification than the reflex, rectified, and connected to a triggering circuit. The FCR H reflex was largely facilitated at the onset of voluntary wrist flexion. Since the susceptibility of an H reflex to facilitation and inhibition varies with the size of the reference (unconditioned) reflex (Crone et al. 1990), the intensity of the test stimulus was decreased so that the size of the unconditioned reflex was the same as at rest. Filled symbols in Fig. 2 show an example of the increase in the ulnar-induced facilitation at the onset of voluntary wrist flexion. The supplementary facilitation (size of the conditioned reflex during contraction minus its size at rest) was highly significant ($p < 0.01$). The facilitation began 0.5 ms earlier than at rest, but the time course was similar in other respects (Fig. 2A). The threshold of the facilitation was lowered to $0.5 \times$ MT, but the supplementary facilitation disappeared when the conditioning stimulus was increased to $0.9 \times$ MT (Fig. 2B). Such an increase in the ulnar-induced facilitation at the onset of voluntary wrist flexion was observed in 45 of 54 experiments, and the mean value of the supplementary facilitation was 12% of the control reflex size. The same arguments as at rest (low threshold, long central delay, brief duration) and, in addition, the depression of the facilitation (Fig. 2B) when the afferent input was increased (see Malmgren and Pierrot-Deseilligny 1988b) strongly suggest that the supplementary ulnar-evoked facilitation during contracting reflects a facilitation of the transmission in the pathway of propriospinal-like excitation. Moreover, because of the characteristics of this facilitation, involvement of other known interneuronal pathways can be excluded, for example, oligosynaptic group I excitation mediated through *segmental* pathways (Jankowska et al. 1981) should be disclosed at earlier intervals (1–2 ms), and depression of presynaptic inhibition of Ia afferents mediating the test volley should last longer (Baldissera et al. 1981; Nakashima et al. 1990, their Fig. 3).

Changes in Presynaptic Inhibition of Afferents to PNs During Movement

The question arises of whether the facilitation of the transmission of ulnar-evoked propriospinal excitation to FCR MNs during movement results from excitation of PNs or from a decrease in presynaptic inhibition of primary afferents projecting to these neurones. Changes in presynaptic inhibition of afferents synapsing with PNs have never been investigated in the cat. Nevertheless the possibility must be considered because there is a decrease in presynaptic inhibition of monosynaptic Ia projections to motoneurones at the onset of voluntary contractions (Hultborn et al. 1987b; Meunier and Pierrot-Deseilligny, 1989). It was therefore investigated whether primary afferents projecting to PNs are subject to presynaptic inhibition, and whether changes in presynaptic inhibition contribute to the increase in excitability of the propriospinal-like system associated with a voluntary contraction (Burke et al. 1992b).

Evidence for Presynaptic Inhibition of Afferents to PNs. Radial nerve stimulation inhibits the FCR H reflex, and the available evidence favors the view that the phase of inhibition at conditioning test intervals of 10–30 ms is due to presynaptic inhibition of the Ia afferent projection to FCR motoneurones responsible for the test reflex (Berardelli et al. 1987). The possibility that this radial stimulation also evokes presynaptic inhibition of ulnar afferents to PNs projecting onto FCR MNs was investigated. Thus the amount of propriospinal excitation of the FCR H reflex evoked by ulnar stimulation (0.7 × MT, 5 ms conditioning test interval) was compared when not preceded (Fig. 2C) and when preceded (Fig. 2D) by a stimulus to the radial nerve (0.95 × MT) 10 ms before the ulnar stimulus. Radial nerve stimulation decreased the ulnar-evoked propriospinal excitation, and the amount of this depression (i.e., D minus C) was investigated varying the time interval between radial and ulnar stimuli. The time course of the radial-induced depression of propriospinal excitation had a long duration (200–300 ms), much as has been described for presynaptic inhibition in the cat hind limb (see Eccles 1964) and in the human lower limb (Hultborn et al. 1987a). This suggests that the depression is not due to postsynaptic inhibition of PNs but to presynaptic inhibition of afferents projecting to them. The absence of radial-induced depression of propriospinal excitation evoked by cutaneous afferents in the superficial radial nerve (Fig. 2E,F) also supports this view, since these cutaneous afferents and mixed afferents in the ulnar nerve have a readily demonstrable convergence onto common PNs (Burke et al. 1992a). Under these conditions it appears unlikely that complete or nearly complete abolition of the mixed nerve evoked facilitation, as was seen sometimes at rest and often during contraction (Fig. 2H), could occur postsynaptically at the PN level without a significant reduction in the cutaneous facilitation. Hence, it appears likely that the radial-induced depression of the propriospinal excitation evoked by ulnar afferents reflects presynaptic inhibition of those afferents.

Increase in Presynaptic Inhibition of Afferents to PNs During Voluntary Movement. In Fig. 2C–J the comparison is drawn between the effects of radial nerve stimulation on the amount of propriospinal excitation evoked by ulnar nerve (C and D, and G and H) and cutaneous (superficial radial, E and F, and I and J) at rest (C–F) and at the onset of voluntary wrist flexion (G–J). At rest, presynaptic inhibition of ulnar afferents to PNs reduced by half the amount propriospinal excitation (C, D). At the onset of contraction, although the ulnar-induced excitation was significantly increased (G), it was completely abolished by radial nerve stimulation (H). By contrast, the cutaneous-induced facilitation was not modified either at rest (E, F) or at the onset of contraction (I,J). Similar results were observed in all seven subjects so explored. Propriospinal excitation evoked by ulnar nerve volleys is thus more depressed by radial stimulation at the onset of voluntary contraction than at rest, and this implies that at least some excitatory inputs to PNs are gated during contraction. This finding is not compatible with the possibility that the facilitation of the transmission in the pathway of propriospinal excitation during voluntary movement (see above) is due to a decrease in presynaptic inhibition of afferents to PNs. Hence, this facilitation reflects an increased excitation of PNs.

Evidence for a Descending Facilitation of PNs

In these experiments the conditioning volley was triggered by the first electromyogram (EMG) potential of the contraction, i.e., in advance of the spindle (Vallbo 1971), group Ib (Binder et al. 1977), and cutaneous (Hulliger et al. 1979) afferent discharges evoked by the active movement. Under these circumstances the increased reflex facilitation observed here at the onset of voluntary wrist flexion is likely to reflect descending facilitation of PNs. It is tempting to assume that this descending facilitation is due to activation during movement of the corticospinal pathway evidenced in experiments using magnetic brain stimulation.

Relative Contribution of Peripheral and Descending Inputs to Excitation of PNs

Contrary to effects mediated through human segmental interneuronal systems (e.g., Ib inhibition or recurrent inhibition; for references see Fournier and Pierrot-Deseilligny 1989), at rest, propriospinal excitation was small (Fig. 2A,B) or absent (Fig. 3F,G), which suggests that the peripheral excitatory input to PNs is weak. The significant reflex facilitation disclosed at the onset of contraction (or after cortical stimulation) may thus be ascribed to the descending excitation of these neurones (since it does not reflect a change in presynaptic inhibition of afferents to PNs). This suggests that, as with the C3–4 PNs in the cat (see Alstermark and Lundberg 1992), PNs receive their main excitatory input from descending tracts. However, several factors could lead to underestimation of the descending facilitation, as assessed here by the effects of a conditioning volley: (a) occlusion at the level of PNs between descending excitation and conditioning volley, (b) the fact that peripheral volleys evoke both excitation and inhibition of PNs, inhibition being also facilitated during contraction (Burke et al. 1992a), and

(c) as seen above, afferent volleys to PNs are gated by presynaptic inhibition during contraction (Burke et al. 1992b).

Evidence that the Descending Command of Movement Is Transmitted Partly to MNs Through PNs

Functionally speaking, the crucial question was then to know whether the descending excitation of PNs is strong enough to contribute significantly to the transmission of the cortical command. This was approached by investigating the effects upon the voluntary MN discharge of a peripheral volley supposed to inhibit PNs. Indeed, if a significant part of the descending command is passing through the propriospinal system, one can expect the voluntary EMG to be depressed by this inhibitory volley (see the schematic diagram of Fig. 3A).

Cutaneous Inhibition of PNs

Cutaneous stimuli, which have been shown to depress PNs with a very specific pattern (Nielsen and Pierrot-Deseilligny 1991), were used to inhibit PNs. An example of cutaneous depression of propriospinal excitation is shown in Fig. 3B–D. The PSTH method was used, and the firing probability of the unit was plotted against the latency measured from the monosynaptic

Fig. 3. **A** Schematic diagram of cutaneous inhibition of homonymous propriospinal pathways to ECR MNs. **B–D** Cutaneous inhibition of the propriospinal excitation evoked in an ECR motor unit by radial nerve stimulation. *Left*, time histograms of the discharge of the voluntarily activated motor unit in response to stimulation (*open columns*) and in control conditions (*shaded columns*); *right*, the difference between the two histograms shown on the left. *Ordinate*, number of counts expressed as a percentage of the number of triggers; *abscissa*, latency measured from the monosynaptic latency. **B** Effect of isolated stimulation of the radial nerve (0.5 × MT). **C** Combined stimulation of the radial nerve and of the palmar side of the fingers. **D** Combined stimulation of the radial nerve and of the dorsal side of the fingers. **E** Rectified and averaged (250 sweeps) EMG of the ECR obtained at the onset of a weak voluntary wrist extension and after cutaneous stimulation applied either to the dorsal (●) or to the palmar (○) side of the fingers. EMG, expressed as a percentage of the control EMG (without cutaneous stimulation), is plotted against the time elapsed after the arrival of the cutaneous volley at the spinal level (*vertical arrow*). **F**, **G** Changes in the amplitude of the ECR H reflex evoked by stimulation of the triceps nerve are compared at rest (○) and at the onset of a voluntary contraction of the triceps (●). The amplitude of the reflex (expressed as a percentage of its control value) is plotted against the conditioning test interval (**F**, conditioning stimulus intensity: 0.8 × MT) and the intensity of the conditioning stimulus (**G**, 4 ms conditioning-test interval). Each point represents the mean of 40 measurements for a single subject. *Vertical bars*, one standard error of the mean, **B–D**, Adapted from Nielsen and Pierrot-Deseilligny 1991; **F**, **G**, adapted from Mazevet and Pierrot-Deseilligny 1993

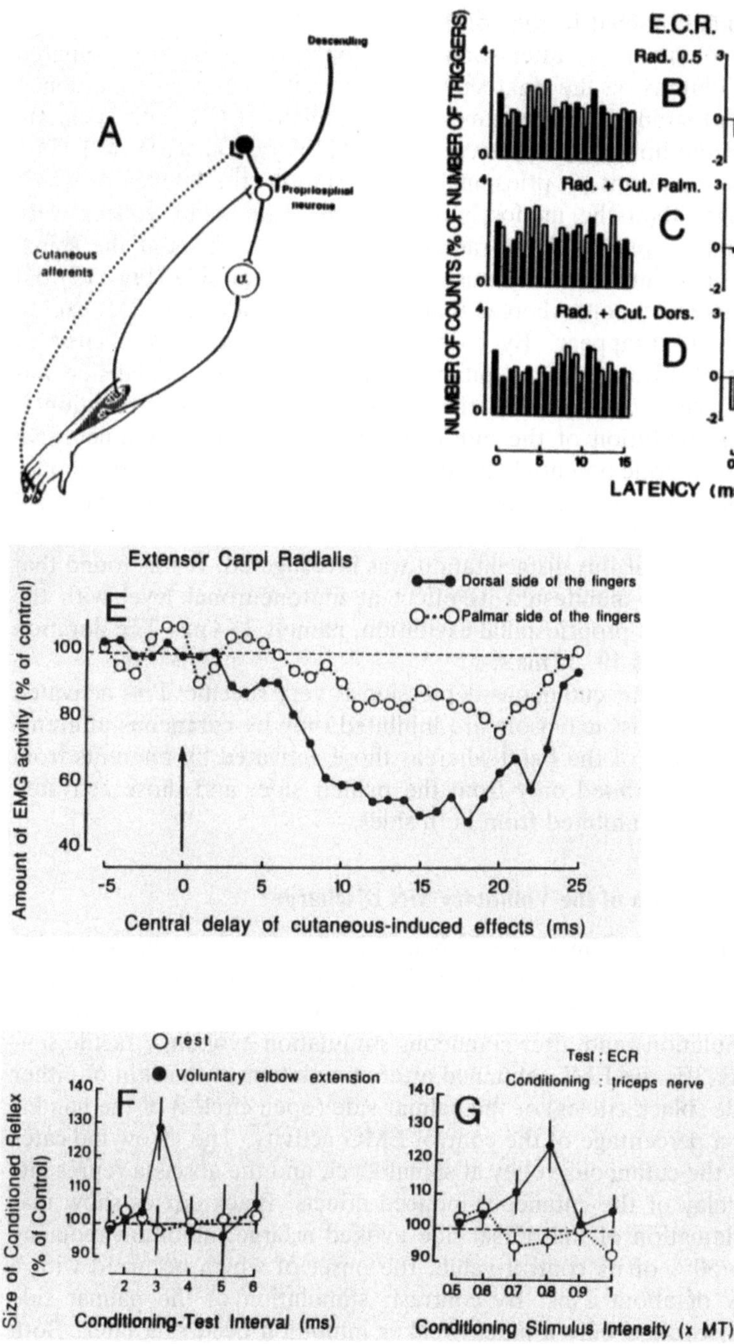
A
Descending
Propriospinal neurone
Cutaneous afferents
α
E.C.R.
Rad. 0.5
B
Rad. + Cut. Palm.
C
Rad. + Cut. Dors.
D
NUMBER OF COUNTS (% OF NUMBER OF TRIGGERS)
LATENCY (ms)
Extensor Carpi Radialis
E
Dorsal side of the fingers
Palmar side of the fingers
Amount of EMG activity (% of control)
Central delay of cutaneous-induced effects (ms)
rest
Voluntary elbow extension
F
Size of Conditioned Reflex (% of Control)
Conditioning-Test Interval (ms)
Test : ECR
Conditioning : triceps nerve
G
Conditioning Stimulus Intensity (x MT)

latency. Results obtained in the control situation (namely without stimulation, shaded columns) and after stimulation (open columns) are compared on the left. Columns on the right show the difference between conditioned and control histograms. In this extensor carpi radialis (ECR) unit, separate stimulation of the homonymous nerve, the radial nerve, at 0.5 × MT (Fig. 3B) evoked a significant propriospinal-like excitation, the latency of which was 3 ms longer than the monosynaptic latency. Cutaneous volleys were timed so that they preceded the radial nerve volley by 1 ms at the spinal level. Cutaneous stimulation administered to the palmar side (Fig. 3C) did not modify this excitation, whereas that of the dorsal side (Fig. 3D) caused the excitation to disappear. By contrast, monosynaptic peak elicited at variable intervals after the same cutaneous volley from the dorsal side was not modified. This indicates that the depression of the late excitation is not due to an inhibition of the motoneurone itself since a motoneuronal inhibitory postsynaptic potential (IPSP) should have also reduced the monosynaptic peak. Thus the inhibition must be exerted on the propriospinal neurones mediating the excitation.

The time couse of this disfacilitation was investigated. It was found that cutaneous inhibition manifested its effect at motoneuronal level with the same central delay as propriospinal excitation, namely 3–4 ms. The duration of the depression was 10–15 ms.

The pattern of this cutaneous depression is very specific: PNs activated by afferents from wrist extensors are inhibited only by cutaneous afferents from the dorsal side of the hand whereas those activated by afferents from wrist flexors are inhibited only from the palmar side, and those activated from the biceps are inhibited from both sides.

Cutaneous Depression of the Voluntary MN Discharge

Effects of cutaneous stimuli were then investigated upon the voluntary EMG of the ECR recorded during phasic, stereotyped movements of wrist extension. The EMG was rectified and integrated in a control situation (without stimulation) and after cutaneous stimulation evoking a tactile sensation. In Fig. 3E, the EMG obtained after stimulation of the skin of either the dorsal side (black circles) or the palmar side (open circles) of the hand is expressed as a percentage of the control EMG activity. The arrow indicates the arrival of the cutaneous volley at spinal level, and the abscissa represents the central delay of the cutaneous-induced effects. Black circles show that cutaneous stimuation of the dorsal side evoked a large inhibition reducing the EMG to 50% of its control value, the onset of which occurred with a central delay of about 3 ms. By contrast, stimulation of the palmar side (open circles) evoked only a much weaker inhibition occurring later. Both the central delay and the pattern (inhibition from the dorsal side) are compatible with an inhibition by the cutaneous volley of the descending command passing through PNs.

Further evidence for such a mechanism was provided by showing that the depression of the EMG reflects an inhibition of neurones mediating exciation to MNs, namely a disfacilitation, and not an inhibition exerted directly to MNs. For this, the H reflex of the same muscle was used. In the case of disfacilitation, the monosynaptic Ia excitatory postsynaptic potential (EPSP) evoked by the test volley should not be modified and the amplitude of the test reflex should be unchanged. By contrast, in the case of direct inhibition of the MNs, the changes in conductance accompanying the motoneuronal IPSP should decrease the test EPSP and thus the reflex amplitude. Contrasting with the depression of the voluntary EMG beginning at the 3 ms central delay, the test reflex size was not modified between 3 and 10 ms. Thus during this first period the depression of the EMG reflects a disfacilitation and not an inhibition exerted directly onto MNs.

We suggest that this disfacilitation, with its particular central delay and pattern, reflects the inhibition by the cutaneous volley of the descending command passing through PNs (Fig. 3A). The important depression of the EMG then implies that PNs may mediate a significant part of the descending command to MNs during normal human movements.

Pattern of Descending Excitation of PNs During Movement

Descending Facilitation of Heteronymous Pathways

Experiments illustrated in Fig. 3A–E deal with homonymous propriospinal pathways to ECR MNs. However, Graciès et al. (1991) have shown that each motor nucleus receives propriospinal excitation from a multimodal and wide-range peripheral input. The question then arises of how the transmission is modified in these heteronymous propriospinal pathways during voluntary contraction. Changes in propriospinal excitation evoked in a given motor nucleus by stimulation of different peripheral nerves were therefore investigated at the onset of voluntary contractions involving various muscles. For example, in Fig. 3F,G, the changes in the ECR H reflex evoked by stimulation of the nerve of the triceps brachii are compared at rest (open circles) and at the onset of voluntary elbow extension (black circles). The size of the test reflex is plotted in Fig. 3F against the conditioning test interval and in Fig. 3G against the intensity of the conditioning stimulation. At rest there was no effect, but a clear facilitation appeared at the onset of triceps contraction. This facilitation had a 3-ms central delay, a very brief duration, and a low threshold (0.7 × MT), and it disappeared when the afferent input was increased (Fig. 3G), i.e., all the characteristics of a propriospinal excitation. A similar result was found in 13 of 16 experiments (seven subjects). The mean value of the supplementary facilitation of the ECR H reflex during contraction (size of the conditioned reflex during contraction minus its size at rest), based upon all the experiments, using the interval 3 or 4 ms, whichever had with the largest facilitation, was 22% of the unconditioned reflex size. Here again, the conditioning volley was triggered in advance of the contraction-induced afferent discharge, and the reflex facilitation must indicate a descending facilitation of PNs.

Selection by Higher Centers of the Subset of PNs Receiving the Afferent Feed-back from the Contracting Muscle

Table 1 compares the supplementary facilitation of the ECR H reflex evoked by stimulation of either the musculocutaneous or the triceps nerve at the onset of a selective voluntary contraction involving either the biceps, or the ECR or the triceps. For each combination (conditioning nerve-contracting muscle) the number of experiments, the number (and percentage) of cases with a significant ($p < 0.05$) supplementary facilitation at the onset of contraction, and the mean value of this supplementary facilitation are given. Table 1 shows that propriospinal excitation evoked in ECR MNs by afferents from the contracting muscle was the only one to be significantly increased. The same has been shown for propriospinal excitation to FCR MNs (Burke et al. 1992a).

This implies that excitations evoked in a given motor nucleus by afferents from various muscles are mediated through different subsets of PNs, as already revealed from their different patterns of cutaneous inhibition (Nielsen and Pierrot-Deseilligny 1991). Different subsets projecting to ECR MNs are sketched in Fig. 4. One subset (right) is selected by motor cortex during elbow extension and receives its main afferent input from the triceps; a second (middle) is selected by motor cortex during elbow flexion and receives its main afferent input from the biceps; and a third (left) is selected by motor cortex during wrist extension and receives its afferent input from

Table 1. Propriospinal-like excitation evoked in ECR MNs by afferents from elbow muscles

Conditioning stimulation	Selective contraction investigated								
	Onset of biceps			Onset of ECR			Onset of triceps		
	n	Fac.	M	*n*	Fac.	M	*n*	Fac.	M
Musculocutaneous	24	19 (79%)	+21.4	7	0 (0%)	+1.9	4	0 (0%)	−7.3
Triceps nerve	4	0 (0%)	−1.7	10	2 (20%)	+6.6	16	13 (81%)	+22

n, Number of experiments performed; Fac., number (percentage) of experiments in which there was a significant ($p < 0.05$) supplementary facilitation at the onset of contraction; M, mean value (calculated from all experiments) of the supplementary facilitation during contraction (value during contraction minus value at rest) expressed as a percentage of the unconditioned reflex size.

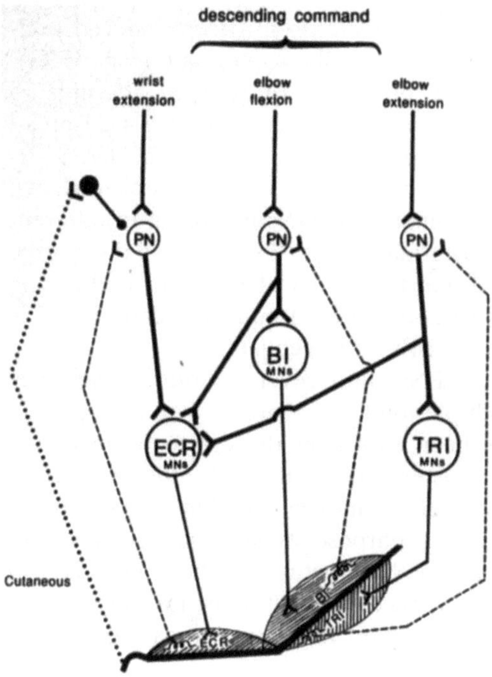

Fig. 4. Hypothetical circuit diagram illustrating the connections of three subsets of PNs. One subset (*right*) is excited by descending tracts during elbow extension, receives its main afferent excitatory input from the triceps, and projects onto triceps and ECR MNs. The second subset (*middle*) is excited by descending tracts during elbow flexion, receives its main afferent excitatory input from the biceps, and projects onto biceps and ECR MNs. The third subset (*left*) is excited by descending tracts during wrist extension, receives its main afferent excitatory input from the ECR, and projects to ECR MNs. Cutaneous inhibition is represented only for the last subset

the extensor carpi radialis. As seen above, each subset receives cutaneous inhibition from a particular skin field (here only cutaneous inhibition of the subset activated during wrist extension is represented).

Thus, *during voluntary contraction, corticospinal excitation and the afferent feedback from the contracting muscle are focused on the same subset of PNs.* Because of this very specific convergence, the afferent feedback may modify and servocontrol the descending activity passing through the propriospinal relay.

The Propriospinal System and Motor Synergies

During selective contraction of one muscle propriospinal excitation evoked by afferents from this muscle is increased in motor nuclei which are not involved in this contraction, for example, increase in triceps-induced excitation of ECR MNs at the onset of voluntary elbow extension (Fig. 3F,G). This divergence may occur either by branching of primary afferents to groups of neurones each supplying its own motor nucleus or by a widespread projection from common neurones to many motor nuclei. The latter explanation is by far the simplest and the most likely since it has been shown in the cat that a single C3–4 PN terminates on MNs belonging to different proximal and distal motor nuclei (Alstermark et al. 1990). Such a divergence is sketched in Fig. 4. It is assumed that the subset of PNs activated during elbow extension (right) or elbow flexion (middle) and projecting in ECR MNs projects also to triceps or biceps MNs, respectively.

These divergent projections of PNs onto several motor nuclei may underlie the co-ordination of multiple-joint movements. For example, in reaching, a movement which has been shown in the cat to be mediated through the propriospinal system (Alstermark et al. 1981), the protraction of the limb is caused by shoulder and elbow movements, but the position of the wrist is essential for the grasping purpose of such movements. We suggest that descending excitation of PNs directed to elbow MNs is also responsible, through their diverging branches to wrist MNs, for the required contractions at wrist level, so that the hand is in an optimal position to perform grasping.

Conclusion

There is evidence for the existence in humans of a neuronal system organized as the C3–4 propriospinal system in the cat: neurones located above C5 transmit disynaptic excitation from the motor cortex to upper limb MNs and also receive weak excitation and strong inhibition from low-threshold afferents from the limb. These neurones, presumed to be PNs, are organized in subsets according to their afferent input. During voluntary contraction, there is descending excitation of the subset(s) of PNs receiving the afferent input from the contracting muscle. It seems that, even during the selective stereotyped contractions explored here, this descending excitation is strong enough to contribute significantly to the transmission of the descending command to MNs. The projections of each subset are divergent, implying that individual PNs project onto diverse motor nuclei, which favors the coordination of multi-joint movement.

Acknowledgements. The authors express their gratitude to Prof. Anders Lundberg for many inspiring discussions. This work was supported by grants from DRED, AP-HP, INSERM (92 80 12), and IRME.

References

Alstermark B, Lundberg A (1992) The C3–4 propriospinal system; target-reaching and food-taking. In: Jami L, Pierrot-Deseilligny E, Zytnicki D (eds) Muscle afferents and spinal control of movement. Pergamon, London, pp 327–354

Alstermark B, Lindström S, Lundberg A, Sybirska E (1981a) Integration in descending motor pathways controlling the forelimb in the cat. viii. Ascending projection to the lateral reticular nucleus from C3–4 propriospinal neurones also projecting to forelimb motoneurones. Exp Brain Res 42:282–298

Alstermark B, Lundberg A, Norrsell U, Sybirska E (1981b) Integration in descending motor pathways controlling the forelimb in the cat. ix. Differential behavioural defects after spinal cord lesions interrupting defined pathways from higher centres to motoneurones. Exp Brain Res 42:299–318

Alstermark B, Lundberg A, Sasaki S (1984a) Integration in descending motor pathways controlling the forelimb in the cat. xi. Inhibitory pathways from higher motor centres and forelimb afferents to C3–4 propriospinal neurones. Exp Brain Res 56:293–307

Alstermark B, Lundberg A, Sasaki S (1984b) Integration in descending motor pathways controlling the forelimb in the cat. xii. Interneurones which may mediate descending feed-forward inhibition and feed-back inhibition from the forelimb to C3–4 propriospinal neurones. Exp Brain Res 56:308–322

Alstermark B, Górska T, Johannisson T, Lundberg A (1986) Hypermetria in forelimb target-reaching after interruption of the inhibitory pathway from forelimb afferents to C3–4 propriospinal neurones. Neurosci Res 3:457–461

Alstermark B, Lundberg A, Pinter M, Sasaki S (1987) Long C3–5 propriospinal neurones in the cat. Brain Res 404:382–388

Alstermark B, Kümmel H, Pinter MJ, Tantisira B (1990) Integration in descending motor pathways controlling the forelimb in the cat. xvii. Axonal projection and termination of C3–4 propriospinal neurones in the C6-Th1 segments. Exp Brain Res 81:447–461

Bach-y-Rita P (1981) Central nervous system lesions: sprouting and unmasking in rehabilitation. Arch Phys Med Rehabil 62:413–417

Baldissera F, Hultborn H, Illert M (1981) Integration in spinal neuronal systems. In: Brooks VB (ed) Motor control. pp 508–595. American Physiological Society, Bethesda, pp 508–595 (Handbook of physiology, sect 1: The nervous system, vol 2)

Baldissera F, Pierrot-Deseilligny E (1989) Facilitation of transmission in the pathway of non-monosynaptic Ia excitation to wrist flexor motoneurones at the onset of voluntary movement in man. Exp Brain Res 74:437–439

Berardelli A, Day BL, Marsden CD, Rothwell JC (1987) Evidence favouring presynaptic inhibition between antagonist muscle afferents in the human forearm. J Physiol (Lond) 391:71–83

Binder MD, Kroin JS, Moore GP, Stuart DG (1977) The response of Golgi tendon organs to single motor unit contractions. J Physiol (Lond) 271:337–349

Burke D, Graciès JM, Mazevet D, Meunier S, Pierrot-Deseilligny E (1992a) Convergence of descending and various peripheral inputs onto common propriospinal-like neurones in man. J Physiol (Lond) 449:655–671

Burke D, Graciés JM, Meunier S, Pierrot-Deseilligny E (1992b) Changes in presynaptic inhibition of afferents to propriospinal-like neurones in man during voluntary contractions. J Physiol (Lond) 449:673–687

Crone C, Hultborn H, Maziéres L, Morin C, Nielsen J, Pierrot-Deseilligny E (1990) Sensitivity of monosynaptic test reflex to facilitation and inhibition as a function of the test reflex size: a study in man and the cat. Exp Brain Res, 81:35–45

Eccles JC (1964) Presynaptic inhibition in the spinal cord. Prog Brain Res 12:65–91

Fournier E, Pierrot-Deseilligny E (1989) Changes in transmission in some reflex pathways during movements in humans. News Physiol Sci 4:29–32

Graciès JM, Meunier S, Pierrot-Deseilligny E, Simonetta M (1991) Pattern of propriospinal-like excitation to different species of human upper limb motoneurones. J Physiol (Lond) (Lond) 434:151–167

Hulliger M, Nordh E, Thelin AE, Vallbo AB (1979) The responses of afferent fibres from the glabrous skin of the hand during voluntary finger movements in man. J Physiol (Lond) 291:233–249

Hultborn H, Meunier S, Morin C, Pierrot-Deseilligny E (1987a) Assessing changes in presynaptic inhibition of Ia fibres: a study in man and the cat. J Physiol (Lond) 389:729–756

Hultborn H, Meunier S, Pierrot-Deseilligny E, Shindo M (1987b) Changes in presynaptic inhibition of Ia fibres at the onset of voluntary contraction in man. J Physiol (Lond) 389:757–772

Illert M, Tanaka R (1978) Integration in descending motor pathways controlling the forelimb in the cat. iv. Corticospinal inhibition of forelimb motoneurones mediated by short propriospinal neurones. Exp Brain Res 31:131–141

Illert M, Lundberg A, Padel Y, Tanaka R (1978) Integration in descending motor pathways controlling the forelimb in the cat. v. Properties of and monosynaptic excitatory convergence on C3–4 propriospinal neurones. Exp Brain Res 33:101–130

Jankowska E, McCrea D, Mackel R (1981) Pattern of "non reciprocal" inhibition of motoneurones by impulses in group Ia muscle spindle afferents. J Physiol (Lond) 316:393–409

Lundberg A (1979) Integration in a propriospinal motor centre controlling the ferelimb in the cat. In: Asanuma H, Wilson VJ (eds) Integration in the nervous system. Igaku Shoin, Tokyo, pp 47–64

Malmgren K, Pierrot-Deseilligny E (1988a) Evidence for non-monosynaptic Ia excitation of wrist flexor motoneurones, possibly via propriospinal neurones. J Physiol (Lond) 405:747–764

Malmgren K, Pierrot-Deseilligny E (1988b) Inhibition of neurones transmitting non-monosynaptic Ia excitation to human wrist flexor motoneurones. J Physiol (Lond) 405:765–783

Mazevet D, Pierrot-Deseilligny E (1993) Pattern of descending excitation of propriospinal neurones at the onset of voluntary movement in man. Acta Physiol Scand (in press)

Meunier S, Pierrot-Deseilligny E (1989) Gating of the afferent volley of the monosynaptic stretch reflex during movement in man. J Physiol (Lond) 419:753–763

Nakashima K, Rothwell JC, Day BL, Thompson PD, Marsden CD (1990) Cutaneous effects on presynaptic inhibition of flexor Ia afferents in the human forearm. J Physiol (Lond) 426:369–380

Nielsen J, Pierrot-Deseilligny E (1991) Pattern of cutaneous inhibition of the propriospinal-like excitation to human upper limb motoneurones. J Physiol (Lond) 434:169–182

Vallbo AB (1971) Muscle spindle response at the onset of isometric voluntary contractions in man. Time difference between fusimotor and skeletomotor effects. J Physiol (Lond) 218:405–431

The Site of Activation of the Corticospinal System by Transcranial Magnetic and Electrical Stimulation of the Human Motor Cortex

D.J. Burke, R. Hicks, J. Stephen, I. Woodforth, and M. Crawford

Introduction

The term "spasticity" refers to the characteristic reflex changes that may be associated with a corticospinal lesion, and as such it is an important component of the upper motoneurone syndrome. However, as argued by others (particularly Landau 1974), the "spasticity" may be a relatively unimportant component of a patient's disability: the attendant weakness is commonly a much greater problem. Until recently, it has not been possible to study corticospinal function in human subjects, other than by making careful measurements of muscle strength. Measurements of strength reflect corticospinal function, in as much as the corticospinal system is the major executive pathway for voluntary motor acts in human subjects, and accordingly, the distribution of corticospinal influences can be inferred from such measurements (see the chapter by Gandevia, this volume). Nevertheless, it is likely that other descending pathways contribute to even the most volitional of motor acts, and this is likely to be so in patients in whom there is a long-standing lesion in the major executive pathway, the corticospinal system.

The demonstration by Merton and Morton (1980) that the corticospinal system could be activated in human subjects by high-voltage capacitively coupled electrical pulses applied through the intact scalp represented a major advance in human neurophysiology. However, pain has limited the widespread adoption of the technique as a diagnostic tool. This obstacle was largely overcome when it was found that brief magnetic pulses could activate the corticospinal system with minimal discomfort (Barker et al. 1985), and this technique has spawned an ever-burgeoning clinical and physiological literature (for review, see Rothwell et al. 1991).

Prince of Wales Medical Research Institute, University of New South Wales, and The Prince Henry, Prince of Wales and Prince of Wales Children's Hospitals, Little Bay N.S.W. 2036, Sydney, Australia

Thilmann et al. (Eds.)
Spasticity: Mechanisms and Management

The D- and I-Wave Hypothesis

There are differences between the two techniques for activating the corticospinal system, other than the pain experienced by the subject. The latency of the EMG response to magnetic stimulation is 1.5–2 milliseconds longer than that for the electrically evoked response (Day et al. 1987, 1989; Hess et al. 1987). Careful analyses of the discharge of single motor units activated by anodal and cathodal electrical stimuli and by clockwise and counter-clockwise magnetic stimuli provided indirect evidence in support of the "D- and I-wave hypothesis", formulated to explain these latency differences (Day et al. 1987, 1989).

In animals, a transient electrical stimulus to the motor cortex using a ball electrode on the pia can evoke multiple descending volleys in the corticospinal system, the initial one termed the "D" wave and the later waves termed "I" waves (Patton and Amassian 1954; Hern et al. 1962; Phillips and Porter 1977; Rothwell et al. 1991). The D-wave is due to direct activation of corticospinal neurones, and the I-waves are due to stimulation of afferent inputs to the corticospinal neurone, thereby producing trans-synaptic re-excitation of some of the corticospinal neurones that responded in the D-wave (see Kernell and Wu Chien-Ping 1967). In human subjects the EMG response of shortest latency occurred with anodal electrical stimulation, and this was therefore attributed to direct activation of corticospinal neurones at or near the initial segment. A response to cathodal stimulation required slightly higher stimulus intensities, and at threshold, the latency was found to be longer, much as was the case with magnetic stimulation. These types of stimulus were thought to produce corticospinal volleys consisting predominantly or exclusively of I waves.

Knowledge of precisely where the corticospinal system is activated is critical for two reasons. First, measurements of "central motor conduction time" are made with an accuracy of tenths of a millisecond: with an overall conduction time of 5–6 ms for the upper limb outflow, an uncertainty of 1–2 ms is unacceptable in a diagnostic test. Second, reservations would need to be retained about the interpretation of physiological studies if their underlying rationale (the D- and I-wave hypothesis) were called into question.

Site of Activation of Corticospinal Neurones with Anodal Electrical Stimulation

With anode at the vertex and cathode 7 cm laterally, stimulus intensities of 150 V are generally sufficient to produce a corticospinal volley that reaches the low-cervical level after a latency of 3–4 ms (Burke et al. 1990). The brevity of the latency and the sensitivity of the threshold volley to volatile anaesthetics (see below) are consistent with its generation at or near the initial segment of the corticospinal neurone, and it is therefore termed the D-wave. However, with modest increases in stimulus intensity (to 225–375 V in different subjects), the latency of the D wave decreases as a component of shorter latency begins to emerge in advance of the low-threshold component (Figs. 1, 2). The original low-threshold component commonly decreases in amplitude and is occasionally completely replaced as the shorter-latency, higher-threshold component emerges. With further increases in stimulus intensity (375–600 V), there is a further latency shortening as a third component of yet shorter latency appears and becomes established, in advance of the lower-threshold components (Fig. 2).

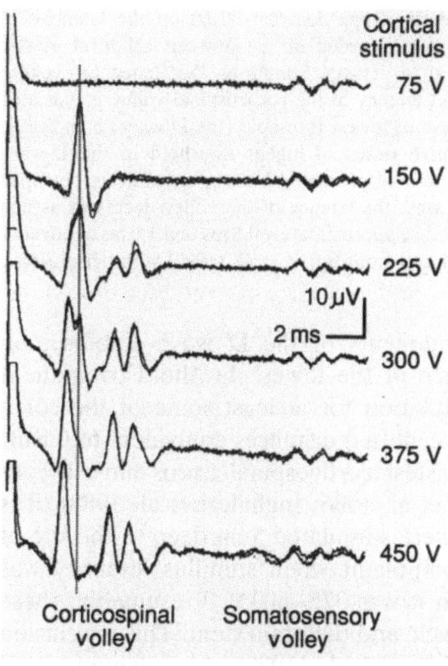

Fig. 1. The effect of increasing stimulus intensity (time constant 50 μs) on the corticospinal volley recorded at the low-cervical level. Electrode configuration: anode at vertex, cathode 7 cm lateral to vertex. Duplicate averages of 25 sweeps are superimposed for each stimulus intensity. The low-amplitude polyphasic deflection to the right of each trace represents an ascending somatosensory volley set up by bilateral stimulation of the tibial nerves in the popliteal fossae. The cortical and somatosensory stimuli were delivered simultaneously, at the onset of the sweeps. The intensity of the cortical stimulus was increased in steps, as indicated on the *right*. The intensity of the somatosensory stimulus remained constant. The *interrupted vertical line* indicates the D wave. Note that this peak is completely replaced at 450 V by an earlier peak that began to appear at 225 V. I waves appear at 225 V and increase with stimulus intensity. Reproduced from Burke et al. (1990)

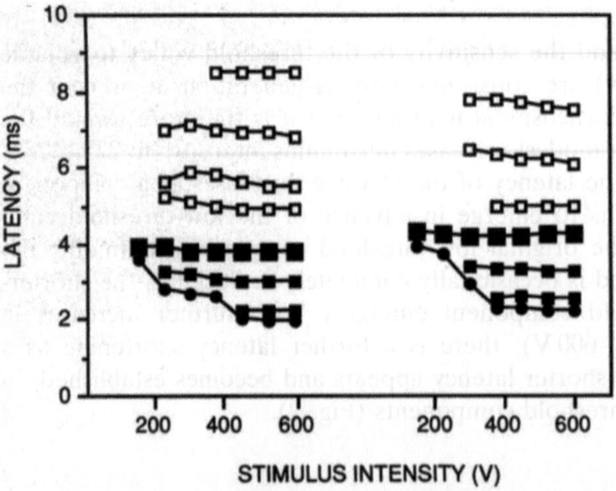

Fig. 2. The effect of increasing stimulus intensity (time constant 50 μs) on the latencies of different components of the corticospinal volley recorded at the low-cervical level in two subjects (anode at vertex; cathode 7 cm lateral to vertex). Stimuli at 75 V failed to evoke a corticospinal volley in either subject. The onset latency of the corticospinal volley is indicated by *round symbols*. The peak latency of the wave of lowest threshold (the D wave) is indicated by *large filled squares*, and the earlier negative peaks of higher threshold in the D-wave complex by *small filled squares*. The peak latencies of presumed I waves are indicated by *open symbols*. Note that as stimulus intensity increases, the latency of the volley decreases as first one, then a second component emerge and stabilise approximately 0.8 ms and 1.6 ms in advance of the component of lowest threshold. Reproduced from Burke et al. (1990) with permission

These two higher-threshold components of the D wave stabilise, on average, 0.8 ms and 1.7 ms in advance of the lowest-threshold component (Fig. 2). The effective point of stimulation for at least some of the corticospinal neurones must therefore have shifted distances equivalent to 0.8 ms and 1.7 ms deep to cortex. If the fastest corticospinal axons have a conduction velocity of 60–65 m/s (Boyd et al. 1986; Inghilleri et al. 1988), it is likely that some corticospinal axons were stimulated 5 cm deep to the site of initiation of the lowest-threshold component when stimulus intensity was 225–375 V, and 10–11 cm deep when it was 375–600 V. Presumably, these sites correspond to the cerebral peduncle and the brain stem. This conclusion is remarkably similar to that of Edgley et al. (1990), who demonstrated that transcranial anodal electrical stimulation could activate corticospinal neurones in the monkey at the medullary pyramids. Such shifts in the site of activation of corticospinal neurones, with comparable latency steps, have been seen with all other stimulus setups so far tested, though they are more readily demonstrated with anode at the vertex and cathode laterally because this setup produces the largest volleys for the least stimulus artifact.

Effects of Changing Motor Cortex Excitability on the Corticospinal Volley Evoked by Transcranial Stimulation

Motor cortex excitability can be depressed by volatile anaesthetics. Accordingly, it should be possible to test whether D waves, I waves or both are sensitive to changes in cortical excitability by withdrawing volatile anaesthetics. As might be expected for potentials dependent on synaptic transmission, I waves proved to be quite sensitive to volatile anaesthetics (Hicks et al. 1992a,b), with the critical level being, for isoflurane, an end-tidal level of 0.5% (Hicks et al. 1992b). The most sensitive were late I waves; there may be little effect on those waves occurring within 3 ms of the D wave (Hicks et al. 1992a). Less expected was a profound depressant effect of volatile anaesthetics on liminal D waves (Hicks et al. 1992a), as illustrated in Fig. 3. Stimuli that produced liminal D waves (mean 2.4 μV) when the end-tidal isoflurane concentration was >0.5% produced much larger D waves (mean 11.8 μV) after withdrawal of the volatile anaesthetic, a change that was significant ($P = 0.037$).

Presumably, the sensitivity of liminal D waves to volatile anaesthetics results from a depressant effect of these agents on the excitability of the corticospinal neurone, an action likely to be both direct and indirect, due to suppression of excitatory inputs. The findings provide support for the view that the liminal D wave arises from the initial segment of the neurone. The implication of the findings is that changes in motor cortex excitability can be reflected in the D wave as well as in I-waves. This raises cautions about studies that have compared changes in the EMG potentials produced by anodal electrical and magnetic stimulation to determine whether a conditioning volley has effects at the cortical level.

Transcranial Stimulation of the Upper-limb Area Using Anodal and Cathodal Stimuli

With one electrode over the upper-limb area of the motor cortex (7 cm lateral to the vertex) and the second 6 cm anteriorly, it should be possible to stimulate selectively the upper-limb area of the motor cortex with threshold stimuli. This setup was used to compare anodal and cathodal stimulation of the upper-limb area in 11 subjects (Burke et al. 1992). With both stimulus polarities, the component of lowest threshold was a D wave which appeared at the low-cervical level with latencies to negative peak of 3.63 $\pm$ 0.16 ms (anodal) and 3.66 $\pm$ 0.5 ms (cathodal). I waves required higher stimulus levels than the D wave with both polarities. On average, cathodal stimulation required a slightly higher increase in stimulus intensity above the D-

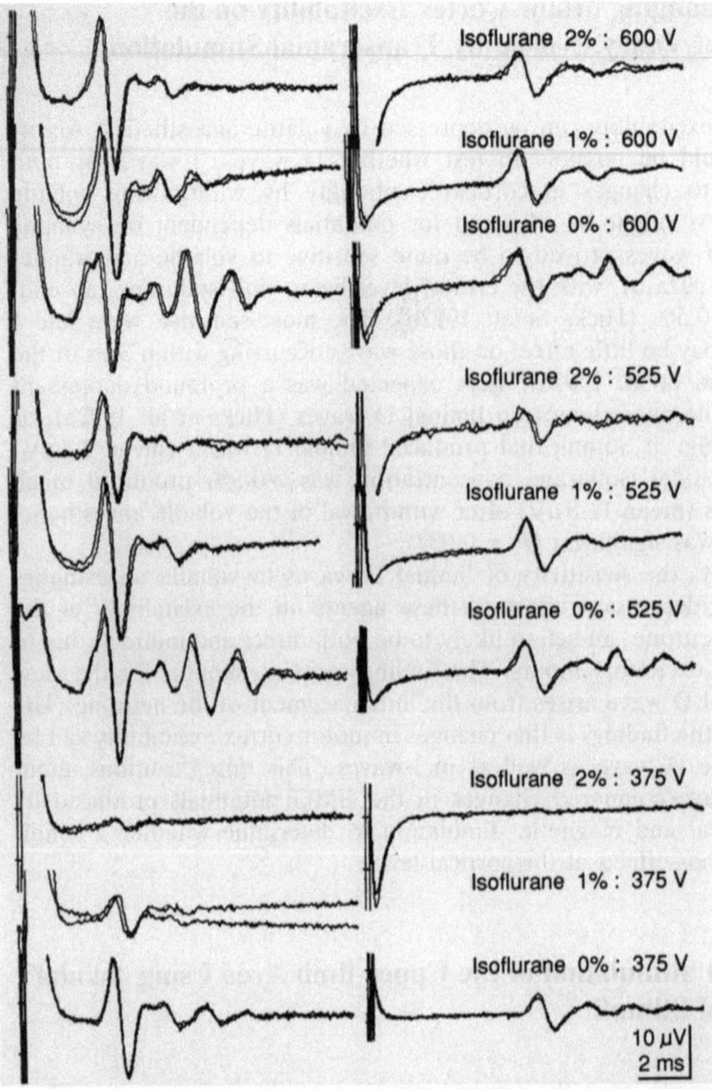

Fig. 3. Effects of isoflurane on corticofugal volleys evoked by vertex anodal stimulation using different stimulus intensities. *Left panels*, Low-cervical recording; *right panels*, corresponding low-thoracic recordings. I waves are more prominent at each stimulus intensity, the lower the isoflurane concentration. Similar changes are seen with the D wave, particularly at 375 V. Reproduced from Hicks et al. (1992a)

wave threshold to produce I waves, but this difference was not significant. In addition, when measured at the same increase in stimulus intensity above the D-wave threshold, there was no difference in the number of I waves with anodal and cathodal stimulation. With both stimulus polarities, strong stimuli resulted in a shortening of the latency of the D wave and the emergence of a high-threshold, short-latency peak, usually 0.8 ms in advance of the component of lowest threshold. Withdrawal of volatile anaesthetics did not change these threshold relationships appreciably.

These results confirm comparable findings for the lower-limb projection in subjects receiving volatile anaesthetics (Burke et al. 1990). They suggest that cathodal stimulation is merely a less effective means of delivering current to motor cortex than anodal stimulation.

Corticospinal Projections from Upper-limb Cortex to Lower-limb Spinal Segments

Using stimulating electrodes over the upper-limb area and 6 cm anterior, as in the previous section, the thresholds were determined for recordable corticospinal volleys at the low-thoracic region as well as at the low-cervical level (Burke et al. 1992). With anodal stimulation, the threshold for a descending volley at the low-cervical recording was 281 ± 33.6 V and at the low-thoracic level 317 ± 39.3 V (mean ± SEM, $n = 10$). In four subjects, a small volley was identified in the low-thoracic recording with the same threshold as at the low-cervical level. It is unlikely that these low stimulus intensities could excite lower-limb motor cortex directly. In addition, it is unlikely that the lower-limb projection was being inadvertently stimulated at some deep site, because this would have resulted in a latency shortening, much as that seen in Fig. 1 and discussed above. The findings imply that there are corticospinal projections to lower-limb segments from upper-limb motor cortex. Furthermore, given the greater difficulty in identifying the smaller volley at the low-thoracic level from noise, the small difference in thresholds for a detectable volley at the two levels is consistent with the view that there may be no true difference: i.e., that whenever upper-limb motor cortex is activated, there will also be a volley directed to the lower-limb segments.

There is ample precedent for a truncal and lower-limb projection from upper-limb motor cortex in cats and monkeys (Sherrington 1889; Phillips and Porter 1977; Shinoda et al. 1976, 1979). Presumably, such a projection serves a "postural" role, necessary to support the volitionally commanded movement. It is possible that a complementary projection exists from the lower-limb motor cortex to upper-limb spinal segments, though this would make more functional sense in a quadruped. It is of interest that a direct lower-limb projection has not been detected in studies that have recorded the EMG response to transcranial stimulation of upper-limb motor cortex (Rothwell et al. 1987). However, this is consistent with the projection being directed not to the motoneurone pool but to spinal interneurones, as might be expected if the projection served a "postural" role.

For those interested in recovery from upper motoneurone deficits, these results are both promising and cautionary. First, given sufficient plasticity, these projections might be functionally beneficial in patients with a deficit restricted to the outflow to one limb (for example, anterior cerebral artery insufficiency with involvement of lower-limb cortex but sparing of upper-limb cortex). Second, improvement in function and the re-appearance of EMG potentials to transcranial stimulation might be due to unmasking of this projection, rather than to recovery of corticospinal transmission or the use of noncorticospinal systems.

Corticospinal Volleys Evoked by Transcranial Magnetic Stimulation

The responses to magnetic stimulation have been recorded in human subjects with neurological deficits during neurosurgical operations by Berardelli et al. (1990; three patients) and Thompson et al. (1991; one patient). The former group identified a D wave in two subjects (but did not report findings with different stimulus intensities), and the latter reported a D wave appearing with stimulus intensities of 100%, above I-wave threshold. We have compared the corticospinal volleys evoked by graded electrical and magnetic stimulation in neurologically normal patients undergoing scoliosis surgery, before and after the withdrawal of volatile anaesthetics (Burke et al. 1993).

At stimulus intensities below 50%, no corticospinal volley was recorded in response to magnetic stimulation. With higher intensities magnetic stimuli produced a D wave, and it was the lowest-threshold component under volatile anaesthesia. Increasing stimulus intensity did not result in significant shortening of the latency of the magnetic D wave. Volatile anaesthetics raised the thresholds for and depressed amplitudes of both D and I waves, but affected I waves more than the D wave, and, following their withdrawal, the threshold for I waves was reduced to that of the D-wave in some but not all subjects. In one subject, stimulus intensities of 100% produced I-waves of $>5\,\mu V$ for a D wave of $<2\,\mu V$, although in others the D wave measured $10\,\mu V$ or more.

Thus, in these anaesthetized patients, transcranial magnetic stimulation did not produce I waves at lower stimulus intensities than the D wave. However, the D wave was commonly smaller than with electrical stimulation, and I waves are therefore likely to contribute more to the evoked muscle twitch with magnetic stimulation than with electrical stimulation.

Conclusions

Our studies are consistent with those of Edgley et al. (1990), who recorded corticospinal volleys in response to anodal electrical and magnetic stimulation of the monkey motor cortex. As they suggested, the difference in latency of the evoked EMG potentials is not due to the fact that the magnetic volley lacks a D wave. Similarly, our data suggest that cathodal stimulation differs from anodal stimulation only in that it is a less effective means of delivering current to motor cortex. These findings are not consistent with the "D- and I-wave hypothesis" as originally formulated.

References

Barker AT, Jalinous R, Freeston IL (1985) Non-invasive magnetic stimulation of human motor cortex. Lancet 2:1106–1107

Berardelli A, Inghilleri M, Cruccu G, Manfredi M (1990) Descending volley after electrical and magnetic transcranial stimulation in man. Neurosci Lett 112:54–58

Boyd SG, Rothwell JC, Cowan JMA, Webb PJ, Morley T, Asselman P, Marsden CD (1986) A method of monitoring function in corticospinal pathways during scoliosis surgery with a note on motor conduction velocities. J Neurol Neurosurg Psychiatry 49:251–257

Burke D, Hicks RG, Stephen JPH (1990) Corticospinal volleys evoked by anodal and cathodal stimulation of the human motor cortex. J Physiol (Lond) 425:283–299

Burke D, Hicks RG, Stephen JPH (1992) Anodal and cathodal stimulation of the upper-limb area of the human motor cortex. Brain

Burke D, Hicks R, Gandevia SC, Stephen J, Woodforth I, Crawford M (1993) Direct comparison of corticospinal volleys in human subjects to transcranial magnetic and electrical stimulation. J Physiol (Lond)

Day BL, Rothwell JC, Thompson PD, Dick, JPR, Cowan JMA, Berardelli A, Marsden CD (1987) Motor cortex stimulation in intact man. 2. Multiple descending volleys. Brain 110:1191–1209

Day BL, Dressler D, Maertens de Noordhout A, Marsden CD, Nakashima K, Rothwell JC, Thompson PD (1989) Electric and magnetic stimulation of human motor cortex: surface EMG and single motor unit responses. J Physiol (Lond) 412:449–473

Edgley SA, Eyre JA, Lemon RN, Miller S (1990) Excitation of the corticospinal tract by electromagnetic and electrical stimulation of the scalp in the macaque monkey. J Physiol (Lond) 425:301–320

Hern JEC, Landgren S, Phillips CG, Porter R (1962) Selective excitation of corticofugal neurones by surface-anodal stimulation of the baboon's motor cortex. J Physiol (Lond) 161:73–90

Hess CW, Mills KR, Murray NMF (1987) Responses in small hand muscles from magnetic stimulation of the human brain. J Physiol (Lond) 338:397–419

Hicks R, Burke D, Stephen J, Woodforth I, Crawford M (1992a) Corticospinal volleys evoked by electrical stimulation of human motor cortex after withdrawal of volatile anaesthetics. J Physiol (Lond) 456:393–404

Hicks RG, Woodforth IJ, Crawford MR, Stephen JPH, Burke DJ (1992b) Some effects of isoflurane on I waves of the motor evoked potential. Br J Anaest 69:130–136

Inghilleri M, Berardelli A, Cioni B, Cruccu G, Meglio M, Manfredi M (1988) The conduction velocity of the corticospinal tract in man. In: Rossini PM, Marsden CD (eds) Non-invasive stimulation of brain and spinal cord: fundamentals and clinical applications. Liss, New York, pp 123–130

Kernell D, Wu Chien-Ping (1967) Responses of the pyramidal tract to stimulation of the baboon's motor cortex. J Physiol (Lond) 191:653–672

Landau WM (1974) Spasticity: the fable of a neurological demon and the emperor's new therapy. Arch Neurol 1:217–219

Merton PA, Morton HB (1980) Stimulation of the cerebral cortex in the intact human subject. Nature 285:227

Patton HD, Amassian VE (1954) Single- and multiple-unit analysis of cortical stage of pyramidal tract activation. J Neurophysiol 17:345–363

Phillips CG, Porter R (1977) Corticospinal neurones. Their role in movement. Academic, London

Rothwell JC, Thompson PD, Day BL, Dick JPR, Kachi T, Cowan JMA, Marsden CD (1987) Motor cortex stimulation in intact man. 1. General characteristics of EMG responses in different muscles. Brain 110:1173–1190

Rothwell JC, Thompson PD, Day BL, Boyd S, Marsden CD (1991) Stimulation of the human motor cortex through the scalp. Exp Physiol 76:159–200

Sherrington CS (1889) On nerve-tracts degenerating secondarily to lesions of the cortex cerebri. J Physiol (Lond) 10:429–432

Shinoda Y, Arnold AP, Asanuma H (1976) Spinal branching of corticospinal axons in the cat. Exp Brain Res 26:215–234

Shinoda Y, Zarzecki P, Asanuma H (1979) Spinal branching of pyramidal tract neurons in the monkey. Exp Brain Res 34:59–72

Thompson PD, Day BL, Crockard HA, Calder I, Murray NMF, Rothwell JC, Marsden CD (1991) Intra-operative recording of motor tract potentials at the cervico-medullary junction following scalp electrical and magnetic stimulation of the motor cortex. J Neurol Neurosurg Psychiatry 54:618–623

Plasticity of the Human Motor Cortex

M. Hallett, L.G. Cohen, A. Pascual-Leone, J. Brasil-Neto, E.M. Wassermann, and A.N. Cammarota

Introduction

Although the role of each part of the brain in voluntary movement is incompletely known, it does appear that the main executor of voluntary movement is the primary motor cortex. Primary motor cortex has monosynaptic and oligosynaptic projections onto spinal motoneurons and a large percentage of cells that are active just prior to movement. Stimulation of these cells produces movement, and averaging of either cellular activity or regional brain potentials from this area shows a clear relationship to muscle activation. Until recently, the predominant view was that the anatomical and physiological relationship between the primary motor cortex and the spinal cord was immutable. Now there are a number of lines of evidence suggesting that the relationship between the motor cortex and its target muscles is modifiable, even in the adult brain. The evidence comes from both animal studies and human studies. This chapter will emphasize the human work and, specifically, information learned using the techniques of transcranial magnetic stimulation (TMS) (Cohen et al. 1991b). With TMS a pulsed magnetic field is applied over the scalp, and the resultant electric current flow in the underlying brain activates neurons. When the neurons activated are in the primary motor cortex, a muscle response or motor evoked potential (MEP) will be produced. By observing the MEPs and varying the stimulus intensity, it is possible to assess excitability of the motor system. Using a figure-8-shaped coil, the stimulus is rather focal, and by systematically moving the coil brain regions can be mapped. Some results will be reviewed using positron emission tomography (PET) with methods for measurement of blood flow. Activity of neurons induces an increase in regional cerebral blood flow, and this effect can be imaged.

In addition to the MEP produced in contralateral muscles by TMS, there is also an inhibitory response, or silent period (SP), that may be observed in the EMG of a voluntarily activated target muscle (Rothwell et al. 1987; Fuhr et al. 1991). SPs have generally been observed to follow

Human Motor Control Section, Medical Neurology Branch, National Institute of Neurological Disorders and Stroke, National Institutes of Health, Bethesda, MD 20892, USA

Thilmann et al. (Eds.)
Spasticity: Mechanisms and Management

MEPs, but, since their threshold is lower than that of the MEP, they may be produced in isolation by low-intensity stimulation (Davey et al. 1991; Wassermann et al. 1991b). Stimuli just slightly above the MEP threshold at the optimal position for producing a MEP elicit small MEPs followed by SPs at a few tightly clustered sites near the optimal position. However, stimuli of the same intensity produce isolated SPs in a more diffuse area extending 1–2 cm laterally from the optimal MEP area. The most prominent isolated SPs are actually elicited from sites lateral to the optimal MEP area.

Although in normal subjects it is rare to record responses to TMS in resting ipsilateral muscles, ipsilateral SPs (iSPs) may be readily observed in voluntarily activated muscles (Fig. 1A) (Wassermann et al. 1991a). These are most prominent in distal muscles. Ipsilateral MEPs on a background of voluntary activation are also seen in some subjects, and these are more prominent in proximal muscles (Fig. 1B).

Blood flow PET studies of voluntary movement show a large activation of contralateral motor cortex, and the appropriate topography can be demonstrated depending on the part of the body moving. This has been identified both with simple repetitive movements (Colebatch et al. 1991) and with tracking movements (Grafton et al. 1991). The supplementary motor area (SMA) also shows consistent activation. Although weak, we have often seen some activation of ipsilateral motor cortex in similar studies (Zeffiro and Hallett 1990, 1992). Other areas of brain are activated as well, the most prominent of which is the ipsilateral cerebellum.

Peripheral Deafferentation

Amputation of a limb produces a dramatic change in the relationship between central cortical representation of movement and effector organs. We studied seven patients with traumatic or surgical amputations and one patient with congenital absence of a limb (Cohen et al. 1991a). Their neurological examinations were normal. MEPs were recorded from biceps and deltoid muscles immediately proximal to the stump and from the same muscles contralaterally. In addition to TMS, electrical and magnetic stimuli were delivered to the brachial plexus at Erb's point to determine the maximal peripheral M responses. For threshold determinations, the lowest intensity of stimulation capable of inducing a 100-μV muscle response in at least five of ten trials was determined. The size of muscle representation areas was determined by delivering stimuli of equal intensities (100% of the output of the magnetic stimulator) to positions on both sides of the scalp, and three MEPs from stimulation of each scalp position were recorded. The average of the three responses was assigned to represent the scalp position stimulated.

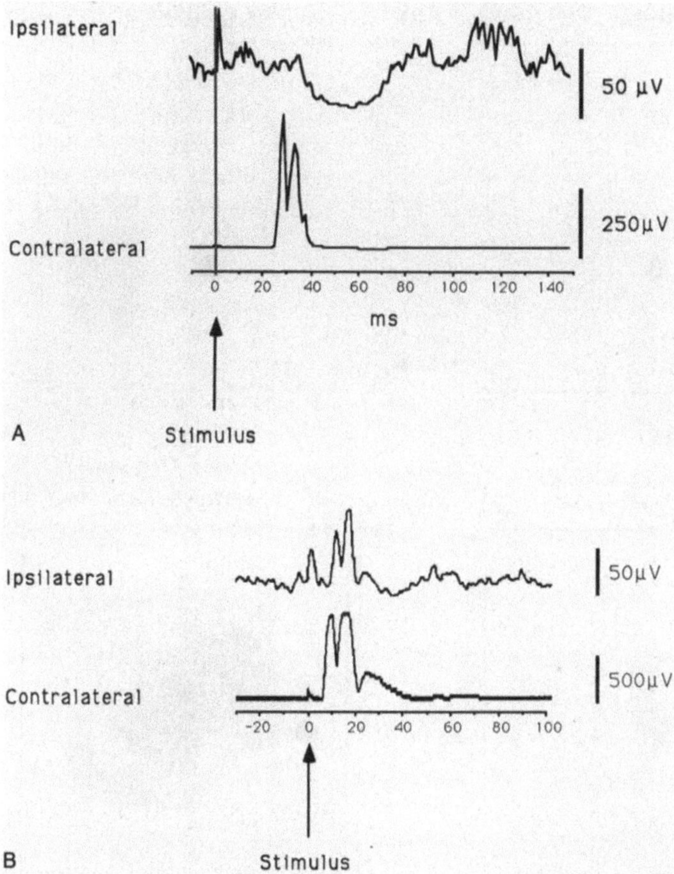

Fig. 1A,B. Averaged surface EMG traces from both contralateral and ipsilateral APBs (**A**) and biceps (**B**) during TMS of one hemisphere in a normal subject. The muscle ipsilateral to the stimulus is voluntarily activated with 20% of maximal force. Traces are from 20 trials of stimulation in **A** and ten trials in **B**.

Motor representation areas targeting muscles ipsilateral and immediately proximal to the stump were larger than those for muscles contralateral to the stump. The side-to-side ratio for the number of excitable scalp positions was larger in the amputees than in a group of normal volunteers. The threshold for activation of muscles proximal and ipsilateral to the stump was decreased compared with that for muscles on the intact side. An example is shown in Fig. 2A. These results are consistent with the idea that the motor cortex for the muscles proximal to the amputation has expanded into the territory of the amputated part.

TMS also induced a sensation of movement in the missing hand in the seven patients with acquired amputations, but not in the patient with

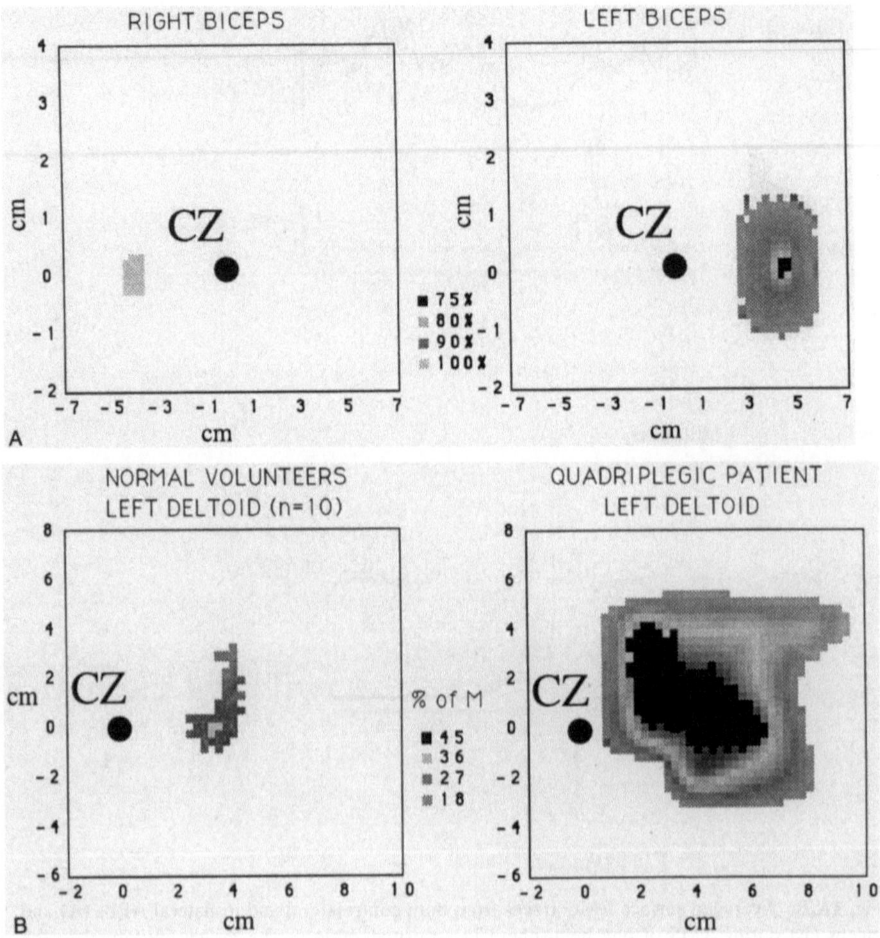

Fig. 2. A Diagrams of the top of the head of a 78-year-old volunteer with a left upper limb amputation. Testing was performed 11 months after amputation. Thresholds for activation of *left* and *right biceps* at different scalp locations are indicated. Note that the muscle representation area for the left biceps is larger than that for the right biceps. Also note that the threshold for activation of the left biceps was lower than that for the right biceps. **B** Diagrams showing views of the top of the head for a group of ten normal volunteers and a patient with complete spinal cord injury at the C5 level that occurred 4 months before testing. The sizes of MEPs from the *left deltoid* (expressed as % of the alpha-motoneuron pool) evoked by magnetic scalp stimulation at different scalp positions are indicated. The group map displays the maximal size of the MEP evoked in the ten control subjects after stimulation of each scalp location. The patient map displays the average size of three MEPs evoked after stimulation of each scalp position. Note that the MEP and motor representation area was larger for the left deltoid in the patient than the largest values of any individual in the control group

congenital lack of a limb. This finding indicates that at least a part of the cortical representation of the amputated limb remains related to the amputated part even years after amputation, a conclusion also supported by the common phenomenon of phantom limb. There are two possible explanations for this phenomenon: either magnetic stimulation activated preserved representation areas in the sensorimotor cortex, or the twitch in stump muscles induced by magnetic stimulation was inappropriately interpreted as originating in the missing limb. We cannot presently differentiate between these possibilities. However, the fact that phantom sensations are present independent of movements or stimulation of the stump areas supports the former possibility.

Experiments in rats have shown that the cortical area from where shoulder movements can be evoked becomes larger after forelimb amputations (Donoghue and Sanes 1988; Sanes et al. 1988). It appears that the functional connection between specific muscle groups and the amount of cortex dedicated to their control can vary and that amputation is an adequate trigger for these changes to occur.

Results from human and animal experiments are compatible with an increased density of physiologically active corticomotoneuronal connections targeting muscles proximal to the stump. However, these results could also be explained by an increased excitability of the alpha-motoneuron pool on the side of the amputation. To evaluate this possibility we studied motor reorganization and excitability of alpha motoneurons to Ia input in six subjects with unilateral lower limb amputation. MEPs were recorded bilaterally from the quadriceps femoris, the first muscle immediately proximal to the stump. Excitability of the spinal alpha-motoneuron pool to Ia afferents was assessed by determining the ratio of the maximal H-reflex to the maximal M-response (H/M ratio) elicited in the quadriceps femoris (Fuhr et al. 1992). As in volunteers with upper extremity amputations, stimuli of equal intensity delivered to optimal scalp positions recruited a larger percentage of the alpha-motoneuron pool in muscles ipsilateral to the stump than in those contralateral to the stump. However, there was no difference in the quadriceps H/M ratios. This absence of changes in the excitability of the alpha-motoneuron pool in the presence of motor reorganization suggests that reorganization occurs proximal to the alpha motoneuron level.

Two possible mechanisms could explain these results: growth of new connections (sprouting) and increased effectiveness of previously existing connections (unmasking). Two different studies were performed in order to see how fast and at which level along the motor pathways modulation of motor outputs could occur following reversible deafferentation of a limb segment in humans. Very rapid modulation would have to be due to unmasking. Reversible deafferentation was accomplished using a blood-pressure cuff.

In the first study (Brasil-Neto et al. 1992), involving three normal subjects, a pneumatic double tourniquet was placed just below the elbow.

An Esmarch bandage was then applied to the hand and snugly wrapped to the level of the tourniquet to exsanguinate the extremity distal to the tourniquet. The proximal cuff of the tourniquet was inflated to 250 mm Hg and the Esmarch bandage was unwrapped. A regional block (Bier block) of the forearm and hand was achieved by administration of lidocaine through a catheter placed in a vein on the dorsum of the hand. The tourniquet remained inflated for up to 70 min. Anesthesia was determined by the absence of voluntary finger and wrist movements, disappearance of MEPs from ipsilateral abductor pollicis brevis (APB), and absence of position sense and tactile sensation below the cuff. The amplitudes of MEPs to TMS from muscles immediately proximal to the temporarily anesthetized forearm increased within minutes after the onset of anesthesia and returned to control values after the anesthesia subsided.

In order to determine the level at which the early modulation of motor outputs took place, we performed a second series of experiments in which we recorded maximal H-reflexes, peripheral M-responses, MEPs to TMS, and MEPs to transcranial electrical stimulation (TES) and spinal electrical stimulation (SES) from a muscle immediately proximal to a limb segment made ischemic by a pneumatic tourniquet (but without using an anesthetic) (Brasil-Neto et al. 1993). In such experiments, which involved six normal subjects, a pneumatic tourniquet was placed just below the subject's right knee with care not to apply pressure over the fibular head. The tourniquet was inflated to a pressure 25%–30% higher than the subject's systolic blood pressure. Effective ischemic nerve block was determined by the absence of voluntary foot and toe movements, the disappearance of MEPs from the extensor digitorum brevis, and the absence of position sense and tactile sensation below the tourniquet. After approximately 50 min, the tourniquet was deflated by releasing the pressure gradually over 5–6 min to minimize postischemic paresthesia. H-reflexes were evoked from the vastus medialis by electrical stimulation of the femoral nerve. The H-reflexes were recorded before, during, and after the induction of ischemia. Supramaximal M-responses were also obtained from both vastus medialis muscles at different times during the experiment. A Digitimer D-180 electrical stimulator was used for TES and SES. The amplitudes of MEPs to TMS, but not to TES and SES, were larger during ischemia. The maximal H/M ratios were unaffected by ischemia, indicating that alpha-motoneuron excitability to segmental Ia inputs remained unchanged. The map of cortical representation areas for the muscle proximal to the ischemic limb segment, obtained with TMS, was also enlarged. TES appears to exert much of its effect by direct stimulation of axons of corticospinal tract neurons, whereas TMS appears to exert much of its effect at a presynaptic level (Amassian et al. 1987; Day et al. 1987; *see* Rothwall et al., this volume; Burke et al., this volume). The fact that MEPs changed after TMS, but not after TES, strongly suggests that cortical neuronal nets targeting corticospinal tract neurons are the sites where most of the modulation takes place.

Taken together, our findings suggest that deafferentation works, at least in part, by unmasking, and that a major component of the unmasking must be cortical. The mechanism of action is likely to be reduction of inhibition at the motor cortical level.

Spinal Cord Injury

To evaluate changes in cortical motor output organization following spinal cord injury, we studied seven patients with thoracic (T9–12) spinal cord injuries (six with complete lesions and one with an incomplete lesion) (Topka et al. 1991). Sensation in all somatosensory modalities and voluntary movement were completely lost below the level of the lesion in six patients and severely impaired in one. Time between injury and testing ranged from 2 to 20 years. Somatosensory evoked potentials to tibial nerve stimulation and MEPs recorded from tibialis anterior to TMS delivered through a coil 9 cm in diameter were absent, indicating lack of conduction in sensory and motor pathways across the lesion. Recordings were made bilaterally using surface electrodes placed on the skin overlying the abdominal wall muscles. TMS was delivered with a Cadwell MES-10 magnetic stimulator through an 8-shaped magnetic coil. Stimulation of ventral roots was done either with supramaximal electrical stimulation over the spine at T3–6, T6–9, and T9–12 (cathode/anode, respectively) using a Digitimer D180 stimulator, or by spinal magnetic stimulation with a circular coil at maximal output intensity with coil positions overlying the spine varying between T5 and T12. Subjects were also asked to describe the sensations felt after each stimulus. If the description was considered incomplete, we inquired about particular aspects of the sensation: location (right vs. left; knee, foot, toes, thigh, hand, forearm, arm, abdomen), quality (dull, sharp, tingling, sensation of movement, sensation of muscle contraction), and duration (very short, several seconds, repetitive). TMS evoked larger MEPs for each muscle level in patients than in controls. The motor representation areas for abdominal muscles were larger in the patients than in the controls. This expansion was present in both dimensions (anteroposterior and side-to-side).

We also studied patients with spinal cord injury leading to quadriplegia. In this case, recordings were from muscles that the patients could move voluntarily and that were immediately proximal to the lesion (usually biceps or deltoid). Our results were similar to those reported after paraplegia, and Levy et al. (1990) have reported similar results. Motor representation areas for muscles immediately proximal to the level of a complete cervical spinal cord injury were larger than those in normal subjects (Fig. 2B). It is interesting that some of these changes may become evident within days following the injury (Streletz et al. 1992). If this is the case, it appears that the anatomical connections underlying these changes are available but inactive under normal conditions and become unmasked after the lesion occurs.

TMS also evoked paresthesias referred to the contralateral leg, which were most commonly in the foot in normal volunteers. Three of the seven patients with paraplegia reported that stimulation evoked sensations of tingling, numbness, touch, or a wave descending along the leg and referred to different parts of the legs and toes (Cohen et al. 1991d). In the patient group, sensations were felt more distally, the closer the site of stimulation was to the midline. The finding of topographically organized paresthesias in patients deafferented by complete spinal cord injury indicates a central origin. Patients with leg paresthesias had less motor reorganization in abdominal muscles than those without paresthesias. These findings suggest that portions of the cortical representation areas for body parts deafferented by a complete spinal cord injury can remain related to those body parts for up to several years after the lesion. Preservation of deafferented cortical representations indicates less reorganization.

Animal experiments also indicate the ability of the developing or adult central nervous system to reorganize. Dramatic somatosensory cortical reorganization has been described in kittens that sustained spinal cord transection at T12 (McKinley and Smith 1990). A significant component of these changes in kittens occurs at synaptic levels below the cortex. If this is the case, the detected cortical reorganization would be a reflection of subcortical changes. The site of reorganization in human beings remains to be determined.

Brain Injury

Studies with PET have given evidence of cortical plasticity after stroke. Chollet et al. (1991) studied six patients with hemiplegic stroke who had recovered full strength, although three of the six had some residual clumsiness. Voluntary finger movement produced significant increases of blood flow in the contralateral sensorimotor cortex, ipsilateral cerebellar hemisphere, contralateral and ipsilateral SMA, contralateral and ipsilateral premotor cortex, contralateral area 40, and contralateral striatum and insula. In recovered stroke patients, in comparison to the normal subjects, there was increased activation of the ipsilateral sensorimotor cortex, insula, and area 40 and contralateral cerebellum. Subsequently, the same investigators (Weiller et al. 1992) studied a group of ten patients with striatocapsular infarctions. The earlier findings were confirmed, and additional areas of activation were noted in bilateral prefrontal cortices and bilateral cingulate cortices.

The importance of ipsilateral control of movement after hemiplegia was demonstrated dramatically by Fisher (1992). He reported two patients with pure motor hemiplegia who had been regaining strength when they developed a subsequent motor hemiplegia on the other side. When the new hemiplegia appeared, the recovering side became re-paralyzed.

Ipsilateral MEPs produced from stimulation of the unaffected hemisphere in patients with stroke have been seen by Iani et al. (1992). These were not found by Palmer et al. (1992; *see* Ashby et al., this volume), but their stimulation intensity may have been too low.

Hemispherectomy is a surgical procedure first described by Lhermite and Dandy for the treatment of cerebral gliomas. It has been used also for treatment of medically intractable seizure conditions. A common observation after this procedure is that motor function in the limb contralateral to the excised hemisphere experiences a substantial degree of recovery, particularly when surgery is performed at an early age (Gardner 1933). To understand the mechanisms underlying this recovery of function, we studied patients with hemispherectomy (Cohen et al. 1991c; Pascual-Leone et al. 1992a). One representative example is a 32-year-old man with a large porencephalic cyst which occupied most of the left hemisphere. At 7 years of age he had a left hemispherectomy for uncontrolled seizures. We studied this patient 25 years after the operation when he was able to move his right arm voluntarily. We recorded MEPs to magnetic scalp stimulation using a Cadwell MES-10 magnetic stimulator and an 8-shaped magnetic coil. Equal stimuli (100% of the stimulator's output) were delivered to a grid of scalp positions 2 cm apart. Recordings were made from biceps and deltoid muscles on both arms. We also measured cerebral blood flow with PET at rest and when he was performing flexion-extension elbow movements.

TMS of the remaining hemisphere induced bilateral activation of deltoid and biceps. Similar findings were obtained by Benecke et al. (1991). Evaluation of MEPs indicated that muscles ipsilateral to the preserved hemisphere were activated by stimulation of scalp positions anterior and lateral to those activating muscles on the normal side (Fig. 3). Similarly, ipsilateral elbow movements were associated with cerebral blood flow increases in an area centered 1.4 cm anterior and lateral to that activated by the same movements on the normal side. These results indicate that ipsilateral and contralateral representations in the remaining hemisphere are topographically differentiated, with ipsilateral representations having a more anterior and lateral scalp distribution.

In our group evaluation, patients who were operated on at a very early age had a more likely chance for substantial recovery of motor function, and this was associated with the development of a distinct (and more anterolateral) ipsilateral representation of proximal arm muscles. Patients who were operated on later had a poorer recovery, and the ipsilateral and contralateral representations in the intact hemisphere were undifferentiated.

These studies of both stroke and hemispherectomy show significant cortical reorganization after brain lesions and suggest that an increased influence of ipsilateral control is one of the mechanisms leading to recovery of motor function.

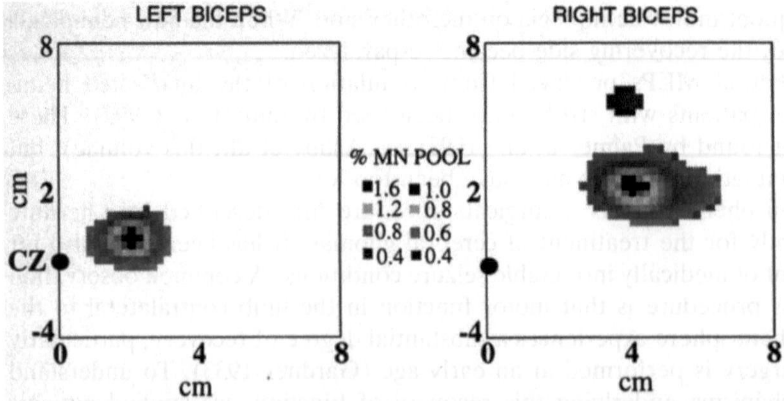

Fig. 3. Diagram of the motor representation areas for the left and right biceps on the right hemisphere of a patient who had a left hemispherectomy. The amplitudes of MEPs evoked by magnetic stimulation (expressed as % of the alpha-motoneuron pool) are indicated. Note that the cortical representation for the right biceps (on the ipsilateral hemisphere) was clearly differentiated and anterolateral to that of the left biceps

Patterns of Use

In the previous sections we have discussed the changes that can take place in the central nervous system to allow recovery of function after injury. The nervous system seems capable of reorganizing, apparently seeking an adaptation to minimize functional deficits. Similar changes may take place in the nervous system according to the patterns of use, therefore providing a substrate for the acquisition of new skills. For example, we have discussed the fact that amputation results in enlargement of the sensorimotor representation of adjacent body parts. These changes could represent the modification of the nervous system, attempting to provide a substrate for compensation of function of the amputated body part. If so, increased sensorimotor demand for one body part may result in the enlargement of its representation, possibly at the cost of the representation of neighboring body parts. If such increased demand is coupled with a significant functional gain for the individual, these changes may be more readily implemented.

Indeed, the sensorimotor representation of the preferred hand in monkeys is more elaborate than that of the nonpreferred hand, and training can result in distortions of body surface and movement representations that lead to behavioral gains (Jenkins et al. 1990). Motor cortical representation of a body part expands following selectively increased activity (Sanes et al. 1992). The vibrissal representational zone in adult rats reorganizes following vibrissa trimming, resulting in enlargement of the representation of the remaining vibrissae, and differential increased stimulation of a restricted skin surface in a finger pad of adult monkeys leads to an enlargement of its somatosensory cortical representation (Merzenich 1989).

The development of noninvasive imaging and neurophysiological techniques enables us to pursue the study of changes in human beings. A recent PET study (Seitz et al. 1990) showed that learning of a complicated finger sequence of voluntary movements is associated with increases in regional cerebral blood flow in the cerebellum, and that once the motor skill has been learned there is an increase in the striatum, suggesting a role for cerebellar and striatal circuits in acquisition and storage of motor skills. Grafton et al. (1992) have shown that regional relative cerebral blood flow increases in primary motor area, supplementary motor area, and thalamus as subjects learn a pursuit rotor task. Changes in slow cortical negative DC-potentials during the acquisition of a complex finger motor task (Niemann et al. 1991) also suggest a reorganization of the cortical representation of movement control during skill learning.

We have compared the sensorimotor representation of the reading index finger of ten blind, proficient braille readers with that of left and right index fingers of ten age-matched untrained subjects using somatosensory evoked potentials (SEPs) (Pascual-Leone and Torres 1992). The braille readers had been blind since childhood (before the age of 10), learned braille in childhood (before age 15), used braille daily for at least 2 h, and had no neurological or systemic diseases. Subjects who self-reported using more than one finger or both hands to read braille were excluded. SEPs were elicited by a focal electrical stimulus to the tip of the index finger and recorded from a contralateral 4×4 grid of scalp electrodes centered around C3′ or C4′. The areas over which the amplitudes of N20 and P27 components of the SEPs decayed by one third of their maximum were significantly larger in SEPs evoked from reading fingers (average difference $8\,cm^2$) than in SEPs evoked for the nonreading index finger in braille readers or for either index finger in controls. Quantitative measurements of sensory threshold to touch, vibration, and two-point discrimination did not reveal differences between subjects and controls.

These findings suggest that learning braille expands the somatosensory cortical representation of the reading finger. However, in order to increase the accuracy in the detection of raised dots, braille readers scan the text while continuously moving the reading index finger from side to side. Movement between the raised symbols and the skin decreases the sensory discrimination threshold, and, in fact, in highly skilled braille readers the frequency of the side-to-side movements of the reading finger can be correlated with the passage difficulty. This suggests that proficient braille reading may be associated not only with a reorganization of the sensory cortex representation of the finger pad of the reading finger, but also with changes in its motor representation to allow the appropriate movement strategy.

To study this possibility, we performed detailed mapping of the motor cortical areas targeting the first dorsal interosseous (FDI) and the adductor digiti minimi (ADM) bilaterally in braille readers and blind controls with focal TMS (Pascual-Leone et al. 1992b). Subjects and controls were matched for age and age of blindness. Subjects had learned braille at age 8–14 years and used it daily for 5–10 h. Controls had not learned braille until age 17–21 years and used it daily for less than 1 h. Motor threshold of subjects and controls was similar. In the controls, motor representations of right and left FDI and ADM were not significantly different. However, in the proficient braille readers, the representation of the FDI in the reading hand was significantly larger than that in the nonreading hand or in either hand of the controls (Fig. 4). Conversely, the representation of the ADM in the reading hand was significantly smaller than that in the nonreading hand or in either hand of the controls. These results suggest that the cortical representation for the reading finger in proficient braille readers is enlarged at the expense of the representation of other fingers.

Braille reading can be considered an example of a highly specialized sensorimotor skill which represents an enormous functional gain for the blind readers. The marked differential sensory stimulation of the finger pads

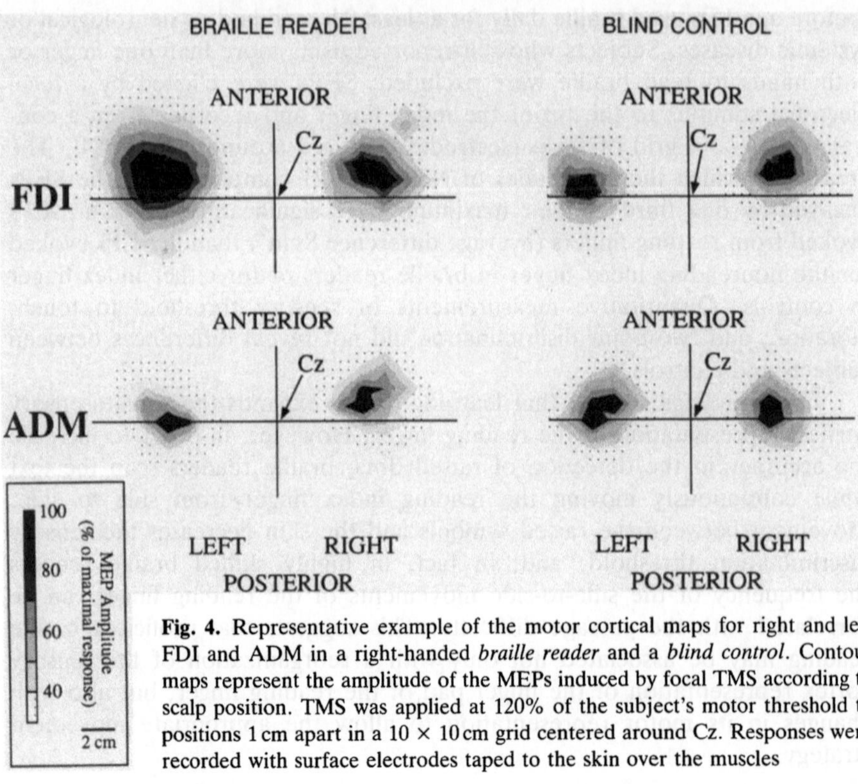

Fig. 4. Representative example of the motor cortical maps for right and left FDI and ADM in a right-handed *braille reader* and a *blind control*. Contour maps represent the amplitude of the MEPs induced by focal TMS according to scalp position. TMS was applied at 120% of the subject's motor threshold to positions 1 cm apart in a 10 × 10 cm grid centered around Cz. Responses were recorded with surface electrodes taped to the skin over the muscles

may lead to identifiable changes in the sensory cortical representation of the reading finger. However, in order to maximize their performance, braille readers adopt particular movement strategies of the reading hand and finger, and, indeed, these new motor demands are associated with changes in the motor cortical representation of the fingers.

Since recovery from brain injury seems to involve plastic changes, and since changes in activity induce plastic changes, it would seem logical that physical therapy should be useful for recovery of function. On the other hand, evidence for the efficacy of this modality is mixed. The majority of hard evidence shows only a statistically small benefit at most (Ernst 1990). Moreover, there is no evidence that one method of therapy has any advantage over any other.

Drug therapy might have a role, particularly drug therapy coupled with physical therapy. Noradrenergic influences on recovery of function have been emphasized by Feeney and colleagues (Feeney and Sutton 1987). They have shown that a single dose of *d*-amphetamine can speed up recovery of coordinated walking in rats after sensorimotor cortex ablation. Moreover, benefit of the single dose can still be seen after 1 month. The *d*-amphetamine has no effect, however, if the animal is restrained. Hence, it is only a combination of drug and activity that works. Haloperidol blocks the useful drug effect. This concept has undergone a small trial in human patients following stroke (Crisostomo et al. 1988). Four patients were treated with amphetamine and another four with placebo in a double-blind trial. All patients had physical therapy, but only those with the drug had improved function on the next day. Clearly, further trials are indicated.

Conclusion

Plasticity of the motor cortex can be readily demonstrated in human beings in a number of circumstances. The process is expressed both by expansion and contraction of the relevant area in the hemisphere contralateral to a body part and by changing influences of the ipsilateral hemisphere. This plasticity is critical for recovery of function after brain injury and may provide a substrate for skill learning. Although still not proven, it may be possible to intervene favorably in the process.

References

Amassian VE, Stewart M, Quirk GJ, Rosenthal JL (1987) Physiological basis of motor effects of a transient stimulus to cerebral cortex. Neurosurgery 20:74–93

Benecke R, Meyer BU, Freund HJ (1991) Reorganization of descending motor pathways in patients after hemispherectomy and severe hemispheric lesions demonstrated by magnetic brain stimulation. Exp Brain Res 83:419–426

Brasil-Neto JP, Cohen LG, Pascual-Leone A, Jabir FK, Wall RT, Hallett M (1992) Rapid reversible modulation of human motor outputs after transient deafferentation of the forearm: a study with transcranial magnetic stimulation. Neurology 42:1302–1306

Brasil-Neto JP, Valls-Solé J, Pascual-Leone A, Cammarota A, Amassian VE, Cracco R, Maccabee P, Cracco J, Hallett M, Cohen LG (1993) Rapid modulation of human cortical motor outputs following ischemic nerve block. Brain 116:511–525

Chollet F, DiPiero V, Wise RJS, Brooks DJ, Dolan RJ, Frackowiak RSJ (1991) The functional anatomy of motor recovery after stroke in humans: a study with positron emission tomography. Ann Neurol 29:63–71

Cohen LG, Bandinelli S, Findley TW, Hallett M (1991a) Motor reorganization after upper limb amputation in man. A study with focal magnetic stimulation. Brain 114:615–627

Cohen LG, Bandinelli S, Topka HR, Fuhr P, Roth BJ, Hallett M (1991b) Topographic maps of human motor cortex in normal and pathological conditions: mirror movements, amputations and spinal cord injuries. Electroencephalogr Clin Neurophysiol [Suppl] 43:36–50

Cohen LG, Roth BJ, Wassermann EM, Topka H, Fuhr P, Schultz J, Hallett M (1991c) Magnetic stimulation of the human cerebral cortex, an indicator of reorganization in motor pathways in certain pathological conditions. J Clin Neurophysiol 8:56–65

Cohen LG, Topka H, Cole RA, Hallett M (1991d) Leg paresthesias induced by magnetic brain stimulation in patients with thoracic spinal cord injury. Neurology 41:1283–1288

Colebatch JG, Deiber MP, Passingham RE, Friston KJ, Frackowiak RS (1991) Regional cerebral blood flow during voluntary arm and hand movements in human subjects. J Neurophysiol 65:1392–1401

Crisostomo EA, Duncan PW, Propst M, Dawson DV, Davis JN (1988) Evidence that amphetamine with physical therapy promotes recovery of motor function in stroke patients. Ann Neurol 23:94–97

Davey NJ, Romaiguere P, Ellaway PH (1991) Inhibition of voluntary contraction by transcranial magnetic stimulation of the brain subthreshold for excitation in man (abstract). J Physiol (Lond) 446:447P

Day BL, Thompson PD, Dick JP, Nakashima K, Marsden CD (1987) Different sites of action of electrical and magnetic stimulation of the human brain. Neurosci Lett 75:101–106

Donoghue JP, Sanes JN (1988) Organization of adult motor cortex representation patterns following neonatal forelimb nerve injury in rats. J Neurosci 8:3221–3232

Ernst E (1990) A review of stroke rehabilitation and physiotherapy. Stroke 21:1081–1085

Feeney DM, Sutton RL (1987) Pharmacotherapy for recovery of function after brain injury. Crit Rev Neurobiol 3:135–197

Fisher CM (1992) Concerning the mechanism of recovery in stroke hemiplegia. Can J Neurol Sci 19:57–63

Fuhr P, Agostino R, Hallett M (1991) Spinal motor neuron excitability during the silent period after cortical stimulation. Electroencephalogr Clin Neurophysiol 1991;81:257–262

Fuhr P, Cohen LG, Dang N, Findley TW, Haghighi S, Oro J, Hallett M (1992) Physiological analysis of motor reorganization following lower limb amputation. Electroencephalogr Clin Neurophysiol 85:53–60

Gardner WJ (1933) Removal of the right hemisphere for infiltrating glioma. J Am Med Assoc 101:823–826

Grafton ST, Woods RP, Mazziotta JC, Phelps ME (1991) Somatotopic mapping of the primary motor cortex in humans: activation studies with cerebral blood flow and positron emission tomography. J Neurophysiol 66:735–743

Grafton ST, Mazziotta JC, Presty S, Friston KJ, Frackowiak RSJ, Phelps ME (1992) Functional anatomy of human procedural learning determined with regional cerebral blood flow and PET. J Neurosci 12:2542–2548

Iani C, Desiato MT, Rossini PM, Bernardi G, Caramia MD (1992) Neurophysiological evidence of motor recovery in brain-damaged patients: a study with magnetic stimulation (abstract). Abstracts of the IXth international congress of electromyography and clinical neurophysiology, Jerusalem

Jenkins WM, Merzenich MM, Recanzone G (1990) Neocortical representational dynamics in adult primates: implications for neuropsychology. Neuropsychologia 28:573–584
Levy WJ, Amassian VE, Traad M, Cadwell J (1990) Focal magnetic coil stimulation reveals motor cortical system reorganized in humans after traumatic quadriplegia. Brain Res 510: 130–134
McKinley PA, Smith JL (1990) Age-dependent differences in reorganization of primary somatosensory cortex following low thoracic (T12) spinal cord transection in cats. J Neurosci 10:1429–443
Merzenich MM (1989) Representational plasticity in somatosensory and motor cortical fields. Biomed Res 10 [Suppl 2]:85–86
Niemann J, Winker T, Gerling J, Landwehrmeyer B, Jung R (1991) Changes in slow cortical negative DC-potentials during the acquisition of a complex finger motor task. Exp Brain Res 85:417–422
Palmer E, Ashby R, Hajek VE (1992) Ipsilateral fast corticospinal pathways do not account for recovery in stroke. Ann Neurol 32:519–525
Pascual-Leone A, Torres F (1993) Plasticity of the sensorimotor cortex representation of the reading finger in braille readers. Brain 116:39–52
Pascual-Leone A, Chugani HT, Cohen LG, Brasil-Neto J, Wassermann EM, Valls-Sole J, Fuhr P, Hallett M (1992a) Reorganization of human motor pathways following hemispherectomy (abstract). Ann Neurol 32:261
Pascual-Leone A, Cohen LG, Cammarota A, Hallett M (1992b) Reorganization of motor cortical outputs in braille readers (abstract). Ann Neurol 32:232
Rothwall JC, Thompson PD, Day BL, Dick JPR, Kachi T, Cowan JMA, Marsden CD (1987) Motor cortex stimulation in intact man. 1. General characteristics of EMG responses in different muscles. Brain 110:1173–1190
Sanes JN, Suner S, Lando JF, Donoghue JP (1988) Rapid reorganization of adult rat motor cortex somatic representation patterns after motor nerve injury. Proc Natl Acad Sci USA 85:2003–2007
Sanes JN, Wang J, Donoghue JP (1992) Immediate and delayed changes of rat motor cortical output representation with new forelimb configurations. Cerebral Cortex 2:141–152
Seitz RJ, Roland E, Bohm C, Greitz T, Stone-Elander S (1990) Motor learning in man: a positron emission tomographic study. Neuroreport 1:57–60
Streletz LJ, Jones SM, Belevich JKS, Verhagen-Metman L, Barber MD, Herbison GJ (1992) Rapid reorganization of motor pathways in acute spinal cord injury revealed by transcranial magnetic stimulation (abstract). Electroencephalogr Clin Neurophysiol 83:85P
Topka H, Cohen LG, Cole RA, Hallett M (1991) Reorganization of corticospinal pathways following spinal cord injury. Neurology 41:1276–1283
Wassermann EM, Fuhr P, Cohen LG, Hallett M (1991a) Effects of transcranial magnetic stimulation on ipsilateral muscles. Neurology 41:1795–1799
Wassermann EM, Pascual-Leone A, Valls-Solé J, Cohen LG, Hallett M (1991b) Muscle inhibition from transcranial stimulation in normal humans (abstract). Soc Neurosci Abstr 17:1022
Weiller C, Chollet F, Friston KJ, Wise RJS, Frackowiak RSJ (1992) Functional reorganization of the brain in recovery from striatocapsular infarction in man. Ann Neurol 31:463–472
Zeffiro T, Hallett M (1990) Regional cerebral blood flow changes in ipsilateral and contralateral motor areas during unilateral movements (abstract). Soc Neurosci Abstr 16:152
Zeffiro T, Hallett M (1992) Regional cerebral blood flow changes during active and passive finger movements: a PET study (abstract). Soc Neurosci Abstr 18:938

Discussion Summary

E. Jankowska

Following Dr. R. Lemon's lecture, the discussion focused on task-related actions of corticospinal neurons during voluntary movements. Most comments were made on the conclusion that different subpopulations of cortical neurons that synapse on motoneurons "may be recruited according to the particular combination of muscles employed for a specific task or phase of movement." There has been general agreement that this is one of the important outcomes of the recent studies of the corticospinal system. However, the possibility that task-related actions of corticomotoneuronal neurons might, in addition, depend on task-dependent feedback from muscle, skin, and other receptors was pointed out by Dr. D. Burke. Attention was also drawn by Dr. S. Miller and Dr. H. Hultborn to the occurrence of parallel actions of corticospinal neurons: by directly exciting motoneurons and by selecting reflex neuronal systems most appropriate to the task.

Of the three motor pathways implicated in responses investigated by Dr. J. Rothwell and his colleagues, the cortico-, reticulo-, and vestibulospinal pathways, the last was the most extensively discussed. Dr. D. Burke presented additional experimental evidence in support of involvement of the vestibulospinal pathways in responses analyzed by Dr. J. Rothwell. Questions were asked on factors which might influence the latency of these responses in different subjects (such as their height; Dr. M. Hallett) and comments were made on sites at which the selection of the postural correction strategies occurred (brain stem and spinal cord; Dr. J. Munson and Dr. E. Jankowska). With respect to the startle response to unexpected sounds, questions were asked on the conduction velocity of the involved reticulospinal neurons (by Dr. Z. Rymer) and on its presence in patients after cortical lesions (by Dr. M. Emre). With respect to responses evoked by transcranial magnetic stimulation, the value of the jitter of single motor units or single motor fibers as an index of monosynaptic activation of motoneurons by corticospinal tract neurons was discussed by Dr. J. Noth.

Dr. E. Pierrot-Deseilligny's lecture raised several questions on the relationship between the propriospinal and other spinal systems in man. One of the questions (put by Dr. S. Miller) concerned the relationship between

Dept. of Physiology, University of Göteborg, Medicinaregartan 11, 41390 Göteborg, Sweden

Thilmann et al. (Eds.)
Spasticity: Mechanisms and Management

patterns of multi-joint movements mediated by propriospinal neurons and patterns of synaptic actions of Ia afferents assisted by gamma fusimotor neurons. The latter did not appear likely to be consistent with the former, but the paucity of heteronomous connections of group-Ia afferents of the human upper limb muscles, found by Dr. Pierrot-Deseilligny's group, suggests that the multi-joint movements are steered to a greater extent via the propriospinal than via the gamma neurons. Another question (put by Dr. M. Illert) concerned co-activation of muscles stabilizing the wrist by propriospinal neurons and the interpretation of its use in preparation for grasping; possible differences in the function of muscles stabilizing the wrist in the man and in the cat have been pointed out. Other questions concerned the relative contribution of the excitatory and inhibitory servocontrol of the propriospinal neurons in man compared with in the cat by peripheral stimuli (by Dr. M. Illert and Dr. D. Burke). Inhibitory actions of cutaneous stimuli appeared to be as strong, or stronger, in man than in the cat, while the facilitatory actions of muscle afferents appeared to be much more efficient. Questions were also raised on the relative contribution of propriospinal neurons to the control of various groups of muscles, in particular the intrinsic hand muscles (Dr. J. Rothwell), and on the relative contribution of the fast- and slow-conducting corticospinal tract neurons to their activation (Dr. D. Kernell). Finally, Dr. R. Lemon, Dr. E. Jankowska, and Dr. H. Hultborn commented on the strikingly complementary actions of descending commands produced by direct corticomotoneuronal connections and those produced by other, parallel, channels, including the propriospinal and segmental interneuronal pathways. The discussants pointed out that these parallel channels may contribute to similar and similarly task-related patterns of muscle activity. Accordingly, too much emphasis should not be put on the importance of only one of them, although the contribution of the various neuronal systems might vary, depending on the kind of movement and according to which muscle groups are to be used in this movement.

Discussion Summary

J. Noth

Percutaneous electrical or magnetic stimulation of the human motor cortex constitutes a major advance in clinical neurophysiology. For the first time it enables the quantitative evaluation of the corticospinal tract in healthy human subjects and in patients with various kinds of central motor disorders. The principle mode of action of electrical stimulation differs little from that used in earlier animal experiments, in which the exposed motor cortex was stimulated with graded anodal or cathodal currents. Thus, the pioneer work of Phillips (1969) on nonhuman primates remains very influential for the interpretation of the more recent human data. Electromagnetic brain stimulation, on the other hand, was introduced at about the same time in human and animal studies, and our understanding of the complicated excitatory and inhibitory cortical and subcortical events elicited by this new stimulation technique has lagged behind its clinical application. The present symposium has provided a unique possibility to discuss recent developments in this field with experts in animal research and in the field of clinical neurology.

The discussion of Dr. Burke's paper was opened by Dr. Eyre. She doubted the common view that the longer motor latency of the electromagnetic brain stimulation is due to transsynaptic excitation of corticospinal neurons, in contrast to the electrical stimulation, which is thought to excite the neurones themselves. Recent recordings of single corticospinal axons in the monkey, rather, suggest that both kinds of stimulation excite corticospinal neurones directly (Edgley et al. 1990). With anodal electrical stimulation, however, corticospinal fibers can be activated at subcortical sites, giving rise to shorter latencies of electrically evoked motor potentials. A related question was posed by Dr. Kernell, who wondered why increasing the stimulus strength of the anodal electrical stimulation shortens the latency of the evoked motor response in distinct jumps and not in a gradual fashion. Dr. Burke suggested that there are sites of preferential accessibility of the axons to the electrical stimulus; possible sites are the internal capsule, the cerebral peduncles, and the pyramids. In man the exact localization of these sites is difficult to assess because the conduction velocity of corticospinal

Neurologische Klinik der Rheinisch-Westfälischen Technischen Hochschule, Pauwels Str., D-52074 Aachen, Deutschland

Thilmann et al. (Eds.)
Spasticity: Mechanisms and Management

fibers within the cranial cavity is not known. Dr. Miller's comment to this question was deduced from his monkey experiments (Edgley et al. 1990). It was found that the corticospinal axons conduct between the cortex and the pyramids with the same velocity as within the dorsolateral funiculus. Calculations based on these data and comparisons with human data suggest that the site of preferential excitability resides in the pedunculus. Dr. Burke concluded this part of the discussion by stating that all kinds of electrical stimulation and stimulus montages produce the distinct latency jumps while electromagnetic stimulation shows a small and monotonic shortening of latencies, indicating a clear difference in the mode by which corticospinal neurones and their axons are excited by the two kinds of stimulation.

Dr. Lemon, finally, asked whether the observed volley recorded at thoracic segments in response to stimulation of "upper limb" cortex in humans undergoing scoliosis operation could be caused by subcortical excitation of fibers from the "lower limb region of the cortex." This possibility should be considered because in monkeys deep fibers can be excited with transcranial electrical stimulation even at threshold. Dr. Burke considered this unlikely, because (a) the volley appears at the thoracical level at exactly the time expected (with deep stimulation, the latency should be shorter), (b) the volley appears at the stimulus strength at which the liminal upper limb response could be obtained, and (c) there is evidence, going back to Sherrington, that branching of corticospinal fibers exists in mammals.

Following Dr. Hallett's talk, Dr. Kernell asked whether differences between the dominant and nondominant arms can be found in normal subjects in the motor or sensory cortical maps. Dr. Hallett responded that he has made the comparison in several subjects, and that he knows results from other groups. Some have found larger representation areas for the dominant hand while others have not; he himself has observed both. The varying results may be related to the different use of the dominant hand in different subjects. Furthermore, whether there is a difference at all, it is a fairly subtle one.

Dr. Munson raised the interesting question of whether in hemispherectomized patients the ipsilateral (to the intact cortex) movements are strictly unilateral, or whether the contralateral side also moves. Dr. Hallett responded that the movements are strictly isolated and are not performed in a mirror-movement fashion, but the movements are altered and fractionated.

Dr. Hultborn turned the discussion to the important question whether the cortical plasticity – as judged from the motor maps following acute ischemic deafferentation of the arm – is due simply to the changes of the excitability of neurones within the cortex, because the cortex is provided with continuous sensory feedback. Dr. Hallett supported this view because the only change that could occur during the short period of ischemic deafferentation is the removal of afferent input to the cortex; he considered it to be a semantic question whether this is called "plasticity." However, he maintained that the side difference in motor maps of patients with limb amputation is a true cortical phenomenon because the H reflex elicited in proximal muscles of the amputated limb shows no side differences, indicating a comparable excitability of α-motoneurones on the two sides. Whether this is a sufficient argument was not resolved because the H reflex estimates merely the fast mono- or oligosynaptic segmental pathway and not the excitability of propriospinal systems. Dr. Rothwell addressed the question of which parameter of the motor map changes in patients with amputations – the threshold or the size. Dr. Hallett answered that the threshold clearly decreases in the center of the motor map on the affected side, but that it cannot be determined whether the increase in area is due to a lowering in threshold.

Dr. Lemon remarked that the motor maps demonstrated by Dr. Hallett look quite simple, with a single hot spot in the middle. This is quite different in the monkey, in which corticospinal axons supplying one muscle are clustered in different areas. Furthermore, maps cannot be described adequately in two dimensions because of the folding of the cortex. Dr. Hallett agreed that the motor maps obtained with transcranial magnetic stimulation are a simplified version of the complicated spatial representation of corticospinal neurones. This is supported by his own activation studies using positron emission tomography which show that the larger amount of activation is seen in the depth of the sulci and not at the surface of the brain, which in fact makes the maps much more complicated.

In the final discussion, questions were addressed relating to more general aspects of plasticity, and how this can be defined. Dr. Hultborn expressed the opinion that in the human studies presented so far evidence for real changes in hardware wiring is weak or missing, and that most of the changes can be explained by changes in adaptation not necessarily accompanied by structural reorganization. All discussants agreed that rewiring in the CNS can occur over very small distances of 1 mm or so, but that plasticity when defined in this way is difficult to confirm with transcranial electrical or magnetic brain stimulation. Plasticity in a wider sense covers a much larger spectrum, such as unmasking of silent synapses, alterations of receptor sensitivity, and changes in the postsynaptic membrane properties. It seems difficult to distinguish between these possibilities using methods which can be applied in human studies. We therefore need animal experiments for a better interpretation of the results obtained in man with this fascinating new method of motor cortex stimulation.

References

Edgley SA, Eyre JA, Lemon RN, Miller S (1990) Excitation of the corticospinal tract by electromagnetic and electrical stimulation of the scalp in the macaque monkey. J Physiol (Lond) 425:301–320

Phillips CG (1969) Motor apparatus of the baboon's hand. Proc R Soc Lond [Biol] 173:141–174

Part II
Disturbance of Voluntary Movement

The Role of the Corticospinal Tract in Spasticity Studied by Magnetic Brain Stimulation

R. Benecke

Introduction

The aim of this study was to elucidate the role of the fast-conducting part of the corticospinal tract in the genesis of spasticity in man. Magnetic brain stimulation was performed in patients with spastic syndromes. The severity of corticospinal tract lesions and the clinically established degree of spasticity were compared in these patients.

The Classical View

The role of the various motor-competent cortical areas and their descending pathways for the genesis of spasticity is still discussed controversially. Earlier ablation studies in monkeys showed that excisions restricted to area 4 leads to an initially flaccid paresis later resolving into a state of approximately normal tonus. Hines (1926) observed spasticity only after excisions involving the anterior part of area 4. For the mediation of this effect McCulloch et al. (1946) suggested a separate extrapyramidal tract projecting to brainstem areas. Kennard et al. (1934) associated spasticity with lesions of the premotor cortex (area 6).

Meanwhile there is, however, clear evidence in monkeys and humans that spasticity can be observed after lesions of the primary motor cortex. Small experimental lesions in monkeys and excisions in patients who had been operated on for hyperkinetic movement disorders were followed by paresis and/or spasticity. The issue of close linkage between paresis and spasticity in such lesions remained unsettled. Despite the considerable variance in symptoms that followed isolated lesions of the primary motor cortex, spasticity in such cases was usually less severe than that which follows damage to the internal capsule or the spinal cord. The pronounced spasticity in such lesions has been attributed to damage of fibers not originating from the primary motor cortex or not passing through the medullary

Neurologische Universitätsklinik, Moorenstr. 5, 40225 Düsseldorf, Deutschland

Thilmann et al. (Eds.)
Spasticity: Mechanisms and Management

pyramid since transection of the pyramids is followed by a hypotonic paresis both in monkeys and humans (Chapman and Wiesendanger 1982; Gilman and Marco 1971). In summary, the classic view of the central motor pathways involved in spasticity is that lesions of fibers which originate from the primary motor cortex or the premotor cortex can be followed by spasticity. These fibers, however, do not pass through the pyramid and therefore belong to the extrapyramidal system. It is still an open question whether these fiber systems have direct access to the spinal cord (corticospinal), or whether their influnces are mediated exclusively via brainstem nuclei.

Magnetic Brain Stimulation

Transcranial magnetic stimulation (TMS; Barker et al. 1985) of the human brain to evoke motor responses is now widely used in research and in clinical practice (Hess and Mills 1986; Hess et al. 1986, 1987; Rothwell et al. 1987; Berardelli et al. 1988; Benecke et al. 1988, 1991; Day et al. 1989; Edgley et al. 1990). Such stimulation induces short-latency motor-evoked potentials (MEPs) in contralateral muscles which are thought to be transmitted by the fast-conducting compartment of the corticospinal tract (Boyd et al. 1986; Rothwell et al. 1987; Day et al. 1989). Recent investigations in anesthetized monkeys with direct recordings from the spinal dorsolateral funiculus after TMS have confirmed this assumption (Edgley et al. 1990). The vast majority of investigations dealing with TMS in intact man and patients with motor deficits have concentrated on evaluating motor compound action potentials and discharges of single motor units which result from descending D and I waves (Patton and Amassian 1954; Kernell and Wu 1967; Boyd et al. 1986; Amassian et al. 1987).

In 1956 Creutzfeldt and coworkers demonstrated that electrical stimulation of the sensorimotor cortex in cats evokes not only excitatory but also inhibitory actions. Extracellular recordings of pyramidal cells showed primary discharges up to 20 ms after the stimulus, which were followed by a constant silent period lasting 150–400 ms. It was suggested that the stimulation-induced silent period in the activity of pyramidal cells was caused by intracortical inhibitory interneurons, similar to the action of Renshaw's cells in the spinal cord. Meanwhile, our knowledge of inhibitory actions on pyramidal cells of the sensorimotor cortex of various species has considerably expanded. γ-Aminobutyric acid (GABA) has been proposed to be a major inhibitory neurotransmitter in the cerebral cortex (Krnjevic 1974, 1987). There are specialized neurons that synthesize, store, and release GABA (Iverson et al. 1971; Ribak 1978; Ottersen and Storm-Mathisen 1984), GABA mimics the actions of the activity of local circuit interneurons (Avoli 1986; Connors et al. 1988; Tseng and Haberly 1988), and the intrinsic inhibitory postsynaptic potentials can be blocked by specific GABA antagonists (Connors et al. 1988). A recent study by McCormick (1989) demonstrated that electrical cortical stimulation results in an initial excitatory postsynaptic potential followed by a fast and a slow inhibitory postsynaptic potential mediated by $GABA_A$ and $GABA_B$ receptors, respectively.

A number of inhibitory actions can also be induced by TMS in awake human subjects. A transcallosally mediated inhibition is observed between the motor cortices (Ferbert et al. 1990), stimulation of the leg-associated motor area decreases the excitability of the ipsilateral arm-hand-associated motor area (surround inhibition), and tonic voluntary muscle activity can be inhibited by ipsilateral motor cortex stimulation, obviously by transcallosal influences (ipsilateral inhibition; Wassermann et al. 1991). The most powerful inhibitory action can be observed when the primary motor cortex is stimulated, and recordings from tonically activated contralateral muscles are performed. In this situation the short-latency response is followed by complete depression of voluntary activity which, especially in the small hand muscles, can last up to 250 ms (postexcitatory inhibition, PI; Fuhr et al. 1991; Roick et al. 1993).

In a recent study by our group (Roick et al. 1993) we analyzed the PI in more detail and provided experimental evidence that PI is mainly of cortical rather than spinal origin. PI was found more pronounced in distal hand muscles than in proximal arm and leg muscles. As is the case with short-latency MEPs, the strongest responses were obtained when the primary motor cortex was stimulated. Double stimulation of motor cortex with delivery of the second stimulus during PI and measurements of H reflex excitability during PI did not reveal a decrease in the excitability of spinal motoneurones during PI. These findings suggest that segmental inhibitory actions cannot be primarily responsible for PI, and that most probably inhibitory actions within the primary motor cortex itself play a major role in the genesis of PI. These results profoundly expand the diagnostic role of TMS in man because not only the excitability of pyramidal cells of the motor cortex and the integrity of the corticospinal tract but also the actions of local circuit interneurons become accessible.

Methods

Subjects

TMS was performed in a total of 30 spastic patients (18 men, 12 women) with defined lesions of the primary sensorimotor cortex ($n = 8$), lesions of the internal capsule (anterior or posterior limb; $n = 11$), or lesions of the premotor cortex ($n = 11$). Of these, 24 patients suffered from a cerebrovascular accident, and in 6 a primary brain tumor was diagnosed. TMS findings in patients were compared with those of 34 healthy volunteers (20 men, 14 women) which had been obtained in a previous study under the same experimental conditions (Roick et al. 1993).

The site of brain lesions was defined by routine computed tomography and magnetic resonance imaging. Clinical examination and TMS examinations were performed during a stabile neurological state. Patients with completely paralyzed limbs were excluded because of their inability to perform tonic activation of muscles.

All patients were examined by a senior resident and scored according to the degree of clinically established spasticity in individual muscles. All patients and subjects gave informed consent for the study, which was approved by the local ethics committee.

Stimulations

TMS was delivered using a Novametrix Magstim 200 HPR with a maximum output of 2.0 T. Capacitors are rapidly discharged through a standard flat circular coil (outer diameter 12 cm). Technical information about the Novametrix stimulator and the physics of TMS are described elsewhere (Barker et al. 1985; Benecke et al. 1988). A stimulation strength of 1.5 threshold level was used. This threshold was defined as the stimulation strength which induces a just visible MEP in at least half the trials when the muscle under study is at rest. The threshold was assessed with stimulation of the intact hemisphere and recordings of the intact contralateral body side. Clockwise and counterclockwise coil currents were used for the stimulation of the right or left motor cortex, respectively. For evoking hand muscle responses, the coil was centered over the vertex, while for leg muscles the coil was moved 4 cm anteriorly along the midline. During stimulation all subjects and patients were asked to perform a standardized isometric voluntary contraction (10% of maximal power) of the biceps brachii, flexor digitorum profundus, vastus medialis, and gastrocnemius muscles.

Recordings

Surface electromyogram (EMG) recordings were taken from muscles on both sides. The signals were amplified using a Toennies Myograph IIR with bandpass filtering between 20 and 3000 Hz. EMG data were collected and evaluated on an IBM compatible personal computer using Aútolab data-collection programs.

The amplitude of MEP was determined on a personal computer screen as the difference between the maximum and minimum of each trial. The onset latency of the MEP was determined as the first deflection from the baseline. In each experimental condition five trials were analyzed. Central motor conduction time (CMCT) was calculated by subtracting the peripheral conduction time (root stimulation) from the onset latency of the fastest cortically evoked potential (Hess et al. 1987; Ingram et al. 1988; Britton et al. 1990). Amplitudes of MEPs were measured as percentages of corresponding maximal M waves which were obtained after maximal stimulation of the peripheral nerves (musculocutaneous, median, femoral, and tibial nerves). Duration of PI was defined as the time from onset of the MEP to

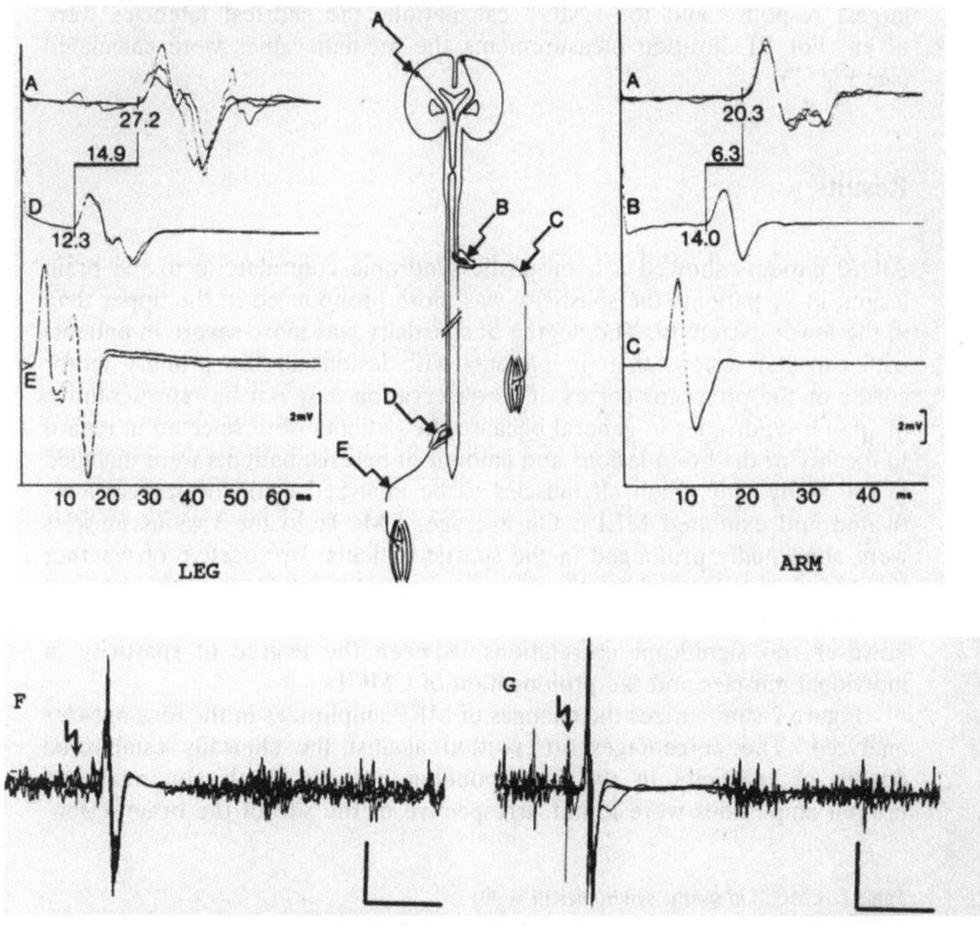

Fig. 1. Routine analysis of MEP parameters and PI. The primary motor cortex was stimulated magnetically to obtain MEPs in arm and leg muscles (*A*). Magnetic root stimulations (*B*, *D*) were performed for measurements of the peripheral conduction times. MEP amplitudes were measured as percentage values of the maximal M wave elicited by electrical stimulation of the peripheral nerves (*C*, *D*). Central motor conduction time was calculated by subtracting the peripheral conduction time from the total latency of MEPs. *F*, *G*, original registrations of MEPs and subsequent PI in gastrochemius and flexor digitorum profundus muscles. PI was measured from time of magnetic stimulation to reccurrence of tonic voluntary activity. Calibrations in *F*, *G*, 100 ms and 2 mV

the reccurrence of tonic voluntary EMG activity. Measurements of PI from the actual onset of EMG silence occurring after the short-latency MEP seemed to be less appropriate, since abnormal polyphasic and prolonged responses would have artificially affected PI duration. In the analysis of PI duration tonic activity amounted to 10% of maximal isometric EMG activity. The degree of EMG activity was monitored on an oscilloscope. For the evaluation of MEP amplitudes, CMCT, and PI durations, five consecutive registrations were analyzed. For amplitude assessments the largest response and for CMCT calculations the shortest latencies were taken. For PI duration measurements the median values were calculated (see Fig. 1).

Results

All 30 patients showed a hemispastic syndrome contralateral to the brain lesion; in 21 patients the spasticity was more pronounced in the upper than in the lower extremity. The degree of spasticity was more severe in patients with capsular lesions than in patients with lesions of the primary motor cortex or the premotor cortex. This observation may not be representative of spastic syndromes in general because the patients were selected in regard to focality of the brain lesions and amount of paresis; patients were included in the study only when all muscles to be analyzed could be tonically activated and exhibited MEPs. On average, CMCTs to the various muscles were abnormally prolonged in the spastic patients, irrespective of whether the lesion was located to the primary sensorimotor cortex, the internal capsule, or the premotor cortex (Table 1; $p > 0.5$, t test). There were, however, no significant correlations between the degree of spasticity in individual muscles and the prolongation of CMCTs.

Figure 2 summarizes the changes of MEP amplitudes in the four muscles analyzed. The percentages are plotted against the clinically established degree of spasticity in the corresponding muscles. Both abnormal and normal amplitudes were found, irrespective of the site of the brain lesion.

Table 1. CMCTs in spastic syndromes ($n = 30$)

Muscle	Normal subjects	Spastic patients
Biceps brachii	4.5 ± 0.6	6.5 ± 1.2
Flexor digitorum profundus	4.8 ± 0.7	6.9 ± 1.3
Vastus medialis	11.1 ± 2.5	14.9 ± 2.8
Gastrocnemius	13.4 ± 2.8	16.9 ± 2.5

For all muscles the latencies were significantly prolonged in spastic patients ($p > 0.05$).

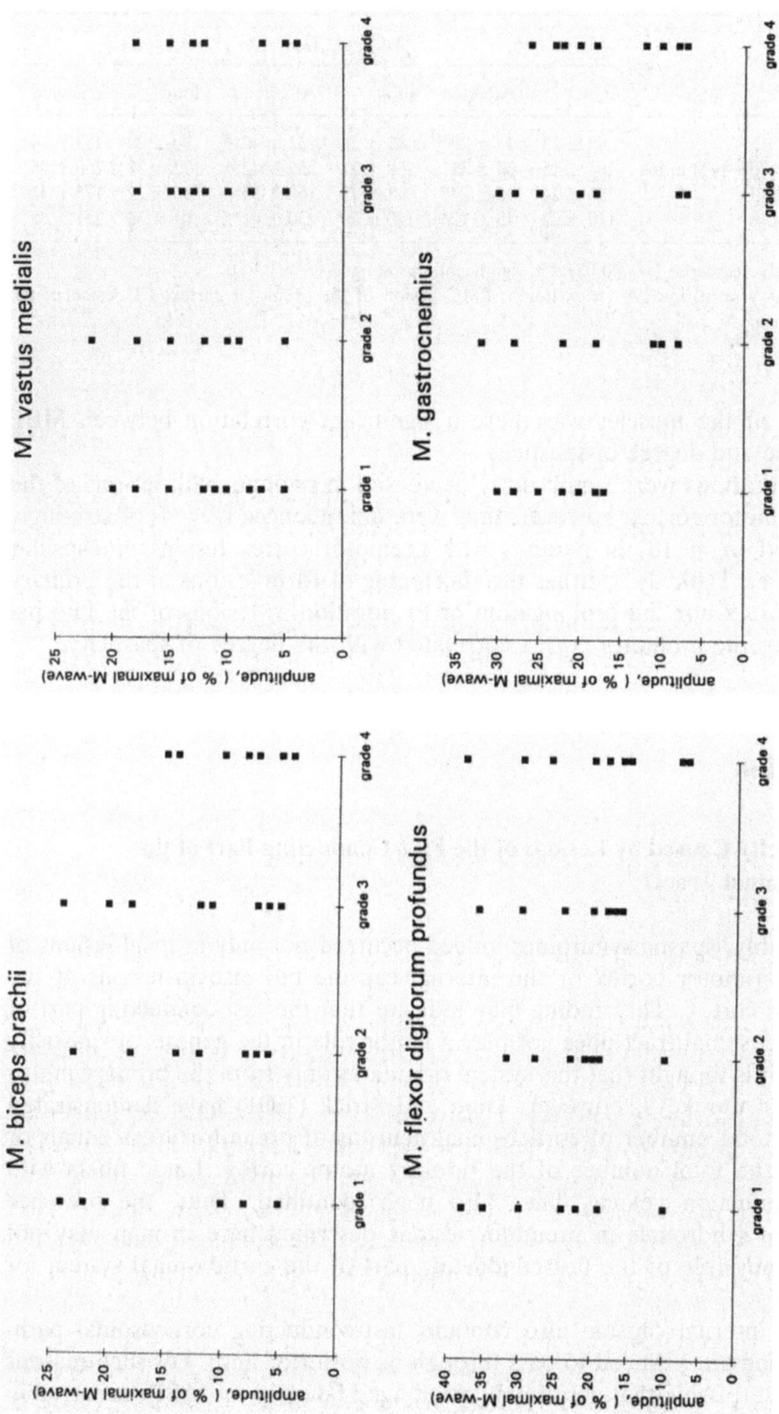

Fig. 2. Correlation of MEP amplitudes and degree of spasticity in 30 individual patients with a spastic hemiparesis. For each muscle the degree of spasticity was scored as follows: *grade 1*, only pathological increase of stretch reflex; *grade 2*, slight increase of muscle tone; *grade 3*, moderate increase of muscle tone; *grade* 4, severe increase of muscle tone. Note the lack of correlation in all muscles

Table 2. Mean PI durations (ms) after stimulation of the intact and the affected hemisphere

Muscle	SMC (n = 8)		PMC (n = 11)		CI (n = 11)	
	Intact	Affected	Intact	Affected	Intact	Affected
Biceps brachii	80 ± 23	45 ± 29*	83 ± 28	132 ± 40**	90 ± 22	152 ± 34**
Flexor digitorum profundus	160 ± 15	60 ± 32*	159 ± 21	255 ± 42**	172 ± 31	280 ± 45**
Vastus medialis	83 ± 24	35 ± 24*	94 ± 19	160 ± 35**	90 ± 22	175 ± 42**
Gastrocnemius	110 ± 22	45 ± 12*	120 ± 29	250 ± 41**	101 ± 33	261 ± 35**

* Significantly decreased ($p < 0.05$); **, significantly prolonged ($p > 0.05$).
SMC, Primary sensorimotor cortex lesion; PMC, lesion of the premotor cortex; CI, lesion of the internal capsule.

In none of the muscles was there a significant correlation between MEP amplitude and degree of spasticity.

PI durations were significantly decreased in patients with lesions of the primary motor cortex; however, they were uninfluenced (n = 4) or strikingly prolonged (n = 18) in patients with premotor cortex lesions or capsular lesions (see Table 2). Neither the shortening of PI in lesions of the primary motor cortex nor the prolongation of PI duration in lesions of the internal capsule or the premotor cortex correlated with the degree of spasticity.

Discussion

Is Spasticity Caused by Lesions of the Fast-Conducting Part of the Corticospinal Tract?

Remarkably, spastic syndromes indeed occurred not only in focal lesions of the sensorimotor cortex or the internal capsule but also in lesions of the premotor cortex. This finding may indicate that the fast-conducting part of the corticospinal tract does not play a major role in the genesis of spasticity because it is thought that this system originates only from the primary motor cortex. In monkeys, however, Dum and Strick (1991) have demonstrated that the total number of corticospinal neurons of premotor areas equals or exceeds the total number of the primary motor cortex. Large fibers with high conduction velocity have also been identified. Thus, the presence of spastic syndromes in premotor lesions described here in man may not exclude any role of the fast-conducting part of the corticospinal system for spasticity.

The internal capsule also contains fast-conducting corticospinal pathways which are believed to pass through its posterior limb. For such reasons it is not surprising that abnormally prolonged latencies of MEPs occurred in all three patient groups. Compared to patients suffering, for example, from multiple sclerosis the extent of latency prolongation is relatively mild. Similarly as in other studies (Hömberg et al. 1991), in most patients with hemispherical brain lesions abnormal amplitudes or a loss of MEPs were observed. In individual patients, however, MEPs may be normal despite the presence of an upper motor neuron syndrome. These considerations suggest that a lesion of the fast-conducting part of the corticospinal system is not a perequesite for a spastic syndrome but may often co-occur due to the close anatomical relationships of the corticofugal systems. This view is further substantiated by the lack of any correlation between the degree of spasticity and MEP amplitudes measured in the present study (Fig. 2). Furthermore, normal MEP amplitudes and latencies have been described in the majority of patients with familial spastic paralysis, a degenerative disease with pronounced spasticity in the lower extremities (Thompson et al. 1987).

Is Spasticity Combined with Decreased Intracortical Inhibition?

There is growing experimental evidence that the EMG silent period usually following the MEP in contralateral tonically activated muscles (PI), especially in its later part, is of cortical origin (Fuhr et al. 1991; Cantello et al. 1992; Uozumi et al. 1992; Roick et al. 1993). Recent studies have described changes in PI in patients with central motor disorders. Uozumi and coworkers (1992) compared durations of PI in normal subjects with those seen in patients with spastic hyperreflexia due to chronic cerebral infarction. In the abductor pollicis brevis muscle the duration of PI was significantly shorter in patients than in normal controls (127 versus 76 ms; $p < 0.01$, t test). It was not mentioned in this study whether the ischemic infarcts were cortically or subcortically located. Haug and coworkers (1992) investigated patients who had suffered from either a cortical or a subcortical ischemic infarct. Their patients had recovered almost fully from hemiparesis at the time of examination. In contrast to the study of Uozumi et al. (1992), in these patients a prolongation of PI durations was observed. Changes in PI duration have also been described in Parkinson's disease (Cantello et al. 1991; Haug et al. 1992). A decrease in PI duration was observed particularly in rigid patients. In Huntington's disease a pronounced prolongation of PI has been demonstrated (Roick et al. 1992) (Fig. 3).

The present investigation demonstrates that both abnormal decreases and increases in PI durations may indeed occur in patients with unilateral cerebral lesions. The primary sensorimotor cortex plays a central role in changes in PI duration. An abnormal short PI was observed when this area was directly lesioned. In the majority of lesions of the premotor cortex or the internal capsule PI was abnormally prolonged. These findings demonstrate that decreased activity of intracortical interneurons cannot be the origin of spasticity; although spasticity was most pronounced in lesions of

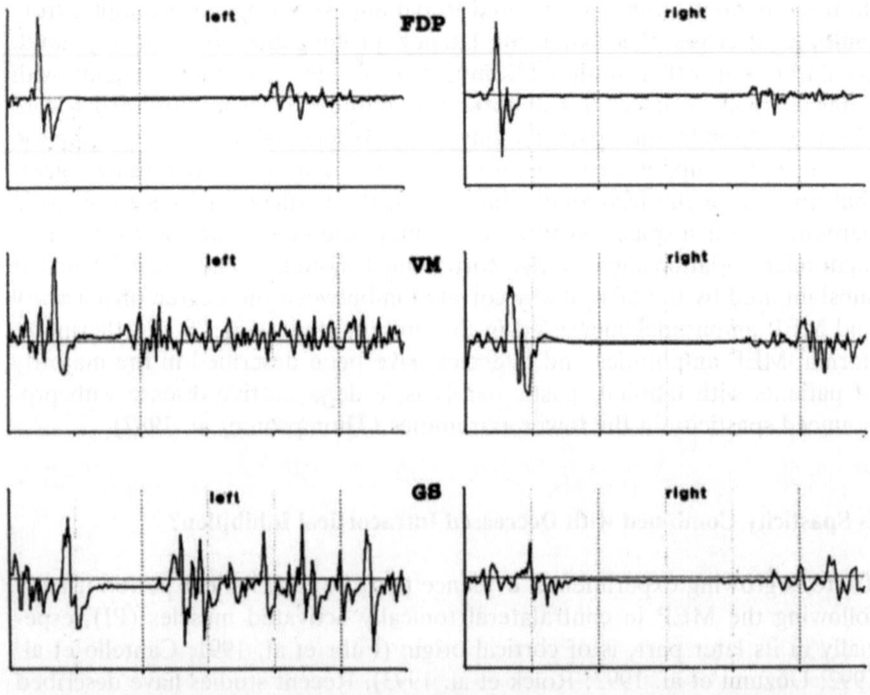

Fig. 3. Prolongation of PI in muscles contralateral to an operated astrocytoma in the left premotor cortex. *FDP*, Flexor digitorum profundus; *VM*, vastus medialis; *GS*, gastrocnemius. Measurements 3 months after the operation. The patient presented with a stabile spastic hemisyndrome on the right body side. Stimulation strength was 1.5 T. Note the pronounced prolongation of PI in the right VM and GS. Calibrations are 50 ms (*between dotted lines*) and ±2 mV in all registrations

the premotor cortex and the internal capsule, PI was prolonged rather than shortened in these patients.

References

Amassian VE, Stewart M, Quirk GJ, Rosenthal JL (1987) Physiological basis of motor effects of a transient stimulus to the cerebral cortex. Neurosurgery 20:74–92

Avoli M (1986) Inhibitory potentials in neurons of the deep layers of the in vitro neocortical slice. Brain Res 370:165–170

Barker AT, Jalinous R, Freeston IL (1985) Non-invasive magnetic stimulation of the human motor cortex. Lancet I:1106–1107

Benecke R, Meyer BU, Schönle P, Conrad B (1988) Transcranial magnetic stimulation of the human brain: responses in muscles supplied by cranial nerves. Exp Brain Res 71:623–632

Benecke R, Meyer BU, Freund HJ (1991) Reorganisation of descending motor pathways in patients after hemispherectomy and severe hemispheric lesions demonstrated by magnetic brain stimulation. Exp Brain Res 83:419–426

Berardelli M, Inghilleri M, Gruccu G, Fornarelli M, Accornero N, Manfredi M (1988) Stimulation of motor tracts in multiple sclerosis. J Neurol Neurosurg Psychiatry 51:677–683

Boyd SG, Rothwell JC, Cowan JMA, Webb PJ, Morley T, Asselman P, Marsden CD (1986) A method of monitoring function in corticospinal pathways during scoliosis surgery with a note on motor conduction velocities. J Neurol Neurosurg Psychiatry 49:251–257

Britton TC, Meyer BU, Herdmann J, Benecke R (1990) The clinical use of the magnetic stimulator in the investigation of peripheral conduction time. Muscle Nerve 13:396–406

Cantello R, Gianelli M, Bettucci D, Civardi C, De Angelis MS, Mutani R (1991) Parkinson's disease rigidity: magnetic motor evoked potentials in a small hand muscle. Neurology 41:1449–1456

Cantello R, Gianelli M, Civardi C, Mutani R (1992) Magnetic brain stimulation: the silent period after motor evoked potnetial. Neurology 42:1951–1959

Chapman CE, Wiesendanger M (1982) Recovery of function following unilateral lesion of the bulbar pyramid in the monkey. Electroencephalogr Clin Neurophysiol 53:374–387

Connors BW, Malenka RC, Silva LR (1988) Two inhibitory postsynaptic potentials, and $GABA_A$ and $GABA_B$ receptor-mediated responses in neocortex of rat and cat. J Physiol (Lond) 406:443–468

Creutzfeldt O, Baumgartner G, Schoen L (1956) Reaktionen einzelner Neurone des sensomotorischen Cortex nach elektrischen Reizen. I. Hemmung und Erregung nach direkten und kontralateralen Reizen. Archiv für Psychiatrie und Zeitschrift Neurologie 194:597–619

Day BL, Dressler D, Maertens de Noordhout A, Marsden CD, Nakashima K, Rothwell JC, Thompson PD (1989) Electric and magnetic stimulation of human motor cortex; surface EMG and single motor unit responses. J Physiol (Lond) 412:449–473

Dum RP, Strick PL (1991) The origin of corticospinal projections from the premotor areas in the frontal lobe. J Neurosci 11:667–689

Edgley SA, Eyre JA, Lemon RN, Miller S (1990) Excitation of the corticospinal tract by electromagnetic and electrical stimulation of the scalp in the macaque monkey. J Physiol (Lond) 425:301–320

Ferbert A, Priori A, Rothwell JC, Colebatch J, Day BL, Marsden CD (1990) Trans-callosal effects on motor cortical excitibility in man. J Physiol (Lond) 429:p38

Fuhr P, Agostino R, Hallett M (1991) Spinal motor neuron excitability during the silent period after cortical stimulation. Electroencephalography and clinical Neurophysiology 81:257–262

Gilman S, Marco LA (1971) Effect of medullary pyramidotomy in the monkey. I. Clinical and electromyographic abnormalities. Brain 94:495–514

Haug BA, Schönle PW, Knobloch C, Köhne M (1992) Silent period measurement revives as a valuable diagnostic tool with transcranial magnetic stimulation. Electroencephalogr Clin Neurophysiol 85:158–160

Hess CW, Mills KR (1986) Low threshold motor units in human hand muscles can be selectively activated by magnetic brain stimulation. J Physiol (Lond) 380:p62

Hess CW, Mills KR, Murray NMF (1986) Measurement of central motor conduction in multiple sclerosis using magnetic brain stimulation. Lancet II:355–358

Hess CW, Mills KR, Furray NMF (1987) Responses in small hand muscles from magnetic stimulation of the human brain. J Physiol (Lond) 388:397–419

Hines M (1929) On cerebral localization. Physiol Rev 9:462–574

Hömberg V, Stephan KM, Netz J (1991) Transcranial stimulation of motor cortex in upper motor neurone syndrome: its relation to the motor deficit. Electroencephalogr Clin Neurophysiol 81:377–388

Ingram DA, Thompson AJ, Swash M (1988) Central motor conduction in multiple sclerosis: evaluation of abnormalities revealed by transcutaneous magnetic stimulation of the brain. J Neurol Neurosurg Psychiatry 51:487–494

Iverson LL, Mitchell JF, Srinivasan V (1971) The release of γ-aminobutyric acid during inhibition in the cat visual cortex. J Physiol (Lond) 212:519–534

Kennard MA, Viets HR, Fulton JF (1934) The syndrome of the premotor cortex in man: impairment in skilled movements, forced grasping, spasticity and vasomotor disturbances. Brain 57:69–84

Kernell D, Wu CP (1967) Responses of the pyramidal tract to stimulation of the baboon's motor cortex. J Physiol (Lond) 191:653–672
Krnjevic K (1974) Chemical nature of synaptic transmission in vertebrates. Physiol Rev 54:418–540
Krnjevic K (1987) GABAergic inhibition in the neocortex. J Neurol Sci 8:537–547
McCulloch WS, Graf C, Magoun HW (1946) A cortico-bulbo-reticular pathway from area 4-s. J Neurophysiol 9:127–132
McCormick DA (1989) GABA as an inhibitory neurotransmitter in human cerebral cortex. J Neurophysiol 62:1018–1027
Ottersen OP, Storm-Mathisem J (1984) Glutamate and GABA-containing neurons in the mouse and rat brain, as demonstrated with a new immunocytochemical technique. J Comp Neurol 229:374–392
Patton HD, Amassian VE (1954) Single- and multiple-unit analysis of cortical stage of pyramidal tract activation. J Neurophysiol 17:345–363
Ribak CE (1978) Aspinous and sparsely-spinous stellate neurons in the visual cortex of rats contain glutamic acic decarboxylase. J Neurocytol 7:461–478
Roick H, Giesen HJV, Lange HW, Benecke R (1992) Postexcitatory inhibition in Huntington's disease. Mov Disord 7[Suppl 1]:27
Roick H, Giesen HJV, Benecke R (1993) On the origin of the postexcitatory inhibition seen after transcranial magnetic brain stimulation in awake human subjects. Exp Brain Res 94:489–498
Rothwell JC, Thompson PD, Day BL, Dick JPR, Kachi T, Cowan JMA, Marsden CD (1987) Motor cortex stimulation in intact man. I. General characteristics of EMG response in different muscles. Brain 110:1173–1190
Thompson PD, Day BL, Rothwell JC, Dick JPR, Cowan JMA, Asselmann P, Griffin GB, Sheehy MP, Marsden CD (1987) The interpretation of electromyographic responses to electrical stimulation of the motor cortex in diseases of the upper motor neurone. J Neurolog Sci 80:91–110
Tseng GF, Haberly LB (1988) Characteristics of synaptically mediated fast and slow inhibitory processes in piriform cortex in an in vitro slice preparation. J Neurophysiol 59:1352–1376
Uozumi T, Ito Y, Tsuji S, Murai Y (1992) Inhibitory period following motor potentials evoked by magnetic cortical stimulation. Electroencephalogr Clinical Neurophysiol 85:273–279
Wassermann EM, Fuhr P, Cohen LG, Hallett M (1991) Effects of transcranial magnetic stimulation on ipsilateral muscles. Neurol 41:1795–1799

Do Rapidly Conducting Ipsilateral Corticospinal Pathways Contribute to Recovery in Adult Hemiplegia?

P. Ashby, E. Palmer, and V.E. Hajek

Recovery of Voluntary Movements Following Lesions of the Corticospinal System in Adult Human Beings

Remarkable recovery of function can follow lesions of the corticospinal system in man (e.g., Nathan and Smith 1973; Ropper et al. 1979). In some instances, the recovery of voluntary movement may be accomplished by the use of pre-existing alternative pathways from the motor cortex on the affected side to the contralateral motoneurons. Which pathways are available will depend on the site of the lesion (Fig. 1). For example, following a lesion of the lateral column of the spinal cord, say on the left, there may be rapid return of movements of the initially paralysed left limbs. This appears to depend on fibers descending in the right lateral column, as a second lesion, affecting the right lateral columns, abolishes the recovery (Nathan

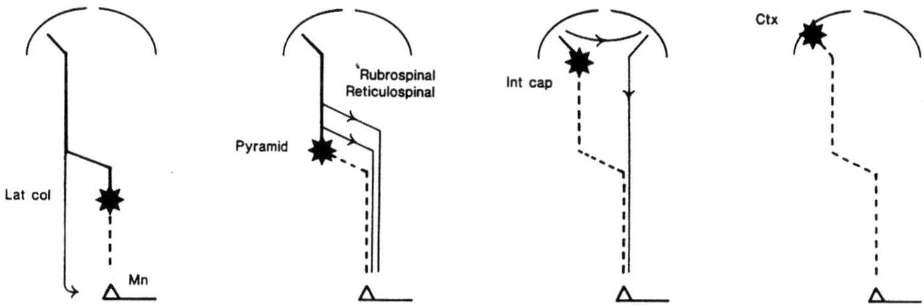

Fig. 1. Alternative pathways from the motor cortex (*Ctx*) to contralateral motoneurons (*Mn*) that could contribute to recovery following lesions of the fast corticospinal system at various sites. Fibers in the opposite lateral column (*Lat col*) may be able to compensate for a unilateral lesion of the spinal cord, the rubrospinal and reticulospinal pathways may be able to compensate for a lesion of the pyramid, and the contralateral cortex for a lesion of the internal capsule (*Int cap*). None of these options are available when the lesion affects the cells of origin of the corticospinal system

Playfair Neuroscience Unit, The Toronto Western Hospital, Edith Cavell Wing, 8-023 399 Bathurst Street, Toronto, Ontario, Canada M5T 2S8

Thilmann et al. (Eds.)
Spasticity: Mechanisms and Management

and Smith 1973). Following a lesion of the right pyramid this option would not be availables, but the right motor cortex could still influence contralateral spinal motoneurons via the rubrospinal or reticulospinal systems. Following a lesion of the right internal capsule there are fewer alternative, but the right cortex could communicate to motoneurons on the left through transcallosal and uncrossed left corticospinal fibers. A lesion in the motor cortex affecting the cells of origin of the corticospinal tract would appear to remove all of these options, yet recovery can still occur. How are voluntary movements carried out in this case?

Recovery of Voluntary Movements Following Lesions of the Motor Cortex in Adult Human Beings

Three possible mechanisms of recovery can be considered (Fig. 2):

Recovery of the Affected Cortex

The function of the affected cortex may return because there has been resolution of edema or other metabolic disturbances affecting cortical neurons, remyelination or regeneration of corticospinal fibers, or reorganization of the adjacent cortex.

Movements Executed by Subcortical Systems

The corticospinal system developed late in evolution. It is present only in advanced mammals, and the monosynaptic connections to motoneurons are present only in primates. It is quite possible that some of the phylogenetically older pathways are still present in man and that voluntary movement can be executed by neural pathways that do not include the motor cortex.

Spared Cortex Develops Projections to the Ipsilateral Motoneurons

There are two lines of evidence for this hypothesis. First, there have been instances in which the contralateral corticospinal system appeared to be responsible for recovery. Fisher (1992) described a patient who had a lesion of the right internal capsule producing severe hemiparesis. The weakness had partly recovered when, 26 days later, an internal capsule lesion on the opposite side caused a new hemiplegia and re-paralysis of the original side. A second patient had hemiparesis from a lesion of the right pyramid. This had recovered when, 7 months later, a second lesion producing infarction of the left pyramidal resulted in tetraplegia.

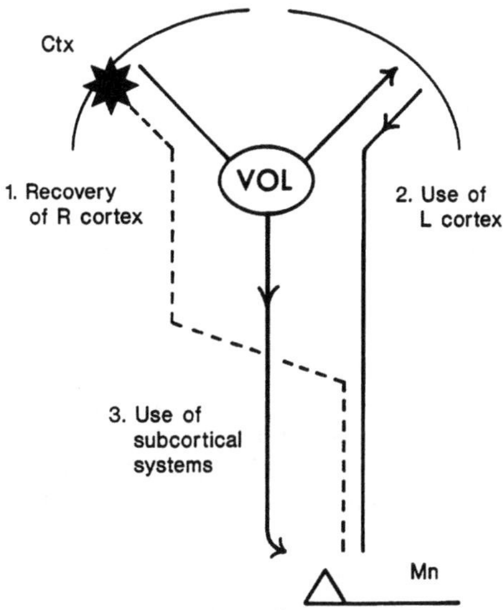

Fig. 2. The recovery of voluntary movements following lesions affecting the motor cortex could be due to (*1*) recovery of the neurons in the affected cortex, (*2*) the emergence of functioning projections from the contralateral cortex, or (*3*) the use of noncortical systems

Second, in patients recovering from hemiplegia there are changes in blood flow over the ipsilateral hemisphere during attempted movements of the recovering hand. Chollet et al. (1991) used positron emission tomography to study regional blood flow in six patients recovering from hemiplegia due to cerebral lesions. Voluntary movements of the spared hand were associated with an increase in blood flow in the contralateral sensorimotor cortex and ipsilateral cerebellar hemisphere. Attempted movements of the recovering hand were accompanied by a bilateral increase in blood flow in the sensorimotor cortices, cerebellar hemispheres, and parietal, insular and premotor cortices. The hypothesis that the spared cortex develops projections to ipsilateral motoneurons can be tested by transcranial stimulation of the cortical neurons.

Testing the Hypothesis that the Unaffected Corticospinal System Gains Access to Ipsilateral Motoneurons Following Lesions of the Corticospinal System in Adult Human Beings

Introduction

Neurons in the human cortex can be excited by transcranial electrical or magnetic stimulation. Stimulation over the motor cortex results in short-latency activation of contralateral motoneurons, which has been attributed to the monosynaptic projections of the fast corticospinal system (Day et al. 1989; Burke et al. 1990; Palmer and Ashby 1992). The hypothesis that the unaffected cortex develops projections to ipsilateral motoneurons can be tested by stimulating over the spared hemisphere and recording from ipsilateral muscles.

Such studies provide evidence that ipsilateral projections *are* present when damage to the cortex is sustained during development. Benecke et al. (1991) examined two patients with porencephaly, one with subsequent hemispherectomy. Magnetic stimulation over the preserved hemisphere produced short-latency activation of ipsilateral as well as contralateral muscles. Farmer et al. (1991) examined four children with hemiplegia and mirror movements from cerebral lesions sustained at birth. Magnetic stimulation over the unaffected hemisphere resulted in short-latency EMG in the small muscles of both hands. Bilateral late-reflex responses (considered to be transcortical) were seen in response to stimulation of the digits of the spared hand, and cross-correlations between motor units in the two hands indicated a common synaptic input. All this evidence suggests that the spared cortex can develop short-latency projections to ipsilateral motoneurons when lesions occur during development.

In contrast, when the lesion to the corticospinal system occurs in adulthood, stimulation over the spared cortex does not result in short-latency contractions of ipsilateral muscles (Thompson et al. 1987; Yang et al. 1990; Benecke et al. 1991; Fries et al. 1991; Homberg et al. 1991), although longer-latency ipsilateral responses have been reported following strong electrical (Thompson et al. 1987) or magnetic (Benecke et al. 1991; Homberg et al. 1991) stimulation.

Technical Problems

There are, however, a number of technical problems with these studies. Electrical stimuli over the scalp excites corticospinal axons and may do so as far caudal as the medulla (Edgley et al. 1990). Stimulation over the spared cortex could therefore excite remaining fibers from the affected motor cortex or fibers in the brainstem with bilateral projections. Similarly, with strong magnetic stimulation there could be spread of a current to the opposite hemisphere, exciting any spared neurons in the affected cortex. In each case, the observer might conclude, falsely, that there were ipsilateral connections. Unless there has been a hemispherectomy, it is important to exclude the presence of residual fibers from the affected cortex. Conversely, connections from the spared cortex to the ipsilateral motoneurons might exist, but if the descending volley produced only a small EPSP in motoneurons and/or if the motoneurons were hyperpolarized, then none of the motoneurons would be brought to threshold. In this case, the observer would conclude, wrongly, that there were no ipsilateral connections. A more sensitive test for the presence of short-latency projections would be to look for subthreshold excitation of motoneurons following stimulation of the cortex.

A Study of Subthreshold Events Produced by Cortical Stimulation

Subthreshold events in individual human spinal motoneurons following cortical stimuli can be derived using post-stimulus time histograms (PSTH) of single motor units. A needle electrode is inserted into the muscle of interest and positioned close to a single, voluntarily activated motor unit. The spike train of the motor unit is recorded during over 100 cortical stimuli, and the changes in firing probability are used to reveal facilitation or inhibition by the cortical volley. In normal subjects it can be shown with this technique that magnetic stimulation generates composite EPSPs with short latency and rise time in spinal motoneurons. These EPSPs presumably represent the monosynaptic facilitation of motor neurons from the fast corticospinal neurons. The composite EPSPs are largest in the motoneurons of distal hand muscles, moderately large in biceps motoneurons and smaller in the motoneurons of proximal muscles (Palmer and Ashby 1992).

We used this technique to look for ipsilateral projections in patients recovering from hemiplegia. Three recordings were made on each patient (Fig. 3). In the first, stimuli were given over the unaffected (say left) cortex

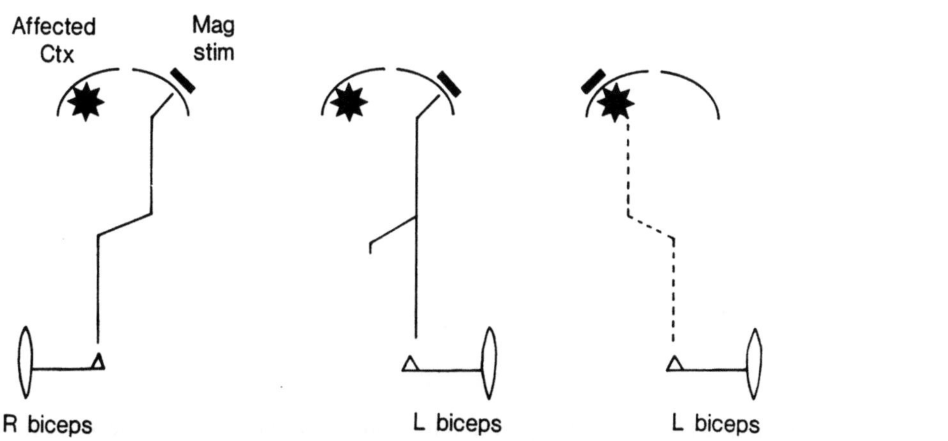

Fig. 3. Method for testing the hypothesis that the spared cortex develops projections to ipsilateral motoneurons. *Left*, Magnetic stimuli are delivered over the spared cortex and subthreshold events in contralateral motoneurons detected from post-stimulus time histograms of single motor units. *Middle*, With the identical stimulus, subthreshold events are sought in ipsilateral motoneurons. *Right*, Sparing or recovery of neurons in the affected corticospinal system is sought by stimulating over the affected hemisphere

and recordings made from single units in the unaffected (right) biceps. The appearance of typical short-latency facilitation indicated that the fast corticospinal neurons were being stimulated. Next, using exactly the same stimulus site and intensity, recordings were made from the affected (left) biceps to look for subthreshold events that would indicate ipsilateral connections. Finally, the stimulator was moved to the affected (right) cortex and recordings were made from the affected (left) biceps to determine whether there were any spared or recovered corticospinal fibers. For this run, the stimulus intensity was increased 5%, as preliminary studies indicated that this would take into account any possible threshold differences between the two sides.

Results

Observations were made on nine patients recovering from hemiparesis (reported in detail by Palmer et al. 1992). The lesions had occurred 7–57

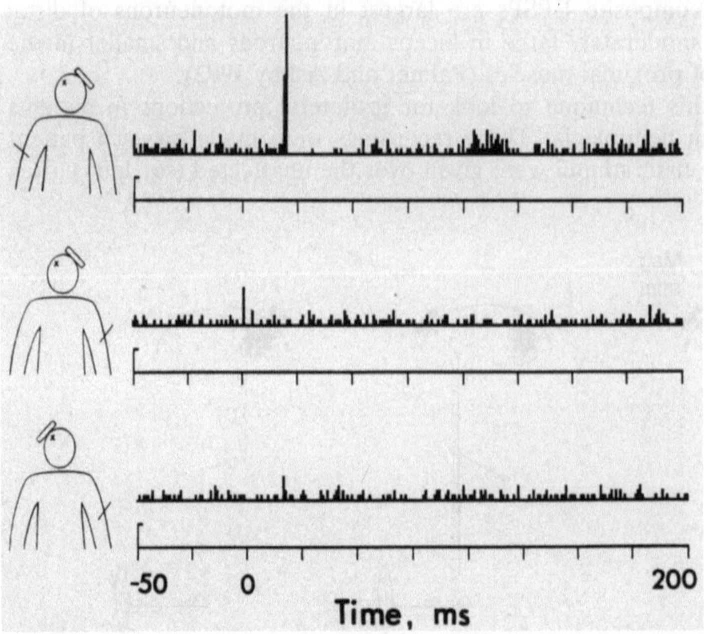

Fig. 4. Result of a typical study in a patient with a lesion in the left internal capsule, giving rise to right hemiparesis with good recovery (biceps strength 5). *Top*, Stimulation over the unaffected right cortex (65% maximum output of stimulator) produced an increase in the firing probability of a contralateral biceps motor unit, representing the effects of a short-latency composite EPSP. *Middle*, The identical stimulus did not produce any changes in firing probability of an ipsilateral biceps motor unit (the peak at time 0 is caused by stimulus artifact). *Bottom*, Stimulation over the affected left cortex (70% maximum output of stimulator) produced a small increase in the firing probability of a contralatetal biceps motor unit. (From Palmer et al. (1992), with permission)

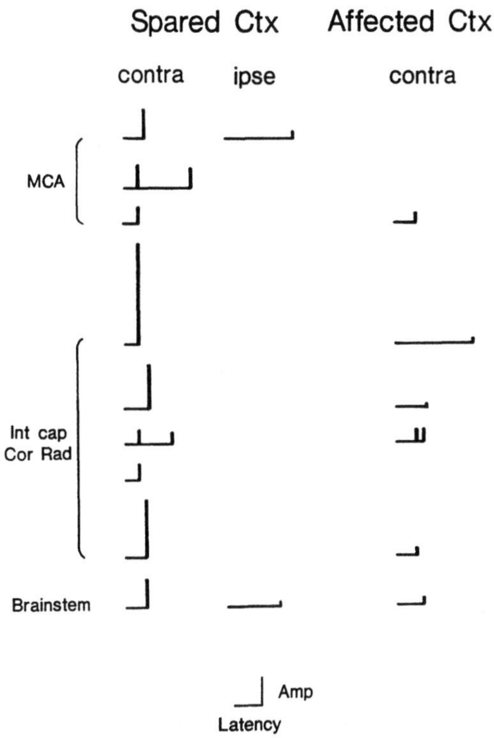

Fig. 5. Composite diagram representing the latencies (*abscissa*) and peak areas (*ordinate*) of the changes in firing probability of single biceps motor units following magnetic stimulation over the cortex (*Ctx*). Stimulation over the spared hemisphere resulted in short-latency facilitation of contralateral biceps motor units in every instance. There was no short-latency facilitation of ipsilateral biceps motor units. A late facilitation was seen in two. Stimulation over the affected hemisphere produced early facilitation of contralateral biceps motor units in five patients, although the peak amplitudes were lower and the latencies longer than on the opposite side. The calibrations represent 20 ms (*abscissa*) and 200 extracounts (*ordinate*). *MCA*, middle cerebral artery; *Cor Rad*, corona radiata; *Int cap*, internal capsule

months previously and included infarcts in the middle cerebral artery territory in three, lacunes in the corona radiata or internal capsule in five and a brain stem lesion in one. In each case there was recovery of function of the affected upper limb; voluntary power in the biceps was between 3 and 5 on the MRC scale, although the return of hand function was less. In every case magnetic stimulation over the unaffected cortex resulted in short-latency facilitation of motor units in the unaffected biceps. An example is shown in Fig. 4 and the results of nine studies are represented diagramatically in Fig. 5. The mean latency of the peaks of increased firing probability was 12.6 ms, the mean duration of the facilitation 2.7 ms (this is related to the rise time of the underlying composite EPSP) and the mean number of displaced counts in the PSTH (which is related to the amplitude of the underlying EPSP) was 233 per 1000 stimuli.

It was surprisingly easy to record regularly discharging single units from the biceps on the affected side. The firing rate was the same as on the unaffected side. No short-latency facilitation was recorded from ipsilateral biceps motor units during stimulation over the unaffected cortex in any subject, even though the stimulus had been shown to excite fast corticospinal neurons in this hemisphere.

The preservation or recovery of the affected corticospinal system was then tested by stimulating over the affected hemisphere and recording from the contralateral biceps. In four patients there was no response. In five, there was a short-latency facilitation but the amplitude was smaller (by a mean of 148 counts) and the latency longer (by a mean of 3 ms) when compared with the data obtained from the unaffected corticospinal pathway of the same five patients.

Discussion

Stimulation of the uninvolved cortex was sufficient to excite fast corticospinal neurons and produce short-latency depolarization of contralateral biceps motoneurons in all patients. The findings were similar to those in normal subjects (Palmer and Ashby 1992). The identical stimulus did not produce any short-latency depolarization of ipsilateral biceps motoneurons. Thus, there is no evidence from these recordings for the emergence of a short-latency ipsilateral corticospinal projection.

It is possible that an ipsilateral fast pathway exists, but that the evoked EPSPs were too small to be observed. The method, however, should be capable of detecting EPSPs as low as 35 μV in amplitude (Palmer and Ashby 1992), and if the EPSPs were smaller than this, the contribution of ipsilateral projections to the recovery of function could be questioned. The EPSPs evoked in biceps motoneurons by stimulating over the normal hemisphere (at the stimulus intensities used) had an estimated mean amplitude of 230 μV.

It is possible that the population of cortical neurons providing short-latency ipsilateral projections have higher thresholds than those providing contralateral projections. This cannot be excluded, but there was no ipsilateral facilitation even when the stimulus intensity was increased by 5%. It is possible that the ipsilateral pathway does not arise from the same area of cortex as the contralateral one, but a large-diameter coil was used and this would likely stimulate cortical cells within several centimeters.

Recovery could be mediated by slowly conducting or polysynaptic ipsilateral pathways and this might be revealed by later periods of facilitation in the PSTH. Stimulation of the unaffected cortex resulted in late (about 50 ms) facilitation in the contralateral biceps in two subjects. This is the same proportion as in normal subjects (Palmer and Ashby 1992). Stimulation of the unaffected hemisphere also resulted in late facilitation of the ipsilateral (affected) biceps motor units in two subjects, but before we attribute recovery to this long-latency ipsilateral pathway it must be noted that the other seven subjects did not show such facilitation although they had also recovered.

From the present data, therefore, we conclude that the recovery of voluntary movements of the previously paretic limb in adults recovering from hemiplegia is not primarily due to the development of an ipsilateral fast-conducting corticospinal pathway. In some cases there is recovery of the damaged corticospinal system; in others, neither cortex appears to project to the motoneurons, and movements must be carried out by noncortical pathways. In patients with lesions of corticospinal axons more caudal in the nervous system, other mechanisms of recovery (see introduction) may also be available.

Acknowledgements. We thank the Canadian Medical Research Council (6727) and the National Institutes of Health (27873).

References

Benecke R, Meyer BU, Freund HJ (1991) Reorganisation of descending motor pathways in patients after hemispherectomy and severe hemispheric lesions demonstrated by magnetic brain stimulation. Exp Brain Res 83:419–426

Burke D, Hicks RG, Stephen JPH (1990) Corticospinal volleys evoked by anodal and cathodal stimulation of the human motor cortex. J Physiol (Lond) 425:283–299

Chollet F, DiPiero V, Wise RJS, Brooks DJ, Dolan RJ, Frackowiak RSJ (1991) The functional anatomy of motor recovery after stroke in humans: a study with positron emission tomography. Ann Neurol 29:63–71

Day BL, Dressler D, Maertens De Noordhout A, Marsden DC, Nakashima K, Rothwell SL, Thompson PD (1989) Electrical and magnetic stimulation as the human motor cortex: surface EMG and single motor unit responses. J Physiol (Lond) 412:449–473

Edgley SA, Eyre JA, Lemon RN, Miller S (1990) Excitation of the corticospinal tract by electromagnetic and electrical stimulation over the scalp in the macagu monkey. J Physiol (Lond) 425:301–320

Farmer SF, Harrison LM, Ingram DA, Stephens JA (1991) Plasticity of central motor pathways in children with hemiplegic cerebral palsy. Neurology 41:1505–1510

Fisher CM (1992) Concerning the mechanism of recovery in stroke hemiplegia. Can J Neurol Sci 19:57–63

Fries W, Danek A, Witt TN (1991) Motor responses after transcranial electrical stimulation of cerebral hemispheres with a degenerated pyramidal tract. Ann Neurol 29:646–650

Homberg V, Stephen KM, Nelz J (1991) Transcranial stimulation of motor cortex in upper motor neurone syndrome: its relation to the motor deficit. Electroencephalogr Clin Neurophysiol 81:377–388

Nathan PW, Smith MC (1973) Effects of two unilateral cordotomics on the motility of the lower limbs. Brain 96:471–494

Palmer E, Ashby P (1992) Corticospinal projections to upper limb motoneurones in humans. J Physiol (Lond) 448:397–412

Palmer E, Ashby P, Hajek VE (1992) Do ipsilateral corticospinal pathways contribute to recovery in stroke? Ann Neurol 32:519–525

Ropper AH, Fisher CM, Kleinmann GM (1979) Pyramidal infarction in the medulla: a case of pure motor hemiplegia sparing the face. Neurology 29:91–95

Thompson PD, Day BL, Rothwell JC, Dick JPR, Cowan JMA, Asselman P, Griffen GB, Sheehy MP, Marsden CD (1987) The interpretation of electromyographic responses to electrical stimulation of the motor cortex in diseases of the upper motor neurone. J Neurol Sci 80:91–110

Yang BJ, Brouwer B, Hajek VE, Ashby P (1990) Magnetic stimulation of short-latency corticospinal pathways following unilateral cerebral infarction. Can J Rehabil 4:39–45

Strength Changes in Hemiparesis: Measurements and Mechanisms

S.C. Gandevia

This chapter is concerned, first, with describing the changes in voluntary muscle strength which occur following hemiparesis and, second, with a brief consideration of some of the central and peripheral factors which contribute to the absolute forces measured in hemiparesis.

Voluntary Muscle Strength in Hemiparesis

Changes in voluntary muscle strength of limb muscles following a unilateral cerebral lesion have been debated for over a century (e.g., Broadbent 1866; Wernicke 1889; Gowers 1893; Mann 1895). At least three issues have aroused particular interest: first, the distribution of strength between proximal and distal muscles; second, the predilection for weakness to occur in extensor muscles in the upper limb and flexor muscles in the lower limb (Mann 1895; Wernicke 1889); and third, the extent to which weakness develops on the side ipsilateral to the lesion. While it has frequently been stated that in upper motoneuron lesions muscle strength is particularly diminished in extensors of the upper limb and flexors of the lower limb, these opinions, now widely disseminated (e.g., Lance and McLeod 1981) have derived from clinical impressions of muscle strength rather than from formally quantified assessments.

Recent quantitative measurements made by Bohannon and colleagues using a hand-held myometer (Bohannon and Smith 1987; Bohannon 1987, 1988, 1989) and in our laboratory using isometric myographs designed for specific muscle groups (Colebatch et al. 1986; Colebatch and Gandevia 1989; Adams et al. 1990) have allowed these questions to be addressed. Patients were studies who had hemiparesis due to single vascular lesions which were usually well lcoalized radiographically. Both research groups used standardized postures for the measurement of maximal strength in each muscle group. We measured at least three maximal efforts (2–3 s duration),

Department of Clinical Neurophysiology, The Prince Henry Hospital and Prince of Wales Medical Research Institute, University of New South Wales, High St, Randwick, NSW 2031 Sydney, Australia

Thilmann et al. (Eds.)
Spasticity: Mechanisms and Management

during which visual feedback and verbal encouragement were provided continuously. In our studies, particular emphasis was placed upon comparison of strength between the strong side (tested first) and the affected side (tested second) and also between the strength on the strong side (in hemiparesis and hemiplegia) and that in a control group of normal subjects.

When data from the hemiparetic subjects were pooled, and relative strength was expressed as a fraction of that on the unaffected side, there was a statistically significant trend for weakness to be relatively greater for distal muscles than proximal ones in both the upper and lower extremity. In the upper limb the first dorsal interosseous was the most distal muscle examined, while the flexors and extensors of the hallux were studied in the lower limb. Table 1 shows data for relative strength of the most and least affected muscle groups. This trend is consistent with data from others (e.g., Bohannon and Smith 1987; Bohannon 1990). While flexors and extensors of the shoulder were relatively spared, it is clear that there is a differential effect on shoulder rotators, with the internal rotators being relatively spared compared with the external ones (Bohannon 1990). However, a surprising feature of the results, given the rather uniform description of strength loss in hemiparesis, was the variability between individual patients in the proximal-to-distal pattern of weakness. An example of this is shown in Fig. 1. One patient had a marked distal component to his weakness, whereas the other had a more uniform distribution of weakness. Our studies have not revealed a correlation between the site of lesion (e.g., cortical versus subcortical) and the particular pattern of strength reduction. Such correlations are probably present, but large numbers of patients will need to be formally tested.

An alternative way to present our results, which takes into account the possibility that the "unaffected" side may not have completely normal

Table 1. Muscle actions of upper limb affected in hemiparesis. Data from two studies

From Colebatch and Gandevia 1989	From Bohannon and Smith 1987
Wrist flexion (32%)[a]	Shoulder external rotators (46%)[a]
Hand grip (34%)	Wrist extensors (50%)
Finger flexion (36%)	Elbow flexors (50%)
Thumb flexion (41%)	Shoulder flexors (54%)
Thumb extension (42%)	Shoulder extensors (57%)
Wrist extension (43%)	Elbow extensors (58%)
Elbow flexion (47%)	Shoulder abductors (61%)
Finger extension (50%)	Shoulder internal rotators (62%)
Abduction of index (52%)	
Elbow extension (53%)	
Shoulder abduction (57%)	
Shoulder abduction (67%)	

[a] The percentages show the residual strength as a fraction of that on the "unaffected" side. Different methods of muscle testing were used in the two studies.

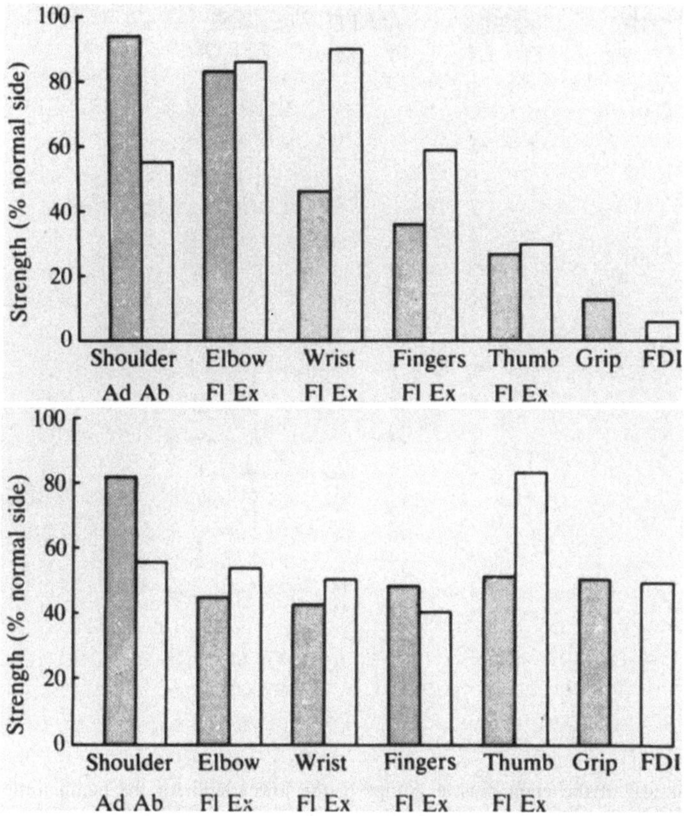

Fig. 1. Distribution of muscle weakness for two of the paretic subjects to reveal the variability encountered. The strength on the paretic side has been expressed as a percentage of the strength of the opposite (spared) side. The muscles are shown below the bars. For the subject shown in the *upper* half of the figure, distal muscles were more severely weakened than proximal ones. The subject shown in the *lower* half of the figure had a similar degree of involvement of all muscle groups. Both subjects show that flexors were usually relatively weaker than extensors, although this difference was often small. *Ad*, Adduction; *Ab*, abduction; *Fl*, flexion; *Ex*, extension. (Reproduced from Colebatch and Gandevia 1989, permission sought)

strength (see below), expresses the weakness for each muscle group on the affected side as a percentage of the overall strength reduction averaged for all muscle groups in the patient's affected limb. Thus, muscles which are preferentially weakened show a relatively greater strength reduction compared with the mean strength reduction for the whole limb. Figure 2 shows this data for strength in the lower limb and also emphasizes the variability on the affected side in the sixteen hemiparetic patients.

A related feature revealed in Table 1 and Fig. 1 is that agonist and antagonist muscles usually showed corresponding degrees of weakness. Such a correspondence is less exceptional, given that strokes which produce

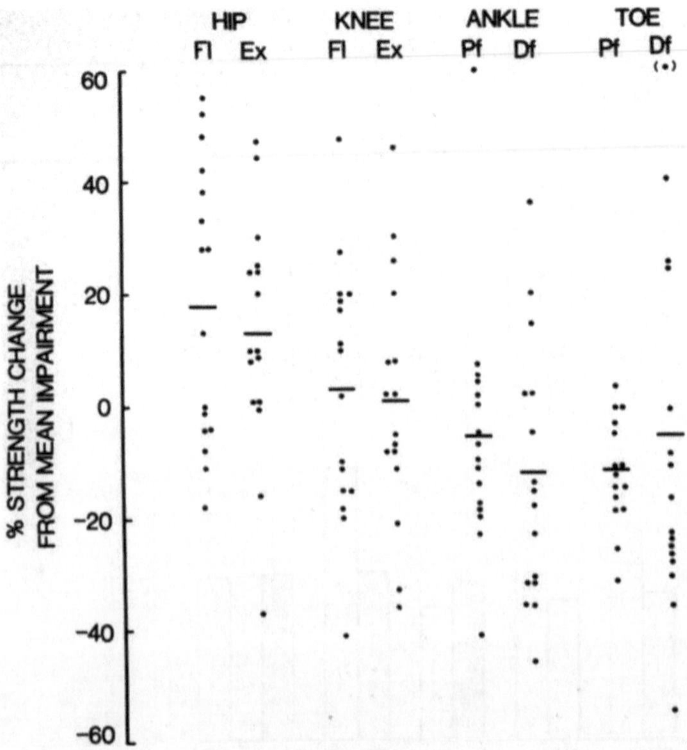

Fig. 2. Relative weakness of the eight muscle groups in the lower limb for the hemiparetic subjects. The degree of weakness has been calculated (as a percentage) by using the strength on the spared side as a reference. To permit comparison between subjects, the weakness of each muscle group has been expressed relative to the overall mean weakness for each subject. Thus for each subject the more severely affected muscles lie below zero, the less affected above it. One subject does not have data points for knee flexion or extension. A value of +88% for hallux plantar flexion in one subject is displaced. There is a tendency for distal muscles to be more severely affected than proximal muscles, but there is no clear difference between the relative effects on flexors and extensors. (Reproduced from Adams et al. 1990, permission sought)

severe hemiparesis or monoparesis usually produce weakness in all muscle groups in the limb. In a statistical sense, the severity of the stroke can be predicted from examination of the weakness in only a few muscle groups (e.g., Bohannon 1990; Warabi et al. 1990). Given the topographical organization of the motor cortex, corticofugal fibers destined for a particular limb or part of a limb are likely to be anatomically close together or even intermingled, thus rendering them susceptible to focal damage. At least after the first few days following the initial vascular episode, absolute strength seems usually to be similarly affected in agonist/antagonist pairs including those at the interphalangeal joint of the thumb, the metacarpophalgeal joint of the fingers, the wrist, elbow, hip, knee, ankle and metatarsophalangeal joint of the great toe. For the shoulder, however, abduction and external rotation are probably affected significantly more than the antagonist movement (see Table 1). For a given population of patients, when the residual strength is expressed relative to that on the contralateral side, the strength differences for these agonist/antagonist pairs are at least 10%.

Warabi and colleagues (1990) recently undertook a detailed correlation between the degenerative shrinkage detected in the cerebral peduncles of stroke patients using computerized tomography and the functional motor abilities of the patients. The major finding was that motor performance (of upper limb, hand or the trunk) deteriorated, the greater the degree of axonal loss. This correlation was stronger for the upper limb and hand movements, but it did not appear to depend on the distribution of shrinkage in three sections of the peduncle. Figure 3 shows their main finding based on 89 hemiparetic subjects with single supratentorial lesions studies at least 1 year following the stroke. In addition, although not specifically addressed by the authors, their data are pertinent to another frequently mentional deficit in stroke – namely, a slowing in the rate of force production. The simple reaction time for a precision grip between index finger and thumb was usually prolonged on the affected side (by 40%–50%) and the absolute rate of force production diminished. However, if we allow for the reduction in absolute strength, then the deficit is much less obvious. This is indicated by

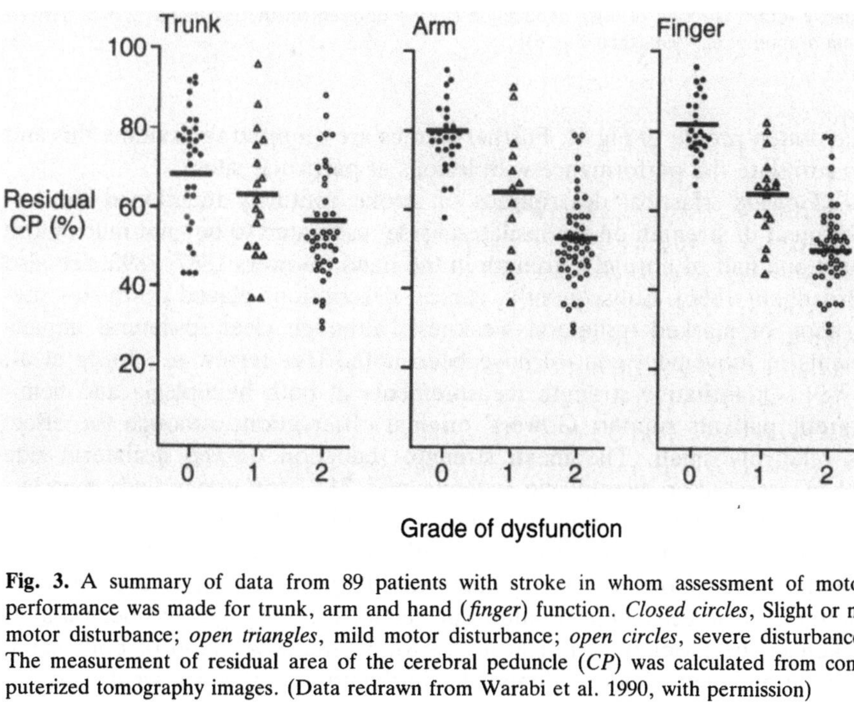

Fig. 3. A summary of data from 89 patients with stroke in whom assessment of motor performance was made for trunk, arm and hand (*finger*) function. *Closed circles*, Slight or no motor disturbance; *open triangles*, mild motor disturbance; *open circles*, severe disturbance. The measurement of residual area of the cerebral peduncle (*CP*) was calculated from computerized tomography images. (Data redrawn from Warabi et al. 1990, with permission)

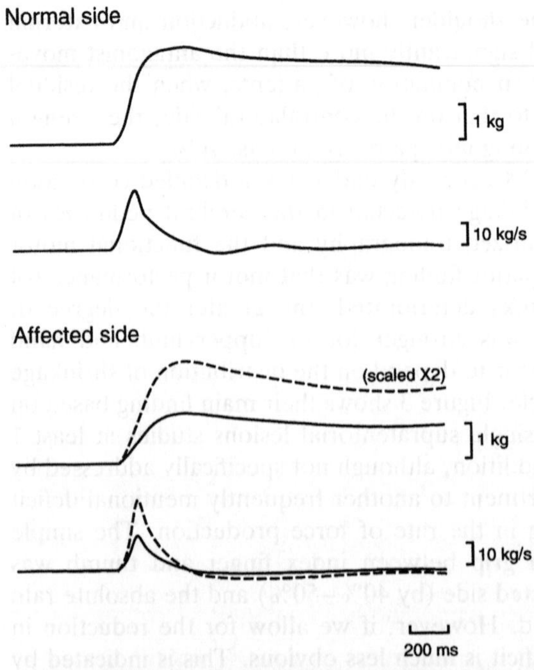

Fig. 4. An example of maximal pinch grip forces in a patient with hemiparesis. *Dotted line* indicates the expected profile when values on the affected side were doubled. Expressed in relative terms, the rate of force production is little different in the two hands. [Data derived from Warabi et al. 1990 (their Fig. 6)]

the dotted profile in Fig. 4. Further studies are required to examine this and to correlate the performance with lesions at particular sites.

Gowers' classical descriptions of stroke routinely mentioned the impairment of strength on the ipsilateral side, estimated to be "not much more than one half of normal" strength in the hand (Gowers 1887, 1893; see also Broadbent 1866). Subsequently, clinical descriptions played down this suggestion of marked ipsilateral weakness, although clear ipsilateral impairments in movement control have been noted (for review see Jones et al. 1989). Quantitative strength measurements in both hemiplegic and hemiparetic patients support Gowers' original observation, although the effect is relatively small. The mean strength reduction on the ipsilateral side when assessed in hemiplegic patients was 21% for upper limb muscles, being prominent for shoulder abduction and adduction (62% and 70%, respectively, of strength in control subjects) and absent for muscles acting on the digits (Colebatch and Gandevia 1989). Quantitative assessment of motor cortical cells with ipsilateral projections to the digits has been undertaken in the monkey (Tanji et al. 1988). In the lower limb of both hemiparetic and hemiplegic patients, the ipsilateral muscles are weakened (by

7% and 26%, respectively), and this weakness occurs in both proximal and distal muscles (Adams et al. 1990).

What factors may have contributed to the impression that the extensor muscles in the upper limb are more greatly affected than the flexors? The upper limb extensors are usually weaker than the flexors by a factor of about 2 to 1 at the elbow and about 3 to 1 more distally. Thus, it will be much easier for the examiner to overcome the attempted maximal contraction of these extensors than flexor muscles, and this will appear as "weakness". A relevant property of skeletal muscle is the force-velocity relationship: this dictates that a greater force (than under isometric conditions) must be applied to lengthen a maximally contracting muscle (e.g., Katz 1939; Joyce et al. 1967; Edman et al. 1978). An interesting corollary to this is that neurologists must themselves be extremely strong if they wish to assess precisely maximal voluntary forces. For example, given that the maximal voluntary elbow torque is about 60 ± 14 Nm (mean ± SD) in adult men and 35 ± 7 Nm in women, and if one assumes that a 20% greater force is required to lengthen the contracting muscles, then the likelihood of being able to assess maximal torque can be calculated. Thus, if the two muscle groups are studied under ideal conditions, a male examiner of normal strengh will be able to lengthen the flexors of subjects with a maximal isometric torque of 50 Nm: this corresponds to 98% of female subjects but to only 26% of male subjects.

In the lower limb, the larger gravitational torques at joints will contribute to the prominent clinical impression about relative strength loss in flexor muscles: a large torque is required simply to elevate the leg in hip flexion when lying supine. Conservative estimates for the weight of the lower limb indicate that for many subjects up to 30% of maximal torque of the "unaffected" hip flexors is required to lift it. The same considerations, but to a quantitatively smaller degree, apply to the torques required for shoulder abduction. Finally, the posture used for strength testing will also have a role in determining the distribution of absolute torques, particularly for the ankle plantar flexors.

Relationship Between Strength Changes and Descending Inputs

Ever since the discovery of the motor functions of the precentral cortex in animals and the realization that lesions of it in primates produced the classical clinical picture of hemiparesis, it has been traditional to equate the deficits of hemiparesis with the lack of excitation arising in descending motor cortical pathways. Thus, the deficits in stroke have often been considered to mirror the positive effects of the major descending outflow from the motor cortex, namely the pyramidal tract. Yet there is not a simple link between motor cortical outflow and the deficits in upper motoneuron lesions. Indeed, some recent studies have emphasized the complex role in premovement planning, not only in motor areas "upstream" of the motor cortex but also in the motor cortex itself (Georgeopoulos et al. 1989). To correlate the motor cortical output with the deficit in strength in stroke makes a number of assumptions, some of which are listed below.

To assess theoretically the influence of the motor cortical output on limb muscles requires knowledge of the number of relevant motor cortical neurons, the degree to which they can be driven by volition, the distribution of the output of the cortical neurons onto motoneuron pools and, finally, the strength and distribution of connections onto a single motoneuron pool. To this should be added the effects reaching the motoneuron pool via paths other than the corticospinal one, including for example, the corticoreticulospinal pathway and those acting through propriospinal interneurons. The complexity of these other influences is easily overlooked, given the somewhat stereotyped view of upper motoneuron lesions, yet it was well recognized by Hughlings Jackson (1931, p. 419).

When set out formally, there are so many steps required to assess quantitatively that it may seem unreasonable to attempt any correlation between motor cortical output and the deficit in stroke. However, evidence from several sources, although incomplete, suggests that the attempted correlation should not be abandoned.

First, the pattern of movement on cortical stimulation following pyramidal tract section in monkeys is somewhat similar to that when the tract is intact (Lewis and Brindley 1965), suggesting that at least the "extrapyramidal" pathways accessed by cortical stimulation reach a similar (not necessarily identical) distribution of motoneuron pools. Second, the properties of one class of motor cortical cells have a well-characterized relationship with motoneurons. These are the corticomotoneuronal cells which make monosynaptic connection with motoneuron pools. Divergence to several pools by a single corticospinal axon is probably a function of the usual movement repertoire, as it is less marked for the intrinsic muscles of the hand than more proximal muscles (e.g., Shinoda et al. 1981). The connection with an individual motoneuron may show marked temporal summation (e.g., Landgren et al. 1962; Lemon and Mantel 1989). While the divergence within a motoneuron pool is not easy to quantify, both electroanatomical and electrophysiological evidence suggests that many (perhaps all) motoneurons are contacted by a single corticospinal axon (Mantel and Lemon 1987, see also Lawrence et al. 1985). Furthermore, both slow and fast pyramidal tract neurons in the monkey evoke post-spike facilitation in the motoneuron pools of hand muscles (Lemon et al. 1992). Third the discharge of many of the motor cortex cells in the monkey occurs phasically with movement and is charcterized by a steep relationship between discharge frequency and muscle force (e.g., Evarts 1968; Smith et al. 1975; Cheney and Fetz 1980). The discharge is also tonically related to static force (Bauswein et al. 1991). Finally, there is direct evidence that the size of the excitatory post-synaptic potentials (EPSPs) of corticomotoneuronal cells is greater for more distal than for proximal muscles in the baboon (Phillips and Porter 1964) and monkey (Lemon 1990). However, for human subjects, who may have evolved monosynaptic corticomotoneuronal connections with more proximal (even truncal) muscles (Gandevia and Rothwell 1987; Gandevia and Plassman 1988; Iwatsubo et al. 1990; cf. Kuypers 1987; Liu and Chambers 1964), this correlation may not exist (e.g., Colebatch et al. 1990; cf. Palmer and Ashby 1992). The uncertainly derives from the difficulty in estimation of the absolute size of EPSPs in a range of human motoneurons using indirect techniques.

When combined with evidence on the evolution of corticospinal pathways (Nathan and Smith 1955, 1982; Heffner and Masterton 1983; cf. Nudo and Masterton 1990), the points above, while not individually conclusive, suggest that the corticospinal and corticomotoneuronal links are likely to be important in the generation of maximal voluntary forces in normal subjects. If so, one would expect the distribution of weakness in an upper motoneuron lesion to reflect, in part, the loss of this input. However, this correlation is complex; it will be dependent upon the properties of residual corticospinal function, the reorganization in motor cortical and other sites following the lesion (e.g., Brouver and Ashby 1991; Weiller et al. 1992), the activation of other motor systems which can influence spinal excitability, and also the properties of the motoneurons and the muscle fibers which they innervate (Edström 1970; Young and Mayer 1982). Overall, intrinsic muscle strength is little affected in stroke (Ismail and Ranatunga 1981) although relative endurance capacity diminishes (e.g., Landin et al. 1977; Miller et al. 1990; see also Edström 1970; Chokroverty et al. 1976). While it is unclear to what extent these changes will influence maximal forces, they are likely to alter the ability to regulate submaximal forces via recruitment and rate coding.

Conclusions

In summary, the strength changes in hemiparesis are predominantly, but not exclusively, distal and they vary between patients. For many limb muscles there may be a significant reduction in strength on the ipsilateral side. Two classes of methanism contribute to the apparent distribution of weakness in hemiparesis. The first involves vagaries of the way in which strength is clinically assessed. Thus, muscles which produce large torques under normal conditions will be more difficult to "overcome" in the clinical examination (e.g., ankle plantar flexion, elbow flexion), whereas muscles required to exert large anti-gravity torques will be more easily overcome (e.g., hip flexors). Measurement of absolute torques, including gravitational components, eliminates the uncertainties. The second involves the distribution of corticofugal influences on the various motoneuron pools. Studies in both human and subhuman primates suggest that EPSPs tend to be larger in distal muscles (particularly those in the hand) than in proximal ones following motor cortical stimulation. This would provide a convenient explanation for the distal strength deficit. However, for more proximal muscles, for example those acting at the elbow, shoulder and even on the trunk, additional factors require consideration. These include the distribution of ipsilateral corticofugal projections, particularly to the shoulder (Colebatch and Gandevia 1989), the relative distribution of corticofugal effects on agonist and antagonists, particularly at the elbow (see Palmer and Ashby 1992) and wrist (see Fetz and Cheney 1980), and the nature of synaptic reorganization following the stroke. Noninvasive studies in human subjects are of special relevance, given that corticospinal projections may reach more of the proximal and axial musculature in them than in subhuman primates.

The goal of finding a single feature of the corticomotoneuronal system which correlates with the distribution of absolute strength in hemiparesis is unlikely to the reached and is arguably unrealistic. Nevertheless, the search will reveal much about the organization of normal movement and its disturbance in hemiparesis.

Acknowledgements. My research is supported by the National Health and Medical Research Council of Australia. I am especially grateful to those colleagues whose studies are discussed in this chapter.

References

Adams RW, Gandevia SC, Skuse NF (1990) The distribution of muscular weakness in upper motoneurone lesions affecting the lower limb. Brain 113:1459–1476

Bauswein E, Fromm C, Werner W, Ziemann U (1991) Phasic and tonic responses of premotor and primary motor cortex neurons to torque changes. Exp Brain Res 86:303–310

Bohannon RW (1987) Relationship between static strength and various other measures in hemiparetic stroke patients. Int Rehabil Med 8:125–128

Bohannon RW (1988) Muscle strength changes in hemiparetic stroke patients during inpatient rehabilitation. J Neurol Rehabil 2:163–166

Bohannon RW (1990) Significant relationships exist between muscle group strengths following stroke. Clin Rehabil 4:27–31

Bohannon RW, Smith MB (1987) Assessment of strength deficits in eight paretic upper extremity muscle groups of stroke patients with hemiplegia. Phys Ther 67:522–525

Broadbent WH (1866) On a case of right hemiplegia, with deviation of the eyes to the left, and aphasia. Lancet i:480–481

Brouwer B, Ashby P (1991) Altered corticospinal projections to lower limb motoneurons in subjects with cerebral palsy. Brain 114:1395–1407

Cheney PD, Fetz EE (1980) Functional classes of primate corticomotoneuronal cells and their relation to active force. J Neurophysiol 44:773–791

Chokroverty S, Reyes MG, Rubino FA, Barron KD (1976) Hemiplegic amyotrophy: muscle and motor point biopsy study. Arch Neurol 33:104–110
Colebatch JG, Gandevia SC (1989) The distribution of muscular weakness in upper motor neuron lesions affecing the arm. Brain 112:749–763
Colebatch JG, Gandevia SC, Spira PJ (1986) Voluntary muscle strength in hemiparesis: distribution of weakness at the elbow. J Neurol Neurosurg Psychiatry 49:1019–1024
Colebatch JG, Rothwell JC, Day BL, Thompson PD, Marsden CD (1990) Cortical outflow to proximal arm muscles in man. Brain 113:1843–1856
Edman KAP, Elzingea G, Noble MIM (1978) Enhancement of mechanical performance by stretch during lengthening contractions of the veterbrate skeletal muscle fibres. J Physiol (Lond) 281:139–155
Edström L (1970) Selective changes in the sizes of red and white muscle fibres in upper motor neurone lesions and parkinsonism. J Neurol Sci 11:537–550
Evarts E (1968) Relation of pyramidal tract activity to force exerted during voluntary movement. J Neurophysiol 31:14–27
Fetz EE, Cheney PD (1980) Postspike facilitation of forelimb muscle activity by primate corticomotoneuronal cells. J Neurophysiol 44:751–772
Gandevia SC, Plassman BL (1988) Responses in human intercostal and truncal muscles to motor cortical and spinal stimulation. Respir Physiol 73:325–338
Gandevia SC, Rothwell JC (1987) Activation of the human diaphragm from the motor cortex. J Physiol (Lond) 384:109–118
Georgeopoulos AP, Kettner RE, Schwartz AB (1988) Primate motor cortex and free arm movements to visual targets in three-dimensional space. II. Coding of the direction of movement by a neuronal population. J Neurosci 8:2928–2937
Gowers WR (1887) Lectures on the diagnosis of diseases of the brain, 2nd edn. Churchill, London, pp 54–55
Gowers WR (1893) A manual of diseases of the nervous system, 2nd edn. Churchill, London (Hafner Reprint, Darien, 1974)
Heffner R, Masterton RB (1983) The role of the corticospinal tract in the evolution of human digital dexterity. Brain Behav Evol 23:165–183
Hughlings Jackson J (1931) Selected writings of John Hughlings Jackson, vol 1. (Ed: J Taylor) Hodder and Stoughton, London
Ismail HM, Ranatunga KW (1981) Isometric contractions of normal and spastic human skeletal muscle. Muscle Nerve 4:214–218
Iwatsubo T, Kuzuhara S, Kanemitsu A, Shimada H, Toyokura Y (1990) Corticofugal projections to the motor nuclei of the brainstem and spinal cord in humans. Neurology 40:309–312
Jackson JH (1865) Lectures on hemiplegia. Clinical lectures and reports by the medical and surgical staff of the London Hospital 2:297–332
Jones RD, Donaldson IM, Parkin PJ (1989) Impairment and recovery of ipsilateral sensory-motor function following unilateral cerebral infarction. Brain 112:113–132
Joyce GC, Rack PMH, Westbury DR (1967) The mechanical properties of cat soleus muscle during controlled lengthening and shortening movements. J Physiol (Lond) 204:461–474
Katz B (1939) The relation between force and speed in muscular contraction. J Physiol (Lond) 96:45–64
Kuypers HGJM (1987) Some aspects of the organization of the output of the motor cortex. In: Bock G, O'Connor M, Marsh J (eds) Motor areas of the cerebral cortex. Wiley, Chichester, pp 63–82 (Ciba Foundation Symposium 132)
Lance JW, McLeod JG (1981) A physiological approach to clinical neurology, 3rd edn. Butterworths London
Landin S, Hagenfeldt L, Saltin B, Wahren J (1977) Muscle metabolism during exercise in hemiparetic patients. Clin Sci Mol Med 53:257–269
Landgren S, Phillips CG, Porter R (1962) Minimal synaptic actions of pyramidal impulses on some alpha motoneurones of a baboon hand and forearm. J Physiol (Lond) 161:91–111

Lawrence DG, Porter R, Redman J (1985) Corticomotoneuronal synapses in the monkey: light microscopic localisation of motoneurons of intrinsic muscles of the hand. J Comp Neurol 232:499–510

Lemon RN (1990) Mapping the output functions of the motor cortex. In: Edelman G, Gall E, Cowan WM (eds) Signal and Sense: local and global order in perceptual maps. Wiley, Chichester, pp 315–356

Lemon RN, Mantel GWH (1989) The influence of changes in discharge frequency of corticospinal neurones of hand muscles in the monkey. J Physiol (Lond) 413:351–378

Lemon RN, Werner W, Bennett KMB, Flament DA (1992) The proportion of slow and fast pyramidal tract neurones producing post-spike facilitation of hand muscles in the conscious monkey. Proc Physiol Soc July: C43

Lewis R, Brindley GS (1965) The extrapyramidal cortical motor map. Brain 88:387–407

Liu CN, Chambers, WW (1964) An experimental study of the cortico-spinal system in the monkey (*Macaca mulatta*). The spinal pathways and preterminal distribution of degenerating fibers following discrete lesions of the pre- and postcentral gyri and bulbar pyramid. J Comp Neurol 123:257–284

Mann L (1895) Über den Lähmungstypus bei der cerebralen Hemiplegie. Sammlung klinischer Vorträge, Leipzig, pp 355–368 (Innere Medizin, no 132; new series, 39)

Mantel GWH, Lemon RN (1987) Cross-correlation reveals facilitation of single motor units in thenar muscles by single corticospinal neurones in the conscious monkey. Neurosci Lett 77:113–118

Miller RG, Green AT, Moussavi RS, Carson PJ, Weiner MW (1990) Excessive muscular fatigue in patients with spastic paraparesis. Neurology 40:1271–1274

Nathan PW, Smith MC (1955) Long descending tracts in man. I. Review of present knowledge. Brain 78:248–303

Nathan PW, Smith MC (1982) The rubrospinal and central tegmental tracts in man. Brain 105:223–269

Nudo RJ, Masterton RB (1990) Descending pathways to the spinal cord, IV: some factors related to the amount of cortex devoted to the cortiospinal tract. J Comp Neurol 296:584–597

Palmer E, Ashby P (1992) Corticospinal projections to upper limb motoneurones in humans. J Physiol (Lond) 448:397–412

Phillips CG, Porter R (1964) The pyramidal projection to motoneurones of some muscle groups of the baboon's forelimb. Prog Brain Res 12:222–245

Phillips CG, Porter R (1977) Corticospinal neurones: their role in movement. Academic, London

Shinoda Y, Yokota JI, Futami T (1981) Divergent projection of individual corticospinal axons to motoneurons of multiple muscles in the monkey. Neurosci Lett 23:7–12

Smith AM, Hepp-Reymond M-C, Wyss UR (1975) Relation of activity in precentral cortical neurons to force and rate of force change during isometric contractions of the finger muscles. Exp Brain Res 23:315–332

Tanji J, Okano K, Sato KC (1988) Neruonal activity in cortical motor areas related to ipsilateral, contralateral, and bilateral digit movements of the monkey. J Neurophysiol 60:325–343

Warabi T, Inoue K, Noda H, Murakami S (1990) Recovery of voluntary movement in hemiplegic patients. Correlation with degenerative shrinkage of the cerebral peduncles in CT images. Brain 113:177–189

Weiller C, Chollet F, Friston KJ, Wise RJS, Frackowiak RSJ (1992) Functional reorganization of the brain in recovery from striatocapsular infarction in man. Ann Neurol 31:463–472

Wernicke C (1989) Zur Kenntniss der cerebralen Hemiplegie. Berl Klin Wochenschr 1889; 26:969–970

Young JL, Mayer RF (1982) Physiological alterations of motor units in hemiplegia. J Neurol Sci 54:401–412

Factors Underlying Abnormal Posture and Movement in Spastic Hemiparesis

J.P.A. Dewald[1] and W.Z. Rymer[2]

Introduction

Hemiparetic stroke is accompanied by abnormalities of muscle tone (identified as spastic hypertonia), weakness for voluntary movement, and disturbances of muscular coordination. The first two disturbances, i.e., spasticity and weakness, are the most visible and the most intensively studied, yet the disturbance of muscular coordination is often the most incapacitating, and the least well understood. For example, it is widely known that in many subjects disturbances of voluntary movement may be masked by abnormalities of muscle tone and/or concurrent muscular weakness. Yet when these physical signs are treated effectively, or when they resolve spontaneously, the resulting incoordination and motor impairment is still severe. It is therefore likely that in many hemiparetic stroke subjects, neither increases in muscle tone nor muscular weakness are the primary sources of movement dysfunction or of their global disability.

The object of this paper is to characterize and quantify those abnormal spatial patterns of muscle activity in the paretic limb of the hemiparetic stroke subject that contribute to movement dysfunction and to evaluate the potential neurophysiological origins of these disturbances.

Disturbances of Voluntary Movement in Hemiparetic Stroke

A detailed description of the abnormal movement patterns in the paretic limb, and the natural history of the evolution of the various components of these abnormal clinical signs was first provided by Twitchell in 1951, who he delineated both the major features of the movement disturbance and the time course of recovery from stroke in some detail. A prominent feature of the disturbed movement patterns was the emergence of "stereotypic"

[1] Sensory-Motor Performance Program, Rehabilitation Institute of Chicago, Programs in Physical Therapy, Northwestern University Medical School, 345 East Superior Ave., Chicago, IL 60611-4496, USA

[2] Departments of Physical Medicine and Rehabilitation, Physiology and Biomedical Engineering, Northwestern University, 345 East Superior Ave., Chicago, IL 60611-4496, USA

Thilmann et al. (Eds.)
Spasticity: Mechanisms and Management

movements in which there appeared to be relatively tight coupling of motion at adjacent joints in the upper and lower limbs. (For the sake of historical accuracy, it should be pointed out that this movement coupling was not examined directly by either kinematic or electromyographic measurements of the abnormal subject but was based on visual clinical assessment.)

Subsequently, Brunnstrom (1970) classified these stereotypic movement patterns into so-called "synergies" which were broadly of either flexor or extensor type. This Brunnstrom classification, which is summarized below, has stood with relatively limited modification by other investigators, and the nature and origins of such muscle synergies form a major component of this discussion:

Extension synergy
- Shoulder girdle: protraction, adduction, internal rotation
- Elbow: extension, pronation
- Wrist: extension
- Finger: flexion

Flexion synergy
- Shoulder girdle: retraction, abduction to 90°, external rotation
- Elbow: flexion, supination
- Wrist: flexion
- Finger: flexion

Origins of Abnormal Muscular Synergies in Hemiparetic Stroke

There are at least two different mechanisms by which these abnormal muscle synergies could be mediated.

The first mechanism is that descending movement instructions, previously transmitted through multiple major dorsolateral spinal pathways such as the corticospinal and rubrospinal projections in the normal subject, and now relayed through other, simpler residual bulbospinal pathways that are undamaged by the stroke. These bulbospinal pathways often lie in ventromedial quadrants of the spinal cord, and probably include major descending spinal systems such as the vestibulospinal, tectospinal, reticulospinal, and perhaps propriospinal pathways. During voluntary movement, these bulbospinal systems are presumably activated either by extrapyramidal projections or by corticobulbar projections not damaged by the cerebral white matter lesion.

There are a number of structural and physiological characteristics of these ventromedial pathways which suggest that movements initiated by them are likely to be quite different from those seen in the normal subject. This is because ventromedial pathways have a different pattern of terminal arborization (compared with the dorsolateral pathways), making most of their connections to spinal interneurons and (indirectly) to motoneurons innervating axial and proximal limb muscles rather than to distal limb and hand muscles (e.g., Kuypers 1964). In addition, ventromedial pathways almost certainly have a more diffuse anatomical pattern of synaptic connectivity, innervating interneurons and motoneurons in many spinal segments. For the case of one key ventromedial pathway, the vestibulospinal system, it is known that efferent axons and their terminal branches may arborize over many spinal segments, including segments innervating both axial musculature and proximal limb muscles, with particular emphasis on extensors and antigravity muscles (Wilson and Yoshida 1969).

In contrast, dorsolateral systems such as the corticospinal or rubrospinal systems show rather more narrowly distributed and focused spinal input, which is directed primarily to premotor interneurons in the intermediate gray matter of the cord and monosynaptically to motoneurons of distal limb and hand muscles (e.g., Kuypers 1981).

The second mechanism (which is not necessarily independent of the first) is that the abnormal synergies are a reflection of augmented excitability of flexor motoneurons, arising as part of a pattern of increased excitability of premotor interneurons of the flexion reflex pathway. This increase in flexor motoneuronal excitability could be attributed in turn to a reduction of descending inhibitory control of spinal interneurons, especially those inhibiting extensors, and activating flexors. These interneurons respond to activation of a broad range of sensory afferents, including those from skin and higher threshold afferents from muscle and are commonly designated as the flexor reflex afferent pathway (FRA; Holmqvist and Lundberg 1961) because their activation leads to excitation of flexor muscles, to inhibition of extensors, and to the motion of generalized flexion of the limb.

Summary Hypotheses

If the *flexor reflex* mechanism predominates, we would expect that the altered synergic patterns observed in voluntary motion of spastic paretic limbs should closely resemble those elicited during the flexion withdrawal reflex in the upper limb. Alternatively, if *abnormal descending commands* (such as would emerge from utilization of residual ventromedial spinal pathways) are responsible, the observed muscle activation patterns might differ substantially from the flexor reflex pattern. We examine these propositions in more detail below (see "Results").

Methods

Our studies were performed using isometric muscle loading, in which the forces generated at the wrist were measured by a three degrees of freedom load cell. To ensure secure transfer of the forces between the limb and the load cell, the lower forearm and wrist of the subject were encased in a closely fitting polymer cast, which was attached in turn to the metal ring supported within the three degrees of freedom load cell. The subject was seated with the shoulder at a 65° abduction angle, the elbow flexed at 90°, and the forearm held in a neutral position (with respect to pronation/supination). The instructions for magnitude and direction of the load were provided to the subject via a three degrees of freedom visual display, using the X-Y plane of the display to identify the relative contributions of flexion/extension and adduction/abduction, and the magnitude of pronation/supination was instructed using rotation of the target about its central axis. Because the shoulder was held at a 65° abduction angle, the display depicted flexion/extension along the line connecting 25° and 205°, and humeral abduction/adduction along the line between 115° and 295°. The subject reached the desired force by acquiring the target cursor, which was also rotated to provide the requisite amount of pronation/supination (see Fig. 1A).

Electromyographic activity from elbow and shoulder muscles was recorded using surface electrodes for some of the larger muscles and intramuscular recordings for several smaller or deeper muscles which were not readily accessed using surface electrodes. The latter muscles included brachialis, pronator teres, anconeus at the elbow, and the different heads of the

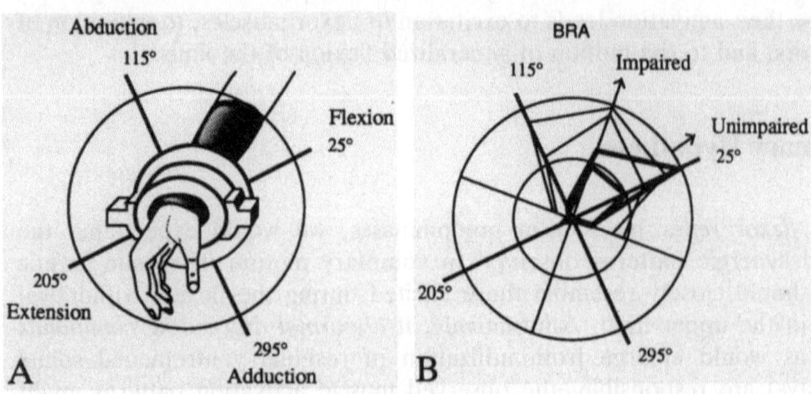

Fig. 1A,B. Experimental set-up and EMG-polar plot. **A** A subject's right wrist inside the three degree of freedom load cell. Also, four of the eight target directions including the corresponding anatomical force exertion directions. The 25° rotation is due to the 65° shoulder abduction angle that was used during the experiments. **B** An EMG polar plot of the brachialis (*BRA*) muscle of a moderately impaired subject. *Thick lines* connect normalized EMG magnitude for the unimpaired limb as a function of load direction, for two different load levels, representing 35% and 50% of maximum force; *thin lines*, the EMG spatial profile for the impaired upper limb, again, for two equivalent load levels (35% and 50% of maximum). *Arrows*, the resulting peak EMG directions for both upper limbs. Note the increase in the angular range of EMG activation as well as the uppward shift in peak EMG activity in the impaired upper limb deltoid muscle at the shoulder. In many experiments in which the relative contribution of the different heads of biceps or triceps were of interest, intramuscular recordings were used there as well. Surface electrodes were used routinely on the brachioradialis, biceps, triceps, pectoralis major, and upper and middle trapezius muscles.

Intramuscular electromyogram (EMG) recordings were taken using 75-μm Teflon-coated stainless steel wires which were inserted using 27-G hypodermic needles. EMGs were recorded differentially through preamplifiers and then further amplified and filtered before acquisition in a Macintosh based data acquisition system. Data was acquired at 200 Hz and low-pass filtered at 100 Hz with an 8-pole filter to prevent aliasing. Surface EMG activity was treated in the same way.

Data were displayed using "polar plots" in which the mean magnitude of normalized rectified EMG activity was plotted on the radial axis, and the direction of the plotted EMG activity was varied according to the angle of load application (see Fig. 1B). Normalized EMG values were obtained by subtracting the baseline EMG and then dividing the residual EMG value by the EMG magnitude obtained during maximum voluntary contractions. The direction of peak EMG activity was then estimated by calculating the vector sum of the individual EMG vectors. Each EMG vector is represented by a magnitude (normalized EMG value) and a direction (one of the target matching directions). The resulting vector angle obtained using vector addition thus determined the direction of peak EMG activity.

In addition, we evaluated the functional status of the impaired upper limb by applying the Fügl-Meyer test, which summarizes in numerical form (0–66) the range of active and passive motion of the joints in the upper limb.

Results

Spatial Patterns of Muscle Activation in Normal Subjects

In recent years studies have characterized the spatial activation profiles of different muscles in the upper limb of normal human subjects during voluntary isometric contractions of differing magnitudes and directions. (For detailed descriptions of the EMG spatial patterns in normal upper limbs see Buchanan et al. 1986; Flanders and Soechting 1990.) The essential features of these EMG spatial patterns are that muscle activity is graded in an orderly fashion with changing load magnitude and changing load direction. The magnitude of the EMG activity is usually greatest in the direction of maximum mechanical advantage of the muscle at the joint it is acting. For the case of the key elbow flexor muscles, biceps, brachioradialis, and brachialis are each activated maximally in (or close to) the pure flexion direction and triceps correspondingly in pure elbow extension.

Figure 1B (heavy traces) shows that for the contralateral limb of the stroke subject, as the direction of the isometric force is varied away from the mechanically optimum one, the magnitude of EMG activity diminishes progressively in a smooth manner. This EMG spatial pattern is usually well fit by a circle, in which the minimum value passes through or close to the origin of the two degree of freedom "polar" plot, and the maximum length chord, equivalent to the diameter, passes through the maximum EMG value. EMG magnitude declines smoothly in between. [Other similar formulations, in which the EMG spatial dependence is characterized by two-cosine functions are often slightly better predictors (Flanders and Soechting 1990); however, the precise form of the EMG directional dependence is not crucial to this study – as is explained further below].

For some arm muscles in either the normal limb or in the contralateral limb of the stroke subject there are often deviations from this "standard" circular pattern, in which EMG activity may achieve a disproportionately large peak value at or near to the optimum mechanical direction. This has the effect that the polar plot becomes elliptical rather than circular in shape. In other muscles EMG tuning may be somewhat broader than predicted by the circular fit, in that there may still be significant EMG activity extending beyond 90° away from the principal EMG axis. This is especially evident at the wrist, where moment arms of the respective muscles are separated by more then 90°. In spite of these differences, the general spatial EMG patterns are surprisingly consistent for quite different muscles in diverse experimental preparations.

Broadly similar EMG spatial findings have also been reported by Flanders and Soechting (1990) for the shoulder, by Buchanan et al. (1993) for the wrist, and for a variety of muscles in legs of quadrupeds (McPherson 1988a,b).

In summary, these observations indicate that as the load direction is oriented progressively further away from the mechanically optimum angle, voluntary EMG activity declines to the point where there is essentially no discernible activation of muscles in directions where no useful mechanical action can result.

Abnormal Spatial Patterns of Muscle Activity in Hemiparetic Stroke Subjects

We have performed a series of EMG and force recordings in our hemiparetic stroke subjects, the results of which were used to describe the EMG spatial patterns in both the paretic-spastic and contralateral upper limbs of each stroke subject. We observed a number of consistent disturbances in spatial EMG patterns in the impaired limbs which varied according to the severity of the neurological impairment. (Aspects of these findings have been reported in abstract form: Dewald et al. 1989, 1992a.)

In the mildly impaired hemiparetic stroke subject (in whom the Fügl-Meyer scores were greater than 40) the pattern of muscle activity in the polar plot was essentially indistinguishable from that of the normal (or of the contralateral limb) for most of the muscles examined in the upper limb and shoulder region.

For stroke subjects with moderate or severe impairment, (i.e., those with Fügl-Meyer scores less than 40) there were consistent and significant disturbances in these spatial, or polar, EMG patterns. The key features of these abnormal polar plots were that the spatial orientations were abnormal, in that the peak value of the electromyogram was often directed well away from the optimum mechanical direction, by 45° or more. The spatial domain, or angular territory occupied by the polar plots, was similarly rotated. In addition, the angular range, or angular extent of EMG activity, was often significantly increased compared with the usual spatial range seen in normal control muscles or in the contralateral limb of the hemiparetic subject.

The net result of these changes in EMG polar plots in the moderately or severely impaired subject is not only that the spatial tuning of the muscles is much broader, but that the muscles may now be activated in directions in which they contribute little useful force, or in which they may even oppose the intended force. This means, in effect, that under some loading conditions there may be partial or even wide spread cocontraction of synergists and even antagonist muscles.

The pattern of change in the polar plots of the impaired limb was also quite consistent between different subjects. For both dominant and nondominant limb impairments, the spatial tuning of the polar plot was altered so that the maximum EMG activity for elbow flexors occurred in a direction which was rotated significantly upwards, towards humeral abduction (as compared with the normal control) even though this direction was not mechanically useful for the generation of elbow flexion. As shown in Fig. 1B (thin lines), brachialis action, which is normally greatest for elbow flexion loads, reaches a maximum 45° or more away from the moment arm of the muscle, in an angular direction where the muscle cannot contribute significant force at the elbow.

For elbow extensors, such as the triceps brachii, the rotation of the polar plot is exactly opposite, in that EMG polar plots are now rotated downwards, towards humeral adduction/humeral internal rotation. Expressed differently, from the viewpoint of an observer facing the subject, with both forearms pointing towards the observer, both the peak EMG and the EMG angular domain are rotated in a counterclockwise direction for the impaired right limb and in a clockwise direction when the left limb is impaired.

Muscle Synergies in Impaired Limbs

These EMG recordings also provide the means to characterize the relations between levels of muscle EMG activation in different muscles of the impaired limb, as load magnitude and direction is varied systematically. EMG

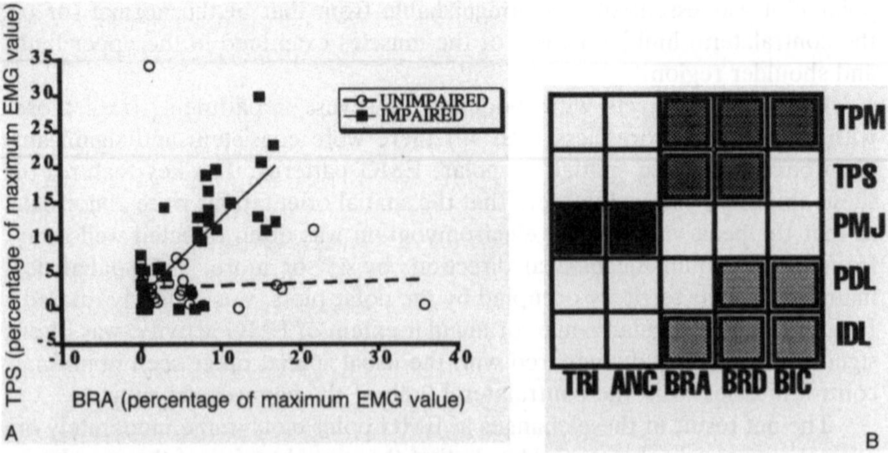

Fig. 2A,B. Muscle coactivation patterns as determined by EMG/EMG scatter plots and straight-line regression analysis. **A** An EMG/EMG scatter plot for the brachialis (*BRA*) and upper trapezius (*TRP*) muscles from a moderately impaired subject. In the impaired limb this pair of muscles is significantly coactivated as a function of load direction and load magnitude as determined by straight-line regression analysis criteria. In the contralateral limb no significant coactivation between the brachialis and trapezius muscles is obtained. **B** Block diagram showing all the significant muscle coactivations of elbow versus shoulder muscles unique to the impaired upper limb. *TRI*, Triceps brachii; *ANC*, anconeus; *BRA*, brachialis; *BRD*, brachioradialis; *BIC*, biceps brachii; *TPM*, middle trapezius; *TPS*, superior trapezius; *PMJ*, pectoralis major; *PDL*, posterior deltoid; *IDL*, intermediate deltoid

relations between different upper limb muscles can be characterized and quantified using "EMG/EMG" plots, in which the electromyographic activity in one muscle is plotted against that of another, regardless of the direction of activation and of load magnitude (see Fig. 2A for example).

Since the EMG activity in the polar plot varies substantially as a function of load direction and load magnitude, EMG/EMG plots drawn from a number of normal limb muscle pairs often show a relatively broad scatter, with no clear signs of linear or even nonlinear functional relation between various muscles (again, see Fig. 2A). In contrast, EMG/EMG plots derived from many elbow versus shoulder muscles in the upper limbs of moderate or severely impaired stroke subjects show a much simpler pattern, in which there is sometimes a strikingly high correlation of EMG-EMG activity between the different muscle groups (see Fig. 2A,B). *In other words, for these muscle groups the severely impaired subject no longer is able to tune the activity of muscles selectively according to load direction, as is the case in the normal upper limb.*

However, our EMG-EMG plots drawn from many other muscles pairs showed significant coactivation patterns, in both the unimpaired limbs of stroke subjects and in normal subjects. These were task dependent, and many were conserved in the impaired limb of the stroke subject. The "abnormal" patterns, which arose in the moderate or severely impaired limb, occurred most frequently between elbow flexors and shoulder abductors/elevators and retractors (see Fig. 2B) and between elbow extensors and shoulder adductors. These new muscle combinations are similar to the flexion and extension synergies as described by Brunnstrom (1970).

These observations suggest that there may be a neurophysiological basis for the abnormal synergies described by Twitchell and Brunnstrom, although our findings are restricted presently to isometric activities only, whereas these authors were describing abnormalities manifested during voluntary motion.

Muscle EMG Activation Patterns During Flexion Withdrawal Reflexes

In unimpaired limbs of the hemiparetic stroke subject, electrical stimulation of the palmar skin of the fingers produces a systematic and orderly activation of certain proximal limb muscles, characterized by shoulder extension and retraction, which is then followed by elbow flexion. In the impaired limb, however, a simultaneous activation of most shoulder muscles and elbow flexors is induced. While these flexor withdrawal responses are to be anticipated (given what we know about the segmental organization of the flexion withdrawal reflex), they also provide a foundation for detailed comparisons between the patterns of muscular response of the flexion reflex, which is clearly an *involuntary* response, and the patterns of *voluntary* activation in the same muscles that serve the limb. (Aspects of these findings have been reported in abstract form: Dewald et al. 1992b.)

It is worth noting, however, that such comparisons are currently limited by the fact that electrical activation of skin does not evoke broad spatial tuning of muscle activity in the limb, such as arises in voluntary activation, but appears simply as muscle activation in one primary angular direction. This direction corresponds essentially to the angle for elbow flexion (0° degrees in Fig. 4). It follows that comparisons between EMG patterns elicited during flexion withdrawal and those seen during voluntary activation are necessarily based upon the rather limited data base of EMG activity profiles in one force direction. Nonetheless, the comparisons have proven to be instructive.

During electrical stimulation of the palmar finger skin in the unimpaired upper limb, at intensities sufficient to induce mild to moderate discomfort, we observed that the stimulation produced a broad activation of muscles in the same limb. This EMG activation began after 45–75 ms, was initiated clearly in proximal muscles, such as the posterior deltoid and trapezius, and was then expressed progressively down the limb, towards the stimulated site (Fig. 3).

In comparison, induction of muscle EMG activity in impaired upper limb muscles of the stroke subject required much less intense electrical stimulation of the skin of the hand in order to be expressed, and muscles were often activated reflexively at lower stimulus intensities. Most striking

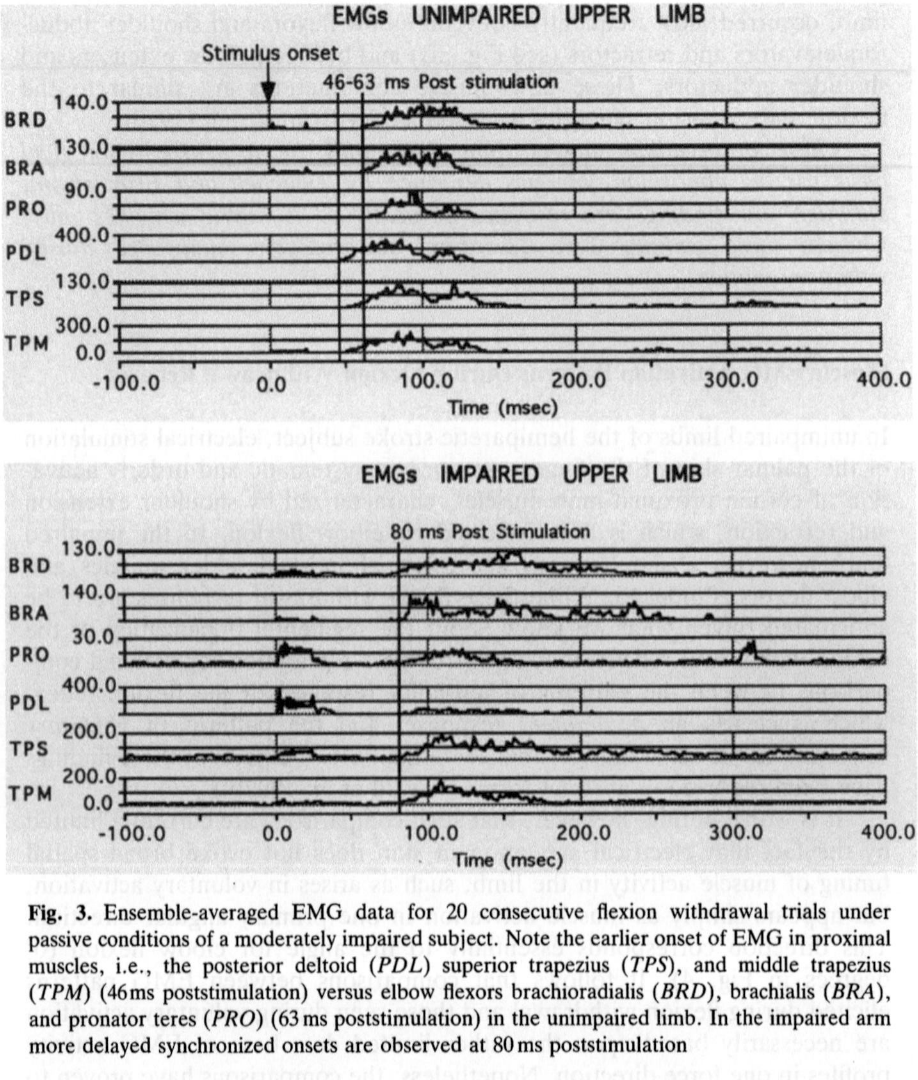

Fig. 3. Ensemble-averaged EMG data for 20 consecutive flexion withdrawal trials under passive conditions of a moderately impaired subject. Note the earlier onset of EMG in proximal muscles, i.e., the posterior deltoid (*PDL*) superior trapezius (*TPS*), and middle trapezius (*TPM*) (46 ms poststimulation) versus elbow flexors brachioradialis (*BRD*), brachialis (*BRA*), and pronator teres (*PRO*) (63 ms poststimulation) in the unimpaired limb. In the impaired arm more delayed synchronized onsets are observed at 80 ms poststimulation

(and unexpected) was the finding that the timing of EMG activation was systematically different in the impaired limb. In particular, the onset of significant EMG activity in the flexion reflex was *delayed* by 20–30 ms (referenced to the contralateral side) even in the earliest activated muscle, so that the onset of EMG activity in proximal and distal muscles was now virtually simultaneous.

The forces generated at the three degrees of freedom load cell by different patterns of skin excitation were also examined (see Fig. 4). The forces elicited in response to skin stimulation were shown to be focused in flexion, without any indication of humeral external rotation or elevation.

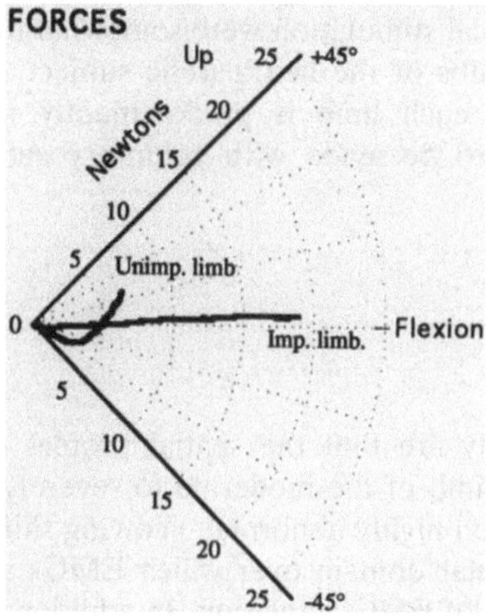

Response times: Unimp. limb 80/170 msec; Imp. limb 100-270 msec

Fig. 4. Averaged force profiles for 20 consecutive flexion withdrawal trials for impaired and contralateral limbs of a severely impaired hemiparetic subject. Note the delayed force onset time the impaired limb as well as the increased onset time to peak force time (see response times). The early down trend observed in the force profile of the unimpaired (i.e., contralateral) limb is believed to be related to the early onset of shoulder muscle activity, most notably in shoulder elevators/retractors (trapezius muscle)

(For the case of Fig. 4, such humeral external rotation or abduction would have induced upward rotation of the force trajectory, towards the +45° line.) However, the selective involvement of particular proximal muscles, such as shoulder extensors and retractors, was especially noticeable in the contralateral limb of the hemiparetic subject, and there was likely to have been a significant (and potentially unmeasured) component of shoulder extension/retraction in these (contralateral) limbs. (The reason that retraction-induced forces were potentially unmeasured is that they lie in the plane orthogonal to the load cell, along the axis of the forearm, a direction not recorded by this load cell.) However, shoulder retraction forces were probably much less important in the impaired limb (see next paragraph).

The time course of force generation was also somewhat different in the impaired limb, with development of a near linear force trajectory, significantly different from that of the contralateral limb (see Fig. 4). In the impaired limb coactivation of all shoulder abductors/elevators and retractor muscles in conjunction with elbow flexors was likely to significantly reduce shoulder extension/retraction forces as compared with the unimpaired (contralateral) upper limb, allowing a relatively clear and linear profile of flexion force to emerge. In contrast, the force profile in the unimpaired limb differs in its time course (see below), and in spatial tuning, in that the force trajectory rotates upwards.

Although the effects of electrical stimulation were somewhat different in the impaired and contralateral limbs of the hemiparetic subject, it is clear that the net force generated in each limb is predominantly in flexion, allowing legitimate comparisons to be made with voluntary activation of elbow muscles.

Discussion

The principal results of this study are that the spatial profiles of muscle activation in the impaired upper limb of the moderate to severely impaired hemiparetic stroke subject are often highly abnormal, showing shifts in peak EMG direction, shifts in the angular domain over which EMGs are active, and increases in the angular range of EMG activation. In addition, there are a number of "abnormal" coactivation patterns most notably between elbow flexors and shoulder abductor (deltoids) and elevator-retractor (trapezius) muscles and a simplification of the relations between EMGs in several different muscle pairs acting in the upper limb. The issue under consideration is whether these abnormal spatial patterns can be explained by enhanced flexor reflex induced changes in motoneuronal excitability, or whether they are best attributed to altered descending commands.

The intent of this electrical stimulation study was to provide a systematic profile of the reflexively activated spatial pattern of muscle activation in the impaired limb, to serve as a basis for comparison with the patterns observed during voluntary isometric contractions. To characterize the EMG profiles during flexor reflex excitation we determined that when skin of the fingers was excited electrically at or near noxious intensities, there was a generalized flexion withdrawal response elicited in which muscles were activated in an orderly proximal to distal sequence while generating forces in the plane of elbow flexion and shoulder extension. Before evaluating these two data sets more regorously a number of issues justify consideration.

Why Should the Flexor Reflex Be Involved in Mediating Abnormal Synergies?

There are number of published observations indicating that flexion withdrawal reflexes are enhanced in stroke-induced himiparesis, primarily on the paretic side. For example, Shahani and Young (1977), among others, have shown increased amplitude and prolonged duration of EMG activity in impaired limbs following the administration of an electrical or mechanical stimulus appropriate to elicit a flexion reflex. Parenthetically, these abnormalities of flexor reflexes are also expressed in the changes in particular cutaneous reflexes, some of which are of established diagnostic value in the neurologically impaired human subject. For example, the plantar reflex (in which the hallux and other toes show plantar flexion after cutaneous pressure on the sole of the foot over the metatarsophalangeal joints) is believed to be broadly analogous to the "plantar cushion reflex" in the quadruped, in which the paws are plantar flexed and the claws extruded on contact to assist with gripping the terrain and (presumably) improving postural stability.

After the occurrence of stroke, there is apparently a loss of inhibition of regional spinal interneurons involved in the generation of the flexion withdrawal reflex, so that these generalized flexion reflexes may override the regionally specific reflexes, such as the normal plantar response, substituting a much coarser flexor reflex onto the whole limb. This substitution is manifested in the appearance of the *Babinski response*, in which plantar cutaneous stimulation elicits a dorsal motion of the hallux, accompanied by generalized withdrawal of the other toes, sometimes even with flexion at the ankle, hip, and knee. This change in the flexor reflex pattern may arise because of disfacilitation of a major descending inhibitory brainstem pathway (the dorsal reticulospinal pathway), or it could arise from changes in the discharge of descending monoaminergic systems including the raphe-spinal and locus ceruleus projections. Whatever the ultimate mechanism, the net result is clearly an increasing flexor reflex responsiveness in subjects with stroke-induced hemiparesis.

In relation to the disturbances of *voluntary* muscle activation, the idea that we are pursuing here is that perhaps the abnormal flexor responsiveness evident in stroke and the abnormal spatial patterns of muscle activity evident in the isometric studies are linked *by virtue of the facilitatory action of tonically discharged cutaneous and muscle afferents on spinal segmental interneurons and flexor motoneurons*. If this were the case, voluntary commands would favor certain flexor muscles because their motoneurons and premotoneurons are biased into a more excitable state. It follows that if this mechanism were in effect, the spatial patterns of muscle activation should be similar under the two conditions, and the latency for activating flexor reflexes should presumably be less than was observed under normal conditions.

The results of our studies on flexor reflexes do not clearly favor the hypothesis of a primary role for flexor reflexes in mediating the abnormal synergies, although this is at least partly for technical reasons in that the data allowing a full comparison of flexor reflex and voluntary muscle activation patterns are not yet available. Specifically, the flexion withdrawal reflexes establish only a single set of torque and EMG directional relations, generating a flexor torque on the lower forearm. There is no spatial EMG tuning to compare under voluntary and reflex conditions, and there is no gradation of motor output available to match the effects of changing load under voluntary conditions. We are left with relatively few data points to use for making comparisons between the two potential hypotheses under examination.

Nonetheless, it appears that the spatial patterns of EMG activation are essentially different in the two experiments. To name a few rather obvious differences, the peak EMG of the impaired limb is rotated upwards in voluntary activation, and the key patterns of EMG coactivation do not correspond closely with those described in the flexion withdrawal reflex of either the normal or even the impaired limb. Furthermore, the delay in onset of flexion reflex responses in the impaired limb (measured with respect to the contralateral side) does not favor significant hyperexcitability of the motoneuron pool. *On the other hand, there are significant points of similarity, such as the strong coactivation of the proximal shoulder muscles and elbow flexors in both voluntary activation of the limb and flexor reflex activation in the impaired upper extremity.*

Where Do Synergies Come From?

Because we are not describing motion but rather EMG patterns during isometric muscle contraction, our data do not provide a good basis for comparison with the traditional movement synergies described by Brunnstrom and others. Nonetheless, it is likely that similar mechanisms are responsible for both the abnormal patterns of isometric contraction and for the altered movement patterns.

If the abnormal spatial patterns visible in voluntary contractions are not *solely* attributable to flexor reflexes, then what is their origin? The spatial patterns observed could be promoted by the residual pathways that were left undamaged by the stoke. Such pathways may originate from many areas of cortex or perhaps subcortical regions, and they are likely to influence spinal motoneurons via pathways that traverse the ventromedial quadrant of the spinal cord. As described earlier, axons of such pathways are likely to have rather extensive branching which may traverse many spinal segments and induce coupling of muscles of the same or distant joints in a rather rigid or stereotypic manner. This anatomical organization could underlie the observed synergic relations.

A related, rather intriguing possibility is that the spatial patterns of muscle activation are induced by residual activity in the C3–4 propriospinal system or its human equivalent. This propriospinal pathway acts as a parallel input to proximal limb muscles from corticospinal systems, by assembling convergent descending and regional segmental input to the upper limb at the higher cervical cord level before projecting down to lower segments regulating limb muscle activation. Although the existence of the C3–4 propriospinal system has not been validated in the primate, strong evidence is presented by Pierrot-Deseignilly and Mazevet (this volume) that there may be an equivalent system present in man. If so, the abnormal synergic patterns described may depend on alternate supraspinal input to this C3–4 propriospinal system, coupled with loss of normally associated corticospinal and rubrospinal input. This suggestion is clearly rather speculative at the moment, but we believe that it is well worth pursuing.

Summary and Final Comment

In the final analysis, it is unlikely that the state of movement dysfunction which exists in the hemiparetic limb is attributable solely to any single mechanism. Instead, *it is quite probable that disturbed descending commands induce changes in the patterns of interneuronal responsiveness and allow altered sensory inflow to interact with and perhaps to distort the preexisting descending commands*. If this is so, it may well prove beneficial to modulate the pattern of muscle activity by changing the sensory inflow that influences proximal muscles.

The particular kinds of physical interventions that are potentially used to modify afferent inflow include warming of skin and muscle, skin vibration or on the other extreme, and reduction or eliminating the skin sensation with local anesthetic. The choice of a particular kind of intervention depends on the particular pattern of motor dysfunction and on the spatial distribution of abnormal muscle activity.

References

Brunnstrom S (1970) Movement therapy in hemiplegia. Harper and Row, New York

Buchanan TS, Almdale DPJ, Lewis JL, Rymer WZ (1986) Characteristics of synergic relations during isometric contractions of human elbow muscles. J Neurophysiol 56:1225–1241

Buchanan TS, Moniz MJ, Dewald JPA, Rymer WZ (1993) A biomechanical analysis of individual muscle forces about the wrist joint during isometric tasks. J Biomech (in press)

Dewald JPA, Buchanan TS, Rovai GP, Rymer WZ (1989) Control of muscle activation during the maintenance of arm postures in the spastic hemiparetic subject. Soc Neurosci Abstr 15:693

Dewald JPA, Pope PS, Given JD, McGuire JR, Rymer WZ (1992a) Evidence for abnormal coactivation patterns in hemiparetic stroke subjects. ACRM/AAPM&R Abstr, p 138

Dewald JPA, McGuire JR, Rymer WZ (1992b) Evidence for flexor reflex alterations in hemiparetic stroke subjects. Soc Neurosci Abstr 18:1410

Flanders M, Soechting JF (1990) Arm muscle activation for static forces in three-dimensional space. J Neurophysiol 64:1818–1837

Fügl-Meyer AR, Jaasko L, Leyman I, Olsson S, Steglind S (1975) The post-stroke hemiplegic patient. I. A method for evaluation of physical performance. Scand J Rehabil Med 7:13–31

Holmqvist B, Lundberg A (1961) Differential supraspinal control of synaptic actions evoked by volleys in the flexion reflex afferents in alpha motoneurons. Acta Physiol Scand 54:1–51

Kuypers HGJM (1964) The descending pathways to the spinal cord, their anatomy and function. In: Eccles JG, Schade J (eds) Organization of the spinal cord. Elsevier, New York, pp 178–200

Kuypers HGJM (1981) Anatomy of the descending pathways. In: Brookhart JM, Mountcastle VB (eds) Handbook of physiology, sect 1, vol 2, part 1. American Physiological Society, Bethesda, pp 597–666

McPherson JM (1988a) Strategies that simplify the control of quadrupedal stance. I. Forces at the ground. J Neurophysiol 60:204–217

McPherson JM (1988b) Strategies that simplify the control of quadrupedal stance. II. Electromyographic activity. J Neurophysiol 60:218–231

Shahani BT, Young RR (1971) Human flexor reflexes. J Neurol Neurosurg Psychiatry 34:616–627

Twitchell TE (1951) The restoration of motor function following hemiplegia in man. Brain 74:443–480

Wilson VJ, Yosida M (1969) Comparison of effects of stimulation of Deiters' nucleus and medial longitudinal fasciculus on neck, forelimb, and hindlimb motoneurons. J Neurophysiol 32:743–758

Disturbances of Voluntary Arm Movement in Human Spasticity: The Relative Importance of Paresis and Muscle Hypertonia

S.J. Fellows[1], C. Kaus[1], H.F. Ross[2], and A.F. Thilmann[1]

Introduction

Spasticity is a term often used loosely, but most workers in the field would agree that among its major features are hyper-reflexia, a velocity-dependent muscle hypertonia and some degree of paresis (Lance 1980). A feature of the first two of these symptoms, however, is that they are defined in the passive state. In clinical practice, hyper-reflexia is assessed by eliciting tendon jerk reflexes with the patient at rest and, similarly, hypertonia is tested by imposed movement of the passive limb. Recent studies have shown, however, that changes in the tendon jerk reflex do not reflect changes in more physiological reflex components elicited by limb displacement (Thilmann and Fellows 1991; Fellows et al. 1993). It is often assumed that spastic hypertonia is a factor in disturbances of voluntary movements commonly associated with spasticity. At first glance, this may seem logical: A voluntary movement which leads to stretch of a hypertonic muscle might be expected to elicit reflex activity which would oppose the attempted movement. The situation in the spinal reflex pathways is, however, more complex during voluntary activity than in the passive state, and it cannot be taken as read that a muscle which shows hypertonia during a passive stretch will behave similarly during stretch occurring as part of a voluntary movement (McLellan 1977). Indeed, although some reports have suggested that voluntary movements in spastic patients may be disrupted by exaggerated reflexes (Corcos et al. 1986), other workers have found no evidence of hyper-reflexia during spastic gait (Dietz and Berger 1983). Furthermore, both these studies concerned the ankle joint, whereas hypertonia is commonly confined to the knee and elbow joints (Burke et al. 1971; Thilmann et al. 1991). At these latter joints earlier studies have also reported conflicting findings, however (Sahrmann and Norton 1977; Mizrahi and Angel 1979). Much of this conflict may have arisen from the heterogeneous nature of the

[1] Neurologische Klinik, Alfried Krupp Krankenhaus, Alfried-Krupp-Straße 21, D-45131 Essen, Deutschland
[2] Department of Physiology, The Medical School, University of Birmingham, Birmingham B15 2TT, UK

Thilmann et al. (Eds.)
Spasticity: Mechanisms and Management

patient groups in these studies, in whom spasticity arose from a variety of causes, such as spinal trauma, tumour, multiple sclerosis and stroke. Spasticity is a syndrome and not a disease, and thus it should not be assumed that its features will be the same irrespective of the underlying cause. Accordingly, it was decided to study a group of patients in whom spasticity arose from a homogeneous cause, namely a single, unilateral ischaemic lesion in the area of the middle cerebral artery. These subjects were asked to make voluntary movements at the elbow, in order to assess the extent to which abnormal antagonist stretch reflex activity was responsible for disruption of voluntary movement. In addition, it was considered whether alternative reasons, such as muscle paresis or inappropriate co-activation of antagonist muscles, might better explain the movement deficits of these patients.

Methods

The study involved 25 patients (age 36–70 years) who had suffered a single, unilateral ischaemic lesion in the region of the middle cerebral artery. All were in stable condition and were tested at least 9 months after their stroke. The patients were selected from the hospital lists by telephone interview on the basis of having some voluntary control of the arm contralateral to the lesion. Fifteen normal subjects (age 23–63 years) served as a control group.

The subject was seated with the arm under study placed in a forearm cast, which was pivoted in alignment with the axis of rotation of the elbow and allowed free movement at this joint. The shoulder was laterally abducted and level with, or slightly above the level of, the elbow; 90° was defined as a right angle between the upper and lower arm. Movements were performed from 150° extension to 60° flexion. A potentiometer was placed on the axis of rotation and provided a position signal which was displayed to the subject on an oscilloscope in order to allow accurate positioning of the arm. A piezoelectric force transducer at the level of the wrist provided a measure of the torque developed at the elbow. Surface electromyogram (EMG) recordings were made from over the biceps brachii (agonist) and triceps brachii (antagonist) muscles. These signals were amplified (band width 20 Hz–1 kHz) and passed, along with the position and force signals, to the ADC board of a computer, which sampled at a rate of 100 Hz and provided an on-line display of the four signals, as well as saving a permanent record on disk for later analysis.

Movements were self-paced over an incremented range up to the subject's maximum velocity. The subjects were instructed to move smoothly and to maintain, as nearly as possible, a constant velocity. Movements were made without load, or with a load of 1.27 or 3.2 Nm (about 5% and 10% of voluntary maximum isometric torque, respectively).

Paresis was assessed with the arm fixed in the middle of the movement range (90°). Subjects were asked to develop a constant flexion or extension torque and then to maintain it for 2–3 s. This was repeated in a series of steps up to the subject's maximum torque. Subsequently, the mean torque and the corresponding EMG activity were determined for each step.

The procedure for analysis of the voluntary movements is illustrated in Fig. 1. This shows a single flexing movement made at 200°/s without load by a normal subject. Each movement cycle was defined using the velocity curve: movement onset was taken as the first visible deflection in this curve (the first vertical line on the velocity trace), and the total movement time (Ttot) was defined as the time between this point and next crossing of the null axis (the second vertical line). The peak velocity (pV) during each cycle was also measured, along with time elapsed since movement onset when it was reached (Tpv). The upper two traces show the rectified EMG records obtained from the agonist and antagonist muscles. The activity occurring during the movement was integrated, and the background EMG activity to be expected over this time interval (measured during the preceding hold phase, indicated by the interval "a") was subtracted. In addition, the integral of the EMG occurring up to the attainment of peak velocity was also obtained and similarly corrected. It should be noted that antagonist EMG activity was considered real, rather than the product of electrode cross-talk, on three grounds: Firstly, needle electrode recordings in a normal subject revealed that surface EMG recordings from the antagonist were accompanied by activity in single motor units in the antagonist. Secondly, as may be seen in Fig. 1, a considerable phase difference was apparent between the agonist and antagonist signals. Thirdly, as will become clear in the "Results" section, agonist and antagonist EMGs were modulated in opposite directions; that is, increasing agonist EMG could be accompanied by a decrease in antagonist activation. The statistical significance of differences between groups was assessed by variance analysis, using Bonferroni/Dunn as the post hoc test.

Results

It was found that the normal subjects were able to move smoothly over the 90° movement range under all load conditions and attained maximum flexing velocities of between 349°/s and 637°/s. It was also found that the mean level of agonist EMG during each movement showed a highly significant linear correlation to the peak movement velocity reached in that movement. The patients showed varied ability in this task, ranging from the nearly normal to the severely impaired subjects who failed to cover the whole 90° range. In order to allow more meaningful analysis, the patients were placed into one of four groups. Group 1 ($n = 9$) consisted of those patients who reached a

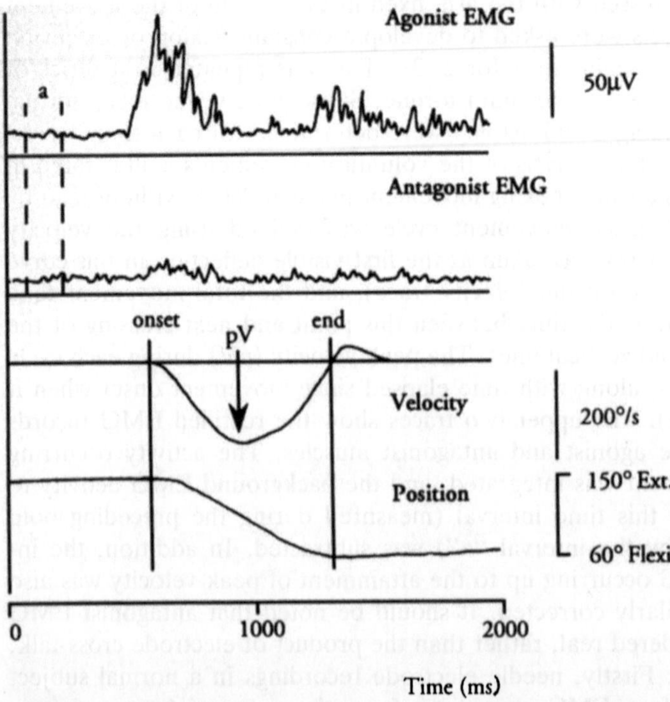

Fig. 1. The rectified EMG recordings obtained from the agonist and antagonist muscles, along with the position and velocity traces, for a 200°/s flexing movement made without load by a representative normal subject. For explanation of terms, see text

maximum flexing velocity within the normal range (i.e., greater than 350°/s) and in whom a linear correlation existed between mean agonist EMG and movement velocity. Group 2 contained those patients ($n = 5$) who did not achieve a maximum velocity within the normal range (i.e., slower than 350°/s) but in whom a linear correlation existed between mean agonist EMG and movement velocity. In these cases a significant correlation was defined when $p < 0.05$. Group 3 consisted of patients ($n = 5$) who did not attain normal maximum velocities and in whom no linear correlation existed between mean agonist EMG and movement velocity. Finally, group 4 contained those patients ($n = 6$) who were unable to complete a 90° flexion at the elbow.

Figure 2a shows the group means (±SEM) of the maximum flexing velocity achieved without load for the normal subjects and the patients. It may be seen that all patient groups capable of 90° flexion were significantly slower than normal ($p < 0.05$, group 1; $p < 0.001$, groups 2 and 3) and that this reduction in maximum flexing velocity became progressively worse across the patient groups.

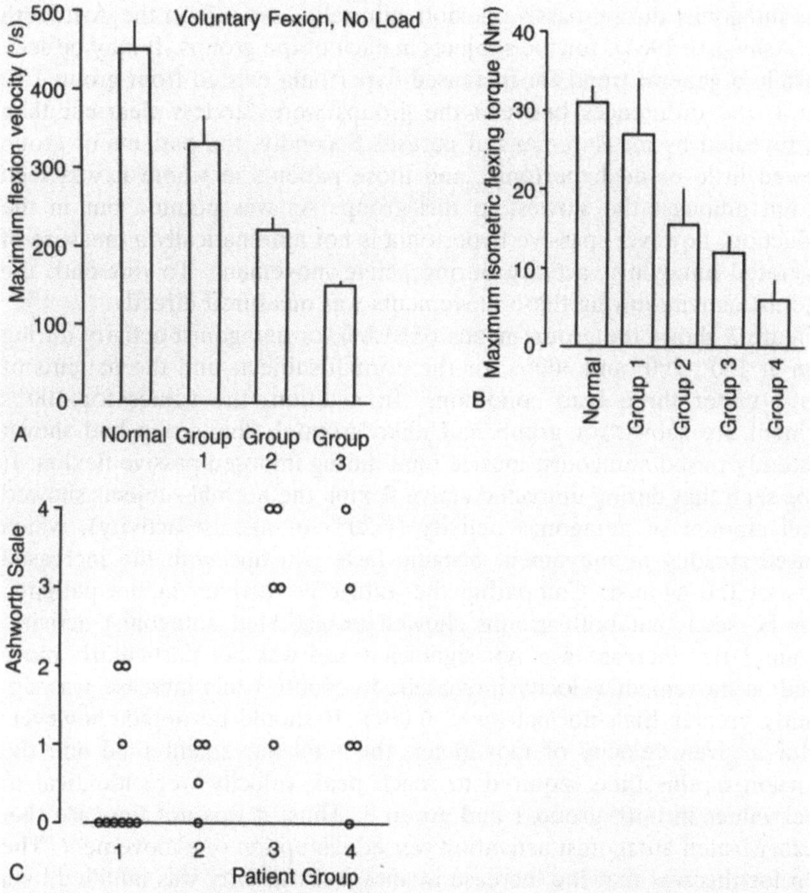

Fig. 2. A The group means (±SEM) for the maximum flexing velocity achieved by a subject while moving without load. **B** The group means (±SEM) for the maximum isometric torque developed by the biceps brachii with the elbow fixed at an angle of 90°. **C** Individual measures, split according to patient group, for the hypertonia of the antagonist triceps brachii during passive flexion, expressed on the Ashworth scale

The cause of this slowing could lie in exaggerated antagonist EMG, in agonist paresis, or in some combination of both. Figure 2b shows the group means (±SEM) for the maximum isometric torque developed by the elbow flexors. Two things in particular should be noted. Firstly, a progressive decrease in isometric torque (that is, an increase in paresis) occurred across the patient groups. This corresponds closely to the decrease in maximum flexing velocity. It is also apparent, however, that while groups 2, 3 and 4 were significantly weaker than normal ($p < 0.05$, group 2; $p < 0.025$, group 3; $p < 0.001$, group 4), the patients of group 1 showed no significant change in agonist strength. It is relevant to ask, therefore, why this group of patients were slower than the normal subjects. Figure 2c shows the hypertonia of the antagonist during passive flexion, clinically assessed on the Ashworth scale (Ashworth 1964), for the subjects in each of the groups. It may be seen that while a general trend for increased hypertonia existed from group 1 to group 4, the differences between the groups were far less clear-cut than those revealed by measurement of paresis. Secondly, the patients of group 1 showed little or no hypertonia, and those patients in whom it was seen were not amongst the slowest in this group. As was pointed out in the introduction, however, passive hypertonia is not automatically a measure of exaggerated antagonist activity during active movement. To this end, the antagonist activity during these movements was measured directly.

Figure 3 shows the group means (±SEM) for antagonist activity during flexion at 100°, 200° and 300°/s for the normal subjects and the patients of group 1 under three load conditions. In addition, the values for 100°/s movement are shown for group 3. Unlike group 1, this group had shown consistently raised antagonist muscle tone during imposed passive flexion. It may be seen that during unloaded active flexion the normal subjects showed a small amount of antagonist activity (≈20% of agonist activity), which increased steadily as movement became faster, in line with the increased activity of the agonist. Comparing the antagonist activity in the patients, it may be seen that both groups showed exaggerated antagonist activity. In group 1 this increase was not significant and was not particularly more marked as movement velocity increased. In group 3 this increase was significantly greater than normal ($p < 0.001$). It should be noted, however, that for a given velocity of movement, the total movement time and the proportion of this time required to reach peak velocity were identical to normal values in both group 1 and group 3. Thus, it was not the case that this exaggerated antagonist activation caused disruption of a movement. The reason for this was that the increase in antagonist activity was nullified by a comparable increase in agonist activity.

Returning to Fig. 3, it may be seen that the situation with the increased antagonist activity changed dramatically when a flexing movement was made under load. With a very light load (1.27 Nm or ≈5% voluntary maximum torque), the patients of group 1 showed normal levels of antagonist EMG with movement velocities of 100° and 200°/s. Although the antagonist activity at 300°/s was slightly exaggerated, this activity was not significantly greater than normal. With a moderate load (3.2 Nm or ≈10% maximum), it may be seen that antagonist EMG activity in group-1 patients was completely normal. Indeed, with this loading even the patients of group 3, in whom hypertonia of the antagonist triceps brachii was severe during imposed passive flexion, showed completely normal antagonist EMG activity during active flexion.

Thus, it is apparent that although antagonist EMG levels may be exaggerated during an unloaded flexion, this can be fully overcome by increased agonist activation in all but the most paretic of patients. More importantly,

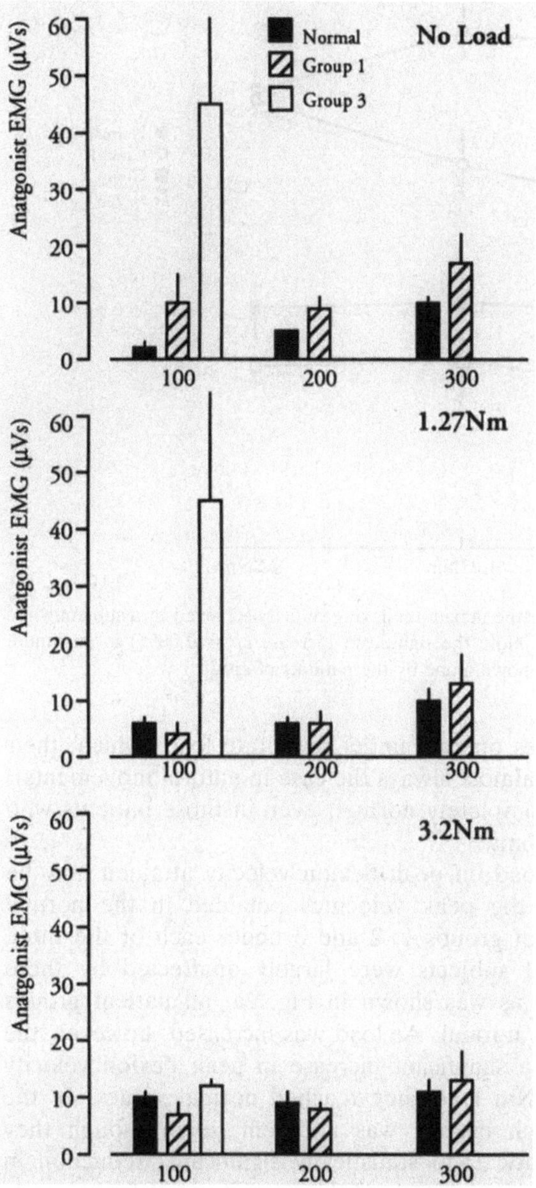

Fig. 3. The group means (±SEM) for the integrated antagonist EMG measured during flexion at 100°, 200° and 300°/s under three load conditions. Note that group 3 means are displayed only for 100°/s flexion, as none of this group were able to attain 200°/s

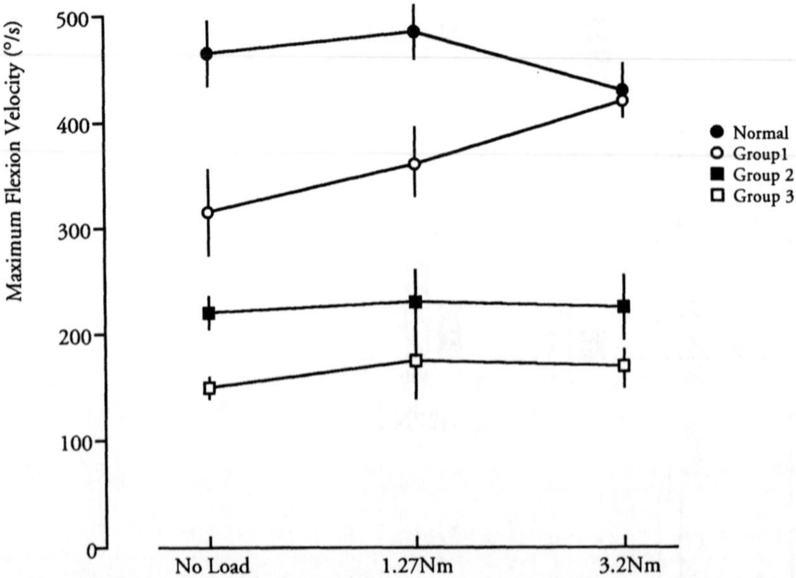

Fig. 4. The group means (±SEM) for the maximum flexing velocity achieved by a subject while moving under three load conditions. Note the significant increase ($p = 0.0055$) in maximum flexion velocity with increasing load shown alone by the patients of group 1

when the elbow flexor muscles operate under moderate load (which, their being anti-gravity muscles, is almost always the case in natural movements), antagonist EMG activity is completely normal, even in those patients who showed severe passive hypertonia.

The effect of increased load on peak flexion velocity attained may be seen in Fig. 4. This shows the peak velocities obtained in the normal subjects and in the patients of groups 1, 2 and 3 under each of the three load conditions. The normal subjects were largely unaffected by these load changes. Without load, as was shown in Fig. 2a, all patient groups were significantly slower than normal. As load was increased, however, the patients of group 1 showed a significant increase in peak flexion velocity ($p = 0.0055$), and with 3.2 Nm load they reached normal values. In the other patient groups no such change was apparent, even though they showed a much more dramatic (and statistically significant) reduction in antagonist EMG activity with increasing load than was the case for group 1.

Discussion

Despite their having a homogeneous cause of spasticity, namely a unilateral ischaemic lesion in the area of the middle cerebral artery, the patients in this study showed considerable variation in their ability to perform voluntary flexion movements at the elbow. From this it would seem sensible that future studies, if they are to yield meaningful data, should restrict themselves to using patients whose spasticity arises from a single cause, at least until such time as common features shared by different subtypes of the spastic syndrome have been identified.

The finding of the present study, that flexion of the elbow without an opposing load leads to abnormal EMG activity, the extent of which corresponds to the passive hypertonia of the antagonist muscle, would seem to support earlier reports that reflex activation of the antagonist muscle causes disruption of voluntary movements in spastic man (Corcos et al. 1986; Mizrahi and Angel 1979). It has also been shown, however, that under normal operating conditions for an anti-gravity muscle (i.e., the elbow flexors moving against a load) antagonist EMG was reduced to normal levels. This supports the finding of McLellan (1977) for his patients with mild spasticity and underlines that parameters assessed under passive conditions (i.e., spastic muscle hypertonia) cannot simply be assumed to operate in the active state. In contrast to the experience of McLellan, however, the present study showed that antagonist EMG activity during flexion against an opposing load was normal even in patients with severe passive hypertonia in the triceps brachii muscle, whereas McLellan reported that antagonist stretch reflex activity was enhanced during voluntary flexion in his subjects with severe spasticity. This difference may arise from the fact that his patients all had spinal spasticity, and it provides a further indication that spasticity should not be regarded as a single entity.

The physiological mechanism responsible for the modulation of the antagonist EMG with load in the patients cannot be inferred directly from the present data. However, it is known that reciprocal inhibition between an agonist-antagonist muscle pair may be absent at rest (Tanaka 1974; but see also Iles 1986) and expressed during activity at a level comparable to agonist activation (Tanaka 1974). As this reciprocal inhibition is present or even enhanced during voluntary activity in spastic man (Ashby and Wiens 1989), it would seem reasonable to assume that this is the mechanism responsible for the observed reduction in antagonist EMG with increasing load. Further indirect evidence for this may be provided by the bilateral reciprocal muster observed in the reflex responses to imposed elbow displacement (Thilmann et al. 1990).

Although the movement deficits observed could be attributed only poorly to antagonist EMG activity, the progressively decreased ability to perform voluntary flexion from group 1 to group 4 reflected well the increasing paresis of the elbow flexors seen across these groups. Only the patients of group 1 were not significantly weaker than the normal subjects. Paresis is not, however, simply a matter of reduced isometric torque, but rather a reduction in the voluntary control of a muscle, which also involves a loss of fine control of motor unit firing properties. Thus, the lower, but almost normal, peak velocities achieved by the patients of group 1 may reflect an inability on the part of these patients to produce motor unit firing rates high enough to achieve the degree of synchronisation of the agonist burst rquired to reach normal peak velocities: a reduction of peak firing rate during voluntary activation has been reported for a variety of paretic muscles (Duchateau and Hainaut 1990; Fitts et al. 1989; Rosenfalck and Andreassen 1980).

The present study has shown that under normal operating conditions, stretch of the antagonist muscles during elbow flexion does not ordinarily produce abnormal antagonist EMG activation. This does not, however, rule out an involvement of antagonist hyper-reflexia in disruption of voluntary movement in spastic patients. Although the movement itself does not, under moderate load, produce abnormal stretch reflex activity, it remains possible that an external disturbance, which is often encountered under natural conditions, might elicit inappropriate reflex responses which then cause disruption of a movement. Whether this is the case must be established in further studies.

Acknowledgements. This work was supported by the *Deutsche Forschungsgemeinschaft*. We also wish to thank the members of the BSB e.V. for their participation in this study.

References

Ashby P, Wiens M (1989) Reciprocal inhibition following lesions of the spinal cord in man. J Physiol (Lond) 414:145–157

Ashworth B (1964) Preliminary trial of carispodol in multiple sclerosis. Practitioner 192:540–542

Burke D, Gillies JD, Lance JW (1971) Hamstrings stretch reflex in human spasticity. J Neurol Neurosurg Psychiatry 34:231–235

Corcos DM, Gottlieb GL, Penn RD, Myklebust B, Agarwal GC (1986) Movement deficits caused by hyperexcitable stretch reflexes in spastic humans. Brain 109:1043–1058

Dietz V, Berger W (1983) Normal and impaired regulation of muscle stiffness in gait: a new hypothesis about muscle hypertonia. Exp Neurol 79:680–687

Duchateau J, Hainaut K (1990) Effects of immobilisation on contractile properties, recruitment and firing rates of human motor units. J Physiol (Lond) 422:55–65

Fellows SJ, Ross HF, Thilmann AF (1993) The limitations of the tendon jerk as a marker of pathological stretch reflex activity in human spasticity. J Neurol Neurosurg Psychiatry 56:531–537

Fitts SS, Hammond MC, Kraft GH, Nutter PB (1989) Quantification of gaps in the EMG interference pattern in chronic hemiparesis. Electroencephalogr Clin Neurophysiol 73: 225–232

Iles JF (1986) Reciprocal inhibition during agonist and antagonist contraction. Exp Brain Res 62:212–214

Lance JW (1980) Pathophysiology of spasticity and clinical experience with baclofen. In: Feldman RG, Young RR, Koella WP (eds) Spasticity: disordered motor control. Year Book Medical Publishers, Chicago, pp 185–203

McLellan DL (1977) Co-contraction and stretch reflexes in spasticity during treatment with baclofen. J Neurol Neurosurg Psychiatry 40:30–38
Mizrahi EM, Angel RW (1979) Impairment of voluntary movement by spasticity. Ann Neurol 5:594–595
Rosenfalck A, Andreassen S (1980) Impaired regulation of force and firing pattern of single motor units in patients with spasticity. J Neurol Neurosurg Psychiatry 43:907–916
Sahrmann SA, Norton BJ (1977) The relationship of voluntary movement to spasticity in the upper motor neuron syndrome. Ann Neurol 2:460–475
Tanaka R (1974) Reciprocal inhibition during voluntary movements in man. Exp Brain Res 21:529–540
Thilmann AF, Fellows SJ (1991) The time-course of bilateral changes in the reflex excitability of relaxed triceps surae muscle in human hemiparetic spasticity. J Neurol 238:293–298
Thilmann AF, Fellows SJ, Garms E (1990) Pathological stretch reflexes on the "good" side of hemiparetic patients. J Neurol Neurosurg Psychiatry 53:208–214
Thilmann AF, Fellows SJ, Garms E (1991) The mechanism of spastic muscle hypertonus: variation in reflex gain over the time course of spasticity. Brain 114:233–244

Spastic Movement Disorder: Similarities and Differences in Children and Adults

W. Berger[1] and V. Dietz[2]

Introduction

This contribution is restricted to findings in spastic movement disorder during stance (following perturbations) and gait performed on a moving treadmill. Nevertheless, these functional automatic movements will provide evidence for new interpretations of this pathological state and therefore new conclusions for therapy.

For further discussions it should be emphasized that the spastic movement disorder is usually made apparent by three characteristic symptoms: movement disorder, spasticity, and exaggerated tendon reflexes. To use only the term "spasticity" has led to confusion, insofar as there exists a widely accepted hypothesis that exaggerated reflexes cause spasticity and spasticity causes impaired functional movement. It will be shown in the following that this causality is incorrect, as has been argued by only a few authors to date (Landau 1980; Young 1982; Feldkamp 1988). Despite these few critical reports, there is a great amount of literature about evaluating the so-called spastic gait only by looking at reflex behavior, either in animal experiments or in patients, thus interpreting drug-induced depressed reflexes as an improvement of gait. This approach, however, is inappropriate.

The results will be discussed in four parts: (a) adult patients with a spastic movement disorder and (b) children with cerebral palsy, with the crucial difference that the latter suffer their supraspinal lesion at a time when the normal gait pattern is not yet established. This is compared with (c) undisturbed development of gait in children. (d) The final conclusions to be drawn from these findings will give further insight into the underlying abnormalities for both groups of patients.

[1] Abt. Klinische Neurologie und Neurophysiologie, Universität Freiburg, Hansa Str. 9, D-79104 Freiburg, Deutschland
[2] Schweizer Paraplegikerzentrum, Klinik Balgrist, Forch Str. 340, CH-8008 Zürich, Schweiz

Thilmann et al. (Eds.)
Spasticity: Mechanisms and Management

Spastic Movement Disorder in Adult Patients

Human gait, as a complex, automatic functional movement, is achieved by central programming on a spinal level and adapted to actual conditions by multisensory afferent input (for review see Dietz 1992a,b). In contrast to quadrupedal animal locomotion, the spinal programming as well as the reflex activity are governed by strong supraspinal control. This close interaction provides electrical leg muscle activity, which is transferred to a functionally modulated muscle tension by mechanical muscle fiber properties (Gollhofer et al. 1984). "Stiff" gait and exaggerated tendon tap reflexes are clinical observations which led to the hypothesis that co-activation of antagonistic leg muscles and the enhanced reflexes result in muscle hypertonia and, consequently, in movement disorder. On the other hand, it is well known that exaggerated reflexes and enhanced muscle tone may differ in their time course following a supraspinal lesion. The gait pattern of patients with a spastic hemi- or paraparesis does not confirm the above-mentioned suggestions: Neither the timing nor the reciprocal mode of activation of the leg muscles differs between healthy subjects and adult patients. There is no evidence for enhanced alpha-motoneuron activity. The exaggerated stretch reflexes with short latency (according to the tendon tap reflexes) form only a small part of the overall activity, which is reduced compared with that of healthy subjects, together with a reduced modulation (Berger et al. 1984b; Dietz and Berger 1983; Dietz et al. 1981). Corrective reactions following perturbations during stance and gait on a treadmill further reveal the altered behavior of the different reflex systems: in patients with spastic hemiparesis a reduction of the functionally essential part in the form of the polysynaptic reflexes was obvious, while only the short-latency reflexes were enhanced on the affected side (Berger et al. 1988). Tension development of the triceps surae muscle, as measured by a strain gauge fixed at the Achilles tendon, showed basic differences between the affected and unaffected side in patients with spastic hemiparesis: correlation with the modulation of EMG activity on the normal side, and correlation with the passive stretching of the tonically activated triceps surae muscle of low amplitude on the affected side (Berger et al. 1984b). The conclusion was that exaggerated monosynaptic reflexes did not contribute to tension development, but that a transformation of motor units occurred which led to enhanced muscle tone by the tonically activated, low-level triceps surae EMG activity (Dietz et al. 1981). Histological muscle fiber changes confirmed these suggestions (Dietz et al. 1986).

Spastic Movement Disorder in Children with Cerebral Palsy

In children with cerebral palsy, too, it is a widely accepted hypothesis that motor impairment is due to increased muscle tone and that muscle hypertonia is the consequence of enhanced motoneuron activity, co-activation, and increased stretch reflexes, as well as enhanced postural reflex activity (Walshe 1923). The results of EMG recordings in children with spastic hemi- or paraparesis indeed showed co-activation of the tibialis and gastrocnemius muscles during the stance phase of gait. The electrical extensor activity, however, as in the adult patients, was strongly reduced and poorly modulated (Berger et al. 1984a). Biphasic stretch reflexes were prominent, together with a digitigrade stepping, following muscle stretch with a short latency, compatible with monosynaptic pathway. Again, as in adult patients, this enhanced stretch reflex activity contributed only a small part of the overall reduced activity. This was parallel to the visible impairment of gait with irregularities, slow speed, and poorly adapated corrective responses following perturbations of the treadmill. The conclusion was that, as in adult patients, movement disorder was due to reduced extensor activity, while the exaggerated stretch reflexes did not play a major role (Berger et al. 1982).

Development of Normal Gait in Children

Newborn stepping, as described by Forssberg and Wallberg (1980), is characterized by an innate locomotor program with co-activation of all antagonistic leg muscles during ground contact of the feet. This pattern is normally lost 10 weeks later. The results recorded in children from a period of free stance (7–9 months of age) to independent gait (about 7 years) were as follows: In the youngest children these steps were induced by the moving treadmill but did not differ substantially from the free ambulatory pattern studied in some subjects. The typical features in this early stage can be characterized by a co-activation pattern of antagonistic leg muscles, by low and poorly modulated extensor EMG activity, and by biphasic potentials arising with a short latency of 20–30 ms after stretch of gastrocnemius muscle at the beginning of the stance phase of gait. This is achieved in the early stages of stepping in a digitigrade manner with a sharp decrease of ankle joint angle followed by ground contact of the whole foot. From 2 to 3 years on, this pattern is transformed to a reciprocal mode of muslce activation, suppression of the biphasic reflexes and significant increase of the low extensor EMG. The most impressive changes take place during the second year of life, while adjustment to the adult gait pattern continues up to 7–8 years of age (Berger et al. 1984a). These changes in EMG activity and reflex behavior are paralleled by the development of free and independent walking, with the ability to react to ground irregularities and, too, by the compensatory EMG activity seen in the stumbling reactions: phasic behavior of the extensor EMG with longer latency, suppression of short-latency reflexes, reciprocal mode of activation (Berger et al. 1987). In addition, developmental changes in interlimb coordination can be demonstrated by these reflex changes: unilateral displacements (on split-belt) elicit long-latency responses also on the contralateral side. These were larger, and the short-latency reflexes were smaller, in the older children (Berger et al. 1990). Mechanically evoked cerebral potentials provide further evidence that during development, changes in afferent pathway control take place which can be interpreted as follows: Together with a suppression of monosynaptic reflexes, a blocking of group-I afferents and an increase of polysynaptic EMG responses take place. It is suggested that maturation of normal gait pattern and compensatory EMG responses is achieved by the descending control of group-I afferents and facilitation of polysynaptic spinal reflexes via group-II afferents.

Conclusions

These results are presented in the time course they were recorded: adult patients with spastic movement disorder, children with cerebral palsy, and normal development gait. If we had started by studying children with spastic movement disorder we would have been in agreement with the common opinion about spasticity and impaired movement: that enhanced stretch reflex excitability and co-activation cause the motor disability. Only the overall reduced motoneuron activity would have been a divergent finding. As we worked in the other way direction, it is obvious that this interpretation can be discarded. We can demonstrate that this pattern is the physiological one during development without signs of muscle hypertonia. Co-activation seems to be a mechanism to support body weight for equilibrium control.

In children with cerebral palsy the normal maturation is inhibited due to an impairment of supraspinal centers. In adult patients, however, a partial regression to this early gait pattern follows the supraspinal lesion, along with the preservation of the reciprocal mode of muscle activation, once established. In both groups of patients muscle hypertonia seems to be the consequence of altered muscle fiber properties.

Consequently, the so-called antispastic drug therapy, measured in the form of reduced reflex excitability, cannot be regarded as an adequate tool to improve the voluntary force and the capacity for functional movements. A physiotherapy should represent the most conclusive mode of treatment to train residual functions or to elicit and influence residual spinal locomotor activity, via special treadmill training assisted by monoaminergic drugs (Barbeau and Fung 1992).

References

Barbeau H, Fung J (1992) New experimental approaches in the treatment of spastic gait disorders. In: Forssberg H, Hirschfeld H (eds) Movement disorders in children. Med Sport Sci 36:234–246

Berger W, Quintern J, Dietz V (1982) Pathophysiology of gait in children with cerebral palsy. Electroencephalogr Clin Neurophysiol 63:538–548

Berger W, Altenmüller E, Dietz V (1984a) Normal and impaired development of children's gait. Hum Neurobiol 34:163–170

Berger W, Horstmann GA, Dietz V (1984b) Tension development and muscle activation in the leg during gait in spastic hemiparesis: the independence of muscle hypertonia and exaggerated stretch reflexes. J Neurol Neurosurg Psychiatry 47:1029–1033

Berger W, Quintern J, Dietz V (1987) Afferent and efferent control of stance and gait: developmental changes in children. Electroencephalogr Clin Neurophysiol 66:244–252

Berger W, Horstmann GA, Dietz V (1988) Spastic paresis: impaired spinal reflexes and intact motor programs. J Neurol Neurosurg Psychiatry 51:568–571

Berger W, Horstmann GA, Dietz V (1990) Interlimb coordination of stance in children: divergent modulation of spinal reflex responses and cerebral evoked potentials in terms of age. Neurosci Lett 116:118–122

Dietz V (1992a) Human neuronal control of automatic functional movements: interaction between central programs and afferent input. Physiol Rev 72(1):33–69

Dietz V (1992b) Spasticity: exaggerated reflexes or movement disorder? In: Forssberg H, Hirschfeld H (eds) Movement disorders in children. Med Sport Sci 36:225–233

Dietz V, Berger W (1983) Normal and impaired regulation of muscle stiffness in gait: a new hypothesis about muscle hypertonia. Exp Neurol 79:680–687

Dietz V, Berger W (1984) Interlimb coordination of posture in patients with spastic paresis. Impaired function of spinal reflexes. Brain 107:965–978

Dietz V, Quintern J, Berger W (1981) Electrophysiological studies of gait in spasticity and rigidity. Evidence that altered mechanical properties of muscle contribute to hypertonia. Brain 104:431–449

Dietz V, Ketelsen UP, Berger W, Quintern J (1986) Motor units' involvement in spastic paresis: relationship between leg muscle activation and histochemistry. J Neurol Sci 75: 89–103

Feldkamp M (1988) In: Feldkamp M, Matthias MH (eds) Diagnose der infantilen Zerebralparese im Säuglings- und Kindesalter. Thieme, Stuttgart, p 22

Forssberg H, Wallberg H (1980) Infant locomotion: a preliminary movement and electromyographic study. In: Berg K, Eriksson BO (eds) Children and exercise. IXth international series on sport sciences. University Park Press, Baltimore, pp 32–40

Gollhofer A, Schmidtbleicher D, Dietz V (1984) Regulation of muscle stiffness in human locomotion. Int J Sport Med 5:19–22

Landau WM (1980) What is it? What is it not? In: Feld RG, Young RR, Loella WP (eds) Spasticity: disordered motor control. Symposia Specialists, Miami, pp 17–24

Walshe FMR (1923) On certain tonic or postural reflexes in hemiplegia, with special reference to the so-called associated movements. Brain 46:1–37

Young RR (1982) The assessment of drug treatment of spastic gait. J Neurol Neurosurg Psychiatry 45:p 760 (letter to the Editor)

Hyper-reflexia and Disordered Voluntary Movement

G.L. Gottlieb[1] and B.M. Myklebust[2]

Introduction

In 1980, J.W. Lance offered the following operational definition of spasticity: "Spasticity is a motor disorder characterized by a velocity-dependent increase in tonic stretch reflexes ('muscle tone') with exaggerated tendon jerks, resulting from *hyper-excitability of the stretch reflex*, as one component of the upper motor neuron syndrome" (Lance 1980) [italics added]. This has come to be the most widely used definition of this collection of clinical signs. Ten years later Lance reiterated the definition with the added remarks that 'spasticity does not include impaired voluntary movement and an abnormal posture. These are often features of upper motoneuron lesions . . ." (Lance 1990).

The issue to be addressed here is whether, or to what degree, when upper motoneuron lesions produce spasticity and impaired voluntary movement together, the one is contributory to the other. To this end, we wish to restate Lance's dictum in the following way: Spasticity is a motor disorder in which a *failure to actively inhibit velocity-sensitive stretch reflexes* can lead to exaggerated muscle resistance to both externally and self-imposed muscle stretch and, consequently, to impaired voluntary movement. This is not incompatible with the former definition, but rather is an expansion and a change of emphasis from the purely static or quasi-static manifestations of hyper-reflexia to those aspects involved in the control of voluntary movement.

[1] Dept. of Neurological Sciences and Physiology, Rush Medical Center, 2242 West Harrison St., Chicago, IL 60612, USA

[2] Dept. of Neurology, Physical Medicine and Rehabilitation, Medical College of Wisconsin and VA Medical Center, 5000 West National Ave., Milwaukee, WI 53295, USA

[3] We use the term "stretch reflex" to refer to the segmental reflex arc, which is in part monosynaptic. There is an extensive body of knowledge concerning longer loop stretch reflex pathways extending to higher levels of the CNS (Desmedt 1978). There is also a basic theory of voluntary movement based upon stretch reflex pathways longer than the short segmental loop (Feldman et al. 1990; Latash and Gottlieb 1991). These longer pathways are not considered here

Thilmann et al. (Eds.)
Spasticity: Mechanisms and Management

Some Properties of the Normal Stretch Reflex

The normal stretch reflex[3] has been exhaustively studied and reported throughout this century (Gordon and Ghez 1991; Houk and Rymer 1981). The point we wish to make here is that the stretch reflex is not a static structure, its well-documented mono- and oligosynaptic connections notwithstanding. The stretch reflex is under profoundly powerful, dynamic modulating influences, both during postnatal development of the individual and from moment to moment throughout life.

The stretch reflex pathway in the normal human infant is functionally quite different from what is commonly described in the adult (Myklebust and Gottlieb 1993; Myklebust et al. 1986b). The normal adult pathway consists of a monosynaptic, excitatory connection from primary afferent fibers to homonomous motoneurons, short oligosynaptic excitatory connections to close synergistic muscle motoneurons, and a disynaptic inhibitory connection to close antagonists. Figure 1A shows adult electromyographic (EMG) responses in four muscles of the leg: soleus (SOL), tibialis anterior (TA), quadriceps (QUAD), and hamstrings (HAM) following an Achilles tendon tap. The expected response in SOL is seen at a monosynaptic latency, with little EMG activity observed elsewhere. The same stimulus delivered to a neonate will produce responses such as those in Fig. 1B. The responses in the various muscles are not simultaneous, denoting different conduction pathways, but they are all in the monosynaptic or short oligosynaptic latency range (Myklebust and Gottlieb 1993; Myklebust et al. 1986b).

The observation of simultaneous SOL and TA responses to an Achilles tendon tap was originally termed "reciprocal excitation" (Gottlieb et al. 1982; Myklebust et al. 1982) and can be simply characterized by the ratio of two responses (peak-to-peak TA/SOL reflex ratio). It was first observed in

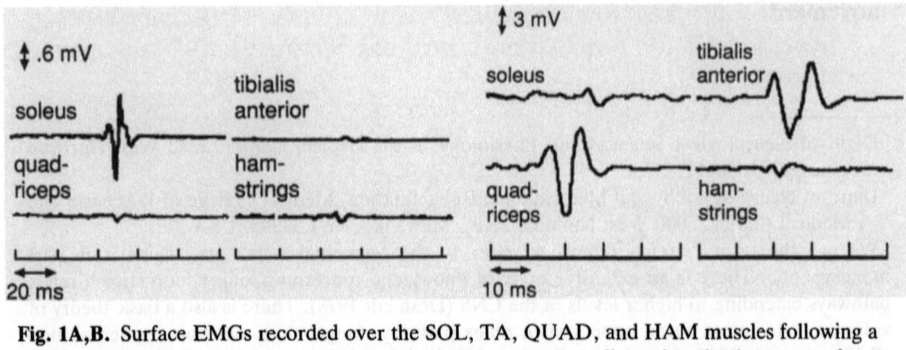

Fig. 1A,B. Surface EMGs recorded over the SOL, TA, QUAD, and HAM muscles following a tap to the Achilles tendon. **A** Normal adult shows a strong, well-synchronized response from the stretched SOL, with only very small voltages measured at other sites. **B** Normal neonate (2 days old) shows responses in all four muscles. The response in the tapped muscle is not necessarily the largest patients with cerebral palsy (see below) and later in normal infants and children. Further studies showed that reflex responses were not localized to the tapped SOL muscle and its antagonist TA. Responses also appeared in muscles of the knee (QUAD and HAM) (Myklebust and Gottlieb 1993). Similar findings have been described by O'Sullivan et al. (1991) in the upper limb. The evolution of this ratio with maturation is shown in Fig. 2. The interpretation of these data is that in the normal newborn infant there is an irradiation of excitatory reflex connections ("reflex irradiation" is an extension of "reciprocal excitation") from primary afferents to the motoneuron pools of many (if not all) of the muscles at their segment of entry, or even at neighboring segments. These may be monosynaptic connections, although a short oligosynaptic path cannot be excluded. Normal maturation processes lead to the diminished expression of these connections, either because of a withering of synapses not use during normal function or because of their masking by maturing inhibitory mechanisms. We favor the latter hypothesis, as discussed below.

The moment-to-moment modulation of reflex excitability is a much older phenomenon (Davenport 1982; Lombard 1887, 1889). The prior voluntary, steady contraction of a muscle such as the SOL enhances its stretch reflex and its H-reflex, while tonic contraction of its antagonist TA inhibits both reflexes (Gottlieb and Agarwal 1971, 1979). Phasic contraction has a similar and much stronger effect on the H-reflex, but the stretch reflex is inhibited in both agonist and antagonist muscles during rapid contraction (Gottlieb and Agarwal 1980; Gottlieb et al. 1970). A possible reason for the difference in modulation of agonist Hoffmann and stretch reflexes during rapid voluntary contraction may be that the H-reflex is better synchronized

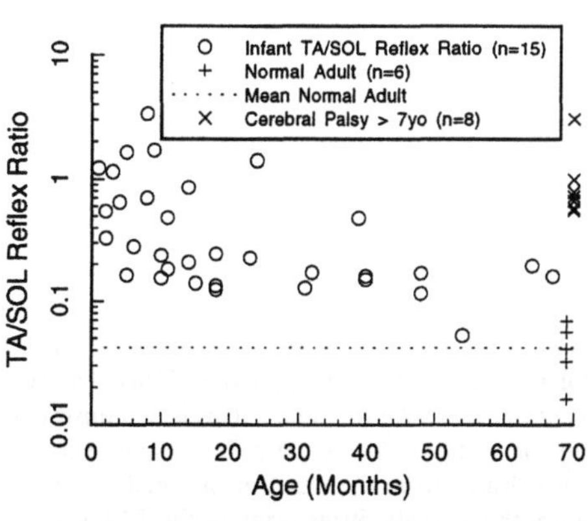

Fig. 2. The mean ratios of peak-to-peak EMG responses in TA and SOL following Achilles tendon taps. The three populations illustrated are normal infants and children (*circles*), normal adults (+), and patients with cerebral palsy (X). The ages of the infants and children, represented by the ordinate axis, range from 2 days to 6 years. Some were followed longitudinally for many months. Even at 6 years of age, the ratio in children has not reached the level of the normal adult. All the cerebral palsy group have ratios similar to those of neonates

and probably better localized to the Ia afferents than the stretch reflex. A dorsiflexing torque takes tens of milliseconds to move the foot, while an electrical pulse requires a millisecond or less. Stretch excites the Golgi tendon organs as well as the muscle spindles, so that there is a convergence of Ia excitation and Ib inhibition during a stretching stimulus. This will of course also be true during any voluntary contraction. Hence, stretch is probably a better functional test of reflex behavior during voluntary activity than is an H-reflex.

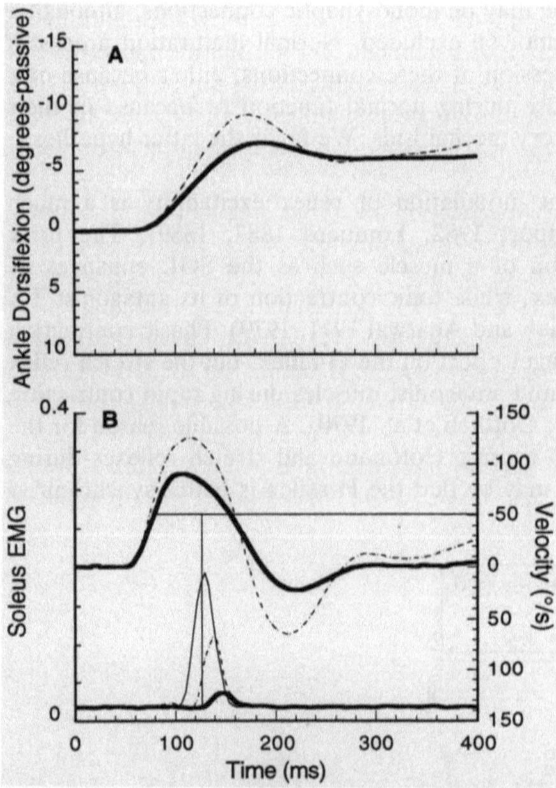

Fig. 3A–D. Reflex and voluntary movements of a normal subject. **A, B** Ankle joint angle, velocity, and the rectified EMG following a dorsiflexing torque step applied at 50 ms. The two torque steps are not far above the threshold necessary to evoke the reflex. This is demonstrated by the third, largest EMG burst (*thin solid line at bottom*) that is the reflex response to the largest torque pulse used. The *vertical line in* **B** is drawn from the onset of the smallest reflex. The *horizontal line* with which it intersects shows the velocity 50 ms prior to the EMG burst, about 50°/s. **C, D** Ankle joint angle, velocity, and the rectified EMGs following two voluntary dorsiflexions of about 10° and 20°. The larger two EMGs are from the TA muscle, while the smaller two are from the SOL. Note that the velocities reached are greater than those produced by the torque in **A, B**. The horizontal line drawn in **D** marks where the velocity reaches 50°/s and the *vertical line* shows where the reflex would occur, were there a threshold for it at that velocity

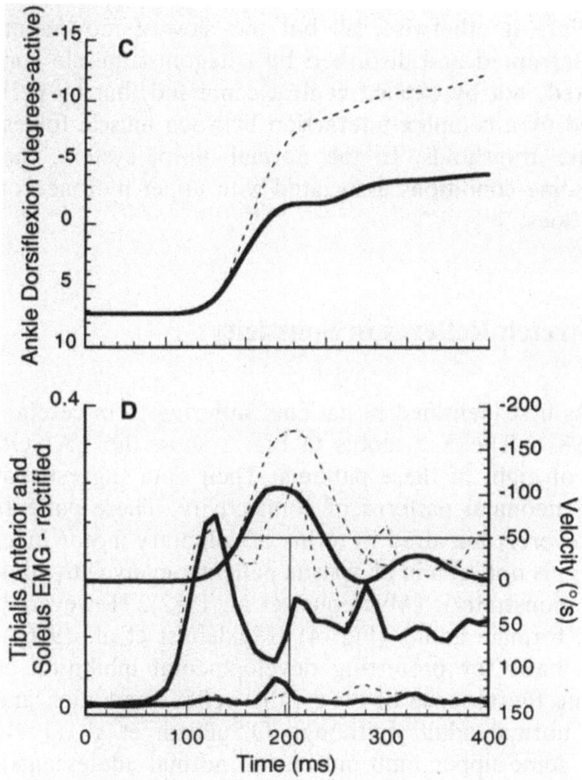

Fig. 3. C, D

Normal stretch reflexes, tested in the relaxed ankle, are quite sensitive to dorsiflexing perturbations. Figure 3A shows the angular displacement of an ankle to two different levels of dorsiflexing torque steps that are not far above threshold for the stretch reflex. Figure 3B shows the angular velocity and the SOL EMG response. There is little angular displacement prior to reaching reflex threshold. We can infer from this figure that the velocity threshold of the reflex, reached about 40 ms before the EMG burst, is about 50°/s. Tonic contraction of SOL against an opposing torque would reduce this threshold significantly, but even in the relaxed leg the reflex can be seen to be very sensitive. Figure 3C and D shows the same kinematic and EMG measures made during a voluntary dorsiflexion. The voluntary movements are of the same order of magnitude as the passive movements. Although the velocities are larger, the voluntary dorsiflexion does not evoke a reflex response in SOL.

Results of studies in which perturbing torques were delivered during movement are consistent with this observation that stretch reflexes are not easily evoked during a movement. This leads us to the conclusion that those reflexes are actively inhibited as part of the normal control process in voluntary movement. Were it otherwise, all but our slowest movements would be continually interrupted and disturbed by antagonist muscle contractions that were elicited, not by explicit central command, but by peripheral events determined by a complex interaction between muscle forces, external loads, and reflex thresholds. In the normal motor system, that does not occur. Under some conditions associated with upper motoneuron lesions, it unfortunately does.

Some Properties of Stretch Reflexes in Spasticity

Reciprocal excitation was first identified in patients suffering from cerebral palsy (Gottlieb et al. 1982). The X symbols in Fig. 2 show the TA/SOL reflex ratio for a group of eight of these patients. Their data suggest that they have retained their neonatal patterns of connectivity. These patients were all moderately to severely impaired in terms of voluntary motor function. Reciprocal excitation is not seen in all spastic patients; many with adult onset injuries do not demonstrate it (Myklebust et al. 1982). However, it can reappear in a labile form in adults (Fig. 4) (Myklebust et al. 1986a). This observation is the basis for preferring developmental inhibition as the mechanism underlying the absence of observable reflex irradiation and reciprocal excitation in normal adults [although O'Sullivan et al. (1991) report its persistence in some upper limb muscles in normal adolescents]. The physical connections remain and can re-emerge following some neurological insults.

Although reflex irradiation and reciprocal excitation cannot be considered a unique causal contributor to impaired voluntary movement in the spastic patient, the existence of this phenomenon, and the nature of these

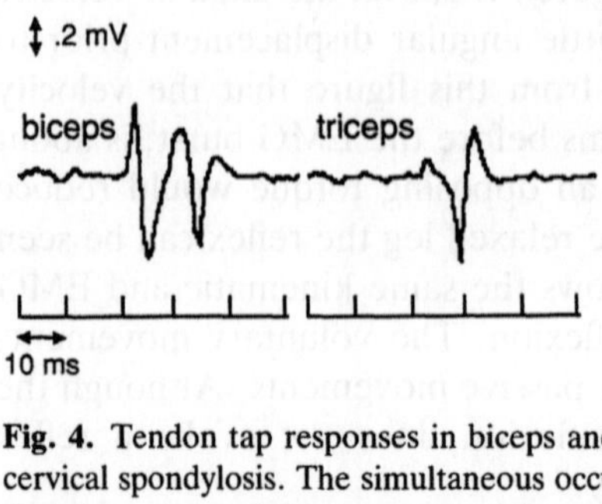

Fig. 4. Tendon tap responses in biceps and triceps of a mildly spastic 80-year-old woman with cervical spondylosis. The simultaneous occurrence of responses in antagonistic muscles suggests that the return of this immature reflex pattern is due to the loss of normal adult inhibitory mechanisms, coupled with the preservation of previously masked connections

reflexes in the neonate, highlight the crucial role that the proper modulation of reflex excitability plays in the production of normal movement.

The normal voluntary and reflex behavior demonstrated above can be contrasted with that of a patient with spinal compression and myelopathy shown in Fig. 5A. The first part of this figure illustrates a "fast," 20° dorsiflexion of the ankle, tested prior to treatment. The voluntary ankle movements are very slow, not reaching 50°/s at peak velocity. There is no sign of antagonist (SOL) activity, either voluntary or reflexive. That absence is not unusual for a movement this slow (Lestienne 1979). This patient was nonambulatory and showed sustained clonus to torque perturbations.

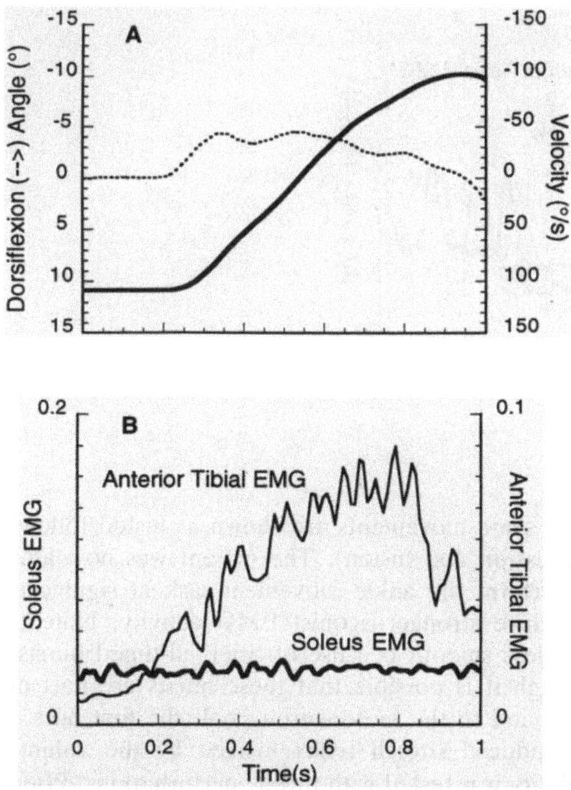

Fig. 5A–D. Ankle joint angle, velocity and SOL and TA EMGs (rectified and filtered) during voluntary movements by a patient with cervical myelopathy before (**A**, **B**) and after decompression laminectomy (**C**, **D**). In **A** and **B** is a slow, 20° dorsiflexion following the instruction to move as fast as possible. The antagonist muscle (SOL) is slightly and tonically active throughout. **C** and **D** show that after surgery, the movements were much faster but less smooth. The brief, phasic activations of SOL are more typical of stretch reflex bursts than the patterns that would be associated with normal antagonist braking of a voluntary movement (**C**, **D**, see p. 162)

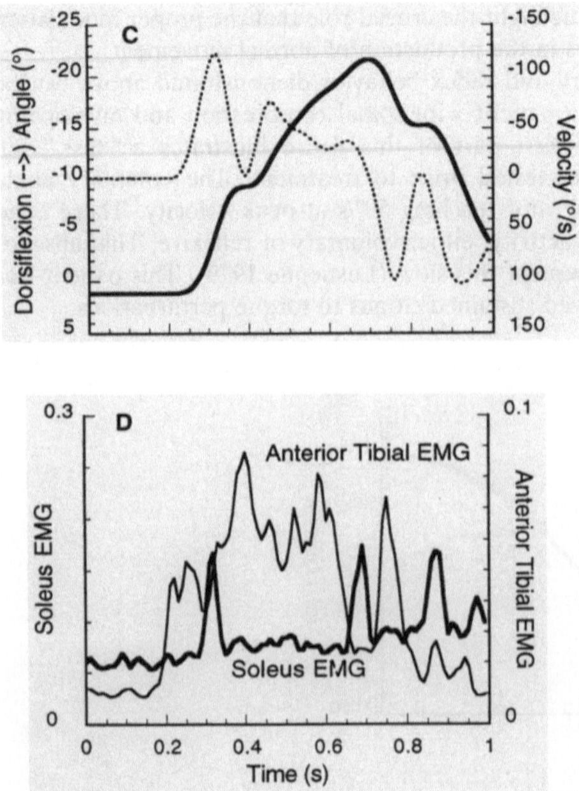

Fig. 5. C, D

Above in Fig. 5B, the same movements are shown as tested following surgery (T10–12 decompression and fusion). The patient was now ambulatory and was able to perform the ankle movement task at significantly higher speeds and to produce stronger agonist EMG activity. Note that the movement is now far less smooth because of brief, ill-timed bursts of antagonist activity. Although it is possible that these bursts are part of a defective descending command to the motoneuron pool, the first burst (at least) is probably a self-induced stretch reflex evoked by the voluntary movement. Reflex excitability was tested with torque perturbations. Prior to surgery, a torque step produced sustained clonus. The patient was still hyper-reflexic after surgery, but less so, since torque perturbations now induced only a damped clonus, as shown in Fig. 6A. The nature of the reflex state is complex, however. Figure 6B shows the unrectified SOL EMG bursts for the first 120 ms following the onset of the torque, both pre- and post-operatively. The postoperative latency of the reflex was significantly longer than preoperatively, and significantly longer than the normal subject in Fig. 4.

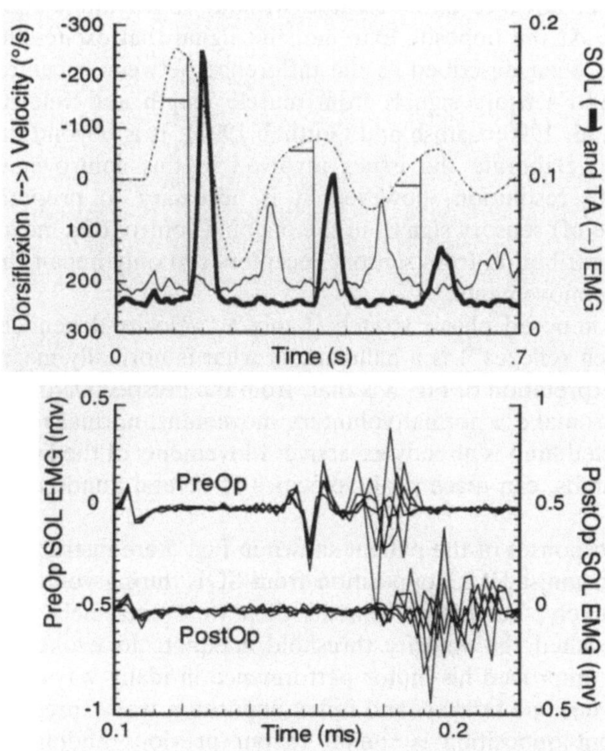

Fig. 6A,B. Stretch reflex responses of the subject illustrated in Fig. 5. **A** Ankle velocity following a dorsiflexing torque step, postoperatively. The SOL and TA rectified EMGs are plotted below. *Vertical lines* denote the onset of the SOL reflex EMG burst. *Horizontal lines* are 40 ms long to mark the velocity that occurred 40 ms prior to the reflex. **B** The SOL EMG responses to five successive, identical dorsiflexing torques applied preoperatively (*top*) and postoperatively (*bottom*). The time origin corresponds to the onset of the perturbing motor's voltage. The rise time of the torque is about 10 ms, so the preoperative reflex latency is about 40 ms, while postoperatively it is about 70 ms

Discussion

Most normal movements are brief and are made relatively automatically. That is, they require little or none of our conscious attention but are controlled at motor centers specifically concerned with the integration of sensory signals with motor commands. There is a long-standing and unresolved controversy concerning the role of sensory information in the initial creation and continuous adjustment of descending central command signals as they act at segmental levels. The sensory role for controlling fast movements may be negligible. This would be true if commands are programmed entirely in advance and executed without reflex modulation (Gottlieb et al. 1989). At the opposite extreme, the signal that excites the motoneuron pools has been described as the difference between a central reference command and sensory signals from muscle length and velocity receptors (Feldman et al. 1990; Latash and Gottlieb 1991). It is beyond the scope of this work to elaborate the issues involved in this controversy. Whatever the ultimate resolution, however, it is necessary to properly use (or prevent the use of) sensory signals in the ongoing control of a movement. An excessive contribution from sensory receptors can only impair the quality and skill of our movements.

Hyper-reflexia to imposed phasic stretch (Lance's "velocity-dependent increase in tonic stretch reflexes") is a hallmark of what is normally meant by spasticity. One interpretation of Fig. 3 is that, from the perspective of the conditions necessary to make a normal voluntary movement, normal reflex excitability of the relaxed limb is already excessive. Movements of the limbs, especially the upper limbs, can reach peak velocities of several hundreds of degrees per second.

In this light, the responses of the patient shown in Fig. 5 are instructive. He did not evoke antagonist EMG opposition from SOL during voluntary dorsiflexion preoperatively, but his movements then were extremely slow and may not have reached the velocity threshold adequate to evoke the reflex. Surgery clearly improved his motor performance in many ways. His voluntary movements became brisker, and reflex opposition was expressed. This velocity-dependent opposition is similar to our previous findings in another patient population (Corcos et al. 1986).

The characterization of this patient as "hyper-reflexic" requires further consideration, however. Hyper-reflexia has been described both as a lowering of the threshold at which the reflex is elicited (Katz and Rymer 1989), or as an increase in the gain – the slope of the relationship between displacement angle and EMG (Thilman et al. 1991). The former is a translation of the stimulus-response curve and the latter is a rotation. The data shown in Fig. 6 suggest that the situation is more complex. Preoperatively, the latency of the reflex was about 50 ms, which we may presume is about the normal, monosynaptic latency for this patient. The postoperative latency was closer to 80 ms. Since nerve conduction velocity had not changed, this indicates that the threshold velocity at which the reflex was initiated became higher. The angular velocity 40 ms before the first EMG was about 150°/s. However, 40 ms prior to the second SOL burst (see Fig. 6A), the velocity is only about 50°/s and prior to the third burst it is close to zero.

These observations do not let us think of the reflex as having a fixed velocity threshold. It is possible that the fall in the velocity threshold from beat to beat is due to the change in joint angle (dorsiflexion) produced by the step of torque. Considering only the first SOL burst, the reflex is not obviously abnormally excitable if the velocity threshold is the measure of excitability. The presence of clonus is sufficient to demonstrate reflex "hyperexcitability," but these data suggest both an elevated threshold and an elevated gain (see also Hultborn, this volume).

We can speculate that the changes in apparent reflex excitability are mediated by interneuronal mechanisms. For example, Delwaide and Oliver (1988) showed that the Ib inhibitory effect is functionally absent in spasticity. The observed differences between normal H-reflexes and stretch reflexes during voluntary movement might be explained by the concurrent production of both Ia and Ib inputs only by the latter stimulus. In spastic subjects, the failure of Ib inhibition allows exaggeration of the reflex response, even if the Ia pathway remains at its normal level of excitability. The failure of higher motor control centers to modulate and amplify inhibition of this pathway during voluntary movements could produce the observed reflex opposition for movements fast enough to exceed the reflex threshold. It is possible that normal or even elevated velocity thresholds could co-exist with an ineffective interneuronal inhibitory mechanism that would increase the slope of the stimulus-response relationship. In this case, slower movements would not produce reflex opposition, while faster movements, either passive or active (which a patient might be unable to make because of the original upper motoneuron injury), would reveal a significant hyperexcitability.

We conclude that stretch reflex hyperexcitability can directly or indirectly play a causal role in movement impairment. Hyper-reflexia will be directly expressed only if movements are sufficiently fast. The threshold need not be continuously lower than nomal. Our understanding of the role of the stretch reflex, both in normal and in injured nervous systems, must be considered in terms of continually modulated thresholds and gains. The failure of the nervous system to perform this modulation, rather than a static hyperexcitable state of the reflex arc, is one key to understanding spastic disordered movement.

Acknowledgments. This work was supported in part by NIH grants RO1-AR33189, RO1-NS28176 and by the Department of Veterans Affairs Research and Development Service.

References

Corcos DM, Gottlieb GL, Penn RD, Myklebust BM, Agarwal GC (1986) Movement deficits caused by hyperexcitable stretch reflexes in spastic humans. Brain 109:1043–1058

Davenport HW (1982) Physiology 1850–1923, the view from Michigan. Physologist [Suppl] 25:1–96

Delwaide PJ, Oliver E (1988) Short-latency autogenic inhibition (IB inhibition) in human spasticity. J Neurol Neurosurg Psychiatry 51:1546–1550

Desmedt JE (1978) Progress in clinical neurophysiology, vol 4. Cerebral motor control in man: long loop mechanisms. Karger, Basel

Feldman AG, Adamovitch SV, Ostry DJ, Flanagan JR (1990) The origin of electromyograms – explanations based on the equilbrium point hypothesis. In: Winters J, Woo S (eds) Multiple muscle systems: biomechanics and movement organization. Springer, Berlin Heidelberg New York, pp 195–213

Gordon J, Ghez C (1991) Muscle receptors and spinal reflexes. In: Kandel E (ed) Principles in neuroscience. Elsevier, New York, pp 564–580

Gottlieb GL, Agarwal GC (1971) The effect of initial conditions on the Hoffmann reflex. J Neurol Neurosurg Psychiatry 34:226–230

Gottlieb GL, Agarwal GC (1979) Response to sudden torques about the ankle in man. I. Myotatic reflex. J Neurophysiol 42:91–106

Gottlieb GL, Agarwal GC (1980) Response to sudden torques about the ankle in man. III. Suppression of stretch-evoked responses during phasic contraction. J Neurophysiol 44: 233–246

Gottlieb GL, Agarwal GC, Stark L (1970) Interactions between voluntary and postural mechanisms of the human motor system. J Neurophysiol 33(3):365–381

Gottlieb GL, Myklebust BM, Penn RD, Agarwal GC (1982) Reciprocal excitation of muscle antagonists by the primary afferent pathway. Exp Brain Res 46:454–456

Gottlieb GL, Corcos DM, Agarwal GC (1989) Strategies for the control of single mechanical degree of freedom voluntary movements. Behav Brain Sci 12(2):189–210

Houk JC, Rymer WZ (1981) Chap 8: Neural control of muscle length and tension. In: Brooks V (ed) Handbook of physiology, section 1: the nervous system. American Physiological Society, Bethesda, pp 257–324

Katz RT, Rymer WZ (1989) Spastic hypertonia: mechanisms and measurement. Arch Phys Med Rehabil 70:144–155

Lance JW (1980) Symposium Synopsis. In: Feldman RG, Young RR, Koella WP (eds) Spasticity: disordered motor control. Symposia Specialists, Miami, pp 485–500

Lance JW (1990) What is spasticity? Lancet 335(1):606

Latash ML, Gottlieb GL (1991) An equilibrium-point model for fast single-joint movement. I. Emergence of strategy-dependent EMG patterns. J Mot Beh 23(3):168–178

Lestienne F (1979) Effects of inertial load and velocity on the braking process of voluntary limb movements. Exp Brain Res 35:407–418

Lombard WP (1887) The variations in the normal knee-jerk and their relations to the activity of the central nervous system. Am J Psychol 1:1–67

Lombard WP (1889) The nature of the knee-jerk. J Physiol (Lond) 10:122–148

Myklebust BM, Gottlieb GL (1993) Reciprocal excitation and reflex irradiation of short-latency reflexes in the healthy neonate. Child Dev 64:1036–1045

Myklebust BM, Gottlieb GL, Penn RD, Agarwal GC (1982) Reciprocal excitation of antagonist muscles as a differentiating feature in spasticity. Ann Neurol 4:367–374

Myklebust BM, Cusick JF, Myklebust JB, Gottlieb GL (1986a) Modulation of reciprocal excitation in cervical spondylotic myelopathy. In: 16th annual meeting of the Society for Neurosciences, 12, no 186.16, Washington DC, p 602

Myklebust BM, Gottlieb GL, Agarwal GC (1986b) Stretch reflexes of the normal infants. Dev Med Child Neurol 28:440–449

O'Sullivan JC, Eyre JA, Miller S (1991) Radiation of phasic stretch reflex in biceps brachii to muscles of the arm in man and its restriction during developement. J Physiol (Lond) 439:529–543

Thilman AF, Fellows SJ, Garms E (1991) The mechanism of spastic muscle hypertonus. Brain 114:233–244

Discussion Summary

M. Hallett

Clinical Pattern of Strength

Following Dr. Gandevia's presentation of patterns of strength after stroke, the concern of a number of the participants was whether the pattern might change with time. Certainly, the magnitude of strength changes with time. Additionally, with the passage of time, there might be confounding changes such as patterns of disuse and deconditioning. Dr. Gandevia noted that while patients in his studies had been seen from 2 weeks to 2 years following their strokes, there was no clear difference in the pattern of weakness at the different intervals. Patients were not studied serially, and, while this might be worthwhile to do, major changes of pattern would not be expected. Deconditioning probably does not play a significant role, since the pattern was seen as early as 2 weeks. Moreover, the weakness ipsilateral to the brain lesion was more prominent in hemiplegic than in hemiparetic patients, and the hemiplegic patients would be likely to be using their ipsilateral extremities more.

There was considerable interest in the finding about weakness in the extremities ipsilateral to the stroke. Dr. Gandevia noted that patients, in general, did not complain about this weakness. The one exception was sometimes the shoulder. What happens to ipsilateral weakness in the circumstance of full recovery of the contralateral side is not known.

Contralateral Motor Evoked Potentials

Questions were also raised for Dr. Benecke regarding the possible changes of motor evoked potentials (MEPs) with time from the onset of a stroke. Dr. Miller recounted his own experience that patients with delayed or absent MEPs soon after a stroke may well show improved responses in subsequent months. Dr. Benecke agreed that this might well be seen and

National Institute of Neurological Disorders and Stroke, National Institute of Health, Bethesda, MD 20892, USA

Thilmann et al. (Eds.)
Spasticity: Mechanisms and Management

that it represents an important phenomenon that deserves further study. Such changes may be useful in regard to the critical issue of recovery. His study, on the other hand, was aimed at looking at the relationship at specific times between MEP results and clinical measurements. Whether there might be changes in this relationship at different stages of recovery would also be worth investigation. MEP studies might be useful for analysis of recovery, and this has received only limited attention so far.

Ipsilateral Motor Evoked Potentials

While Dr. Benecke noted the presence of ipsilateral MEPs in patients with hemispherectomy, both he and Dr. Ashby reported evidence against ipsilateral MEPs being present in patients after stroke. This was difficult to understand, given Dr. Gandevia's clinical observation about loss of strength in ipsilateral muscles. Also, the observations seemed incompatible with Dr. Hallett's presentation in an earlier session, where he noted ipsilateral MEPs even in normal subjects and gave strong evidence for involvement of ipsilateral pathways in stroke recovery. Dr. Hallett noted that Dr. Benecke had studied all his patients at rest, and that, at least for the normal ipsilateral MEPs, some background contraction was needed. Dr. Hallett also pointed out that ipsilateral responses may require much higher stimulation intensities than contralateral responses, and, although Dr. Ashby used some higher intensities than those needed for contralateral responses, they may not have been high enough. Dr. Ashby pointed out that if this were true, then the population of cortical neurons would have to be different for the ipsilateral responses.

There was some discussion about whether ipsilateral responses could be expected equally in all muscles, or whether there might be a proximal or distal preference. From an anatomical point of view, proximal muscles would be more likely, and Dr. Hallett confirmed that in his studies ipsilateral MEPs had been seen mainly in proximal muscles.

Dr. Lemon raised the issue that muscle activation might be accomplished differently in different tasks. He wondered whether bilateral contractions, for example, might bring out ipsilateral projections better.

Role of the Flexor Reflex in Voluntary Movement

Dr. Gottlieb asked Dr. Rymer to expand on his views about how abnormal flexor reflexes might play a role in disordered voluntary movement. Dr. Rymer replied that a continuous running of the reflex pathways probably provides a background upon which voluntary movement is superimposed.

Following along this point, a number of participants wondered what would happen to the reflexes if they were studied during voluntary movement rather than at rest. Dr. Rymer said that the patterns of flexor reflexes were similar at rest and with background force. On the other hand, he has not investigated how they might look during dynamic voluntary movement.

Dr. Burke asked Dr. Rymer about the relevance of nociceptive stimuli for ordinary voluntary movement; he noted that sensory input was certainly important, but that nociceptive influences were not common. Dr. Rymer replied that the threshold for the flexor reflex was markedly reduced on the abnormal side of his patients and that, for this reason, flexor reflexes could be produced with stimuli that were not nociceptive.

Discussion Summary

J. Eyre[1] and S. Miller[2]

In the study by S. Fellows, C. Kaus, H. Ross and A.F. Thilmann, the experimental paradigm constrained the elbow so that flexion and extension movements could be made with or without load and without the influence of gravity. In response to technical questions, Fellows replied that subjects understood the instructions and the experimental paradigm and that they made smooth movements in 10–12 velocity steps to the point of moving as fast as possible. The EMG did not distinguish brachioradialis or brachialis from biceps brachii. No attempt was made to investigate alternating movements, but Fellows commented that unidirectional movements about the elbow include a component of reaching or retrieving.

In the condition without loading and in the absence of gravity, there would be less afferent information, particularly from Ib afferents, to shape the outputs of the motoneuronal pools of biceps and triceps. Fellows agreed, but did not think that this had influenced the results. He commented, in fact, that the movements and EMG activities in the stroke patients were remarkably close to those of normal subjects, and that the abnormal activity in biceps had very little effect on bursts of EMG from triceps. He then added that in voluntary isometric contraction, a set of experiments not reported in the present paper, there were no abnormal levels of antagonist activity.

In reply to the question whether stroke patients' ability to make ballistic movements correlates with assessments of activities of daily living, Fellows stated that this had not been studied, but that it would be of interest with regard to rehabilitation regimes.

The first question addressed to W. Berger was whether comparison had been made of different patterns of gait between adults with infarction of the middle cerebral artery and babies with infarction. Berger replied that her infant subjects were largely diplegic, with a few hemiplegic, but that the nature of the brain damage was mostly unknown. In a few the lesion was intracerebral haemorrhage, and these subjects had the same gait pattern as

[1] Dept. of Child Health, The Medical School, Framlington Place, Newcastle upon Tyne ME2 4HH, England, UK
[2] Division of Clinical Neuroscience, The Medical School, The University of Newcastle upon Tyne, Newcastle upon Tyne NE2 4HH, England, UK

Thilmann et al. (Eds.)
Spasticity: Mechanisms and Management

children with diplegia. In adult subjects the gait patterns differed little whether the lesion was cervical or cerebral.

It was suggested that it would be extremely rare to observe an adult with the equivalent of periventricular leucomalacia, a common cause of hemiparesis in neonates. Berger replied that a young adult of 18 years with the hemiplegic form of cerebral palsy had the same infantile pattern, and that his gait had clearly not undergone normal maturation.

The next part of the discussion focused on whether heel strike or toe strike might be responsible for the differences of reflex excitability; heel strike evoking little stretch of triceps surae, but toe strike producing stretch. Berger commented that they had searched for children who walk voluntarily on tiptoe. At the age of 4 years they did not have exaggerated stretch reflex excitability.

Asked whether the spastic patients exhibited co-contraction, Berger replied that she had not observed it in adults, but noted that at the slow rate of stepping (1–2 km/h) this might be common.

Several participants questioned the timing of the reflexes and the EMG burst of the gait patterns. Berger stated that the interval between the monosynaptic reflex and the resumption of muscle activity was not prolonged in spastic children as compared with adults, but that it had not been systematically studied. She did concede that the intervals in the changes between agonist and antagonist activity were longer in children that in adults, although the inter-subject variability was large.

Miller commented that in the cat the silent period of a muscle during gait is due to the motoneurones receiving strong reciprocal inhibition, and that a knowledge of the maturation of this inhibition during human development would provide a further basis on which to interpret the results presented. Berger replied that inhibitory transmission had not been studied.

In the first part of the discussion of the paper presented by G. Gottlieb, Dietz questioned whether the paired EMG bursts in agonist and then antagonist muscles constitute a reflex or represent part of the triphasic EMG bursts seen in the performance of fast ballistic movements. Gottlieb replied that there was some evidence to support the view that the EMG sequence is in fact programmed and not a reflex, but that further work was required.

The second part of the discussion concerned the nature of irradiation of reflexes observed in neonates which become restricted during development to adulthood. On the one hand, Burke cited evidence that a tap to a muscle or its tendon generates travelling waves in the limb which can excite spindles in antagonist muscles, for example, soleus to tibialis anterior. Gottlieb replied that H reflex testing in spastic patients circumvents the problem of mechanical spread of the stimulus, and that the results provide confirmation of reflex irradiation.

Attention was drawn to Eyre and Miller's studies on babies and children similar to those in Gottlieb's paper, in which excitatory radiated reflexes were recorded in muscles of the shoulder, arm and hand, involving all the segments of the brachial plexus (O'Sullivan et al., J Physiology 1991, 439:529–543). Adequate controls of mechanical spread had been performed. From the latencies of the responses it had been concluded that at least the early part of the radiated responses had to have an intraspinal neural substrate. In response to criticisms of the findings both of Gottlieb and of Eyre and Miller, it was noted that in experiments in foetal, newborn and adult animals non-homonymous and non-synergistic monosynaptic projections between primary afferents and α-motoneurones had been reported (Eccles et al., J Physiology 1957, 136:527–546 and 1957, 137:22–50; Eccles et al., J Physiology 1963, 165:392–402; Saito, J Physiology 1979, 294:581–594).

In the general discussion of Session II, Rothwell posed two questions on which the ensuing discussion took place. The first was:

1. How do abnormal reflexes contribute to the movement disorder of spasticity?

In the papers presented in this session some confusion existed as to whether stretch reflexes in an antagonist muscle prevent the agonist from moving the limb, whereas in normal movement, such as walking, the movements are not held up by abnormal stretch reflexes. Thus, Rothwell continued, should reflexes be studied in the relaxed state and the conclusion drawn that the reflexes so measured contribute to the disorders of movement in patients with spasticity?

Rymer answered that spasticity is a syndrome and not a lesion, and that no single simple answer can be expected. He suggested that spasticity, as it is defined, involves different patients with different lesions, and the answers will therefore be different. Others supported this view. Rothwell thought this an easy way out of the problem and asked which particular lesions might result in a movement disorder caused by abnormal reflexes. Cerebral palsy and some spinal cord compressions were given as possible candidates. However, Rymer emphasized that, despite being able to characterise small changes in stretch reflexes, we can treat a subject with baclofen or phenolic blockade and reduce the severity of the spastic state, but that movement control may not improve. Thilmann commented that we still have the responsibility to treat patients with spasticity, and perhaps we need to select more variables than a single reflex to assess the nature of the spasticity and to guide intervention.

The particular task being performed carries important consequences for the performance of movement in the spastic limb. A case was cited in which, in a ballistic task without preloading, a subject must develop a greater agonist burst to achieve a velocity in the presence of antagonist activity. Gravity may also play a role in helping to withdraw the unwanted antagonist activity. Benecke commented that patients with spasticity often avoid unwanted antagonist activity by making movements of the limbs slowly. Finally, Burke suggested that normal subjects modulate reflex activity according to the task and can suppress that activity. In spastic subjects he suggested that the manner in which reflex disturbance contributes to the motor disorder is that the reflexes can no longer be modulated to suit the task. By administering drugs to reduce involuntary reflex activity, it would not be surprising that function is not improved. This does not mean that a disturbance of the reflex does not contribute to the movement disorder.

Up to this point the discussion had concentrated on adults with spasticity. Eyre commented that the characteristics of spasticity, described for adult subjects in this workshop, are in fact typical of the responses of the normal newborn baby, with hyperreflexia, radiated excitatory responses, co-contraction, a high threshold for activating the corticospinal pathway with abnormal responses and abnormal durations, and upgoing extensor plantar responses. But the newborn do *not* have spasticity. This raises important questions. Even babies with very extensive lesions in the central nervous system, who subsequently develop spasticity at 6 months do not have it at birth. Thus you can measure all these characteristics in the newborn baby, exactly as you can in the spastic adult, but the babies are not spastic.

2. *What is the role of ipsilateral (undamaged) cortex in recovery from stroke?*

Hallett commented that PET and cerebral blood flow studies indicate that the undamaged, ipsilateral hemisphere has a higher cerebral blood flow than that in normal subjects. The observations are interesting but need rigorous study. Another commentator noted that if the right wrist is moved, the left shoulder is stabilised, and one should not be surprised to find ipsilateral cortical projections, although it does not follow that ipsilateral projections are available for, or account for, the recovery of the contralateral paretic side.

Lemon raised the observations of Ingram and Farmer of mirror movements in patients with hemiplegic spasticity. One these subjects was a remarkable lady with very precise branch connections from one cortex to the same groups of spinal motoneurones on each side. She could not move one finger without moving the same finger of the contralateral hand. But what did she do for a living? She was a typist – and she could suppress the mirror movements when typing!

Part III
Spinal Pathophysiology: Animal Models

Regulated Properties of Motoneurons and Primary Afferents: New Aspects on Possible Spinal Mechanisms Underlying Spasticity

J. Nielsen and H. Hultborn

Introduction

The classical view of the mammalian spinal motoneuron emerged from the laboratories of Eccles and Granit during the 1950s and 1960s. They held the view that the cell membrane in areas of synaptic contact was essentially passive, allowing a linear summation of synaptic inputs at the spike-initiating region. The stronger the net excitatory current, the higher the discharge rate. This situation would lead to a rather simple input-output relationship. It is now known that there are several active membrane properties that contribute to the electro-responsiveness of motoneurons. In addition, a great number of reports have appeared during the past 5 years showing that such properties are often strongly transmitter modulated. Taken together, this implies not only that motoneurons actively take part in information processing, but also that the input-output relation of *individual* motoneurons seems to be a flexible variable that may be adapted according to the external requirements.

In order to describe the transmission across a *pool* of motoneurons it is also necessary to consider the distribution of intrinsic properties and external synaptic input among the individual neurons. These factors will determine both recruitment *order* and recruitment *gain* of the motoneurons during a graded contraction.

When the input-output relation is extended to include the presynaptic elements, i.e., the Ia afferents in the case of the monosynaptic part of the stretch reflex, it is necessary to take into account presynaptic inhibition of their terminals as well as the blend of facilitation/depression of transmitter release that follows activity of these afferents.

This chapter covers some recent work in this field, with particular emphasis on the possible contribution of these mechanisms to the symptoms of spasticity. Other reviews from our laboratories covering similar topics have recently been published (Eken et al. 1989; Kiehn 1991; Hultborn and Illert 1991; Brownstone and Hultborn 1992; Hultborn and Kiehn 1992).

Division of Neurophysiology, Department of Medical Physiology, The Panum Institute, Blegdamsvej 3, DK-2200 Copenhagen N, Denmark

Thilmann et al. (Eds.)
Spasticity: Mechanisms and Management

The Post-Spike Afterhyperpolarization

The dependence of firing frequency on somatic current, mimicked by current injected through an intracellular microelectrode, is known as the frequency-current relationship. The post-spike afterhyperpolarization (AHP) is an essential mechanism in the transduction of net current to spike trains of different frequencies. For this reason, much attention has been given to the AHP, which in motoneurons is generated by a Ca^{++}-dependent K^+ conductance (Krnjevic et al. 1978). A reduction of this AHP current, by blocking either Ca^{++} entry (by Mn^{++}; Walton and Fulton 1986) or the AHP K^+ conductance (by apamin; Hounsgaard and Mintz 1988), increases the steepness of the f–I relation. Administration of serotonin (5-hydroxytryptamine; 5-HT) effectively reduces the slow AHP in the lamprey (Wallén et al. 1989), the turtle (Hounsgaard and Kiehn 1989), the rat (Wu et al. 1991), and the cat (White and Fung 1989). 5-HT appears to decrease the AHP conductance "directly" and not secondary to a decreased Ca^{++} availability (Wallén et al. 1989; Hounsgaard and Kiehn 1989). With the exogenous administration used in the above studies it was not possible to obtain a physiological time course for 5-HT action. This aspect has been addressed by electrical stimulation of the 5-HT-containing raphe-spinal projection in cats, as illustrated in Fig. 1 (R.M. Brownstone, H. Hultborn and T. Toth, unpublished data). The AHP reduction is seen within 100 ms from the onset of stimulation (mainly conduction delay), and the effect lasts less than 1 s after cessation of stimulation. Concomitant with the AHP reduction, the slope of the frequency-current relation becomes steeper (not illustrated). This mechanism therefore opens up the possibility for a rapid control of the gain of the motor neuron's input-output relation.

Although the AHP is a transmitter-regulated property, which may be adjusted following acute spinal lesions secondary to an interrupted monoaminergic innervation from the brain stem, there are still no experimental findings that demonstrate slow adaptive changes following spinal cord lesions that may correlate to the development of spasticity. However, that possibility should not be excluded. It is known from experiments in which fast and slow muscles are cross-innervated that the intrinsic properties (also the AHP) are "appropriately" adapted to the newly innervated muscle (Foehring and Munson 1990).

Plateau Potentials

Many neurons in the central neurvous system can generate plateau potentials, i.e., prolonged maintained depolarized states. These are initiated by transient depolarizations, e.g., a short barrage of excitatory postsynaptic potentials

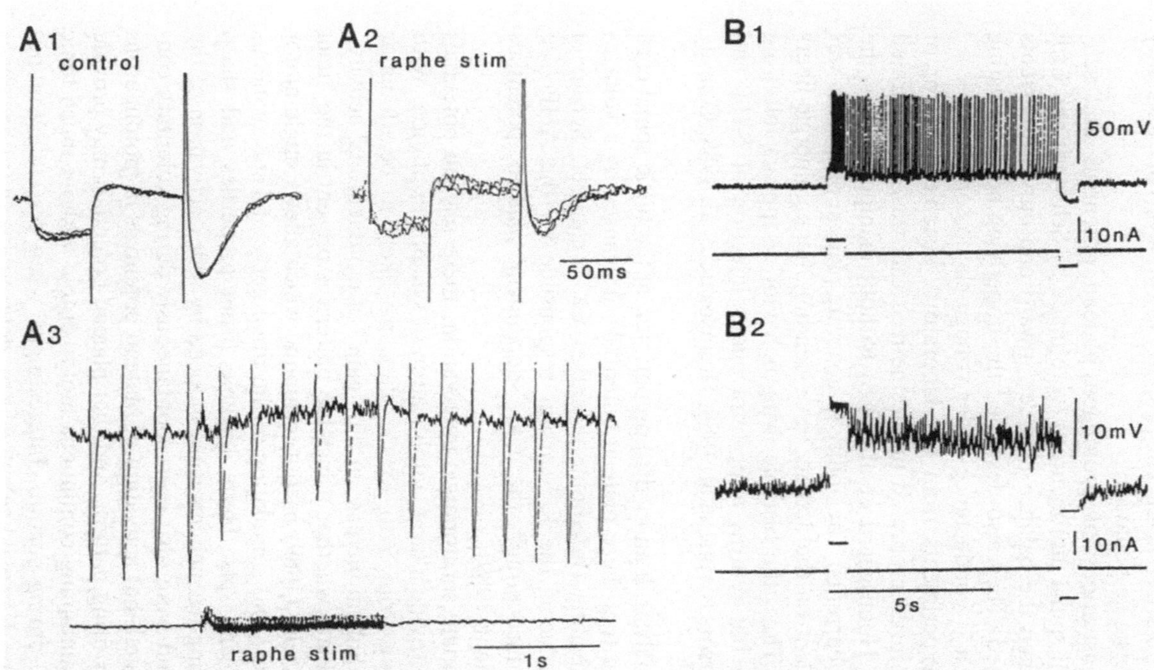

Fig. 1. Depression of the afterhyperpolarization (*A*) and presence of plateau potentials (*B*) in motoneurons. The motoneuron in A was recorded under nembutal anesthesia. Intracellular current pulses were used to measure input resistance and to evoke action potentials (and the following afterhyperpolarization) in the control situation (*A1*) and during tetanic (20 Hz) stimulation of the caudal raphe nuclei (*A2*). In *A3* only spikes were elicited at a frequency of 5 Hz. The record illustrates the fast onset and disappearence of depression of the AHPs in relation to a short train of stimuli to the raphe nuclei. The motoneuron in *B* was recorded in an unanesthetized decerebrate preparation. In *B1* a brief depolarizing current pulse (lower trace) triggered a self-sustained firing which was terminated by a hyperpolarizing pulse. When the spikes were inactivated (in *B2*) similar current pulses recruited and terminated an all-or-none plateau potential. (A records by Hultborn and Toth, published in Brownstone and Hultborn 1992; B modified from Hounsgaard et al. 1988)

(EPSPs) or an intracellular current injection, and may remain for a long period of time before they terminate spontaneously or are actively turned off by a brief inhibition. The plateau potential therefore allows for two quasi-stable states of activity, one around "resting" membrane potential and another at a more depolarized level. The property of two levels of excitability is referred to as a "bistable behavior".

The first demonstrations of prolonged plateau potentials and self-sustained firing in cat spinal motoneurons were made by Schwindt and Crill (1980a,b,c, 1984) when the balance between inward and outward currents was artifically changed. They also described the underlying noninactivating voltage-dependent Ca^{++} conductance under voltage-clamp conditions.

In a series of experiments from our laboratory, bistable behavior of spinal motoneurons in decerebrated unanesthetized cats was demonstrated (Crone et al. 1988; Hounsgaard et al. 1988) without manipulation of the inward and outward currents. The ability to generate plateau properties was shown to be dependent on the activity in descending monoaminergic fibers in this preparation. Thus, the bistable firing behavior disappears following a spinal transection but returns after intravenous injection of 5-HT and noradrenaline precursors (Crone et al. 1988; Hounsgaard et al. 1988; Conway et al. 1988).

A subsequent analysis in an in vitro preparation of the turtle spinal cord led to the conclusion that these plateau potentials are caused by a voltage-dependent noninactivating (nifedipine-sensitive) Ca^{++} channel. Serotonin acts primarily by reducing the K^{+} current responsible for the AHP (see above), thereby "uncovering" the Ca^{++}-dependent plateau potentials (Hounsgaard and Kiehn 1989).

With an appropriate serotonergic innervation, these plateau potentials can be expected to provide a large amplification of synaptic excitation. With an amplitude ranging from 5 to 15 mV, they are likely to be of major importance in providing an increase in the "gain" of motoneuronal activity.

In order to evaluate whether a bistable pattern is present in the intact animal, Eken and Kiehn (1989) used a technique which allows single motor unit recording (soleus) and stimulation of the tibial nerve and subcutaneous tissue in freely moving rats. Here, bistable firing properties and sharp discontinuities in firing frequencies were induced by reflex activation of the motoneuron pool and were also seen spontaneously during voluntary contraction. It is suggested that activation of plateau potentials in motoneurons may account for this firing pattern, and that plateau potentials may provide a mechanism for maintaining continuous motor output without much tonic synaptic input, e.g., during posture. Interestingly, when descending monoaminergic fibers were destroyed selectively by neurotoxins (intrathecally at the lumbar spinal cord level) the total tonic activity in the soleus muscle decreased markedly with a simultaneous increase in more burstlike events of short duration (Kiehn et al. 1992). This pattern is aparently similar to the electromyographic pattern seen in 8- to 15-day-old rats (Eken et al. 1990).

The immature pattern is replaced by an adult pattern at the age of about 20 days, i.e., at the time when the motoneurons receive an adult pattern of serotonergic innervation (Tanaka et al. 1992), and it is tempting to suggest a causal relationship.

To what extent can plateau potentials be responsible for the increased reflex activity following chronic lesions? There is strong evidence that the tonic stretch reflexes seen in the decerebrate cat are dependent on plateau potentials in the activated motoneurons (Crone et al. 1988), but that may be a very particular case without much significance for the spasticity which develops slowly after central lesions in man. In this respect, it is important that a spastic syndrome *can* be reproduced following complete chronic (3–4 weeks) spinal transection in cats (Bailey et al. 1980 and unpublished observations from our laboratory). In such preparations short-term synaptic excitation by vibration or trains of group-I stimuli can result in a long-lasting motor discharge (Crone et al. 1983 and unpublished) reminiscent of the response in the decerebrate preparation (Crone et al. 1988). This may suggest that the exaggerated reflexes seen in the chronic spinal state are related to the development of plateau potentials in motoneurons or interneurons.

Shortly after the spinal transection only phasic stretch reflexes (tendon taps) were evoked, but after about 2 weeks more tonic stretch reflexes were regularly recorded. This is exemplified in Fig. 2A, which shows the triceps surae EMG during a manual flexion of the ankle joint. The lower part of the figure shows intracellular responses in a motoneuron of the same cat the following day. Figure 2B and C illustrate that the firing outlasts the stimulating current pulse. With slightly stronger depolarizing bias current the firing after the pulse was maintained (see in Eken et al. 1989). These findings in the chronic spinal state are in sharp contrast to the lack of any signs of bistability in the acute spinal cat (Hounsgaard et al. 1988). However, the percentage of clearly "bistable" motoneurons in the chronic spinal state is small in comparison with the decerebrate preparation. Proof of the possible causal relation between development of stretch reflexes and plateau potentials in motoneurons must await further experiments.

In the chronic spinal preparations monoaminergic innervation of the spinal cord from the brain stem is lacking (Andén et al. 1964), and the seeming reappearance of plateau potentials under such circumstances may then be attributed to slow changes at the level of intracellular messenger systems.

Baldissera et al. (1991) recently described some patients in whom muscle cramps were triggered by volleys in Ia afferents and terminated by trains of antidromic volleys of the innervating motoneurons (causing Renshaw inhibition and AHP). They then proposed that the generation of the motoneuronal activity underlying the cramp was due to activation of plateau potentials in the motoneurons. This observation should be followed up in further studies.

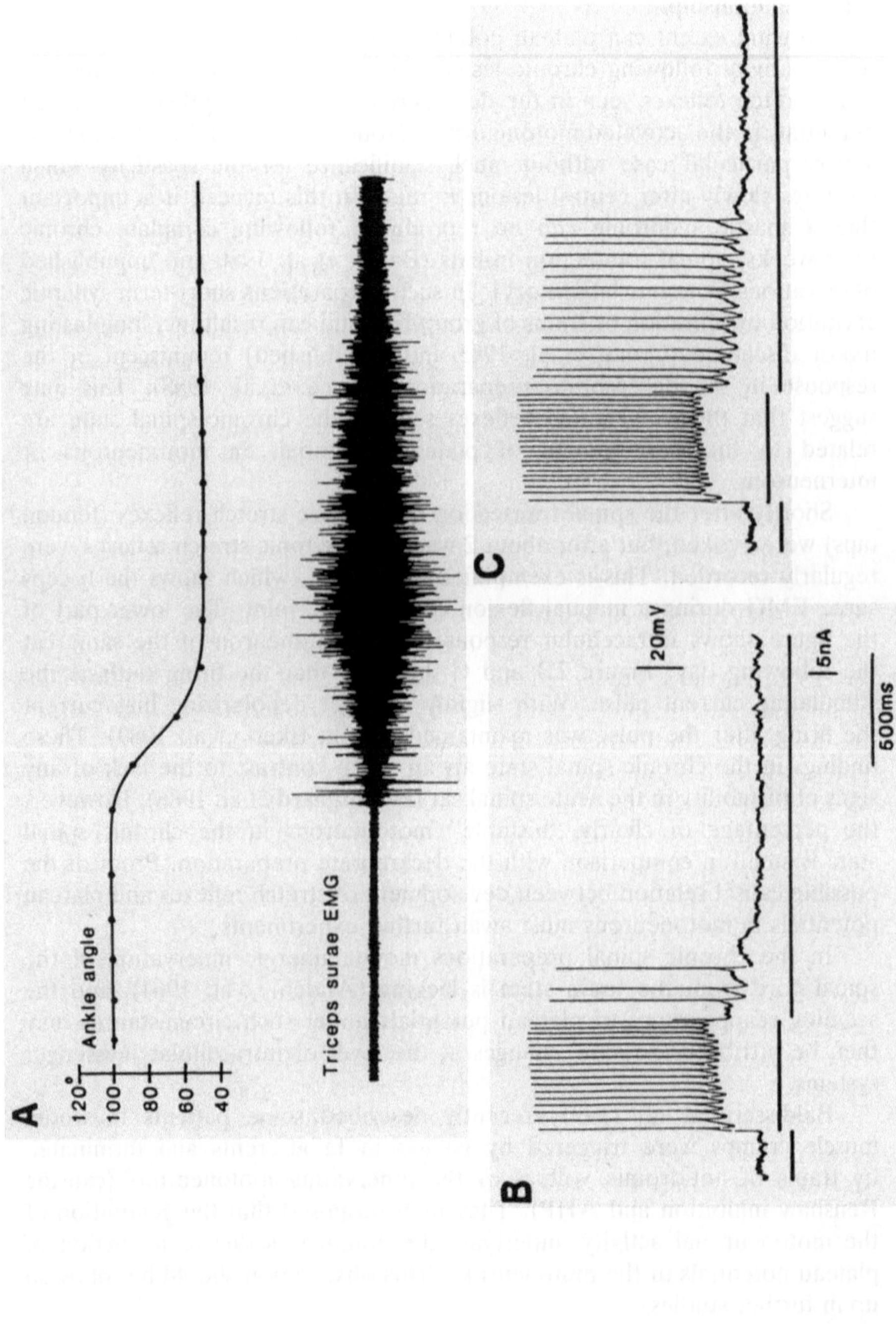
A
120°
100
80
60
40
Ankle angle
Triceps surae EMG
B
20mV
5nA
500ms
C

Renshaw Inhibition

The mechanisms discussed in the previous sections act within individual motoneurons. Renshaw inhibition, however, is a negative feedback system which is driven by the integrated activity of a motoneuronal pool and projects back to most motoneurons in the same pool. It has been argued that recurent inhibition might change the input/output relation across the motor pool, which is otherwise largely determined by the intrinsic properties of the motoneurons (cf. previous sections). With the recurrent pathway in operation one would expect a smaller total output for any given input. As Renshaw cells receive both strong facilitatory and inhibitory control from many motor centers, it was proposed that the central nervous system uses the recurrent pathway to regulate the gain at the output stage (Hultborn et al. 1979). A "low-gain" condition (during facilitation of Renshaw cells) would allow force-generating circuits to play over a considerable part of their working range while causing only small changes in muscular force, whereas a "high-gain" state (during inhibition of Renshaw cells) would allow the central command to generate larger forces for a given drive. Such a variable gain control at a motoneuronal level may thus optimize the resolution of the motor output. Experiments in human beings (Hultborn and Pierrot-Deseilligny 1979) have indeed suggested that the excitability of Renshaw cells changes in parallel to the strength of graded voluntary contractions as predicted from the variable gain hypothesis: it decreases during strong contractions (increasing the motoneuronal output in response to a given input) and increases during weak ones (decreasing the motoneuronal output).

As a reduction of Renshaw inhibition certainly would cause an increased motor activity for a given input, it has often been proposed that recurrent inhibition may be reduced in spasticity. However, using the conditioned H-reflex technique by Bussel and Pierrot-Deseilligny (1977), it has been shown (Katz and Pierrot-Deseilligny 1982) that the excitability of Renshaw cells is unchanged or enhanced at rest in spastic hemiplegia following brain lesions. It therefore is unlikely that changes in recurrent inhibition could contribute to the hyperexcitability of stretch reflexes in this condition. In contrast, in patients developing spasticity following spinal cord lesions, the same

---

Fig. 2A–C. Tonic stretch reflex motor activity and plateau potentials in motoneurons in the chronic spinal cat. **A** Tonic stretch reflex in the triceps surae muscle recorded in the "intact" cat 3 weeks following a complete spinal transection. *Lower trace* shows raw EMG (recorded with surface electrodes) during a manual flexion of the ankle joint as indicated in *Upper trace*. **B**, **C** responses of an extensor α-motoneuron to injection of rectangular current pulses in the same cat illustrated in **A**. *Upper traces* in **B** and **C** illustrate intracellular recordings, *lower traces* the injected current. (Results by Eken, Hultborn, Kiehn and Toth, previously published in Eken et al. 1989) technique suggested that the amount of recurrent inhibition is increased (Shefner et al. 1992). The supraspinal control of Renshaw cell excitability, which occurs during postural or voluntary contractions in normal subjects, was lacking in most patients with supraspinal lesions (Katz and Pierrot-Deseilligny 1982). This could be partly responsible for the difficulty spastic patients have in graduating contraction and in regulating reciprocal Ia inhibition (as the 'Ia inhibitory interneurons' are controlled by Renshaw cells). Recent experiments in our laboratory have indeed demonstrated that reciprocal Ia inhibition between ankle dorsiflexors and plantarflexors is depressed in spastic patients (C. Crone, unpublished data).

Change of "Recruitment Gain" in a Motoneuron Pool

The distribution of intrinsic properties (excitability, firing threshold) and of synaptic excitation/inhibition among the individual neurons of a motoneuron pool will determine not only the recruitment *order* during a gradually increasing excitatory drive, but also the ease of recruitment gradation, i.e., the recruitment *gain* (Kernell and Hultborn 1990). A motoneuron may be said to have an "intrinsic excitability," which could be defined as the amount of current required to discharge that motoneuron in the absence of background synaptic activity. At any given time, the ability of a motoneuron to generate an action potential will depend on both its intrinsic excitability and the sum of its synaptic inputs (background activity). Kernell and Hultborn (1990) pointed out that synaptic input systems with an uneven distribution may expand or compress the range of functional thresholds among the motoneurons and thereby change the ease with which a more equally distributed input (e.g., command system) will activate the pool. More precisely, the term "recruitment gain" was used to indicate the relation between the synaptic input (drive) to a motoneuron pool and the output from that pool (i.e., the number of active cells), and it predicted that a "biasing" input (with a different distribution to low- and high-threshold motor units than the "drive") would indeed change this gain.

To see whether such changes of the recruitment gain can also be demonstrated experimentally, Nielsen et al. (1990) investigated the effect of a conditioning stimulation of the caudal cutaneous sural nerve on the input/output relation of the medial gastrocnemius motoneuron pool in the cat. They used a graded stimulation of the dorsal roots as the test stimulation. The size of the EPSP in sampled medial gastrocnemius motoneurons was used as a measure of the input and the size of the monosynaptic reflex as a measure of the output of the pool. As the relation between the monosynaptic reflex and the EPSP became steeper following stimulation of the sural nerve, they were able to conclude that the input/output relation of the pool, i.e., the recruitment gain, had indeed increased. Further experiments have indicated that a similar increase in recruitment gain can occur in man (Nielsen and Kagamihara 1993). In this case sural nerve stimulation inhibited low-threshold motor units and facilitated high-threshold motor units in the tibialis anterior muscle, thereby compressing the range of functional thresholds in the pool. As a consequence, it was also possible to demonstrate a steeper input/output relation for the monosynaptic reflex. There is at present no direct experimental evidence suggesting a contribution of an increased recruitment gain to the enhanced reflexes seen in spasticity. Tang and Rymer (1981) have, however, demonstrated that spastic patients require a larger integrated EMG to obtain a given force than healthy subjects (or the healthy control side). This may be explained by an increased recruitment gain in the following way: if the threshold spacings between the motoneurons are reduced (increased recruitment gain), the rate increase of the previously recruited units is not sufficient to attain optimal force output from those units before new units are recruited. Thereby, an increased recruitment gain could contribute to the 'uneconomical' EMG/force ratio in the spastic patients. Obviously, this tentative interpretation requires further experimental analysis.

Modulation of Transmitter Release from Presynaptic Terminals

The physiological significance of primary afferent depolarization (PAD) was first recognized when it became associated with a depression of synaptic transmission (Frank and Fourtes 1957; Eccles et al. 1961). The present understanding is that the PAD is caused by axo-axonal $GABA_A$-ergic synapses, which produce presynaptic inhibition by reducing transmitter release (see Rudomin 1990). The Ia afferent terminals of any hindlimb muscle receive PAD from a very wide receptive field, but the main source is Ia and Ib afferents from ipsilateral flexor muscles (Eccles et al. 1962). The PAD and the related depression of the monosynaptic excitation of motoneurons attain a maximum at about 20 ms following the triggering stimulus and then decay within 150–300 ms.

In addition to this form of synaptic control of the terminals, transmitter release following an impulse is also dependent on the preceding impulse activity of the individual axon. Several independent mechanisms seem to contribute, and the net result is the sum of simultaneous facilitation, potentiation, and depression (for references in relation to Ia afferents, see Lev-Tov and Pinco 1992). In the present context, we will focus on the long-lasting depression that follows single pulses or trains of impulses.

It was originally observed that the size of the monosynaptic reflex in the cat decreases during repetitive stimulation, and a frequency-related depression of this reflex was described at stimulus intervals as long as 10–20 s (Lloyd and Wilson 1957). Similarly, the H-reflex in man decreases during repetitive stimulation (Magladery and McDougal 1950; Paillard 1955; Rothwell et al. 1986; Crone and Nielsen 1989) even with stimulus intervals of up to 10–15 s (Rothwell et al. 1986; Crone and Nielsen 1989). This long-lasting depression cannot be attributed to a long-lasting excitability decrease of the motoneurons themselves, since they can be easily activated from other sources (such as heteronomous Ia excitation in the cat, Beswick and Evanson 1957, or magnetic activation of the corticospinal pathway in man, unpublished findings from our laboratory). Direct intracellular recording from the motoneurons also shows the lack of effects in passive properties and membrane potential (Lev-Tov and Pinco 1992; unpublished findings from our laboratory). On the other hand, the depression is evoked by activation of the primary afferents even when the volley fails to evoke a motoneuronal discharge (Katz et al. 1977; Crone and Nielsen 1989). Studies in the neonatal rat (lev-Tov and Pinco 1992) and the adult cat (Ballegaard et al. 1991, and unpublished by the same authors) demonstrate that the decrease in monosynaptic reflex discharge is related to a decrease of the underlying monosynaptic Ia EPSPs. Since it is unlikely that the long-lasting decrease of the EPSP is due to receptor desensitization (see references in Lev-Tov and Pinco 1992), the most likely interpretation is that the preceding impulse activity causes a prolonged decrease of transmitter release with subsequent impulses. It was demonstrated that this type of presynaptic depression was unaffected by $GABA_A$-antagonists (Lev-Tov and Pinco 1992). Several features, such as (a) the very long duration, (b) limitation of depression to previously activated afferents, and (c) insensitivity to $GABA_A$-antagonists, thus differentiate this form of depression from the synaptic axo-axonal presynaptic inhibition described above.

Crone and Nielsen (1989) introduced the term "post-activation depression" to denote this mechanism, but terms such as homosynaptic depression, frequency depression, and habituation have been used by other groups. The post-activation depression has now been demonstrated following a tendon tap (also when subthreshold for reflex activation; Katz et al. 1977; Crone and Nielsen 1989), following electrical stimulation of the nerve (thus influencing the so-called recovery curves; Crone and Nielsen 1989), after voluntary contraction (Crone and Nielsen 1989; Schiepatti and Crenna 1984), and following passive lengthening of the muscle (Ballegaard et al. 1991, and unpublished). It is thus obvious that the depression of monosynaptic reflexes evoked by vibration of the muscle in which the reflex is recorded is also, at least partly, caused by post-activation depression, although it has generally been assumed to be exclusively caused by $GABA_A$-mediated presynaptic inhibition (Gillies et al. 1969; see, however, their Fig. 3 which demonstrates a time course of the depression of the reflex similar to that of post-activation depression; Delwaide 1973). The interpretation of a decreased vibratory inhibition in spastic patients (Delwaide 1973; Ashby et al. 1980) as being caused by a decreased presynaptic inhibition exclusively clearly also calls for some skepticism.

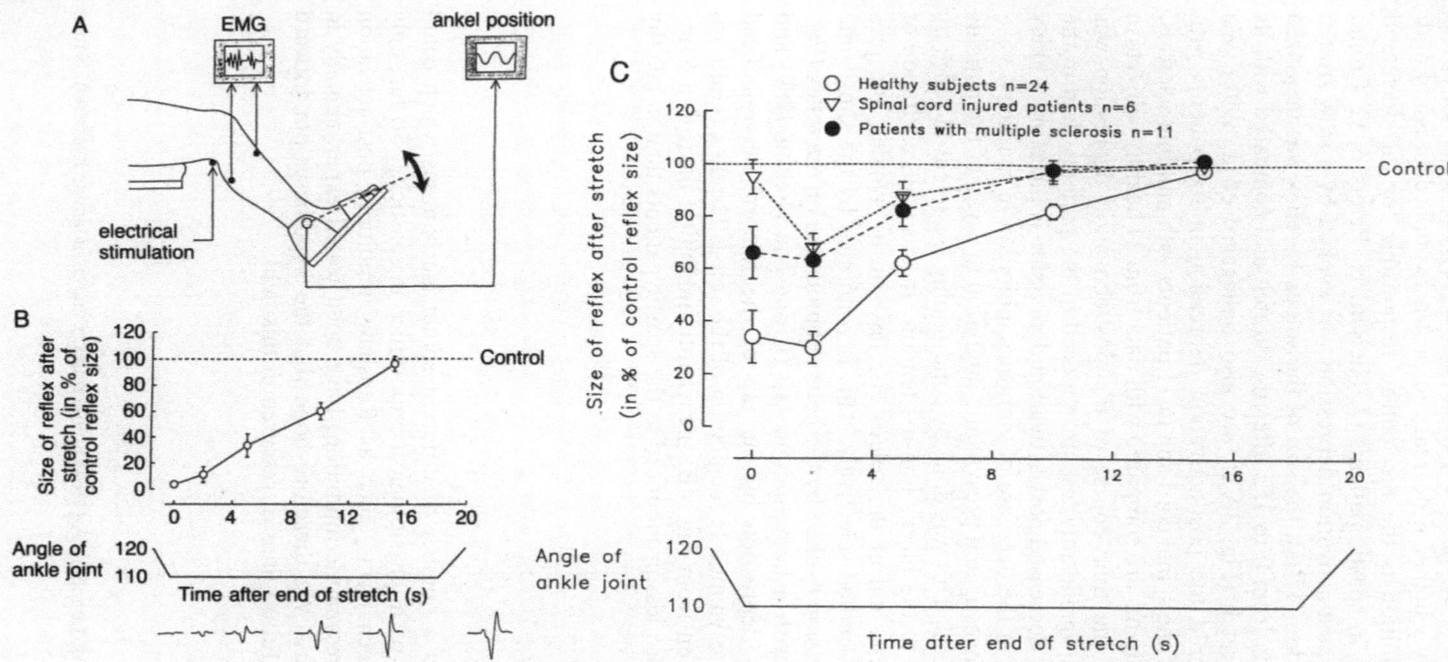

Fig. 3A–C. Post-activation depression in healthy and spastic subjects. **A** Experimental arrangement: The subject was seated in a reclining armchair with the right leg attached to a footplate, which was connected to a potentiometer. The size of the soleus H-reflex was measured after a slow passive dorsiflexion of the foot (from 120° plantarflexion to 110° plantarflexion; 17°/s). The reflex was always measured at an ankle joint angle of 110°. The interval between each reflex measurement was 20 s. The size of the reflex was expressed as a percentage of the size of the reflex in the control situation without stretch. **B** Measurements from one of the healthy subjects: The *traces* at the bottom of the figure show reflexes measured at different times following the stretch, the *rightmost sample* being the control reflex without preceding stretch. Ten reflex measurements for each interval were averaged. The result of these measurements are shown in the graph. **C** The average size of the post-activation depression (expressed as the size of the reflex measured after stretch in percent of the reflex size without stretch) from 0 s to 15 s after the stretch in 24 healthy subjects, six patients with traumatic lesion of the spinal cord, and 11 spastic patients with multiple sclerosis. (Data from Nielsen, Petersen, Ballegaard, Biering-Sørensen, Crone, Kiehn, unpublished)

For this reason, a series of experiments was recently initiated in our laboratory to clarify whether a decrease in post-activation depression and/or presynaptic inhibition is the mechanism responsible for the decreased vibratory inhibition in spastic patients (J. Nielsen, N. Petersen, C. Crone, unpublished). The post-activation depression was evoked by a slow stretch of the Achilles tendon (17°/s); the size of the soleus H-reflex was measured at different intervals from 0 to 15 s after the stretch and compared with its size without the stretch (Fig. 3A,B; see also Ballegaard et al. 1991). We have so far measured the post-activation depression in six patients with traumatic lesions of the spinal cord and in 11 patients with multiple sclerosis. Data from these patients are compared with data from 24 healthy subjects in Fig. 3C. As seen, the time course of the post-activation depression was shorter, and it was significantly less pronounced in the patients than in the healthy subjects. A decreased post-activation depression is therefore a likely contributory factor in the decrease of vibratory inhibition in spasticity.

We have also investigated presynaptic inhibition in seven of the patients with multiple sclerosis. This was done by conditioning the soleus H-reflex by a brief tap applied to the biceps femoris tendon. In this way, elicitation of post-activation depression of the soleus H-reflex can be avoided and presynaptic inhibition studied selectively. This is confirmed by the short time course of the depression evoked by the heteronomous tap on the reflex (Fig. 4A) as compared with a homonomous tap (Crone and Nielsen 1989) and by a simultaneous depression by the tap of the femoral nerve-induced heteronomous monosynaptic Ia facilitation of the soleus H-reflex (Hultborn et al. 1987). As seen from Fig. 4B, presynaptic inhibition measured 60 ms after the tap was also less pronounced in the spastic patients than in healthy subjects.

Conclusion

We have described a number of different spinal mechanisms which could theoretically be of pathophysiological importance in spasticity. So far, only a few of these mechanisms, such as Renshaw inhibition, post-activation depression, and presynaptic inhibition, have been tested experimentally in relation to spasticity. We therefore hope that the hypotheses put forward here may point to future lines of research in this field.

References

Andén NE, Jukes MGM, Lundberg A (1964) Spinal reflexes and monoamine liberation. Nature 202:1222–1223

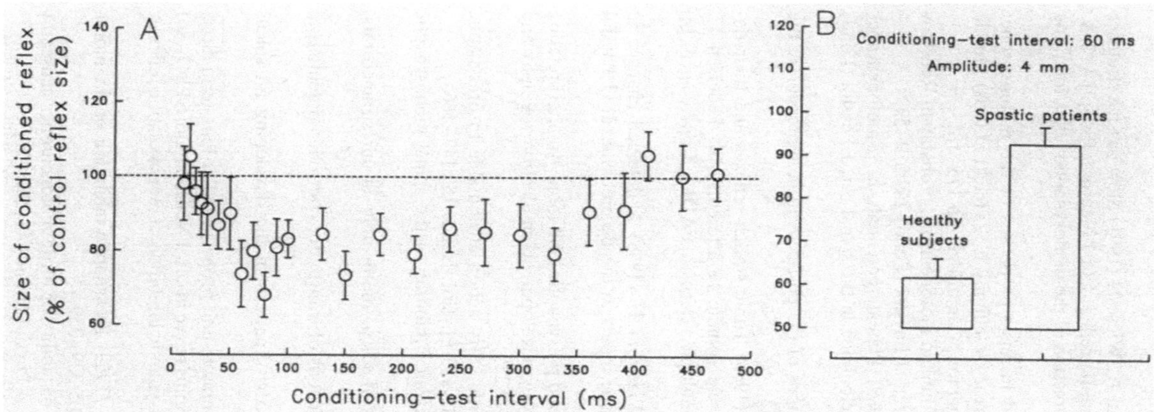

Fig. 4A,B. Presynaptic inhibition in healthy and spastic subjects. The soleus H-reflex was conditioned by a preceding tap applied to the biceps femoris tendon (4 mm amplitude; 4 ms duration). **A** Time course of the size of the conditioned reflex as a percentage of the control (unconditioned) reflex size in a healthy subject. Conditioned and unconditioned reflexes were randomly alternated. **B** Size of the conditioned reflex (in percent of the control reflex size) in seven healthy subjects and in seven spastic patients (multiple sclerosis) at a conditioning-test interval of 60 ms, i.e., when the depression of the reflex was at its peak (as seen in *A*). The amplitude of the tap was 4 mm. In all subjects the amplitude of the tap was graded from 0.5 mm to 4 mm. In the healthy subjects a significant depression of the reflex was seen with an amplitude of 1 mm, whereas a significant depression was not seen in spastic patients unless the amplitude was increased to 4 mm. (Data by Nielsen Petersen and Crone, unpublished)

Ashby P, Verrier M, Carleton S (1980) Vibratory inhibition of the monosynaptic reflex. Prog Clin Neurophysiol 8:254–262

Bailey CS, Lieberman JS, Kitchell RL (1980) Response of muscle spindle primary endings to static stretch in acute and chronic spinal cats. Am J Vet Res 41:2030–2034

Baldissera F, Cavallari P, Dworzak F (1991) Cramps: a sign of motoneurone "bistability" in a human patient. Neurosci Lett 133:303–306

Ballegaard M, Hultborn H, Illert M, Nielsen J, Paul A, Wiese H (1991) Slow passive stretches of a muscle depress transmission of its monosynaptic reflex. Eur J Neurosci [Suppl] 4: 298

Beswick FB, Evanson JM (1957) Homosynaptic depression of the monosynaptic reflex following activation. J Physiol (Lond) 135:400–411

Brownstone R, Hultborn H (1992) Regulated and intrinsic properties of the motoneurone: effect on input-output relations. In: Jami I, Pierrot-Deseilligny E, Zytnicki D (eds) Muscle afferents and spinal control of movements. Pergamon, Cambridge, pp 175–181

Bussel B, Pierrot-Deseilligny E (1977) Inhibition of human motoneurons, probably of Renshaw origin, elicited by an orthodromic motor discharge. J Physiol (Lond) 269:319–339

Conway BA, Hultborn H, Kiehn O, Mintzl (1988) Plateau potentials in α-motoneurones induced by intravenous injection of *l*-dopa and clonidine in the spinal cat. J Physiol (Lond) 405:369–384

Crone C, Nielsen J (1989) Methodological implications of the post-activation depression of the soleus H-reflex in man. Exp Brain Res 78:28–32

Crone C, Hultborn H, Malmsten J, Mazieres L (1983) Tonic stretch reflexes and their dependence of polysynaptic excitation from muscle spindle la afferents. In: Pedersen E, Clausen J, Oades L (eds) Actual problems in multiple sclerosis. FADLs, Copenhagen, pp 99–102

Crone C, Hultborn H, Kiehn O, Mazieres L, Wigström H (1988) Maintained changes in motoneuronal excitability by short-lasting synaptic inputs in the decerebrate cat. J Physiol (Lond) 405:321–343

Delwaide PJ (1973) Human monosynaptic reflexes and presynaptic inhibition. An interpretation of spastic hyperreflexia. In: Desmedt JE (ed) New developments in electromyography and clinical neurophysiology. Karger, Basel, pp 508–522

Eccles JC, Eccles RM, Magni F (1961) Central inhibitory action attributable to presynaptic depolarization produced by muscle afferent volleys. J Physiol (Lond) 159:147–166

Eccles JC, Schmidt RF, Willis WD (1962) Presynaptic inhibition of the spinal monosynaptic reflex pathway. J Physiol (Lond) 161:282–297

Eken T, Kiehn O (1989) Bistable firing properties of soleus motor units in unrestrained rats. Acta Physiol Scand 136:383–394

Eken T, Hultborn H, Kiehn O (1989) Possible functions of transmitter-controlled potentials in α-motoneurones. Prog Brain Res 80:257–267

Eken T, Elder G, Lømo T (1990) Motor-unit activity pattern in the developing rat soleus muscle. Soc Neurosci Abstr 16:331

Foehring RC, Munson JB (1990) Motoneuron and muscle-unit properties after long-term direct innervation of soleus muscle by medial gastrocnemius nerve in cat. J Neurophysiol 64:847–861

Frank K, Fourtes MGF (1957) Presynaptic and postsynaptic inhibition of monosynaptic reflexes. Fed Proc 16:39–40

Gillies JD, Lance JW, Neilson PD, Tassinari CA (1969) Presynaptic inhibition of the monosynaptic reflex by vibration. J Physiol (Lond) 205:329–339

Hounsgaard J, Kiehn O (1989) Serotonin-induced bistability of turtle motoneurones caused by a nifedipine-sensitive calcium plateau potential. J Physiol (Lond) 414:265–282

Hounsgaard J, Mintz I (1988) Calcium conductance and firing properties of spinal motoneurones in the turtle. J Physiol (Lond) 398:591–603

Hounsgaard J, Hultborn H, Jespersen B, Kiehn O (1988) Bistability of α-motoneurones in the decerebrate cat and in the acute spinal cat after intravenous 5-hydroxytryptophan. J Physiol (Lond) 405:345–367

Hultborn H, Illert M (1991) How is motor behavior reflected in the organization of spinal systems? In: Humphrey DR, Freund HJ (eds) Motor control: concepts and issues. Wiley, New York, pp 49–73
Hultborn H, Kiehn O (1992) Neuromodulation of vertebrate motor neuron membrane properties. Curr Biol 2:770–775
Hultborn H, Pierrot-Deseilligny E (1979) Input-output relations in the pathway of recurrent inhibition to motoneurones in the cat. J Physiol (Lond) 297:267–287
Hultborn H, Lindström S, Wigström H (1979) On the function of recurrent inhibition in the spinal cord. Exp Brain Res 37:399–403
Hultborn H, Meunier S, Morin C, Pierrot-Deseilligny E (1987) Assessing changes in presynaptic inhibition of Ia fibres: a study in man and the cat. J Physiol (Lond) 389:729–756
Katz R, Pierrot-Deseilligny E (1982) Recurrent inhibition of alpha-motoneurons in patients with upper motor neuron lesions. Brain 105:103–124
Katz R, Morin C, Pierrot-Deseilligny E, Hibino R (1977) Conditioning of H-reflex by a preceding subthreshold tendon reflex stimulus. J Neurol Neurosurg Psychiatry 40:575–580
Kernell D, Hultborn H (1990) Synaptic effects on recruitment gain: a mechanism of importance for the input-output relations of motoneurone pools. Brain Res 507:176–179
Kiehn O (1991) Plateau potentials and active integration in the "final common pathway" for motor behaviour. TINS 14:68–73
Kiehn O, Erdal J, Eken T, Bruhn T (1992) Activity pattern in the adult soleus muscle after selective depletion of descending monoaminergic fibres. Acta Physiol Scand [Suppl] 608:77
Krnjevic K, Puil E, Werman R (1978) EGTA and motoneuronal after-potentials. J Physiol (Lond) 275:199–223
Lev-Tov A, Pinco M (1992) In vitro studies of prolonged depression in the neonatal rat spinal cord. J Physiol (Lond) 447:149–169
Lloyd DPC, Wilson VJ (1957) Reflex depression in rhythmically active monosynaptic reflex pathways. J Gen Phys 40:409–426
Magladery JW, McDougal DB jr (1950) Electrophysiological studies of nerve and reflex activity in normal man. I. Identification of certain reflexes in the electromyogram and the conduction velocity of peripheral nerve fibres. Johns Hopkins Hosp Bull 86:265–290
Nielsen J, Kagamihara Y (1993) Differential projection of the sural nerve to early and late recruited tibialis anterior motor units in man: change in recruitment gain. Acta Physiol Scand 147:385–401
Nielsen J, Hultborn H, Gossard J-P (1990) Changes of recruitment gain by stimulation of the caudal sural or superficial peroneal nerve in cat. Eur J Neurosci [Suppl] 3: 193
Paillard J (1955) Reflexes et regulations d'origine proprioceptive chez l'homme. Thesis, Libraire Arnette, Paris, pp 1–293
Rothwell JC, Day BL, Berardelli A, Marsden CD (1986) Habituation and conditioning of the human long-latency stretch reflex. Exp Brain Res 63:197–204
Rudomin P (1990) Presynaptic inhibition of muscle spindle and tendon organ afferents in mammalian spinal cord. TINS 13:499–505
Schieppati M, Crenna P (1984) From activity to rest: gating of excitatory autogenetic afferences from the relaxing muscle in man. Exp Brain Res 56:448–457
Schwindt PC, Crill WE (1980a) Properties of a persistent inward current in normal and TEA-injected motoneurons. J Neurophysiol 43:1700–1724
Schwindt PC, Crill WE (1980b) Effects of barium on cat spinal motoneurons studied by voltage clamp. J Neurophysiol 44:827–846
Schwindt PC, Crill WE (1980c) Role of a persistent inward current in motoneuron bursting during spinal seizures. J Neurophysiol 43:1296–1318
Schwindt PC, Crill WE (1984) Membrane properties of cat spinal motoneurons. In: Davidoff RA (ed) Handbook of the spinal cord, vols 2 and 3. Dekker, New York, pp 199–242
Shefner JM, Berman SA, Sarkarati M, Young RR (1992) Recurrent inhibition in patients with spinal cord injury. Neurology 42:2162–2168

Tanaka H, Mori S, Kimura H (1992) Developmental changes in the serotoninergic innervation of hindlimb extensor motoneurons on neonatal rats. Dev Brain Res 65:1–12

Tang A, Rymer WZ (1981) Abnormal force-EMG relations in paretic limbs of hemiparetic human subjects. J Neurol Neurosurg Psychiatry 44:690–698

Wallén P, Buchanan JT, Grillner S, Hill RH, Christenson J, Hökfelt T (1989) Effects of 5-hydroxytryptamine on the afterhyperpolarization, spike frequency regulation, and oscillatory membrane properties in lamprey spinal cord neurons. J Neurophysiol 61:759–768

Walton K, Fulton BP (1986) Ionic mechanisms underlying the firing properties of rat neonatal motoneurones studied in vitro. Neuroscience 19:669–683

White SR, Fung SJ (1989) Serotonin depolarizes cat spinal motoneurones in situ and decreases motoneuron afterhyperpolarizing potentials. Brain Res 502:205–213

Wu SY, Wang MY, Dun NJ (1991) Sertonin via presynaptic 5-HT1 receptors attenuates synaptic transmission to immature rat motoneurons in vitro. Brain Res 554:111–121

Functional Properties of the Neuromuscular System and Force Gradation

D. Kernell

Introduction

When trying to explain motor symptoms that appear after damage to the central nervous system (CNS), each motoneurone (MN) pool with its associated muscle is usually treated as a functional unit that reacts in a stereotyped manner to the net sum of incoming postsynaptic excitation and inhibition. On such a view, disease causes the motor output to become deranged simply because of changes in the total amount of net excitation reaching the various MN pools from upstream structures. One aim of the present chapter is to direct attention to the fact that there are also considerable possibilities for influencing the *working manner of the neuromuscular output interface itself*, and that such alterations may conceivably contribute to motor symptoms in neurological disease (see also Nielsen and Hultborn, this volume).

As a background for the present, fairly general survey I first give an overview of the normal strategies for output gradation in the neuromuscular system. Thereafter, the review deals briefly with some of the possible ways in which the gain and other gradation properties of the system may become altered such as to promote, for instance, the appearance of an increased neuromuscular tone or reactivity. In this context, special attention is devoted to some of the expected consequences of changes in "recruitment gain" (Kernell and Hultborn 1990).

Normal Neuromuscular Gradation Properties

Pool Innervation and Gradation Strategies

Since the beginnings of electromyography (Adrian and Bronk 1929), it has been well known that the force of a mammalian skeletal muscle is controlled by the CNS via two aspects of MN management: (a) by a variation in the

Department of Neurophysiology, Academisch Medisch Centrum, University of Amsterdam, Meibergdreef 15, NL-1105 AZ Amsterdam, The Netherlands

Thilmann et al. (Eds.)
Spasticity: Mechanisms and Management

number of activated MNs (*recruitment gradation*) and (b) by a variation in the discharge rate of already recruited MNs (*rate gradation*; for general references, see Burke 1981; Henneman and Mendell 1981; Kernell 1992). For both strategies the actual force production is, of course, dependent on properties of the MNs as well as on those of their muscle units (*MUs*). The way in which the two strategies for MN pool management work together depends on how the pool is innervated, i.e., on the patterns of premotoneuronal synaptic divergence.

If every individual MN had its own "private line" of synaptic input from command centers of the brain, it would be difficult to predict in what general manner the different members of an MN pool would cooperate in motor functions. Fortunately, this does not seem to be the common principle of organization. The simplest way to guarantee that all the MNs of a given muscle will actually work together in driving a muscle is to provide them with innervation of a "*low-level wide divergence*" such as that shown in Fig. 1A. Here nearly every MN of a pool is reached by branches of practically every individual axon of a given synaptic input. This organization of innervation comes close to that actually observed for the best studied example of synaptic distribution to MNs so far: the Ia monosynaptic connections (Mendell and Henneman 1971; Fleshman et al. 1981; Henneman and Mendell 1981). In the case of a "*high-level wide divergence*," of which Fig. 1B shows one of many possible varieties, subportions of the MN pool might be (but need not always be) influenced relatively independently of each other; such an alternative synaptic organization may be of relevance for cases in which different motor tasks tend to be performed preferentially by different, more or less overlapping "task teams" within individual muscles (e.g., Ter Haar Romeny et al. 1984; Hoffer et al. 1987; Hensbergen and Kernell 1992).

The considerations below concern primarily MN pools (or "task teams" within whole-muscle pools) for which the whole MN population is activated

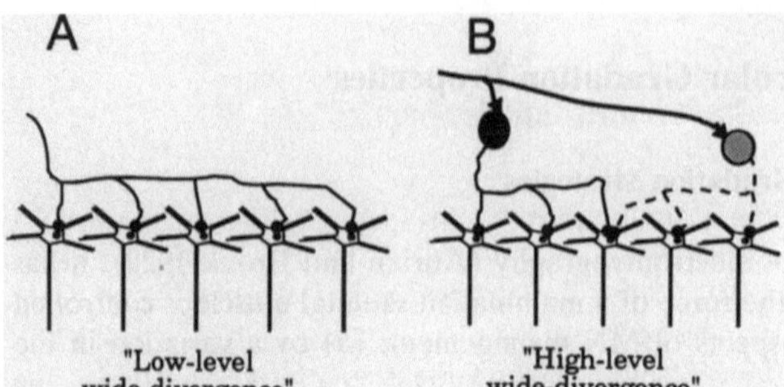

Fig. 1A,B. Schematic drawing illustrating two of the many possible ways in which individual axons may distribute their synapses among the MNs of a given muscle. Distributions of the kind shown in **A** automatically link rate and recruitment gradation

by synaptic systems with a divergence pattern such as that of Fig. 1A. This assumption simplifies the analysis because, under such conditions, the increase in excitatory pool input needed for recruiting a greater number of MNs also increases the excitatory drive (and, hence, discharge rate) for MNs that had already become recruited, i.e., recruitment gradation is parallelled automatically by rate gradation. Such a parallelism has also commonly been observed during motor acts.

Recruitment Gradation

If all the cells of a given functional population had the same membrane properties and were at all times receiving exactly the same excitatory current, they would also all be recruited at the same level of pool excitation, i.e., recruitment gradation would then be impossible. The differences in recruitment threshold that normally exist among the MNs of a given muscle may depend on quantitative differences in synaptic distribution to the various cells as well as on differences in MN intrinsic electrical excitability. The latter factor seems to be of great importance, and this may help to explain why, statistically speaking, MNs tend to become recruited in about the same sequence for many different excitatory inputs (Henneman and Mendell 1981). Thus, at the low end are the most easily recruited MNs which generally have slow, weak, and fatigue-resistant S units. At the other extreme are the least easily recruited MNs, which tend to possesss relatively fast, strong, and fatigue-sensitive FF units. In between there is a continuous variation. Easily recruited MNs have a relatively low electrical current threshold for activation, and such differences in excitability apparently depend largely on the fact that the MNs differ in input resistance and membrane resistivity (see Kernell and Zwaagstra 1981, 1989; Gustafsson and Pinter 1985; Fleshman et al. 1988).

Rate Gradation

MN intrinsic properties are also obviously important for the other gradation strategy: the modulation of discharge frequency. The MNs are typically firing repetitively as a muscle is made to contract, and such rhythmic discharges are normally elicited by relatively steady synaptic currents which are produced by the summation of many asynchronous postsynaptic events. Such activating synaptic currents may be well imitated by maintained currents injected through an intracellular microelectrode (for review see Kernell 1984, 1992). Using such stimuli, the input-output relation of cat MNs shows a fairly definite minimum rate of maintained discharge (F_{min}) below which no regular firing takes place. From F_{min} upwards there is a so-called "primary range" of firing with an approximately linear frequency-current relationship (f-I relationship). At higher rates than those of this linear relationship, the curve bends off upwards ("secondary range").

In anesthetized animals, F_{min} is closely related to the duration of the postspike afterhyperpolarization (AHP); at F_{min}, the impulse intervals tend to be about equal to AHP duration (Kernell 1965). In addition to the AHP, other factors may also contribute to determining the F_{min} (e.g., Carp et al. 1991). Besides being of importance for F_{min}, the changes of potassium conductance that give rise to the AHP (Barrett et al. 1980) are also likely to be of major influence for determining the gain for the relationship between impulse frequency and the intensity of activating current (frequency-current slope, f-I slope). In simple neurone models, the f-I slope for the first impulse interval has been found to be inversely proportional to: (a) the postspike starting value of the AHP conductance, (b) the time constant of decline in the AHP conductance, and (c) the voltage "driving" the AHP current (i.e., the difference between the voltage threshold for spike initiation and the potassium equilibrium potential; see Kernell 1968; Kernell and Sjöholm 1973).

The discharge rate of MNs is time as well as current dependent; MNs display at least two different types of *frequency adaptation*, which may be experimentally observed as a decline in discharge rate during constant-current stimulation. The "initial adaptation" takes place largely during the first few impulse intervals after current onset. This early frequency decline is associated with a marked decrease in the f-I slope, and both phenomena are probably caused mainly by a "summation" of the AHP conductances following consecutive initial impulses (Kernell and Sjöholm 1973; Baldissera and Gustafsson 1974; Barrett et al. 1980; Kernell 1984). The subsequent, more prolonged phase of "late adaptation" proceeds slowly over at least 0.5–1 min of a continuous discharge. This phase of adaptation is not associated with any prominent changes in f-I slope (Kernell and Monster 1982), and the underlying membrane mechanisms are still uncertain.

The maximum and minimum rates of MNs are not determined at random but are related in a functionally corresponding manner to the contractile speed of their muscle fibers. Mammalian MUs have a sigmoid relationship between tension and activation frequency (T-f relationship), whereby the rate gradation of force is effective only within a steep middle region covering a relatively narrow range of frequencies. The rates of this steep region are lower, and the slope of the curve is steeper for slow than for fast muscles and MUs (Cooper and Eccles 1930; Kernell et al. 1983). In fast- as well as slow-twitch MN-MU combinations, the F_{min} of an MN typically corresponds to a rate for which consecutive twitches of its MU barely start to sum (i.e., a frequency at the lower end of the steep T-f range), and the MN primary range typically covers the steep portion of the MU T-f curve. This aspect of "speed-matching" is partly a direct consequence of the fact that there is a similarity in total duration between the AHP of an MN and the twitch of its MU (Bakels and Kernell 1992).

The maximum impulse frequency within the primary range typically corresponds to roughly 80% of the maximum tetanic MU force (as measured at the optimum muscle length for a twitch), and the observed maximum rate within the secondary range is just about sufficient for producing the maximum tetanic tension (Kernell 1965, 1979). Over the high rates of the secondary range, the steep f-I slope compensates for the leveling-off of the T-f curve.

Recruitment and Rate Gradation Working Together

Some general principles concerning the combined use of both gradation mechanisms are discussed here with the help of a very simple pool model (Figs. 2, 3). In this model (see Kernell and Hultborn 1990) all the 100 MNs and their MUs have the same properties (resembling those of average cat

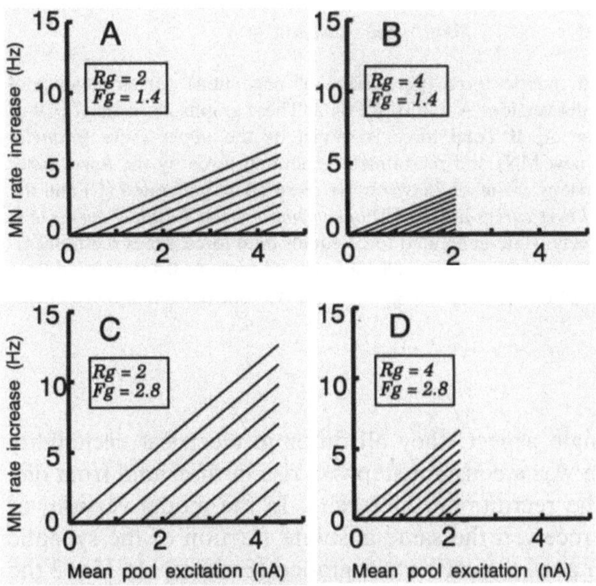

Fig. 2A–D. Motoneuronal discharge behavior in four versions of a simple pool model, illustrating the effects of changes in rate and recruitment gain. In each panel, MN discharge rate (Hz) is plotted versus the mean intensity of a gradable excitatory drive (E drive, nA). Plotted discharge rates are those exceeding the minimum rate for each MN (i.e., the impulse rate at recruitment is subtracted from each MN frequency), and the plotted pool-excitation currents are those exceeding the threshold for the most excitable MN of the pool. Each MN receives the same fraction of the total E drive, i.e., the "mean pool excitation" is here in fact equivalent to the intensity of E drive per MN. The four model versions differ only in their recruitment gain (*Rg*, given in MNs recruited per nA) and/or in their frequency gain (*Fg*, given in Hz per nA). In each panel, MN discharge behavior is shown from pool threshold up to the excitation needed for recruiting the 10th cell

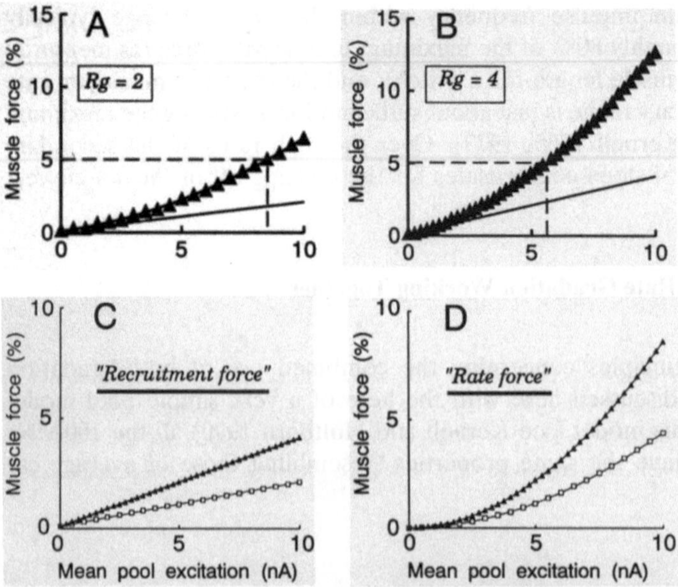

Fig. 3A–D. Plots of calculated muscle force (percentage of maximum) versus mean pool excitation (E drive, nA) for model versions A and B of Fig. 2. These graphs illustrate effects of a doubling of recruitment gain. **A, B** Total force is shown by the *upper curve* (*triangles* indicating recruitment of each new MN) and recruitment-generated force by the *lower curve* (*continuous line*). **C, D** Comparisons are made between the recruitment-generated (**C**) and the rate-generated (**D**) forces in **A** (*lower curves* in **C** and **D** *open symbols*) and **B** (*upper curves* in **C** and **D**, *closed symbols*) respectively. Rate-generated force equals total force minus recruitment-generated force. For further information on force calculations, see text, **A, B** Interrupted lines are drawn through the 5% value of total force; note greater relative contribution of recruitment-generated force in **B** than in **A**

MNs) except for one single aspect: they all differ in electrical excitability (current threshold); there was a constant stepwise rise in threshold from one MN to the next within the recruitment hierarchy. In the model versions of Figs. 2 and 3, each MN received the same absolute fraction of the synaptic excitatory drive by which pool activation was graded (E drive). In Fig. 2 the x-axes of the graphs show the intensity of this E drive, and the y-axes show the resulting suprathreshold discharge rates for consecutively recruited MNs. In the "standard" case of Fig. 1A, the "ease" with which new cells can be recruited is two cells per nA ("recruitment gain;" Kernell and Hultborn 1990), and the f-I slope is 1.4 Hz/nA (see Kernell 1979). For calculations of the total force output of the model muscle (Fig. 3) the further assumption was made that the mere recruitment of an MN (i.e., keeping it discharging at its F_{min} would produce a mean force of 10% of the MU tetanic tension (see Kernell 1979). Furthermore, each Hz rate increase of an MN is assumed to increase the force output by 3% of the tetanic tension of its MU (see

Kernell et al. 1983). Under these conditions, the MN discharges shown in Fig. 2A produce the forces displayed in Fig. 3A.

The data of Fig. 3A demonstrate one important general rule: that the muscle force produced by rate gradation increases relatively more rapidly than that caused by recruitment (see Milner-Brown et al. 1973). Such a result is intuitively understandable: a change in pool excitation from that needed for recruiting MN-1 to that for recruiting MN-2 is parallelled by a rate increase in only *one* unit (MN-1; see Fig. 2A). The corresponding recruitment interval from MN-2 to MN-3 is accompanied by rate increases in *two* units, and so on. Hence, in a simple pool model such as that of Figs. 2 and 3, the input-output relationship shows an increasing gain, thanks to the increasingly large number of MNs participating in the rate modulation. As has also been noted by Matthews (1986), this provides one possible mechanism for the repeatedly observed "gain compensation" of reflexes (higher gain at high than at low levels of contraction; for references, see Matthews 1986; see also Heckman and Binder 1991). It should be stressed that the present simple model illustrations concern only a range of weaker contractions for which the MNs operate within their "primary range" of firing and, for the MUs, within the steep portion of their T-f curves; for more complex and realistic pool models covering a greater force range, see Heckman and Binder (1991).

Changes in Neuromuscular Input-Output Properties

In neurological disease, changes in neuromuscular function are presumably often the result simply of alterations in the *commands* arriving at the MN pool; wrong commands produce wrong actions even if the functional state of the neuromuscular machinery itself is completely normal. Some of the motor symptoms may be intensified and aggravated by simple decreases in *neuromuscular threshold*, as caused by an increase in the total net amount of steady background synaptic excitation (and/or a decrease in background inhibition) arriving, for instance, by pathways descending from the brain. In the long term, shifts in neuromuscular threshold may conceivably also be produced by changes in the intrinsic excitability of the MNs themselves and/or by changes in properties of synapses connecting to the MNs. Besides these relatively simple kinds of influence on neuromuscular function, normal motor commands may conceivably also produce unexpected/pathological reactions because of alterations in the *gradation properties* of MN pools and their MUs. Regarding this latter category, two major classes of possible causes are briefly considered below: (a) alterations in gradation caused acutely by a change in synaptic input to normal-property MN pools and (b) alterations in gradation caused by long-term changes in functional properties of the MNs and/or MUs themselves.

Acute Synaptic Influences on MN Gradation Mechanisms

Changes in the synaptic input to a MN pool may have an effect on both of the main strategies for the gradation of MN activity. An increased *rate-gradation gain* (i.e., a steeper f-I slope for the MNs) can be produced by certain synaptic transmitters, which act by decreasing the conductance changes underlying the AHP (best known for the monoamines serotonine and noradrenaline; e.g., van Dongen et al. 1986; Hounsgaard and Kiehn 1989; Hultborn and Toth 1989). An increased *recruitment gain* is produced if all the activation thresholds within the MN pool are brought closer together. Such an effect may take place, for instance, if the pool is influenced by a tonic synaptic background excitation which delivers more depolarizing current to MNs with a high intrinsic threshold than to those with an intrinsically lower one (for further examples and considerations, including similar effects of background inhibition, see Kernell and Hultborn 1990).

Simplified examples of how the behavior of an MN pool is influenced by changes in rate and/or recruitment gain are shown in Fig. 2, in which panel A shows a provisional "standard" case, B the effect of a doubled recruitment gain (activation thresholds twice as close), C a doubled rate gain (f-I slopes doubled), and D both alterations combined. The effects on force gradation of an *increased rate gain* are relatively straightforward; in this case the total input-output relationship becomes steeper thanks to a selective rise in the amount of force generated by rate gradation (see Fig. 2C versus 2A). As a result, each force level is now produced at a higher average discharge rate of the MNs and, consequently, at a higher average tension level for each one of the MUs. High tension levels would generally be expected to be associated with a smaller degree of endurance; such secondary effects on MU function may be of relevance during more long-lasting contractions.

The consequences of an *increased recruitment gain* are somewhat more complex (Figs. 2, 3, panels A versus B). Firstly, it is evident that with the activation thresholds spaced more closely together there will be less increment of synaptic current (E drive) between the recruitment of each new MN; hence, at a given total *force level* there is now a relatively greater amount of recruitment-generated force (see interrupted lines at 5% force in Fig. 3A versus 3B). On the other hand, a given suprathreshold *current level* now recruits more MNs, and as a result more MNs now participate in the rate increase that takes place in parallel with recruitment. For this reason, an increased recruitment gain secondarily causes an marked increase also in the rate-generated force per unit of excitatory pool drive (Fig. 3D). Hence, a change in recruitment gain is quite an efficient method for altering the gain of the total neuromuscular input-output relationship. In neurological disease, persistent changes in the synaptic background activity arriving at the MNs may conceivably contribute to pathological motor symptoms as a result (partly) of alterations in recruitment gain. It should be noted that changes in recruitment gain may also cause quantitative alterations in the input-output relationship of phasic test reflexes, such as the tendon jerk or the H reflex (e.g., Nielsen and Kagamihara 1991).

Long-Term Influences on MN and MU Gradation Mechanisms

Rather little is to date known about the extent to which the *membrane properties of MNs* may change as an effect of long-term alterations in their use and/or synaptic connectivity. It has been observed, however, that the membrane properties of at least some MNs may alter markedly as a result of spinalization. Evidence for such modifications (Czéh et al. 1978; Munson et al. 1986) has been observed for properties of importance for recruitment gradation (input resistance, current threshold) as well as for rate gradation (AHP time course). Thus, it seems quite possible that CNS lesions may indeed, in the long run, cause changes in rate and/or recruitment gain to occur due to long-term alterations in MN properties.

Besides influences acting on the MNs themselves, it is important not to forget that alterations in the neuromuscular input-output relationship may also occur as a result of changes in *contractile MU properties*. The very presence of neurological motor symptoms implies that the MUs of affected muscles tend to become subjected to other daily patterns of activation than those normally employed. Such long-term changes in activation pattern may produce modifications in the force-production and force-gradation properties of the MUs. Animal experiments with chronic nerve stimulation have shown that even a comparatively modest amount of "extra" activity per day (7.2 min cumulative time; Kernell et al. 1987) may cause a small but significant slowing in the T-f relationship of fast muscle. Moderate to large amounts of chronic stimulation cause greater degrees of slowing and also an increase in muscle endurance and a decrease in maximum muscle force (for further details and references, see Kernell 1992). It is evident that such changes in MU properties will affect the neuromuscular input-output relationship. For instance, even if other muscle properties (including maximum force) remain constant, a slight slowing of the MU T-f relationship, such as that produced by the 7.2 min of "extra" activity per day (see above), would cause significantly larger forces to be produced by MN rates close to F_{min} (see Fig. 3 of Kernell 1987). In the absence of matching modifications in the rate-gradation properties of the MNs (e.g., in their AHPs), such changes in MU speed would at least have a marked effect on the recruitment-related force gradation of the muscle (see Fig. 3C). Other MU alterations (e.g., changes in maximum MU force) might have marked effects on the recruitment- as well as on the rate-related force gradation; many further examples may also be considered.

The Question of "Recalibration"

For an otherwise normal CNS, the effects of (long-term) alterations in neuromuscular input-output relationships would presumably become neutralized by some procedure of central "recalibration" of the relationship between central command and expected force output. Such recalibration procedures may be less readily available in a lesioned CNS than in a normal one

References

Adrian ED, Bronk DW (1929) The discharge of impulses in motor nerve fibres. II. The frequency of discharge in reflex and voluntary contractions. J Physiol (Lond) 67:119–151

Bakels R, Kernell D (1993) Matching between motoneurone and muscle unit properties in rat medial gastrocnemius. J Physiol (Lond) 463:307–324

Baldissera F, Gustafsson B (1974) Firing behaviour of a neurone model based on the afterhyperpolarization conductance time course and algebraical summation. Adaptation and steady state firing. Acta Physiol Scand 92:27–47

Barrett EF, Barrett JN, Crill WE (1980) Voltage-sensitive coutward currents in cat motoneurones. J Physiol (Lond) 304:251–276

Burke RE (1981) Motor units: anatomy, physiology and functional organization. In: Brooks VB(ed) Handbook of physiology 1, vol 2/1. American Physiological Society, Bethesda, pp 345–422

Carp JS, Powers RK, Rymer WZ (1991) Alterations in motoneuron properties induced by acute dorsal spinal hemisection in the decerebrate cat. Exp Brain Res 83:539–548

Cooper S, Eccles JC (1930) The isometric responses of mammalian muscles. J Physiol (Lond) 69:377–385

Czéh G, Gallego R, Kudo N, Kuno M (1978) Evidence for the maintenance of motoneurone properties by muscle activity. J Physiol (Lond) 281:239–252

Fleshman JW, Munson JB, Sypert GW (1981) Homonymous projection of individual group la-fibers to physiologically characterized medial gastrocnemius motoneurons in the cat. J Neurophysiol 46:1339–1348

Fleshman JW, Segev I, Burke RE (1988) Electrotonic architecture of type-identified α-motoneurons in the cat spinal cord. J Neurophysiol 60:60–85

Gustafsson B, Pinter MJ (1985) On factors determining orderly recruitment of motor units: a role for intrinsic membrane properties. Trends Neurosci 8:431–433

Heckman CJ, Binder MD (1991) Computer simulation of the steady-state input-output function of the cat medial gastrocnemius motoneuron pool. J Neurophysiol 65:952–967

Henneman E, Mendell LM (1981) Functional organization of motoneuron pool and its inputs. In: Brooks VB(ed) Handbook of physiology 1, vol 2/1. American Physiological Society, Bothesda, pp 423–507

Hensbergen E, Kernell D (1992) Task-related differences in distribution of electromyographic activity within peroneus longus muscle of spontaneously moving cats. Exp Brain Res 89:682–685

Hoffer JA, Loeb GE, Sugano N, Marks WB, O'Donovan MJ, Pratt CA (1987) Cat hindlimb motoneurons during locomotion. III. Functional segregation in sartorius. J Neurophysiol 57:554–562

Hounsgaard J, Kiehn O (1989) Serotonin-induced bistability of turtle motoneurones caused by a nifedipine-sensitive calcium plateau potential. J Physiol (Lond) 414:265–282

Hultborn H, Toth T (1989) Raphe-spinal depression of motoneurone after-hyperpolarization. Acta Physiol Scand 136:37A

Kernell D (1965) The limits of firing frequency in cat lumbosacral motoneurones possessing different time course of afterhyperpolarization. Acta Physiol Scand 65:87–100

Kernell D (1968) The repetitive impulse discharge of a simple neurone model compared to that of spinal motoneurones. Brain Res 11:685–687

Kernell D (1979) Rhythmic properties of motoneurones innervating muscle fibres of different speed in m gastrocnemius medialis of the cat. Brain Res 160:159–162

Kernell D (1984) The meaning of discharge rate: excitation-to-frequency transduction as studied in spinal motoneurones. Arch Ital Biol 122:5–15

Kernell D (1987) Properties of motoneurones and motor units in relation to problems of sensorimotor integration. In: Struppler A, Weindl A (eds) Clinical aspects of sensory motor integration. Springer, Berlin Heidelberg New York, pp 65–74

Kernell D (1992) Organized variability in the neuromuscular system: a survey of task-related adaptations. Arch Ital Biol 130:19–66

Kernell D, Hultborn H (1990) Synaptic effects on recruitment gain: a mechanism of importance for the input-output relations of motoneurone pools? Brain Res 507:176–179

Kernell D, Monster AW (1982) Time course and properties of late adaptation in spinal motoneurones in the cat. Exp Brain Res 46:191–196

Kernell D, Sjöholm H (1973) Repetitive impulse firing: comparisons between neurone models based on 'voltage clamp equations' and spinal motoneurones. Acta Physiol Scand 87:40–56

Kernell D, Zwaagstra B (1981) Input conductance, axonal conduction velocity and cell size among hindlimb motoneurones of the cat. Brain Res 204:311–326

Kernell D, Zwaagstra B (1989) Dendrites of cat's spinal motoneurones: relationship between stem diameter and predicted input conductance. J Physiol (Lond) 413:255–269

Kernell D, Eerbeek O, Verhey, BA (1983) Relation between isometric force and stimulus rate in cat's hindlimb motor units of different twitch contraction time. Exp Brain Res 50:220–227

Kernell D, Eerbeek O, Verhey BA, Donselaar Y (1987) Effects of physiological amounts of high- and low-rate chronic stimulation on fast-twitch muscle of the cat hindlimb. I. Speed- and force-related properties. J Neurophysiol 58:598–613

Matthews PBC (1986) Observations on the automatic compensation of reflex gain on varying the pre-existing level of motor discharge in man. J Physiol (Lond) 374:73–90

Mendell LM, Henneman E (1971) Terminals of single la fibers: location, density and distribution within a pool of 300 homonymous motoneurons. J Neurophysiol 34:171–187

Milner-Brown HS, Stein RB, Yemm R (1973) Changes in firing rate of human motor units during linearly changing voluntary contractions. J Physiol (Lond) 230:371–390

Munson JB, Foehring RC, Lofton SA, Zengel JE, Sypert GW (1986) Plasticity of medial gastrocnemius motor units following cordotomy in the cat. J Neurophysiol 55:619–634

Nielsen J, Kagamihara Y (1991) Changes in recruitment gain in man induced by stimulation of the caudal sural nerve. Eur J Neurosci 3[suppl]:298

Ter Haar Romeny BM, Denier van der Gon JJ, Gielen CCAM (1984) Relation between location of a motor unit in the human biceps brachii and its critical firing levels for different tasks. Exp Neurol 85:631–650

Van Dongen PAM, Grillner S, Hökfelt T (1986) 5-Hydroxytryptamine (serotonin) causes a reduction in the afterhyperpolarization following the action potential in lamprey motoneurons and premotor interneurons. Brain Res 366:320–325

Development of an Animal Model for the Study of Spinal Spasticity

J.S. Taylor, C.J. Vierck, Jr., and J.B. Munson

Introduction

For over a decade the commonly adopted definition of human spasticity has been "a motor disorder characterized by a velocity dependent increase in tonic stretch reflexes (muscle tone) with exaggerated tendon jerks" (Lance 1980). However, the manifestations of spasticity are heterogeneous, which complicates attempts to develop a test that encompasses all the clinical signs observed in human spinal cord injury (SCI; Wiesendanger 1985). To avoid this problem, Noth (1991) has suggested that the term spasticity be used "only when the typical increase in muscle tone is present." Thus, progress in understanding human spasticity might arise from the study of "one typical subtype of 'spasticity'" (Young 1980). Accordingly, we have operationally defined animal spinal spasticity as augmented muscle tone and stretch reflexes resulting from spinal injury (for a review on the mechanisms and quantitation of spastic hypertonia see Katz and Rymer 1989).

Development of this behavioral animal model of SCI in the cat has involved: (a) formation of a behavioral method to quantify muscle tone and stretch reflexes, using natural means of dorsiflexion of the feet in unanesthetized and anesthetized cats, (b) determination of the minimal spinal lesion which produces unilateral spasticity and observation of the time course of changes in tone and reflexes, and (c) identification of an anesthetic which preserves spinal spasticity. This model is used to correlate the behavioral consequences of SCI with cellular physiological mechanisms and ultimately to investigate the possibility of amelioration of spinal spasticity.

Development of the Behavioral Stretch Reflex Model

Most investigations of spasticity in animal models have involved placement of spinal lateral hemisections in the middle to low thoracic or middle to high lumbar region (Carter et al. 1991; Fujimori et al. 1966; Hultborn and

Department of Neuroscience, University of Florida, Gainesville, FL 32610-0244, USA

Thilmann et al. (Eds.)
Spasticity: Mechanisms and Management

Malmsten 1983; Murray and Goldberger 1974). In these studies behavioral evidence for increases in muscle tone and stretch reflexes were monitored using clinical tests. Hyperreflexia could not be confirmed behaviorally in some of these studies (Hultborn and Malmsten 1983), but various pathological signs, including exaggerated tendon jerk reflexes were obtained in others (Carter et al. 1991; Fujimori et al. 1966; Murray and Goldberger 1974). These contradictions highlight the necessity of relating quantitative behavioral results with anatomical characterizations of lesions to different regions of the spinal cord (Goldberger et al. 1990).

However, with the exception of one study (Brothers et al. 1983), no quantitative analysis of stretch reflex effects following spinal lesions in the awake, behaving animal has been reported. The behavioral SCI model described here permits the delivery of different rates and displacement angles of stretch to generate electromyographic (EMG) and torque measures of reflexes from behaviorally conditioned animals. In addition, it stresses that long-term postoperative/preoperative comparisons are required to identify the development of lesion-induced effects on reflexes following a spinal injury.

Methods

Cats were conditioned with a food reward to accept gentle, quiet restraint in an apparatus (Fig. 1A) in which their hind feet were secured into paired "boots." Stretch reflexes were elicited by producing ramp and hold dorsiflexion of both feet using a DC torque motor. Quantitative stretch reflex activity was measured as EMG activity recorded from the gastrocnemius medialis and soleus muscle groups in the triceps surae, using chronic implanted mono- or bipolar electrodes. Torque was measured by placing specially designed force transducers directly under the plantar pad of each foot (see Fig. 1A). Both the angle and velocity of displacement were controlled to examine both the slope of the input-output relationship of the muscle and the velocity-dependent components of the reflex.

Following amplification, raw EMG and torque signals were converted through a VCR-based analog/digital unit and stored on VHS tape. Using laboratory-written software, these signals were full-wave rectified and filtered (with a 30-ms window), averaged over six trials, and finally analyzed for several reflex attributes, including latency of the reflex at threshold, dynamic amplitude and gain, static amplitude, dynamic amplitude over three angular displacements (the input-output relationship), and the velocity dependence of the dynamic phase. Typical records of averaged EMG (rectified and filtered) and torque signals are shown in Fig. 1B (left column).

Following adaptation to the testing regimen, the cats were run over a 6-week control period before a chosen spinal lesion was administered under anesthesia. After a 1-week recovery period, these cats were run once per

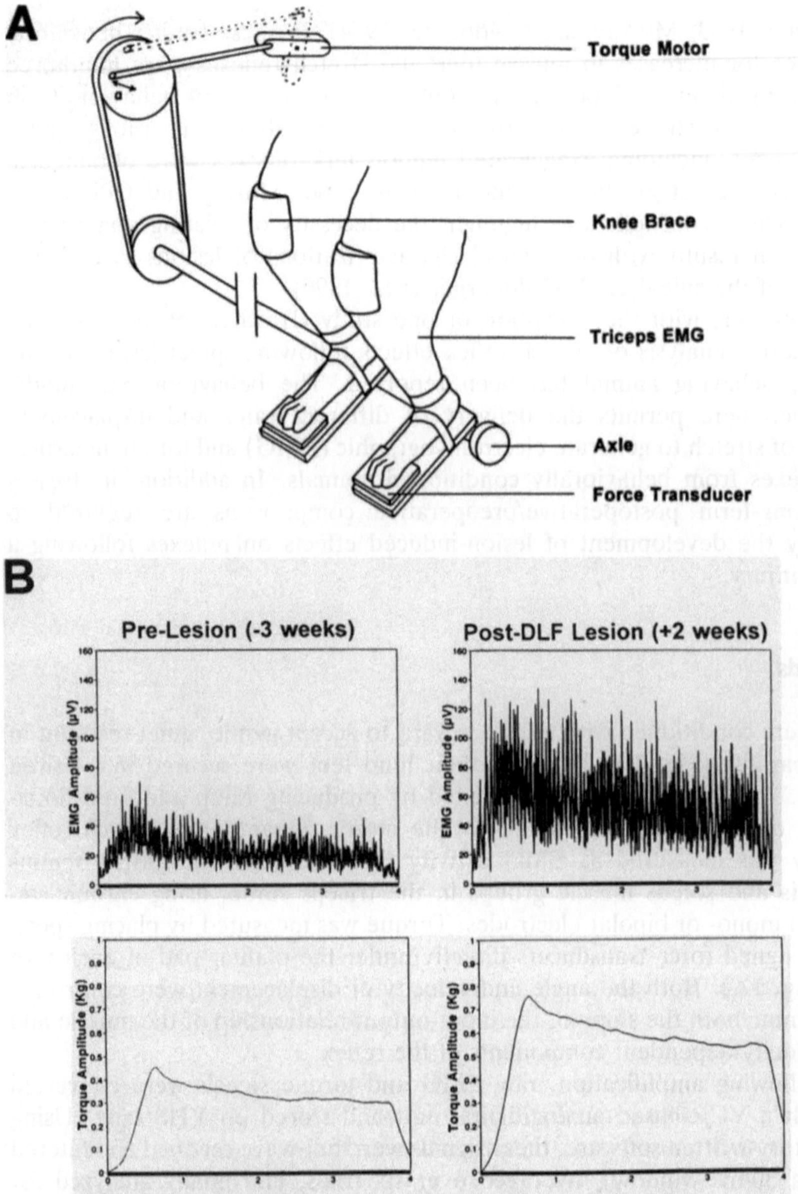

Fig. 1. **A** Schematic representation of the apparatus for quantitative testing of the stretch reflex in conscious or anesthetized cats (see text for details). Electromyographic and torque stretch reflex activity is evoked by dorsiflexion of the feet using a DC torque motor. The angular displacement (a, 20°–43°) and velocity of perturbation ($10°-60° s^{-1}$) can be controlled. **B** The effect of a left DLF low thoracic lesion on the ipsilateral rectified and filtered EMG and torque stretch reflex activity is shown over one complete displacement. Reflex activity over a 30° displacement angle at $60 s^{-1}$ 3 weeks prior to (*left*) and 2 weeks following (*right*) the unilateral DLF spinal lesion

week over a 5-month session. The behavioral apparatus produced EMG and torque measures of stretch reflex activity in the conscious cat over several months following a lesion (10 months to date; Fig. 1B). Accurate assessment of the dynamic amplitude and gain, static amplitude, and velocity-dependent reflexes could be made. This model does not permit accurate assessment of the reflex threshold latency over time.

In parallel with the quantitative testing, several neurological tests of spasticity were performed, including the modified Ashworth test (Bohannon and Smith 1987), the positive weight support test (see Goldberger 1988, 1991), and subjective assessment of the Babinski response and hindlimb muscle tone (in response to passive stretch) while the animal was suspended in air (see Ashby and McCrea 1987). Average hindlimb muscle tone was assessed using the modified Ashworth scale (ranked from 0 to 5, where 2 represents normal tone) by displacing each of the three joints and producing a final summed score. The positive weight support test was quantified using videotape analysis of the final displacement of the ipsilateral limb relative to the intact side (measured in inches).

Definition of a Low-Thoracic Spinal Lesion that Produces Unilateral Spasticity

It is possible that the effect of lateral hemisection on muscle tone depends upon the level of the spinal lesion. Middle lumbar spinal lesions have been shown to produce increased ipsilateral tone in the cat (Carter et al. 1991; Nelson and Mendell 1979). However, other investigators have reported difficulty in detecting hyperreflexia of a hindlimb from lateral thoracic hemisection (Hultborn and Malmsten 1983), or evidence for an ipsilateral hyperreflexia is provided only by comparison of responses from the two levels postoperatively. Given the possibility that contralateral effects on tone could have been produced by lateral hemisection (e.g., Vierck et al. 1990), it is important to relate the development of changes in reflex measures to preoperative responses of the same limb.

Therefore, to identify a unilateral spinal lesion at a thoracic level that would produce unambiguous evidence for ipsilateral spasticity, cats were surgically prepared under halothane anesthesia with a T_{13} partial laminectomy, whereupon a dorsolateral funiculus (DLF) or a dorsal quadrant (DQ) was surgically interrupted.

Responses Following the Dorsolateral Lesion

For two animals the DLF lesion produced variable results. Both cats revealed an early increase in the ipsilateral torque amplitude (muscle tone) over the

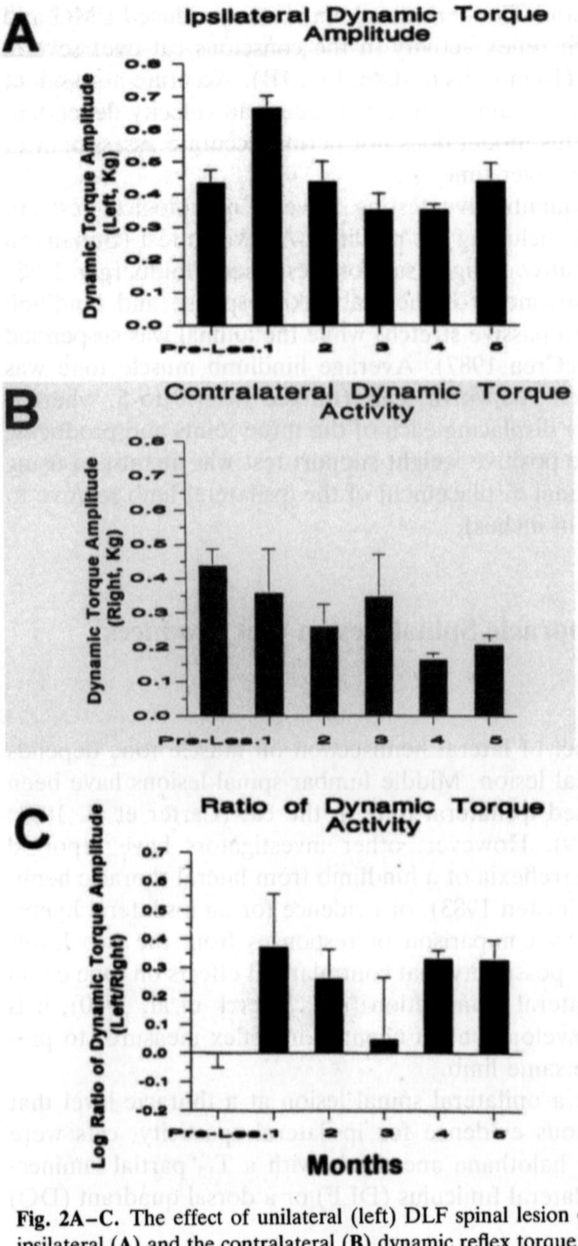

Fig. 2A–C. The effect of unilateral (left) DLF spinal lesion on the absolute amplitude of the ipsilateral (**A**) and the contralateral (**B**) dynamic reflex torque measure 6 weeks prior to and up to 5 months following injury. **C** The log ratio (lesion/intact) of the dynamic torque reflex component relative to the time of DLF lesion application. With reflexes of equal size in both hindlimbs this logarithm is 0. If reflexes are double on the ipsilateral lesioned side (shown here as positive value), the logarithm is +0.3

first month. Figure 1B (right column) shows a dramatic increase in both the dynamic and static components of the ipsilateral triceps surae stretch reflex (hyperreflexia) 2 weeks after the lesion. An increase in gain of the dynamic ipsilateral reflex can also be observed (Fig. 2A). Reflex EMG activity also increased over the 1 month in this one animal (not shown). Note that the immediate increase in ipsilateral reflex torque activity was normalized by the 2nd month. Also note that the DLF injury produced cyclical changes (mostly decreases) in the contralateral dynamic torque amplitude over the course of the 5 months (Fig. 2B).

The ratio measures (lesion/intact Fig. 2C) revealed the subtle effects of the DLF lesion over time. However, unlike the ipsilateral torque records, the ratio measure suggested a biphasic enhancement of the ipsilateral reflex at 1–2 and 4–5 months. This temporal pattern of relatively greater ipsilateral tone reflects effects on reflex amplitude for both the ipsilateral and contralateral limb. Neither the velocity-dependent dynamic torque amplitude nor the dynamic ratio of the ipsilateral stretch reflex changed following the DLF lesion.

With regard to the neurological tests performed, low-thoracic DLF lesions produced some subtle correlates of human spasticity. These included increases in ipsilateral extensor muscle tone and tonic digital extension (which could also be evoked by light pressure applied to the plantar cushion). Of note was that suspension of the cat in the air produced strong extension and immobilization of the affected limb. Turning to the semiquantitative tests, hypermetria of the affected limb could also be demonstrated with the use of the positive weight support test. This test helped in identifying the efficacy of the applied lesion (see below). Although the Ashworth test revealed small increases in ipsilateral muscle tone, in our hands this method of assessment was not useful in following the reflex changes observed with the behavioral stretch model. No gross deficit in locomotion was observed.

Responses Following the Dorsal Quadrant Lesion

Generally the unilateral DQ lesion produced the same quantitative deficits as obtained with a DLF lesion: immediate enhanced ipsilateral extensor tonus, enhanced ipsilateral dynamic and static reflex components, increased gain of the ipsilateral dynamic reflex, and no change in either the dynamic ratio or the velocity sensitive dynamic component (not shown). Preliminary results indicate that the DQ lesion produced the same magnitude of increase in reflex amplitude as the DLF injury, but that this enhancement was sustained over a greater period (at least more than 1 month).

Neurological testing of the deficit also suggests that the DQ lesion is sustained over a greater period than the DLF injury, and that it produces a larger, more reproducible injury. This is based mainly on the results of the positive weight support test and passive stretch of the affected limb. As with the DLF lesion, this larger injury also produced tonic extension of the ipsilateral limb and digits (which could also be revealed as the Babinski sign) during suspension of the animal. Again, no gross deficit in locomotion was noted.

Although the precise extended temporal pattern of the DQ lesion-induced deficit is not yet known, Fig. 5A shows the muscle tone of both the ipsilateral and contralateral leg 4–5 weeks prior to and 4–5 weeks following this injury. The DQ lesion increased the slope of the input-output relationship for the ipsilateral reflex (0.030 versus 0.010 kg/degree). This increase in slope is supported by an increase in the gain of the dynamic phase following the DQ lesion. In addition, the ipsilateral torque amplitude obtained at the larger displacement angles was larger (see "Discussion").

Identification of an Anesthetic that Preserves Asymmetric Spinal Spasticity

The third aim of this research was to define an anesthetic regimen that would permit humane, acute cellular studies in cats, but would preserve the reflex asymmetries produced by a DQ lesion. Early studies in our laboratory indicated that intravenous infusion of ketamine preserved reflex activity and maintained the reflex asymmetry produced by a unilateral lesion of the DQ (Taylor et al. 1990). However, the use of ketamine alone as an anesthetic in small animals is still a controversial issue within the veterinary field (for reviews of the use of ketamine see, Domino 1990; Reich and Silvay 1989; White et al. 1982; Wright 1982). Thus, an important goal of this work was to establish whether ketamine infusion could provide a sufficient level of anesthesia for the purposes of acute electrophysiological experiments on cats.

Several studies have investigated the effect of anesthetics on segmental mono- and polysynaptic responses (for a review see Davidoff and Hackman 1983). Ketamine is unique in that it preserves stretch reflex activity and muscle tone during anesthesia, which is reflected in part by its CNS excitant properties (for a review of this concept see Winters 1976). Electrophysiological evidence suggests that ketamine (1–10 mg/kg) has a negligible effect upon monosynaptic reflexes, although reduction in the polysynaptic pathways has been observed (Chen and Chow 1975; Lodge and Anis 1984; Tang and Schroeder 1973). The effect of ketamine on segmental reflexes could result from the interaction at a number of receptors within the spinal cord or at a more supraspinal level (see Reich and Silvay 1989; White et al. 1982).

There are a small number of tantalizing anecdotal reports which suggest that effects of SCI on reflexes are preserved under anesthesia (particularly chloralose and light pentobarbital; Aoki et al. 1976; Bailey et al. 1980; Fujimori et al. 1966). With regard to Winter's schema of the action of anesthetics, it is interesting to note that chloralose and possibly light levels of pentobarbital attain stage IIc, characterized by CNS excitation (but not by CNS depression; Winters 1976).

The requirements for our behavioral animal model of SCI necessitate the use of an anesthetic that permits (a) repeated evaluation of reflex function before and after SCI with a minimum of side effects to the animal and to the experimenter, (b) comparability of stretch reflex function in the awake and acute electrophysiological state, using the DC torque motor (Fig. 1A), (c) stable maintenance of anesthesia in short (behavioral) and long-term (acute electrophysiological) studies, and (d) preservation of SCI-induced reflex deficits. Within the first three constraints, ketamine is the anesthetic of choice. Thus, we sought answers to the following two questions: (a) At what intravenous infusion rate of ketamine is electroencephalographic (EEG) arousal in response to noxious stimulation blocked in the cat? (b) What are the effects of ketamine anesthesia on stretch reflex EMG and torque activity, relative to the responses in the conscious animal? In addition, what was the effect of ketamine on reflex asymmetries produced by a unilateral, low thoracic DQ lesion?

Methods

To define anesthesia under ketamine we adopted both cortical (EEG) and vital sign (respiration rate, systolic blood pressure, heart rate) measures and quantified these in response to noxious stimulation of the tail (for a review of the assessment of anesthesia for animal surgery see Steffey 1983). Psychophysical calibration of the mechanical stimulus was performed in human subjects so as to define noxious and subthreshold (to pain) intensities over a continuous 2- to 4-min period.

Under halothane anesthesia three cats were implanted with chronic extradural stainless steel screw electrodes to enable monopolar EEG recording from the suprasylvian gyrus. Several weeks later this group was used in two anesthesia experiments, each separated by a 1- to 2-week interval. EEG analysis was performed by measuring absolute total power (0.6- to 30.0-Hz spectral analysis) over 12 20-s windows within 2-min of control (no stimulus) and 2 min of stimulus presentation. Integrated data were normalized for each animal, and the mean change in power for each 20-s window was calculated over all animals.

To characterize EEG and vital sign nociceptive-specific responses, a single intramuscular injection of ketamine (33.0 mg/kg) was given, and the responses were recorded periodically as ketamine was metabolized until the cat exhibited a weak orientation to the stimulus. Following the nociceptive-specific characterization of EEG and vital sign arousal, the ability of intravenous infusion of ketamine (10.0, 16.1, and 22.2 $mg\,kg^{-1}\,h^{-1}$) to block nociceptive-specific EEG and vital sign arousal components was investigated.

Stretch reflex activity was obtained from three cats in both the conscious and the ketamine anesthetized state (10.0, 16.1, and 22.2 $mg\,kg^{-1}\,h^{-1}$) by recording torque responses and triceps surae EMG activity. The absolute amplitudes of dynamic EMG and torque activity were monitored over three intravenous ketamine infusion levels. Finally, the effect of ketamine (13.3 $mg\,kg^{-1}\,h^{-1}$ intravenously) on SCI-induced reflex asymmetry from one animal was determined 4–5 weeks before and 4–5 weeks following a unilateral low thoracic DQ lesion. This was performed by analyzing the absolute amplitude and slope of the input-output relationship of the ipsilateral and contralateral dynamic torque reflex (see Fig. 5A).

Results of the Anesthesia Studies

The raw EEG signals and the response to noxious stimulation of the tail following an im ketamine injection at 33.0 mg/kg is shown in Fig. 3. The nociceptive-specific EEG arousal response is shown in the raw signals (A–C) and the power spectral records (D, E), and the change in EEG at each 20-s window is plotted against the 2-min control (no stimulus) and 2-min stimulus presentation period. Relative to the conscious EEG signals (A), even light ketamine anesthesia produced characteristically large amplitude delta wave activity (B). Noxious stimulation of the tail produced an immediate decrease in the number of delta waves at 74 min after ketamine injection (C). During suprathreshold noxious stimulation, the power within the delta band decreased, in agreement with the visual interpretation of this response (D, E). Thus, noxious stimulation changed primarily the delta EEG band.

As the effects of ketamine wore off, significant changes in total EEG power were observed following noxious pinch (see Fig. 3F at 36 min after ketamine administration). These changes were nociceptive specific and correlated well with increases in vital sign measures (not shown). In the subsequent study (not shown), ketamine at intravenous infusion levels of (10.0, 16.6, and 22.2 $mg\,kg^{-1}\,h^{-1}$) blocked nociceptive-specific EEG and vital sign arousal (Taylor et al. 1992a,b).

Results of the Reflex Studies

Both dynamic EMG and torque activity were maintained over all the intravenous infusion levels (10.0–22.2 $mg\,kg^{-1}\,h^{-1}$, Fig. 4), although there was a dose-dependent inhibition in the reflex. Note that the displacement angle required to elicit the dynamic reflex was increased when ketamine was given at 10.0 $mg\,kg^{-1}\,h^{-1}$ relative to the responses measured in the conscious animal. Both the dynamic and static components were equally affected. Neither the dynamic ratio nor the velocity-dependent stretch reflex components were affected within the range of doses delivered (not shown).

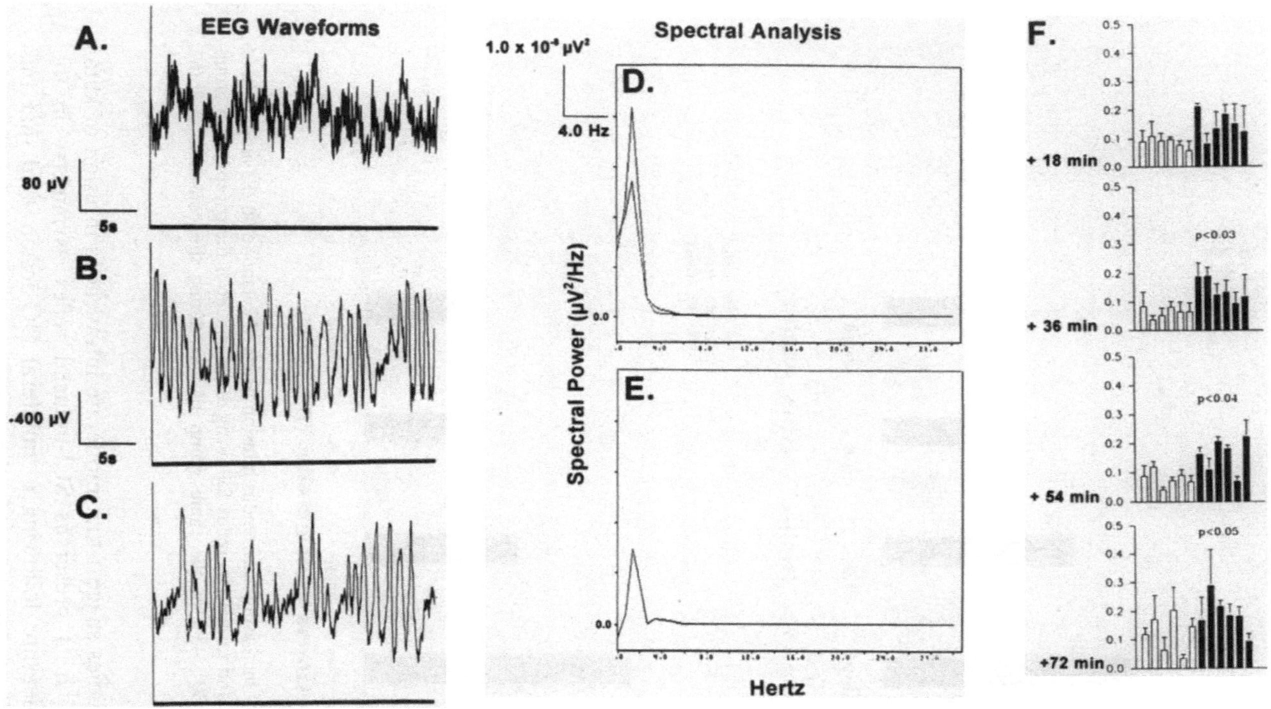

Fig. 3A–F. EEG signals during a 20-s window of activity collected from one cat in the quiet conscious state (**A**), during a 20-s control interval prior to stimulation (**B**), and during the first 20-s period of suprathreshold noxious mechanical stimulation of the tail (**C**) following a single im bolus injection of ketamine at 33.0 mg/kg (stimulus presented at 74 min after injection). Integrated 20-s power spectral records for the corresponding signals shown above (**D**) plotted against frequency (0–30 Hz) of the EEG. **E** The difference in spectral power between the unstimulated control and the noxious stimulation period. Note that noxious stimulation primarily changes the delta EEG band. **F** The histograms represent the change in total EEG power following suprathreshold noxious stimulation of the tail from 18 to 74 min following an intramuscular bolus dose of ketamine given at 33.0 mg/kg. *White bars*, the total power change at each 20-s window relative to the mean of the 2-min control period; *black bars*, each 20-s window in the 2-min stimulus period

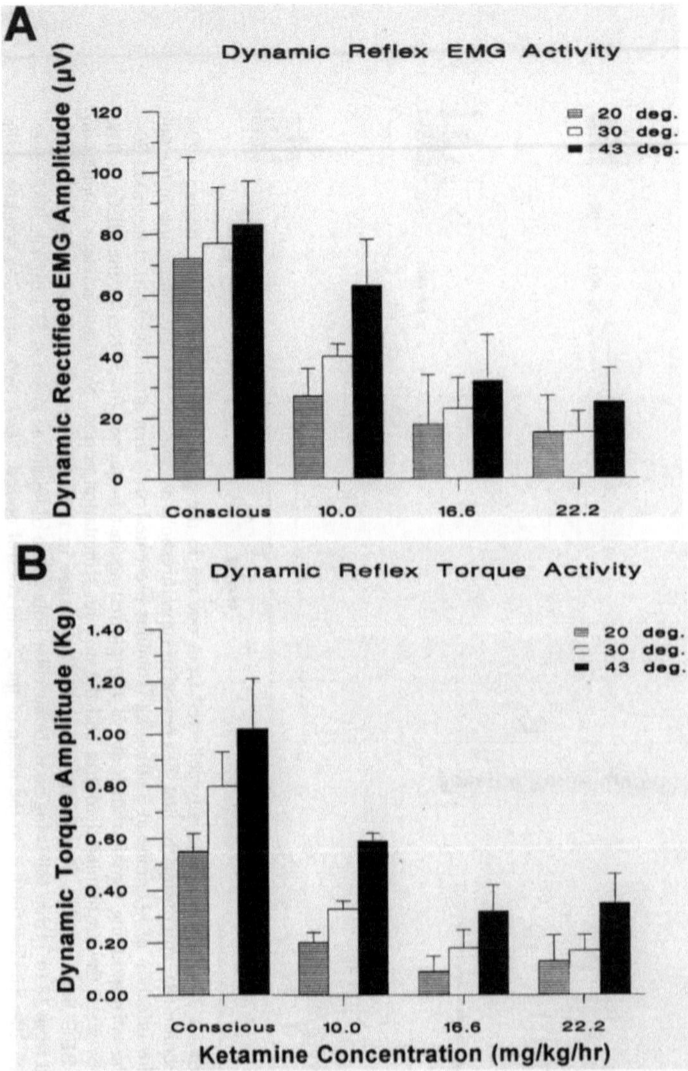

Fig. 4A,B. The effect of increasing intravenous infusion levels of ketamine on the amplitude of the dynamic EMG (**A**) and torque (**B**) reflex following displacement of the triceps surae group at three angles (20°, 30°, 43°) in the same group of cats during the conscious and the anesthetized state

Preliminary studies suggest that moderate intravenous infusion of ketamine ($13.1\,\mathrm{mg\,kg^{-1}\,h^{-1}}$) preserves SCI-induced reflex asymmetry in the dynamic torque measure following a unilateral DQ lesion, and that this unilateral lesion effect was present in the conscious state (for details of the reflex effects of the unilateral DQ lesion in the awake animal, shown in Fig. 5A, see above). The preservation of the SCI-induced ipsilateral hyperreflexia under ketamine anesthesia suggests that the absolute amplitude was

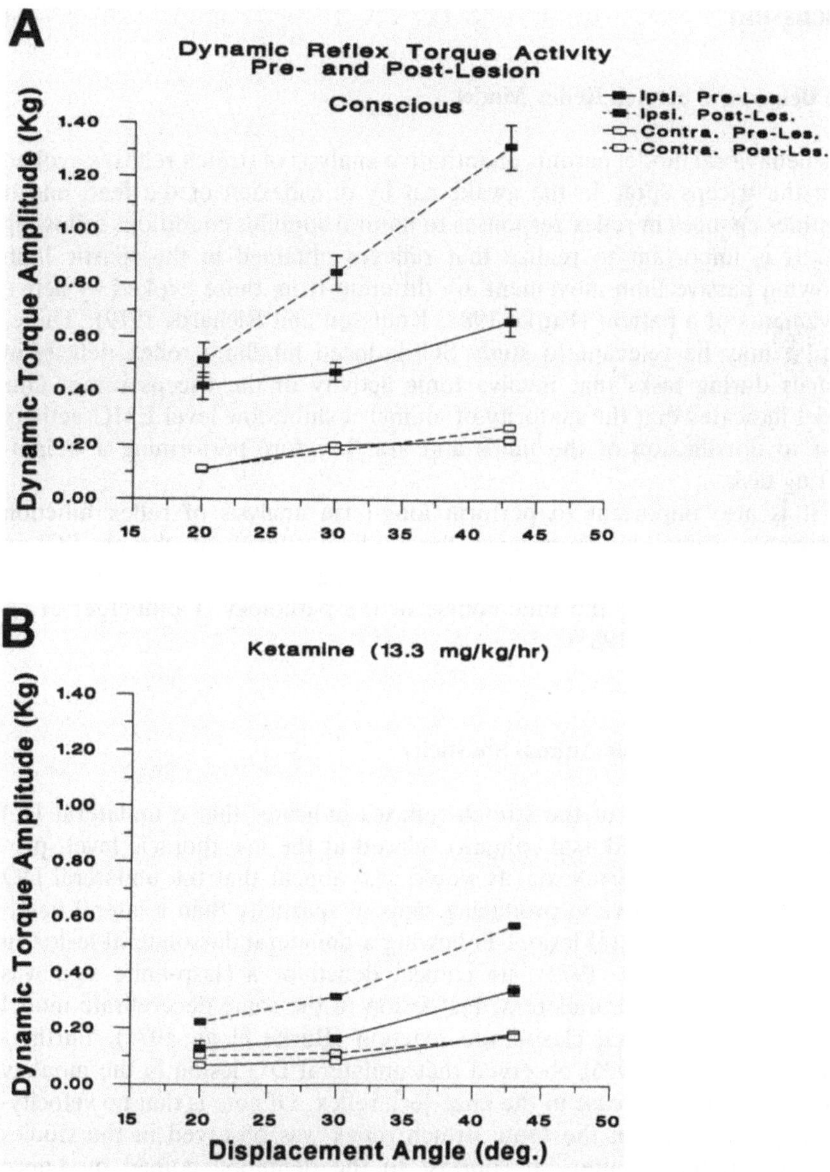

Fig. 5A,B. The effect of a unilateral DQ lesion on the ipsilateral and contralateral dynamic reflex torque amplitude 4–5 weeks before and at the equivalent time following injury, shown both in the conscious (**A**) and the ketamine anesthetized (**B**, 13.3. mg $kg^{-1} h^{-1}$) state. Dynamic stretch reflexes were evoked over three angular displacements (20°, 30°, 43°) at 60° s^{-1}

greater in this limb over all angular displacements (Fig. 5B). In addition, the slope of the input-output relationship was slightly greater compared to that observed with this agent before administration of the DQ lesion (0.016 rather than 0.010 kg/degree).

Discussion

The Behavioral Stretch Reflex Model

This behavioral model permits quantitative analysis of stretch reflexes evoked from the triceps surae in the awake cat by dorsiflexion of the feet, and it identifies changes in reflex responses to natural stimulus conditions following SCI. It is important to realize that reflexes obtained in the spastic limb following passive limb movement are different from those evoked by active movements of a patient (Burke 1988; Knutsson and Richards 1979). Therefore, it may be relevant to study SCI-induced hindlimb reflex deficits in animals during tasks that involve tonic activity of the triceps surae. Our model indicates that the majority of animals exhibit low-level EMG activity prior to dorsiflexion of the limbs and are therefore performing a weight-bearing task.

It is also important to perform long-term analysis of reflex function following a spinal lesion because longitudinal studies of chronic SCI in animals have the double advantage of validating the reliability of the testing method and assessing the time course of the pathology (Goldberger et al. 1990; Wiesendanger 1985).

The Optimal Lesion for Animal Spasticity

Quantitative analysis of the stretch reflexes indicates that a unilateral DQ lesion (including the dorsal column), placed at the low thoracic level, produces ipsilateral hyperreflexia. It would also appear that the unilateral DQ lesion is more effective in producing signs of spasticity than a lateral hemisection or a dorsolateral lesion. Following a unilateral dorsolateral lesion in the cat (Burke et al. 1972), no clinical deficit or a clasp-knife sign was observed. However, a unilateral DQ lesion in the same decerebrate model produced an ipsilateral clasp-knife reaction (Burke et al. 1972). Furthermore, Aoki et al. (1976) observed that unilateral DQ lesion in the monkey produced a weak increase in the knee-jerk reflex. Of note is that no velocity-dependent increase in the tonic stretch reflex was observed in the studies discussed in this chapter, in contrast to the definition coined by Lance (1980). It is possible that the behavioral reflex testing apparatus used in these studies does not permit the examination of the velocity-dependent reflex components in the conscious cat because the highest velocity produced is only around $60° s^{-1}$. Indeed, the crucial angular velocity required to evoke enhanced stretch reflex activity has been suggested to be in excess of $100° s^{-1}$ (for a review see Noth 1991).

The importance of producing a quantitative behavioral model of SCI is highlighted by the variable effects of the unilateral DLF injury (Fig. 2).

Only one animal (one of two) showed a clear increase in the ipsilateral stretch reflex over the 1st month, and the log ratio (lesion/intact) reflex amplitude waxed and waned over succeeding postoperative months. Of interest was the obvious cyclical decreases in the contralateral reflex following the DLF lesion (Fig. 2B). In the hemiparetic patient, reciprocal changes in upper limb stretch reflex patterns have been described in the "unaffected" contralateral limb (see Gandevia, this volume; Thilmann et al. 1990). These crossed effects may be thought of as reflex changes in response to intrinsic compensatory mechanisms. It is also relevant to systematically examine reflex activity over a range of displacement angles (Ashby and McCrea 1987; Katz and Rymer 1989). Figure 5A shows that the effect of the unilateral DQ lesion-induced injury can only be revealed with reflex testing over a wide range of displacement angles.

The data presented here also highlight the necessity of obtaining adequate pre- and postoperative reflex data from each animal. The possibility that late changes in reflex activity following spinal injury can be evaluated only by comparisons to both the preoperative and immediate postlesion reflex levels. In addition, such late changes in tone can be revealed only by comparisons with a stable control period. Normal side-to-side predominance of reflex activity should also be identified before application of a spinal lesion. Figure 5A represents the effect of the unilateral DQ lesion on both the ipsilateral and contralateral dynamic torque reflexes. This animal showed not only an increase in muscle tone after the lesion but also a large reflex asymmetry during the prelesion period. Thus, any quantification of the lesion-induced reflex deficit must account for the normal reflex asymmetry before any conclusion can be made regarding the magnitude of the spinal injury.

Release of spinal circuitry from supraspinal modulation would be expected to produce an immediate enhancement in reflex activity following SCI (Goldberger 1991; Wiesendanger 1991). The most likely candidate for SCI reflex abnormalities following dorsolateral spinal cord injury is damage to the descending reticulospinal pathway (Burke et al. 1972; for a review of this system see Lundberg 1982). It is likely that the reflex deficits observed in this study are due to loss of descending inhibitory modulation. This is supported by the development of the Babinski sign in these cats, a characteristic marker of loss of descending control that is associated with spastic hypertonia (Burke 1988; Katz and Rymer 1989). However, it is important to recognize that descending modulatory pathways typically project bilatarally (Kuypers 1981), and that bilateral effects on segmental circuitry have been described. Using electrical stimulation of the bulbar reticulospinal formation, residual crossed connections reaching the lumbosacral cord ipsilateral to a high lumbar hemisection have been shown (Eidelberg et al. 1986). In addition, the reticulospinal pathway has also been demonstrated to descend in the ventrolateral funiculus (Kuypers 1981). Indeed, inhibitory effects of a ventral reticulospinal pathway have been observed (Jankowska et al. 1968). Collectively these studies suggest the importance in evaluating contralateral reflex changes following unilateral SCI.

Anesthesia and Reflexes Under Ketamine in the Normal and SCI Animal

The present study demonstrates that preservation of reflex responses can occur with doses of ketamine that eliminate arousal following nociceptive stimulation. There is much controversy about the use of ketamine as an maintenance anesthetic and as such is seldom used in major surgical procedures (Report of a National Institutes of Health Workshop 1991). In addition, EEG measures as indicators of anesthesia have been found to be either equivocal or at variance with other findings, and their relationship as signs of conscious awareness during general anesthesia has been questioned (Jessop and Jones 1991). However, problems in the definition of EEG arousal may arise with the use of a poorly defined or inadequate noxious stimulus or from the EEG analysis procedure itself. The successful characterization of noxious stimulation-induced changes in EEG under ketamine anesthesia in the present study probably reflects calibration of the noxious stimulus using psychophysical calibration techniques and the development of a fast Fourier transform software system designed to analyze ketamine EEG waveforms during the stimulus presentation period. Furthermore, systematic analysis of changes of total EEG power in response to different stimulus strengths permits the identification of nociceptive-specific effects as distinguished from changes due to other unidentified factors. This study has established that intravenous ketamine infusion at moderate to high intravenous infusion levels is sufficient to block EEG arousal to pain and is therefore ideal as a maintenance anesthetic during acute electrophysiological experimentation.

Previous investigations have shown that monosynaptic reflexes are maintained in the cat during ketamine anesthesia (Chen and Chow 1975; Lodge and Anis 1984; Tang and Schroeder 1973). The present results (Fig. 5A) support these studies and extend our earlier finding (Taylor et al. 1990) that an intravenous infusion of ketamine preserves a reflex asymmetry that is induced by a dorsal quadrant lesion.

Although ketamine and other *N*-methyl-D-aspartate (NMDA) receptor antagonists have been proposed as neuroprotective agents in animal models of ischemia (for a review see Hall and Murdoch 1990), it is unlikely that such effects are active in this behavioral model of spinal spasticity using the ketamine anesthetized cat. Ketamine has been shown to be effective in ameliorating ischemic damage to rat hippocampal CA1 cells when administered as cumulative doses of 60–210 mg/kg and has been tested up to 8 h after injury (Church et al. 1988). In the study presented in this chapter (see Fig. 5B) the total cumulative intravenous dose of ketamine was less than 30 mg/kg. In addition, this agent was administered in the chronic period following SCI, between 4–5 weeks after the DQ spinal lesion. It should also be noted that other anesthetics such as the inhalant isoflurane or the parenteral barbiturate (sodium pentobarbital or thiopental) agents have also been shown to possess similar neuroprotective properties (Hall and Murdoch 1990).

Using barbiturate anesthesia in the monkey, a phenomenon described as "reflex resistance" to anesthesia has been observed in the hindlimb ipsilateral to a spinal lateral hemisection (d'Aquili et al. 1966). Anecdotal reports in this paper also refer to the "ankle jerk" reflex as being particularly resistant to anesthesia, and that the cat showed weaker "reflex resistance" to anesthesia than the monkey. Although scant, it would appear therefore that there is other evidence that administration of some anesthetics can maintain reflex asymmetries or spastic signs following SCI. Furthermore, the present study demonstrates that preservation of reflex responses can occur with doses of ketamine that eliminate arousal by nociceptive stimuli, and to our knowledge it may be the only study that has attempted to correlate the presence of a neurological deficit under anesthesia with that observed in the conscious state.

Conclusion

The results presented here establish an animal model of spasticity that is based on rigorous quantitative measures of stretch reflex activity in the awake and unanesthetized cat with measures obtained over a prolonged period before and after spinal cord injury (Taylor et al. 1992b).

A reproducible, immediate, and sustained increase of ipsilateral muscle tone and stretch reflex can be demonstrated. Intravenous infusion of ketamine as a maintenance anesthetic (Taylor et al. 1992a) preserves both stretch-evoked activity and SCI-induced reflex asymmetry. This anesthetic provides opportunities to correlate results obtained from the conscious behavioral and acute electrophysiological spastic animal models.

Acknowledgements. We would like to express many thanks to Carolyn Baum, Alex and Boza Radisavljevic, Dr. Dan Theele, Archana Vasanthakumar, and Dr. Alistair Webb. The advice and helpful comments of Dr. David Burke, Dr. Hans Hultborn, and Dr. Zev Rymer are gratefully acknowledged. Funding was provided by NIH grants NS 15913, NS 27511, and MH 15737 and by the State of Florida Spinal Research IDSTF.

References

Aoki M, Mori S, Fujimori B (1976) Exaggeration of knee-jerk following spinal hemisection in monkeys. Brain Res 107:471–485

Ashby P, McCrea DA (1987) Neurophysiology of spinal spasticity. In: Davidoff DA (ed) Congenital disorders and trauma & infections and cancer. Dekker, New York, pp 119–145 (Handbook of the spinal cord, vols 4, 5)
Bailey CS, Lieberman JS, Kitchell RL (1980) Response of muscle spindle primary endings to static stretch in acute and chronic spinal cats. Am J Vet Res 41(12):2030–2036
Bohannon RW, Smith MB (1987) Interrater reliability on modified Ashworth scale of muscle spasticity. Phys Ther 67:206–207
Brothers MF, McCrea DA, Tatton WG, Tasker RR (1983) Stretch reflexes after spinal lesions in the cat (Abstr). Soc Neurosci Abstr 9:528
Burke D (1988) Spasticity as an adaptation to pyramidal tract injury. Adv Neurol 47:401–423
Burke D, Knowles L, Andrews C, Ashby P (1972) Spasticity, decerebrate rigidity and the clasp-knife phenomenon: an experimental study in the cat. Brain 95:31–48
Carter RL, Ritz LA, Shank CP, Scott EW, Sypert GW (1991) Correlative electrophysiological and behavioral evaluation following L5 lesions in the cat: a Model of spasticity. Exp Neurol 114:206–215
Chen CF, Chow SY (1975) Effects of ketamine on synaptic transmission in cat spinal cord. Neuropharmacology 14:147–149
Church J, Zeman S, Lodge D (1988) The neuroprotective action of ketamine and MK-801 after transient cerebral ischemia in rats. Anesthesiology 69:702–709
D'Aquili EG, Chambers WW, Liu CN, McCouch GP (1966) Reflex resistance to anesthesia in partially denervated neurons. Exp Neurol 16:307–315
Davidoff RA, Hackman JC (1983) Drugs, chemicals, and toxins: their effects on the spinal cord. In: Davidoff RA (ed) Pharmacology. Dekker, New York, pp 409–477 (Handbook of the spinal cord, vol 1)
Domino EF (ed) (1990) Status of ketamine in anesthesiology. NPP, Ann Arbor
Eidelberg E, Nguyen LH, Deza LD (1986) Recovery of locomotor function after hemisection of the spinal cord in cats. Brain Res Bull 16:507–515
Fujimori B, Kato M, Matsushima S et al. (1966) Studies on the mechanism of spasticity following spinal hemisection in the cat. In: Granit R (ed) Muscular afferents and motor control. Almquist and Wiskell, Uppsala, pp 397–413 (Nobel symposium I)
Goldberger ME (1988) Spared-root deafferentation of cat's hindlimb: hierarchical regulation of pathways mediating recovery of motor behavior. Exp Brain Res 73:329–342
Goldberger ME (1991) The use of behavioral methods to predict spinal cord plasticity. Restorative Neurol Neurosci 2:339–350
Goldberger ME, Bregman BS, Vierck CJ Jr, Brown M (1990) Criteria for assessing recovery of function after spinal cord injury: behavioral methods. Exp Neurol 107:113–117
Hall R, Murdoch J (1990) Brain protection: physiological and pharmacological considerations. II. The pharmacology of brain protection. Can J Anaesth 37(7):762–777
Hultborn H, Malmsten J (1983) Changes in segmental reflexes following chronic spinal cord hemisection in the cat. Acta Physiol Scand 119:405–422
Jankowska E, Lund E, Lundberg A, Pompeiano O (1968) Inhibitory effects evoked through ventral reticulospinal pathways. Arch Ital Biol 106:124–140
Jessop J, Jones JG (1991) Conscious awareness during general anaesthesia – what are we attempting to monitor? Br J Anaesth 66(6):635–637
Katz RT, Rymer WZ (1989) Spastic hypertonia: mechanisms and measurement. Arch Phys Med Rehabil 70: 144–155
Knutsson E, Richards C (1979) Different types of disturbed motor control in gait of hemiparetic patients. Brain 102:405–430
Kuypers HGJM (1981) Anatomy of the descending pathways. In: Brooks VB (ed) Handbook of physiology, sect 1, vol 2. Williams and Wilkins, Baltimore, pp 597–666
Lance JW (1980) Symposium synopsis. In: Feldman RG, Young RR, Koella WP (eds) Spasticity: disordered motor control. Year Book Publishers, Chicago, pp 485–494
Lodge D, Anis NA (1984) Effects of ketamine and three other anaesthetics on spinal reflexes and inhibitions in the cat. Br J Anaesth 56:1143–1151

Lundberg A (1982) Inhibitory control from the brain stem of transmission from primary afferents to motoneurons, primary afferent terminals and ascending pathways. In: Sjolund B, Bjorklund A (eds) Brain stem control of spinal mechanisms. Elsevier, Amsterdam, pp 179–224

Murray M, Goldberger ME (1974) Restitution of function and collateral sprouting in the cat spinal cord: the partially hemisected animal. J Comp Neurol 158:19–36

Nelson SG, Mendell LM (1979) Enhancement in Ia-motoneuron synaptic transmission caudal to chronic spinal cord transection. J Neurophysiol 42(3):642–654

Noth J (1991) Trends in the pathophysiology and pharmacotherapy of spasticity. J Neurol 238:131–139

Reich DL, Silvay G (1989) Ketamine: an update on the first twenty-five years of clinical experience. Can J Anaesth 36(2):186–197

Report of a National Institutes of Health Workshop (1991) Preparation and maintenance of higher mammals during neuroscience experiments. NIH, Washington

Steffey EP (1983) Concepts of general anesthesia and assessment of adequacy of anesthesia for animal surgery. In: Kitchell RL, Erickson HH, Carstens E et al. (eds) Animal pain perception and alleviation. American Physiological Society, Bothesda, pp 133–150

Tang AH, Schroeder LA (1973) Spinal-cord depressant effects of ketamine and etoxadrol in the cat and the rat. Anesthesiology 39(1):37–43

Taylor JS, Friedman RM, Bao JX, Theele DP, Vierck CJ Jr, Munson JB (1990) Reflex asymmetry resulting from lesions of the spinal cord and effects thereon of anesthesia (Abstr). Soc Neurosci Abstr 16(1):48.1

Taylor JS, Vasanthakumar A, Vierck CJ Jr, Munson JB (1992a) Anesthesia and stretch reflex activity in the ketamine treated cat as measured with EEG and EMG analysis techniques (Abstr). Soc Neurosci Abstr 18(2):1403

Taylor JS, Baum CK, Radisavljevic B, Friedman RM, Vierck CJ, Munson JB (1992b) Stretch reflex activity in the awake behaving cat as a quantitative measure of acute and chronic spinal cord injury. J Neurotrauma 9(4):401

Thilmann AF, Fellows SJ, Garms E (1990) Pathological stretch reflexes on the "good" side of hemiparetic patients. J Neurol Neurosurg Psychiatry 53:208–214

Vierck CJ Jr, Greenspan JD, Ritz LA (1990) Long-term changes in purposive and reflexive responses to nociceptive stimulation following anterolateral chordotomy. J Neurosci 10(7): 2077–2095

White PF, Way WL, Trevor AJ (1982) Ketamine – its pharmacology and therapeutic uses. Anesthesiology 56:119–136

Wiesendanger M (1985) Is there an animal model of spasticity? In: Delwaide PJ, Young RR (eds) Clinical neurophysiology in spasticity. Elsevier, Amsterdam, pp 1–12

Wiesendanger M (1991) Neurophysiological bases of spasticity. In: Sindou M, Abbott R, Keravel Y (eds) Neurosurgery for spasticity. A multidisciplinary approach. Springer, Vienna, New York, pp 15–21

Winters WD (1976) Effects of drugs on the electrical activity of the brain: Anesthetics. Annu Rev Pharmacol Toxicol 16:413–426

Wright M (1982) Pharmacologic effects of ketamine and its use in veterinary medicine. J Am Vet Med Assoc 180(12):1462–1471

Young RR (1980) Summary remarks. In: Feldman RG, Young RR, Koella WP (eds) Spasticity: disorderd motor control. Symposia Specialists, miami, pp 495–500

Monoaminergic Inhibitory Control of Spinal Interneurons

E. Jankowska

Introduction

Several observations indicate that a failing monoaminergic control of spinal interneurons may be one of the main causes of the exaggerated stretch, flexion, and postural reflexes associated with spasticity: (a) The exaggerated stretch reflex may involve responses evoked by group II as well as by group I muscle spindle afferents (e.g., Burke et al. 1970, 1972; Ashby and McCrea 1987; Burke 1985). (b) Responses of group II origin are to a much greater extent evoked via interneurons than by direct actions of group II muscle afferents upon motoneurons (see Lundberg et al. 1977). (c) Interneurons interposed between group II muscle afferents and motoneurons (referred to as "group II interneurons") may mediate stretch-evoked excitation of motoneurons as well as flexion reflexes, crossed extension reflexes, and labyrinthine and neck reflexes, because sensory fibers and neurons that induce these reflexes utilize to a great extent the same interneurons (for references see Jankowska and Edgley 1993; Jankowska 1992). When group II interneurons become hyperexcitable they thus may be responsible for exaggeration of not only stretch, but also flexion and postural reflexes. (d) Two of the drugs that depress transmission from group II muscle spindle afferents (Edgley et al. 1988; Bras et al. 1989, 1990), skin receptors (Barbeau et al. 1987), and nociceptors (e.g., Belcher et al. 1978; Headley et al. 1978; Davies and Johnston 1984; Fleetwood-Walker et al. 1985) also reduce muscle tonus and exaggerated stretch and flexion reflexes in spastic patients. These drugs (the α_2 noradrenaline agonists tizanidine and clonidine) counteract spasticity of various etiologies, whether it is caused by central lesions (e.g., Hassan and McLellan 1980; Knutsson et al. 1982; Meinck et al. 1985) or spinal lesions, and whether the latter are associated with tetraplegia, paraplegia, hemiplegia (see, e.g., Bes et al. 1988; Mathias et al. 1989; Stewart et al. 1991), or multiple sclerosis (e.g., Lapierre et al. 1987; Eyssette et al. 1988). Thus, these drugs must operate, at least in part, at the spinal level. (e) In contrast to transmission from group II muscle afferents, trans-

Department of Physiology, University of Göteborg, Medicinaregatan 11, 413 90, Göteborg, Sweden

Thilmann et al. (Eds.)
Spasticity: Mechanisms and Management

mission from group I muscle afferents is not depressed by monoamines (Wiesendanger et al. 1984; Edgley et al. 1988; Bras et al. 1989, 1990), so that the depression of the exaggerated stretch reflexes by locally acting α_2 agonists cannot be explained by an interaction of these drugs with monosynaptically evoked excitation of motoneurons. (f) A similarly selective depression of reflex actions of group II, but not of group I, afferents may be evoked by stimuli applied within nuclei of origin of the descending monoaminergic pathways (locus ceruleus and medullary raphe nuclei; Skoog and Noga 1991; Noga et al. 1992; Jankowska et al. 1992). The first three of these observations have recently been reviewed (see, e.g., Burke 1985; Pierrot-Deseilligny and Mazières 1985; Jankowska 1992); only the last three will therefore be discussed in the following.

Depressant Effects of Monoamines on Interneurons in Polysynaptic Flexion-Extension Reflex Pathways

Transmission in polysynaptic spinal reflex pathways from group II and III muscle afferents and from high-threshold skin and joint afferents (flexion reflex afferents; see Eccles and Lundberg 1959) is strongly depressed after systemic application of L-dopa and 5-HTP (precursors of noradrenaline and serotonin, respectively). Both the excitatory and inhibitory actions of these afferents on motoneurons are affected (see, e.g., Andén et al. 1966; Schomburg and Steffens 1988; and Marshall 1983 for review). However, the depressive effects of monoamines are restricted to certain categories of neurons with input from flexion reflex afferents and/or nociceptors, since responses of only some neurons evoked by these afferents are depressed by locally applied noradrenaline (NA) and serotonin (5-HT) (Biscoe et al. 1966; Engberg and Ryall 1966; Weight and Salmoiraghi 1966; Jankowska et al. 1967; Jordan et al. 1977; Belcher et al. 1978; Headley et al. 1978; Davies and Johnston 1984; Fleetwood-Walker et al. 1985; Howe and Zieglgänsberger 1987). The depressive effects of monoamines on polysynaptic flexion-extension reflex pathways cannot yet be linked to their actions at any particular sites along these pathways.

Depressant Effects of Monoamines on Interneurons in Disynaptic Pathways from Group II Muscle Afferents

Nerve impulses in group II muscle afferents induce reflex responses not only via interneurons of polysynaptic flexion-extension reflex pathways but also disynaptically, i.e., via single interneurons interposed between group II afferents and motoneurons (Edgley and Jankowska 1987a,b; Cavallari et al.

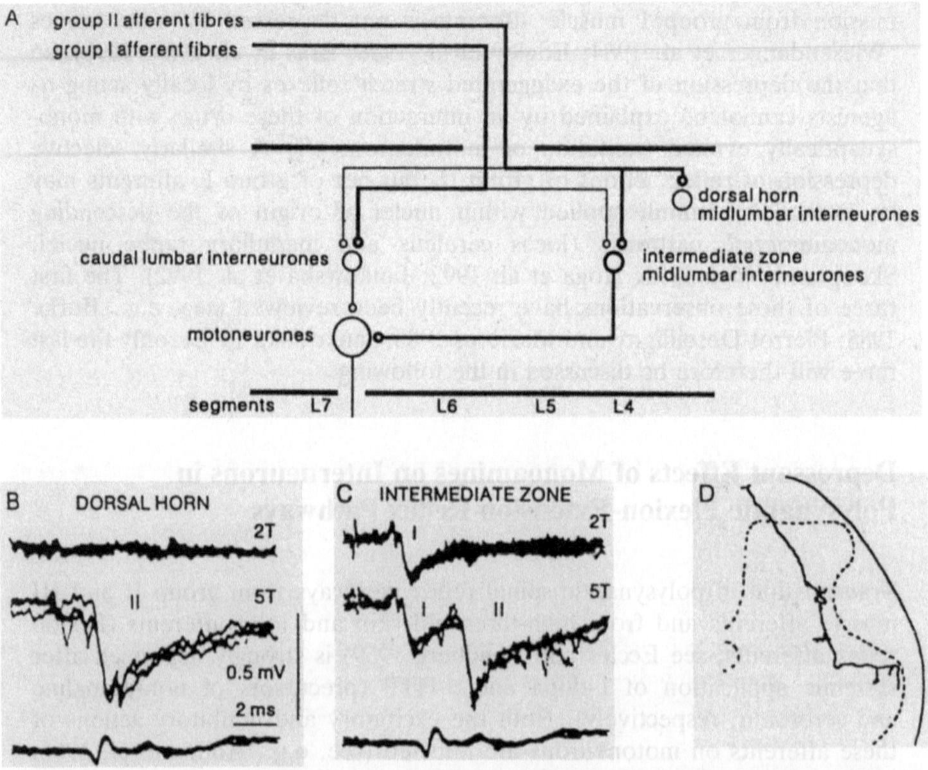

Fig. 1. A Diagram of disynaptic neuronal pathways between group I and II muscle afferents and motoneurons: via interneurons located at the same segmental level as hindlimb motoneurons and via interneurons in more rostral segments. *Thicker* and *thiner lines* indicate stronger and weaker connections, respectively. **B** An example of population EPSPs (*middle traces*) evoked by group II afferents in the dorsal horn. *Top* and *bottom traces* show lack of effects of group I afferents in the dorsal horn and records of the incoming volleys from the surface of the spinal cord, respectively. **C** Examples of population EPSPs in the intermediate zone (top and middle records) with an early component (*I*) evoked by group I afferents and a late component (*II*) evoked by group II afferents. *Bottom trace* shows incoming volleys as in **B**. In traces of **B** and **C** stimulus intensities twice and five times threshold (2T and 5T) were near maximal for group I and group II afferents of the deep peroneal nerve, respectively (modified from Edgley and Jankowska 1987). **D** Reconstruction of a single group II muscle spindle afferent labeled after intra-axonal injection of HRP by Fyffe (modified from Fyffe 1979). Note that the fiber gave off collaterals both in the dorsal horn and in the intermediate zone

1987; see Fig. 1). Effects of monoamines on disynaptic pathways from group II muscle afferents could, accordingly, be investigated at the level of synapses between the afferents and the group II interneurons. Experimental conditions for such an investigation are very favorable, as many group II interneurons are located within a restricted area of the midlumbar segments. Effects of monoamines could therefore be investigated both on monosynaptically evoked responses of single interneurons and on monosynaptically evoked postsynaptic potentials induced in their populations (population EPSPs, or focal field potentials). As shown in Fig. 1B and C, distinct population EPSPs are induced by group II afferents (by electrical stimulation of a muscle nerve with the intensity five times (5T) but not twice threshold (2T) for the most excitable fibers of this nerve). The figure shows that such population EPSPs are induced both in the dorsal horn and in the intermediate zone, although in the intermediate zone they are preceded by shorter latency and lower threshold potentials of group I origin (Edgley and Jankowska 1987a). Reconstructions of axonal projections of single group II muscle afferents made after intra-axonal labeling of functionally identified afferents [with an example of reconstructions made by Fyffe (1979) in Fig. 1D] and in physiological studies (Fu and Schomburg 1974) suggest that these population EPSPs are primarily evoked by axon collaterals of the same sensory fibers. At least there are no indications that different subgroups of group II afferents synapse on neurons in the dorsal horn and on neurons in the intermediate zone.

Figure 2A and C shows that population EPSPs of group II origin are very effectively depressed by monoamines applied at either of these locations, and that the group I components of EPSPs evoked in the intermediate zone are then unchanged (Bras et al. 1988, 1989, 1990). A similarly effective depression of responses of individual interneurons evoked by group II (Fig. 2D), but not of those evoked by group I, afferents is illustrated in Fig. 2B,D and in Fig. 2E, respectively.

It turned out, however, that transmission from group II afferents to neurons in the dorsal horn and in the intermediate zone is under a differential monoaminergic control. In the intermediate zone it is very effectively depressed by a local application of NA, its precursor L-dopa, and its agonists tizanidine, BHT-933, clonidine, isoproterenol, and phenylephrine (Bras et al. 1988, 1989, 1990), as well as by a systemic application of L-dopa (Edgley et al. 1988), while it is hardly or much less affected by serotonin and its specific agonists (Bras et al. 1990). In contrast, serotonin and five of six of its agonists very effectively depressed transmission from group II afferents in the dorsal horn but not, or much less, in the intermediate zone (Bras et al. 1990). Differential effects of two NA and two 5-HT agonists are shown in Fig. 3, in which the mean amplitudes of the population EPSPs recorded in the intermediate zone and in the dorsal horn are plotted against the time of ionophoresis of the drugs.

These observations indicate that transmission via interneurons in disynaptic pathways from group II afferents to motoneurons ought to be under a preferential control of descending noradrenergic pathways, while descending serotonergic pathways would preferentially gate transmission through the dorsal horn interneurons, i.e., in polysynaptic pathways from group II afferents. Neither the noradrenergic nor the serotonergic descending neurons were, on the other hand, expected to influence transmission from group I afferents. These expectations were confirmed in experiments in

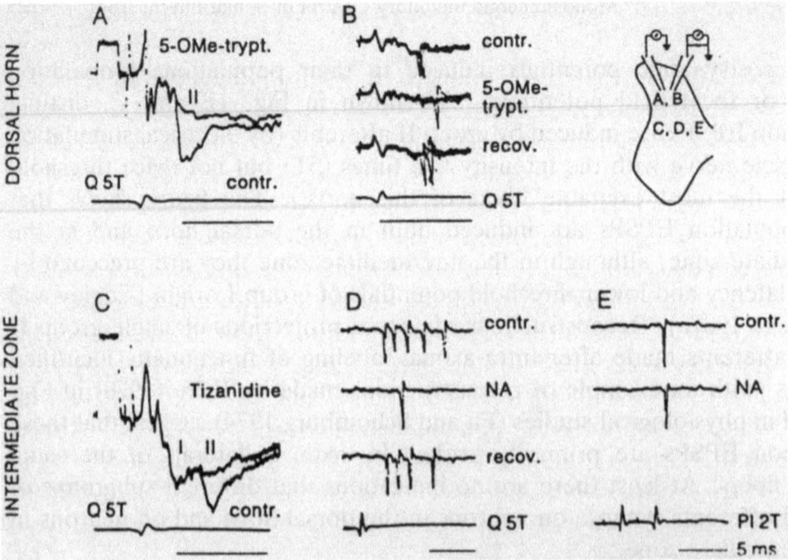

Fig. 2A–E. Examples of depression of responses evoked by group-II afferents by monoamines. In **A** and **C** are examples of the depression of population EPSPs evoked by group II afferents (*II*) by a serotonin agonist, *5-OMe-trypt*amine, in the dorsal horn and by a noradrenaline α_2 agonist, *tizanidine*, in the intermediate zone. The smaller of the two superimposed *upper traces* are potentials recorded after a few minutes of ionophoresis of the drugs, while the larger ones are control records. In **B** and **D** are examples of the depression of responses evoked by group II afferents in two interneurons by the same serotonin agonist and by noradrenaline (*NA*), while records in **E** illustrate a lack of effect of NA on responses evoked by group-I afferents. The records were obtained before (*contr.*), during, and after recovery following application of the two drugs, as indicated. The interneurons were excited by group II afferents of quadriceps (*Q 5T*) and by group I afferents of plantaris (*Pl 2T*). The diagram shows the experimental arrangement and the location of the recording and drug-containing electrodes. (Modified from Bras et al. 1989, 1990)

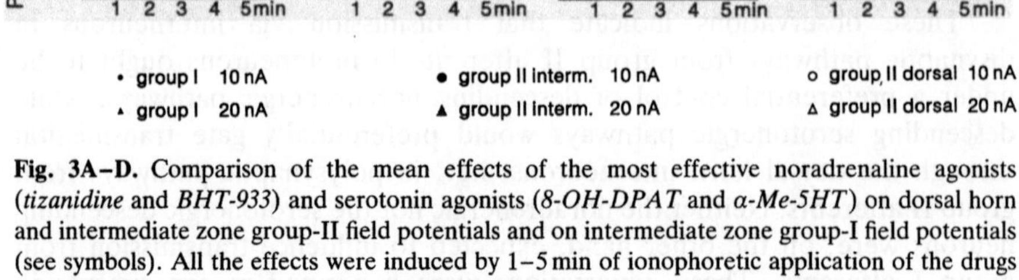

Fig. 3A–D. Comparison of the mean effects of the most effective noradrenaline agonists (*tizanidine* and *BHT-933*) and serotonin agonists (*8-OH-DPAT* and *α-Me-5HT*) on dorsal horn and intermediate zone group-II field potentials and on intermediate zone group-I field potentials (see symbols). All the effects were induced by 1–5 min of ionophoretic application of the drugs using 10 nA or 20 nA DC current. (Modified from Bras et al. 1990)

which effects of conditioning electrical stimulation within nuclei of origin of the descending monoaminergic neurons (the locus ceruleus/subceruleus, Kölliker-Fuse, and raphe magnus, obscurus, and pallidus nuclei) were tested on monosynaptic population EPSPs of group I and group II origin in midlumbar segments (Noga et al. 1992; Jankowska et al. 1993). As illustrated in Fig. 4A, only group II components of these potentials were depressed 100–200 ms after a short train of conditioning stimuli, i.e., after nerve impulses in fibers conducting at less than 5 m/s would reach the explored spinal segments. A similarly selective depression of disynaptically evoked EPSPs and IPSPs of group II origin was found in hindlimb motoneurons following such stimuli (Fig. 4B–D; see Jankowska et al. 1993). Thus, one may hypothesize that when the ceruleo-spinal and/or raphespinal neurons fail to gate transmission from group II muscle afferents to group II interneurons, a continuous bombardment of these interneurons by nerve impulses in secondary endings of muscle spindles (see Matthews 1972) makes them

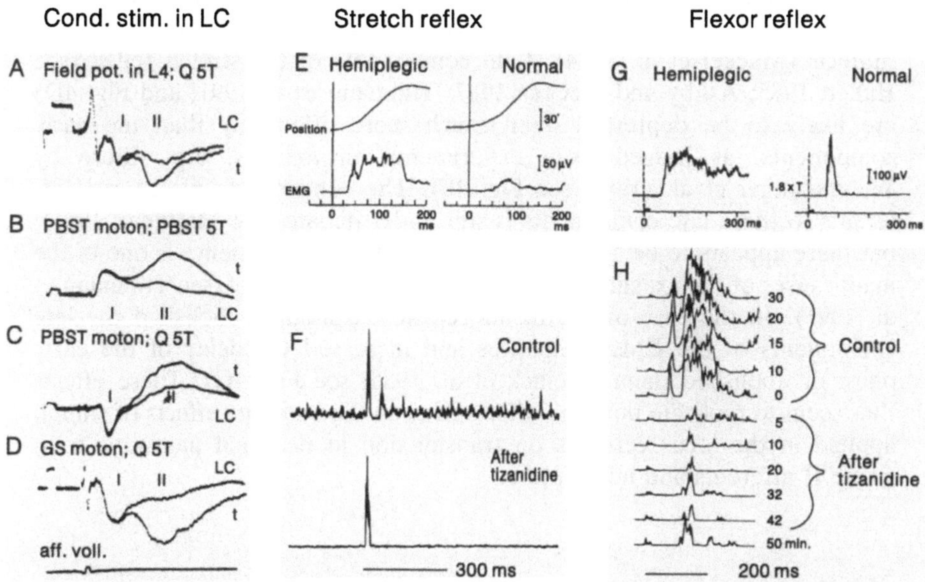

Fig. 4A–H. Examples of the selective depression of late components of responses evoked by 5T stimulation of a motor nerve following a short train of stimuli applied within the locus ceruleus/subceruleus region. **A** Depression of the group II component of a focal field potential in the intermediate zone. **B–D** Depression of the group II components of PSPs evoked in motoneurons. (Modified from Jankowska and Edgley 1992) **E** Comparison of EMG responses evoked on the affected and on the normal side during stretch reflex in a hemiplegic patient. (Modified from Hallett 1985) **F** Effects of tizanidine on early and late components of EMG responses during the stretch reflex in a decerebrate cat. (Modified from Wiesendanger et al. 1984) **G** Comparison of EMG responses evoked on the affected and on the normal side during flexor reflex in a hemiplegic patient. (Modified from Meinck et al. 1985) **H** Effects of tizanidine on EMG responses during flexor reflex in a hemiplegic patient. (Modified from Meinck et al. 1985)

hyperexcitable, so that they then respond much more vigorously to all nerve impulses that reach them. This hypothesis may appear to be at variance with reports that stimuli applied in the locus ceruleus may not only inhibit but also facilitate some spinal reflex actions (for references see Palmeri and Wiesendanger 1990; Noga et al. 1992). However, it is difficult to evaluate to what extent the facilitatory actions of these stimuli were mediated by noradrenergic or non-noradrenergic descending tract neurons, especially in the cat, where fast conducting neurons were reported to be responsible for a great part of the facilitation (see, e.g., Fung and Barnes 1987; and Noga et al. 1992 for further comments). Facilitation of the flexor reflexes has, in addition, been shown to involve α_1 adrenoceptors (Palmari and Wiesendanger 1990). It might thus be secondary to actions of ceruleo-spinal neurons on motoneurons (see Conway et al. 1988), rather than on interneurons, since α_1 agonists depress only synaptic actions of group II afferents (Bras et al. 1990) and have no effect on synaptic actions of nociceptors (Fleetwood-Walker et al. 1985) at the premotoneuronal level.

Tizanidine has been found to reduce EMG responses evoked by muscle stretches both in spastic patients (Hassan and McLellan 1980) and in normal subjects (Mackel et al. 1984). Late components of the stretch reflex (see Hallett 1985; Ashby and McCrea 1987; Thilmann et al. 1991; and Fig. 4D) are likely to be depressed by it much more effectively than the early components, as judged from experiments on animals, e.g., those by Wiesendanger et al. (1984; see Fig. 4F). The contribution of group II afferents to these late components is still under debate (see Matthews 1991), but there appears to be a general consensus that their presence is one of the main causes of the exaggerated stretch reflexes in spasticity (see Thilmann et al. 1991). In the case of the flexor reflexes, tizanidine abolished the late components of the EMG responses and increased the delay of the early ones, or abolished them (Meinck et al. 1985; see Fig. 4G). These effects thus seem to replicate both the effects of tizanidine and the effects of stimuli applied in the locus ceruleus on transmission in neuronal pathways from group II afferents and nociceptors.

Additional Observations on Contribution of Midlumbar Group II Interneurons to Exaggerated Spinal Reflexes Combined with Spasticity

Spinal cord hemisections caudal to the location of midlumbar interneurons (within the L5 segment) have recently been reported by Carter et al. (1991) to induce more consistently exaggerated hypertonia, clonus, and stretch and flexion reflexes than lesions made rostral to the location of these interneurons (within the upper lumbar or lower thoracic segments; see, e.g.,

Hultborn and Malmsten 1983a,b; Ashby and McCrea 1987). However, before the stronger effects of the L5 lesions are attributed to disruption of connections between midlumbar group II interneurons and motoneurons, the relative contribution of these interneurons and of a neuronal system described by Cavallari and Pettersson (1989) ought to be compared. The latter neurons are located more rostral (within the L2–3 segments) than the midlumbar group II interneurons. It should therefore be verified whether chronic hemisections at a caudal L3 level are followed by spastic symptoms as strong as those that follow hemisections at an L5 level, or by similarly weak effects as those following hemisections at a lower thoracic level.

Closing Remarks

Several supraspinal and propriospinal neuronal systems are involved in gating transmission in spinal reflex pathways (see Lundberg 1982 and Jankowska 1992, for review). Lesions affecting a number of these neuronal systems may thus be followed by a similar disinhibition of spinal neurons and the resulting exaggeration of spinal reflexes. It may nevertheless be justified to single out the monoaminergic inhibitory control of transmission in disynaptic pathways from group II afferents, because it is particularly potent and because it involves interneurons that constitute an important final common premotoneuronal path. Another reason for singling out the monoaminergic inhibitory control of transmission from group II afferents is that it provides a much more selective mechanism for counteracting spasticity than the generalized effects of baclofen. It may therefore become a mechanism of choice in the future treatment of spasticity.

However, a failing monoaminergic control of transmission between group II muscle afferents and interneurons may be only one of the spinal mechanisms of the exaggerated reflexes. An enhancement of excitatory actions of serotonergic neurons upon motoneurons (see Hounsgaard et al. 1986; and Hultborn, this volume) may be another, but the two may well operate in concert. For instance, central lesions of the noradrenergic, but not of the serotonergic, descending neuronal systems might cause a parallel disinhibition of raphe neurons (see Sagen and Proudfit 1985, 1986) and of group II intermediate zone spinal interneurons by noradrenergic neurons. Higher discharge rates of raphe neurons and stronger excitatory actions of group II interneurons upon motoneurons, in which self-sustained firing is evoked under the influence of serotonergic neurons, would then jointly lead to the exaggeration of motoneuronal responses.

Acknowledgements. The study was supported by grants from the Swedish Medical Research Council (project 5648), Sandoz Ltd. and Magnus Bergvalls Foundation.

References

Andén NE, Jukes MGM, Lundberg A, Vyklicky L (1966) The effect of dopa on the spinal cord. 1. Influence on transmission from primary afferents. Acta Physiol Scand 67:373–386

Ashby P, McCrea DA (1987) Neurophysiology of spinal spasticity. In: Davidoff RA (ed) Handbook of the spinal cord. Decker, New York, pp 119–143

Barbeau H, Julien C, Rossignol S (1987) The effects of clonidine and yohimbine on locomotion and cutaneous reflexes in the adult chronic spinal cat. Brain Res 437:83–96

Belcher G, Ryall RW, Schaffner R (1978) The differential effects of 5-hydroxytryptamine, noradrenaline and raphe stimulation on nociceptive and non-nociceptive dorsal horn interneurones in the cat. Brain Res 151:303–321

Bes A, Eysette M, Pierrot-Deseilligny E, Rohmer F, Warter JM (1988) A multicentre, double-blind trial of tizanidine, a new antispastic agent, in spasticity associated with hemiplegia. Curr Med Res Opin 10:709–718

Biscoe TJ, Curtis DR, Ryall RW (1966) An investigation of catecholamine receptors of spinal interneurones. Int J Neuropharmacol 5:429–434

Bras H, Cavallari P, Jankowska E (1988) An investigation of local actions of ionophoretically applied dopa in the spinal cord. Exp Brain Res 71:447–449

Bras H, Cavallari P, Jankowska E, McCrea D (1989) Comparison of effects of monoamines on transmission in spinal pathways from group-I and -II afferents in the cat. Exp Brain Res 76:27–37

Bras H, Jankowska E, Noga BR, Skoog B (1990) Comparison of effects of various types of NA and 5-HT agonists on transmission from group-II muscle afferents in the cat. Eur J Neurosci 2:1029–1039

Burke D (1985) Mechanisms underlying the tendon jerk and H-reflex. In: Delwaide PJ, Young RR (eds) Clinical neurophysiology in spasticity. Restorative neurology, vol 1. Elsevier Amsterdam pp 55–62

Burke D, Gillies JD, Lance J (1970) The quadriceps stretch reflex in human spasticity. J Neurol Neurosurg Psychiatry 33:216–223

Burke D, Knowles L, Andrews C, Ashby P (1972) Spasticity, decerebrate rigidity and the clasp-knife phenomenon: an experimental study in the cat. Brain 95:31–48

Carter RL, Ritz LA, Shank CP, Scott EW, Sypert GW (1991) Correlative electro-physiological and behavioral evaluation following L-5 lesions in the cat: a model of spasticity. Exp Neurol 114:206–215

Cavallari P, Pettersson LG (1989) Tonic suppression of reflex transmission in low spinal cats. Exp Brain Res 77:201–212

Cavallari P, Edgley SA, Jankowska E (1987) Post-synaptic actions of midlumbar interneurones on motoneurones of hind-limb muscles in the cat. J Physiol (Lond) 389:675–690

Conway BA, Hultborn H, Kiehn O, Mintz I (1988) Plateau potentials in α-motoneurones induced by i.v. injection of L-dopa and clonidine in the spinal cat. J Physiol (Lond) 405:369–384

Davies J, Johnston SE (1984) Selective antinociceptive effects of tizanidine (DS 103-282), a centrally acting muscle relaxant, on dorsal horn neurones in the feline spinal cord. Br J Pharmacol 82:409–421

Eccles RM, Lundberg A (1959) Synaptic actions in motoneurones by afferents which may evoke the flexion reflex. Arch Ital Biol 97:199–221

Edgley SA, Jankowska E (1987a) Field potentials generated by group I and II muscle afferents in the middle lumbar segments of the cat spinal cord. J Physiol (Lond) 385:393–413

Edgley SA, Jankowska E (1987b) An interneuronal relay for group I and II muscle afferents in the midlumbar segments of the cat spinal cord. J Physiol (Lond) 389:675–690

Edgley SA, Jankowska E, Shefchyk S (1988) Evidence that interneurones in reflex pathways from group-II afferents are involved in locomotion in the cat. J Physiol (Lond) 403:57–73

Engberg I, Ryall RW (1966) The inhibitory action of noradrenaline and other monoamines on spinal neurones. J Physiol (Lond) 185:298–322

Eyssette M, Rohmer F, Serratrice G, Warter JM, Boisson D (1988) Multicentre, double-blind trial of a novel antispastic agent, tizanidine, in spasticity associated with multiple sclerosis. Curr Med Res Opin 10:699–708

Fleetwood-Walker SM, Mitchell R, Hope PJ, Molony V, Iggo A (1985) An α_2 receptor mediates the selective inhibition by noradrenaline of nociceptive responses of identified dorsal horn neurones. Brain Res 344:243–254

Fu TC, Schomburg ED (1974) Electrophysiological investigation of the projection of secondary muscle spindle afferents in the cat spinal cord. Acta Physiol Scand 91:314–329

Fung SJ, Barnes CHD (1987) Membrane excitability changes in hindlimb motoneurons induced by stimulation of the locus coeruleus in cats. Brain Res 402:230–242

Fyffe REW (1979) The morphology of group-II muscle afferent fibre collaterals. J Physiol (Lond) 296:39–40P

Hallett M (1985) Spasticity: assessment and pathophysiology (modification of long-latency reflexes). In: Delwaide PJ, Young RR (eds) Clinical neurophysiology in spasticity: Restorative neurology, vol 1. Elsevier, Amsterdam, pp 83–93

Hassan N, McLellan DL (1980) Double-blind comparison of single doses of DS-103-282, baclofen and placebo for suppression of spasticity. J Neurol Neurosurg Psychiatry 43: 1132–1136

Headley PM, Duggan AW, Griersmith BT (1978) Selective reduction by noradrenaline and 5-hydroxytryptamine of nociceptive responses of cat dorsal horn neurones. Brain Res 145: 185–189

Hounsgaard J, Hultborn H, Kiehn O (1986) Transmitter-controled properties of alpha-motoneurones causing long-lasting motor discharge to brief excitatory inputs. Prog Brain Res 64:39–49

Howe JR, Zieglgänsberger W (1987) Responses of rat dorsal horn neurons to natural stimulation and to iontophoretically applied norepinephrine. J Comp Neurol 255:1–17

Hultborn H, Malmsten J (1983a) Changes in segmental reflexes following chronic spinal cord hemisection in the cat. I. Increased monosynaptic and polysynaptic ventral root discharges. Acta Physiol Scand 119:405–422

Hultborn H, Malmsten J (1983b) Changes in segmental reflexes following chronic spinal cord hemisection in the cat. II. Conditioned monosynaptic test reflexes. Acta Physiol Scand 119:423–433

Jankowska E (1992) Interneuronal relay in spinal pathways from proprioceptors. Prog Neurobiol 38:335–378

Jankowska E, Edgley S (1993) Interactions between pathways controlling posture and gait at the level of spinal interneurones in the cat. Prog Brain Res 97:161–171

Jankowska E, Jukes MGM, Lund S, Lundberg A (1967) The effect of dopa on the spinal cord. 5. Reciprocal organization of pathways transmitting excitatory action to alpha motoneurons of flexors and extensors. Acta Physiol Scand 70:369–388

Jankowska E, Riddell JS, Skoog B, Noga BR (1993) Gating of transmission to motoneurones by stimuli applied in the locus coeruleus and raphe nuclei of the cat. J Physiol (Lond) 461:705–722

Jordan LM, McCrea DA, Steeves JD, Menzies JE (1977) Noradrenergic synapses and effects of noradrenaline on interneurons in the ventral horn of the cat spinal cord. Can J Physiol Pharm 55:399–412

Knutson E, Mårtensson A, Gransberg L (1982) Antiparetic and antispastic effects induced by tizanidine in patients with spastic paresis. J Neurol Sci 53:187–204

Lapierre Y, Bouchard S, Tansey C, Gendron D, Barkas WJ, Francis GS (1987) Treatment of spasticity with tizanidine in multiple sclerosis. Can J Neurol Sci 14:513–517

Lundberg A (1982) Inhibitory control from the brain stem of transmission from primary afferents to motoneurones, primary afferent terminals and ascending pathways. In: Sjölund

B, Björklund A (eds) Brain stem control of spinal mechanisms. Elsevier Biomedical, Amsterdam, pp 179–224

Lundberg A, Malmgren K, Schomburg ED (1977) Comments on reflex actions evoked by electrical stimulation of group-II muscle afferents. Brain Res 122:551–555

Mackel R, Brink EE, Nakajima Y (1984) Action of tizanidine on responses of forearm flexors and extensors to torque disturbances. J Neurol Neurosurg Psychiatry 47:1109–1116

Marshall KC (1983) Catecholamines and their actions in the spinal cord. In: Davidoff RA (ed) Handbook of the spinal cord. Decker, New York, pp 275–328

Mathias CJ, Luckitt J, Desai P, Baker H, El-Masri W, Frankel HL (1989) Pharmacodynamics and pharmacokinetics of the oral antispastic agent tizanidine in patients with spinal cord injury. J Rehabil Res Dev 26:9–16

Matthews PBC (1972) Mammalian muscle spindles and their central action. Arnold, London, pp 1–630

Matthews PBC (1991) The human stretch reflex and the motor cortex. TINS 14:87–91

Meinck HM, Benecke R, Conrad B (1985) Spasticity and the flexor reflex. In: Delwaide PJ, Young RR (eds) Clinical neurophysiology in spasticity. Restorative neurology, vol 1. Elsevier, Amsterdam, pp 41–54

Noga BR, Bras H, Jankowska E (1992) Transmission from group-II muscle afferents is depressed by stimulation of locus coeruleus/subcoeruleus, Kölliker-Fuse and raphe nuclei in the cat. Exp Brain Res 88:502–516

Palmeri A, Wiesendanger M (1990) Concomitant depression of locus coeruleus neurons and of flexor reflexes by an α2-adrenergic agonist in rats: a possible mechanism for an α2-mediated muscle relaxation. Neuroscience 34:177–187

Pierrot-Deseilligny E, Mazieres L (1985) Spinal mechanisms underlying spasticity. In: Delwaide PJ, Young RR (eds) Clinical neurophysiology in spasticity. Restorative neurology, vol 1. Elsevier, Amsterdam, pp 63–76

Sagen J, Proudfit HK (1985) Evidence for pain modulation by pre- and postsynaptic noradrenergic receptors in the medulla oblongata. Brain Res 331:285–293

Sagen J, Proudfit HK (1986) Alterations in nociception following lesions of the A5 catecholamine nucleus. Brain Res 370:93–101

Schomburg ED, Steffens H (1988) The effect of dopa and clonidine on reflex pathways from group-II muscle afferents to alpha-motoneurons in the cat. Exp Brain Res 71:442–446

Skoog B, Noga BR (1991) Do noradrenergic descending tract fibres contribute to the depression of transmission from group II muscle afferents following brain stem stimulation in the cat? Neurosci Lett 134:5–8

Stewart JE, Barbeau H, Gauthier S (1991) The effects of clonidine in the modulation of the locomotor pattern and spasticity in spinal cord injured patients: a preliminary study. Can J Neurol Sci 18:321–331

Thilmann AF, Fellows SJ, Garms E (1991) The mechanism of spastic muscle hypertonia: variation in reflex gain over the time course of spasticity. Brain 114:233–244

Weight FF, Salmoiraghi GC (1966) Responses of spinal cord interneurons of acetylcholine, norepinephrine and serotonin administered by microelectrophoresis. J Pharmacol Exp Ther 153:420–427

Wiesendanger M, Chapman CE, MacPherson J, Berger W (1984) Neurophysiological investigations of tizanidine. In: Conrad B, Benecke R, Bauer HJ (eds) Die klinische Wertung der Spasticität. Schattauer, Stuttgart, pp 39–55

Discussion Summary

D.J. Burke

The first two papers of this session, by Daniel Kernell and Hans Hultborn, addressed factors that could alter the input-output relationship of motoneurones, leading to enhancement of the response to a constant afferent input. The third paper, by John Munson, described attempts to produce a model of spasticity in the cat, by a small cord lesion resulting in a chronic preparation, in which the clinical features of the human disorder would be manifest. Elzbieta Jankowska then presented data on the cat supporting a unifying hypothesis on the role of disturbed monoaminergic control of specific interneuronal systems in the development of the enhanced reflex reactivity typical of spasticity.

In response to the comment that many human voluntary contractions are relatively brief, so that the steady-state firing levels presented in his paper might not be appropriate, Kernell discussed further the time dependence of motoneurone adaptation to a maintained input. There are two phases of adaptation. The first involves a dramatic decrease in gain of the current-frequency relationship for the motoneurone and is complete within the first few intervals. Most voluntary contractions last longer than this. The second phase of adaptation goes on for 0.5–1.0 min or longer and could be relevant in postural contractions.

Kernell dealt with inhomogeneities of recruitment within a motoneurone pool by endorsing the concept of "task teams" within a muscle, each of which could be differently activated in different tasks. This concept (of what constitutes the physiological pool of motoneurones that the nervous system is driving) he considered analogous to task teams from different muscles activated together as part of the movement synergy selected to perform a task. In each case, within the specific task team, there would be homogeneity and appropriate gradation of recruitment, as discussed in his paper.

In a discussion of the limits of the discharge frequencies of motoneurones, questions were raised about the control of the lowest rates at which motoneurones will fire steadily. A number of factors can influence the lower-limit frequency, the minimal maintainable discharge rate.

Dept. of Neurology, University of N.S.W., The Prince Henry Hospital, Little Bay N.S.W. 2036, Sydney, Australia

Thilmann et al. (Eds.)
Spasticity: Mechanisms and Management

In anaesthetised animals, the most important is the duration of the after-hyperpolarization, which also helps to determine the gain of the current-frequency relationship. There are a number of superimposed mechanisms that can produce even longer interdischarge intervals than can be attributed to the after-hyperpolarization. In addition, there is a close relationship between the contractile properties of muscle units and the minimal rates of motoneurones, a property which is maintained despite normally occurring variations in contraction speed, such that, e.g., under dynamic (but possibly not static) conditions, shortening contractions are associated with higher discharge rates (matching the shorter twitch properties) than lengthening contractions which start from the same position. In the discussion a further factor was raised, namely the afferent input from the contracting muscles. Human motor axons appear to maintain lower discharge frequencies in weak, moderate and maximal contractions when the muscle afferent input has been completely blocked.

Two issues dominated the discussion of Hultborn's paper: post-activation depression and the relevance of plateau potentials as a factor in spasticity. Post-activation depression (or low-frequency depression) was clearly a pre-motoneuronal phenomenon but, in Hultborn's view, a manifestation of transmitter depletion, not the result of presynaptic inhibition. Hultborn distinguished between activity-dependent depression and post-tetanic potentiation, which he saw as inversely related. For example, in low [Ca^{++}] a small amount of transmitter is released and there is little post-activation depression but prominent post-tetanic potentiation.

Establishing the relevance of plateau potentials in chronic animal preparations and in human experiments was technically very difficult, and, while some observations were consistent with plateau potentials in these studies, there was insufficient evidence to indicate that they contribute significantly to the enhanced motoneurone response to afferent input typical of spasticity. A doubling of motoneurone discharge rates (from 8–10 Hz to 15–20 Hz) had been seen in the rat preparation. While these rates are still within the range over which motoneurones normally fire, 15–20 Hz would be close to fusion for soleus muscle units. Experiments in human subjects using vibration have been even less rewarding so far. For example, in the cat the tonic vibration reflex evoked after administration of 5-HT continues long after vibration has ceased, but in human subjects, whether spastic or normal, the reflex decays immediately when the vibration ceases. However, in the human situation, vibration excites receptors other than just the primary spindle ending (e.g., cutaneous receptors and other muscle receptors), and an inhibitory input from these sources would complicate the reflex response, perhaps preventing the appearance of the prolonged motor discharge due to the plateau potential.

Discussion of Munson's paper focused on the choice of ketamine as the anaesthetic agent for acute studies on cats with chronic unilateral spinal cord lesions, and on the similarities between the feline model and the human condition. It was pointed out that ketamine was a relatively potent non-competitive NMDA antagonist and therefore had a potentially antispastic action. In addition, there is disturbed afferent input in ketamine-anaesthetised animals, with a dramatic tonic discharge of Ia afferents. This effect is central, presumably due to enhanced fusimotor activity. Munson acknowledged that there was no perfect anaesthetic agent, but, because the asymmetrical increase in stretch reflex activity in his model survived the ketamine anaesthesia, it should still be valid to look for differences in spinal mechanisms between the two sides.

In discussing his previous studies on chronically hemisected cats and rats, Hultborn raised the question of whether such animals were "spastic" and indicated that perhaps he should withdraw the statement in his papers that these animals were not spastic. That the animal models do have some features different from the clinical condition was acknowledged, but there was no consensus that this prevented the use of the term "spasticity". It was pointed out that, under ketamine anaesthesia, there was a maintained response to tonic stretch on the unaffected side of Munson's cats, something that is not seen in human beings whether anaesthetised or awake. Perhaps this indicates that the reflex threshold is lower on both sides in the cat.

Accordingly, Munson's failure to demonstrate velocity dependence on the lesioned side could be analogous to Rymer's demonstration that the velocity dependence of the stretch reflex in spastic human beings was very low when stretch was applied while the patients maintained a steady tonic contraction. Under such circumstances, a 50-fold change in stretch velocity produces a less than twofold change in torque.

Jankowska's hypothesis faced two challenges: to reconcile the different effects of tizanidine in previous animal studies with the spinal actions described in her paper, and to find clinical correlations in which there was degeneration of neurones in the locus coeruleus. The various effects of tizanidine in different studies seem to be related to dosage, route of administration and site of action (see Emre, this volume). For example, it was reported that, on the one hand, systemic tizanidine and tizanidine injected into the locus coeruleus decrease polysynaptic reflexes and systemic tizanidine decreases muscle tone in spastic mutant rats, but that, on the other hand, intrathecal tizanidine increases muscle tone and has no effects on flexor reflexes. Following direct injection into the locus coeruleus, the activity of neurones in the locus decreases, and this could have a number of indirect distant effects, including disinhibition of raphe nuclei. This latter effect could lead to the bistability discussed in Hultborn's paper.

A discrepancy was noted with studies by Wiesendanger's group: tizanidine given to the intact rat reduced reflex activity, and this was attributed to inhibition of a coeruleo-spinal projection, but much stronger doses given to spinal rats produced an increase in reflexes. To explain this discrepancy, Jankowska suggested that dosage might have been the critical factor: at the high dosage used in the latter studies, tizanidine (which is normally considered a selective α_2 agonist) might influence α_1 receptors on motoneurones, leading to a bistability similar to that produced by serotonin. Overall, the published data were considered by Jankowska to "fit" (or at least not to be incompatible) with her hypothesis.

The question of suitable clinical correlates produced no definitive answers. The locus coeruleus degenerates in the Shy-Drager syndrome and in Parkinson's disease, but both are associated with rigidity rather than with spasticity. However, given that her hypothesis concerns enhancement of reflex function (and this need not be restricted to the reflex enhancement occurring in spasticity), Jankowska saw no conflict.

In patients with hemiplegia, degeneration of the locus coeruleus has not be reported but does not appear to have been specifically sought. However, to determine whether function is normal might require more than just simple anatomical techniques. Degeneration of the locus coeruleus was found at autopsy in four dystonic patients by Marden's group in London, but the same group then undertook detailed functional studies (including a search for loss of REM inhibition of muscle tone) in a large group of dystonic patients, both primary and secondary, only to conclude that there was no evidence of loss of locus coeruleus function.

Part IV
Reflux Disturbances in Spasticity: Movement Studies

Pathological Changes in Spastic Muscle Reflexes Evoked by Passive Stretch or Tendon Taps

A.F. Thilmann[1], S.J. Fellows[1], and H.F. Ross[2]

Introduction

Spasticity is a motor disorder which occurs after CNS lesions. Although in diagnostic terms clinicians worldwide would largely agree over the term spasticity, no totally satisfying physiological definition has been agreed upon (see Thilmann, this volume). The most widely accepted definition is that of Lance (1980): 'Spasticity is a motor disorder characterized by the velocity-dependent increase in tonic stretch reflexes ('muscle tone') with exaggerated tendon jerks, resulting from hyperexcitability of the stretch reflex as one component of the upper motor neurone syndrome." This definition contains two assumptions: (a) spasticity is a disorder mainly affecting reflex systems, and (b) it is defined by a passive movement in which increased muscle tone correlates with the velocity of a passive muscle stretch. Implicitly it is often assumed that the disability of the patient is correlated to the degree of disturbance in reflexes and to muscle tone (for literature see Feldman et al. 1980). During voluntary activation of muscles, however, the disability may differ greatly from that seen during passive disturbance. This issue is discussed further in the chapters by Fellows et al. and Dewald and Rymer in this volume.

Despite this difference between the passive and active states, passive movement is nevertheless used in routine clinical diagnosis of the spastic syndrome, and hypertonus is subjectively assessed by the clinicians and rated on the Ashworth scale (Ashworth 1964). To test hyperreflexia neurologists use a reflex hammer to evoke a synchronized reflex contraction of the muscle. Increased tendon jerks are often assumed to be synonymous with spasticity, a practice which should be discontinued, as is shown below. Furthermore, the Hoffmann reflex (H reflex) is often taken as the electrical equivalent of the tendon jerk, although there are important differences between these reflexes (Burke et al. 1984). Attempts have been made with

[1] Alfried Krupp von Bohlen und Halbach Krankenhaus, Neurologische Klinik mit Klinischer Neurophysiologic, Alfried Krupp Strasse 21, D-45131 Essen, Deutschland

[2] Dept. of Physiology, The Medical School, University of Birmingham, Birmingham B152TJ, England

H reflex testing to assess motor function (Bathien and Rondot 1985) and the effects of treatment on spastic patients (Müller et al. 1988). For more detailed discussion of H reflexes, see the chapters by Yanagisawa et al., Ongerboer de Visser et al., Delwaide, and Dressandt et al. in this volume.

While most research groups would agree with the hypothesis that reflex alterations play an important role in the spastic motor disorder, Dietz et al. (1981) have pointed out that for disturbances of the calf muscles during gait "no electrophysiological explanation could be found for the increased muscle tone." This group proposed that "the muscle fibres undergo changes which are responsible for increased muscle tone in spasticity." From this it can be suggested that the main cause of spasticity is not localized in the CNS but in the periphery.

It is the aim of this chapter to show that, in contradiction to the views of Dietz et al., pathologically increased stretch reflexes are the cause of the passive muscle hypertonus seen in spasticity. The pathological reflex disturbances in passive spastic limbs arise from different reflex components which do not behave as in the tendon tap. It is concluded that the tendon tap may present a misleading picture of reflex changes in the spastic syndrome.

Subjects and Methods

Subjects. Experiments were performed on hemiparetic patients ($n = 63$) and control groups of age-matched normal subjects ($n = 47$). The hemiparetic subjects had all suffered from an ischaemic lesion in the area of the left or right middle cerebral artery. All lesions were confirmed by computed tomography or magnetic resonance imaging to be single unilateral cerebral infarcts. The time of examination was – depending on the group – within the 1st month, between 1 and 3 months, or more than 1 year after the stroke. None of the patients was receiving antispastic medication at the time of the study, and patients with other neurological disorders, such as polyneuropathy, were excluded. Patients and normal subjects gave informed consent to all procedures, which had previously been approved by the local ethics committee.

Movements of the Foot. While sitting in a stable chair with the knee flexed (105°), the foot was fixed in an adjustable cast so that the only possible movement of the leg was in the anatomical axis of the ankle joint. From a start position of 90° between the sole of the foot and the tibia the ankle was dorsiflexed 4.7° at velocities of 25, 50, 75, 100, 155 and 175°/s by a torque motor (Servalco MC24P) with a four-quadrant control system (Seidel, series 500), responding to pulses generated by a laboratory computer (DEC, PDP11/73). Subjects were instructed to remain relaxed and not to attempt to resist the imposed displacements. Ten consecutive displacements were applied at each stretch velocity at intervals of 5–8 s.

Movements of the Elbow Joint. Subjects were seated in a stable chair with the arm laterally abducted and the forearm and hand fixed in a plastic mould. The height of the chair was adjusted so that the shoulder was slightly above arm level. With the same technical equipment as described above, extension movements were applied at the elbow. Starting from 75° between forearm and upper arm, a displacement of 30° was applied at velocities of 80°, 115°, 175°, 240° and 300° per second. Ten displacements at a randomized interval between 5 and 8 s were applied. The subjects were asked to relax and not to resist the movements.

Data Acquisition. Joint position was monitored by a light-sensitive detector (United Detector Technology, LSC5D). Electromyographic (EMG) activity was recorded using surface electrodes placed over the belly of the muscles (soleus and tibialis anterior (ankle) or biceps brachii and triceps brachii (elbow)). The electrodes had a fixed interelectrode distance of 2 cm and contained a small pre-amplifier designed to reduce movement artefacts to a minimum. EMG signals were amplified by a custom-built device (bandwidth 20 Hz–1 kHz). All data were sampled by the computer AD board at a rate of 1 kHz per channel. EMG was rectified and averages were constructed from ten consecutive sweeps at each displacement velocity. In an off-line analysis the area of the EMG integral (μVs), the latency (ms), the duration (ms), the average EMG during a movement (μV) and the peak amplitude (μV) were measured. Latencies were determined from the onset of displacement. Velocity and acceleration records were calculated by the computer from the position records.

Tendon Taps. Before and after all movement experiments, tendon taps of the biceps brachii muscle or triceps surae muscle were recorded. In the same apparatus as described above, maximum tendon percussion was produced using a reflex hammer which contained a micro-switch to trigger the sampling of the computer. Three maximal tendon tap responses were averaged.

For detailed description of methods see Thilmann et al. (1990, 1991), Thilmann and Fellows (1991), and Fellows et al. (1993).

Results

Ankle Dorsiflexion

As previously shown (Fellows and Thilmann 1989), dorsiflexion of the relaxed normal ankle resulted in a single synchronized reflex EMG response in the triceps surae muscle, with a duration of 20–25 ms. Reflexes were

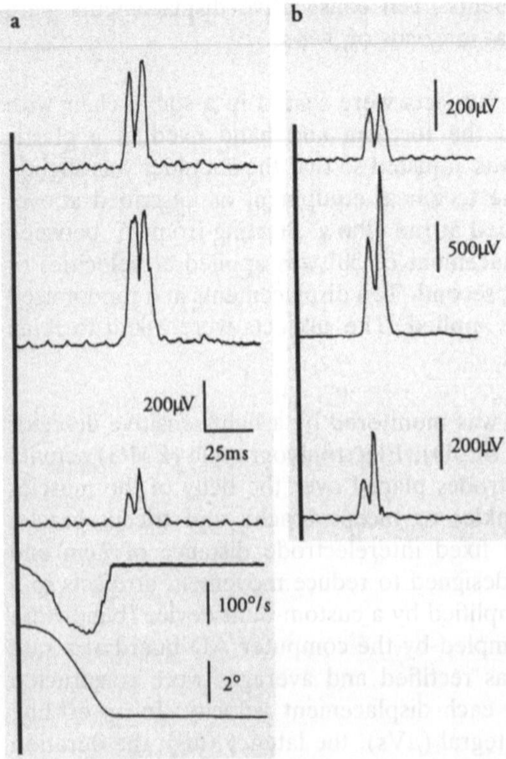

Fig. 1. a The rectified EMG response of the relaxed soleus muscle to dorsiflexion of 4.7° at a velocity of 175°/s. *Upper trace*, response in a normal subject; *middle trace*, contralateral side; *lower trace*, ipsilateral side compared to the cerebral lesion of a hemiparetic subject. Each EMG record is the average of ten rectified sweeps. The position and velocity traces for the displacement are shown below. **b** The maximal Achilles tendon reflex obtained from the same subjects. Note the different scale factor for the middle EMG record. Onset of the EMG traces correspond to the contact of the hammer with the tendon. (From Thilmann and Fellows 1991)

evoked only if velocity of movement exceeded 50°/s. The latency of the stretch reflex was 54 ms (for a stretch of 175°/s).

A representative reflex response for normal subjects is shown in the upper trace of Fig. 1a. In the hemiparetic subjects a similar response was evoked by the dorsiflexion on the side contralateral (middle EMG trace Fig. 1a) and ipsilateral (lower EMG trace Fig. 1a) to the cerebral lesion. Latency and duration of the responses on both sides of the hemiparetic subjects were identical to those of the normals. However, the amplitude of the reflex response differed from that of a normal subject, being increased on the contralateral and decreased on the ipsilateral side. The corresponding tendon jerk responses are shown in Fig. 1b, in which the exaggeration of the contralateral side is visible (note different scale), while the magnitude of the ipsilateral tendon jerk did not differ from normal.

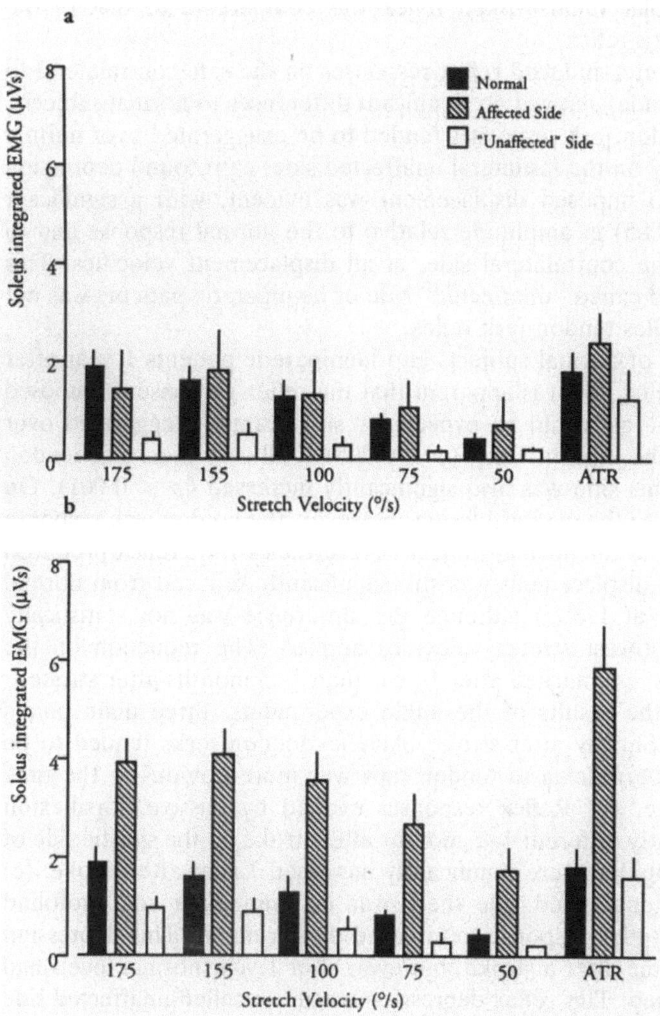

Fig. 2a,b. Group averages (±SEM) of the integrated EMG response in the soleus muscle at five stretch velocities and following percussion of the Achilles tendon (*ATR*). **a** Results for the group 1–3 months after stroke. **b** Results for the group 1 year after stroke. (From Thilmann and Fellows 1991)

Figure 2 shows the group averages (±SEM) for the integrated EMG values of the stretch reflex responses recorded over triceps surae muscle. The results are shown for the normals, the ipsilateral "unaffected" and contralateral "spastic" sides of the hemiparetic subjects tested between 1 and 3 months after the stroke (Fig. 2a) or 1 year after stroke (Fig. 2b). It is apparent that in normal subjects the size of the reflex EMG was proportional to the velocity of the imposed displacement. At high velocities, the magnitude of this displacement-evoked reflex was comparable to that of the Achilles tendon jerk reflex.

In patients, stretch-induced reflex responses on the side contralateral to the lesion (spastic side) showed no significant differences to normal subjects, although their tendon jerk responses tended to be exaggerated over normal values. Surprisingly on the ipsilateral unaffected side, a profound depression of the response to imposed displacement was evident, with a significant reduction ($p < 0.005$) in amplitude relative to the normal response and to the response on the contralateral side, at all displacement velocities. This reduction on the so-called "unaffected" side of hemiparetic patients was not visible in the Achilles tendon jerk reflex.

A comparison of normal subjects and hemiparetic patients 1 year after stroke is given in Fig. 2b. It is apparent that the reflex response to imposed displacement was – as would be expected – significantly exaggerated over normal values on the "spastic" side ($p < 0.005$) at all velocities. The tendon jerk response to this side was also significantly increased ($p < 0.001$). On the side ipsilateral to the cerebral lesion, however, the tendon jerk response remained identical to the normal subjects. Nevertheless, the reflex produced following imposed displacement was still significantly reduced from normal levels ($p < 0.025$ at 175°/s) although this difference was not statistically significant at the lowest stretch velocities applied. The reduction on the ipsilateral side was less marked after 1 year than 1–3 months after stroke.

Summarizing the results of the ankle experiments, three main points are apparent: (a) Shortly after stroke, Achilles tendon jerks tended to be increased, and hyperreflexia in tendon taps was more obvious in the time course after stroke. (b) Reflex responses evoked by passive dorsiflexion were not significantly different 1–3 months after stroke on the spastic side of hemiparetic patients but were significantly increased 1 year after stroke. (c) On the so-called unaffected side there was an immediate and profound depression of the reflex response to imposed displacement. This depression diminished over time after a stroke, but even after 1 year normal levels had not yet been reached. This reflex depression on the so-called unaffected side is not reflected in changes in the tendon jerk response.

Passive Stretch at the Elbow

An extension of 30° at the elbow elicited a reflex response in the relaxed biceps brachii muscle in about 35% of normal subjects.

In Fig. 3a the responses of a normal subject with active reflexes are shown at a variety of stretch velocities. The length of each trace indicates the duration of the displacement, beginning with the start of the movement. In all normal subjects in which a reflex responses was evoked, a single synchronized burst of EMG activity was visible, occurring between 61 and 107 ms after the start of displacement. In the following descriptions this time interval is called the "normal range."

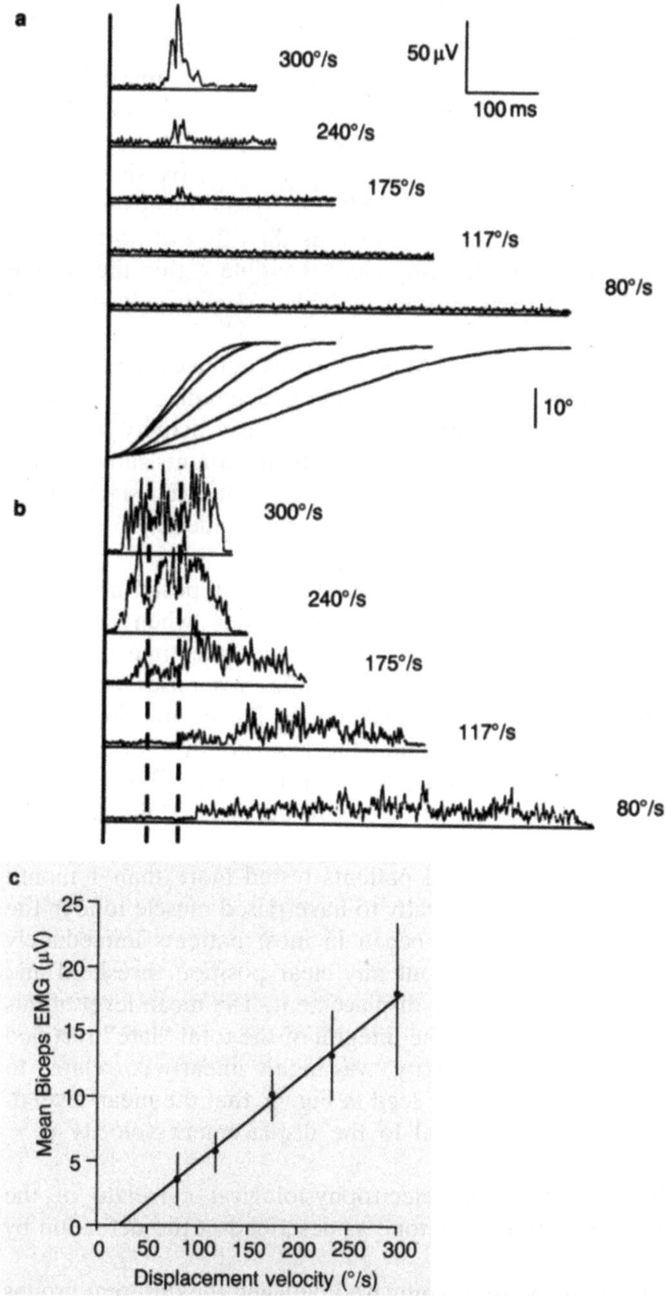

Fig. 3a–c. The response of the biceps brachii muscle of a normal subject (**a**) and the spastic side of a hemiparetic subject (**b**) to a 30° extending displacement applied at the relaxed elbow. Each trace represents the rectified and averaged EMG record obtained from ten displacements at the velocity shown beside it. The length of a trace corresponds to the duration of the displacement. Position records are shown below **a**. *Dotted lines* (**b**), timing of the "normal range" (61–107 ms, see text for definition). **c** Linear correlation between displacement velocity and the mean biceps brachii EMG level of the "late" response (see text for definition). Each point is the average value (±SEM) of a group of patients in whom spasticity had been established for 1 year or more. (Partly adapted from Thilmann et al. 1991)

Figure 3b shows a typical set of EMG responses of a spastic biceps brachii muscle of a hemiparetic subject 1 year after stroke. The dotted lines define the normal range (61–107 ms, see above), and, as for the normals, the length of the EMG trace reflects the duration of dispacement. It is evident that the response pattern in the EMG of spastic biceps muscle is more complicated and contains more components than that of normal subjects. As in normals, a considerable component is visible within the normal range, additionally earlier activity is seen at higher stretch velocities and later activity at all velocities.

"*Early activity*" (occurring before 61 ms) was (as defined) never visible in normals, but was seen in 15%–90% of patients depending on the time after stroke. While in the 1st month after stroke early activity was only rarely elicited, more than 1 year after stroke nearly all patients showed such early activity. The amplitude of this early activity did not differ significantly over the time course after a stroke, but there was a tendency for it to be largest after 1–3 months.

The *normal-range activity* of the biceps EMG after imposed movement did not show any significant differences in magnitude when compared between normal subjects and patient groups or over the time course of spasticity. The percentage of subjects showing reflex responses within this normal range was higher in these patients than in the normal group and increased from 75% of the subjects within the first month after stroke to 100% of hemiparetic patients more than 1 year after stroke. The early and normal-range components are described in more detail in Thilmann et al. (1990) and Fellows et al. (1993).

The "*late activity*" appeared in all patients tested more than 1 month after stroke who were considered clinically to have raised muscle tone in the biceps brachii. The late reflex EMG began in most patients immediately after the normal range (107 ms) without any clear position threshold and terminated at or around the end of the displacement. The mean level of this EMG activity, calculated by dividing the integral of the total "late" rectified EMG by the duration of the late activity, was highly linearly correlated to the velocity of displacement. It may be seen in Fig. 3c that the mean level of biceps EMG was directly proportional to the displacement velocity ($r = 0.998$, $p = 0.0012$).

This late EMG activity is the electrophysiological correlate of the velocity-dependent increase in muscle tone as described in the definition by Lance (1980).

When this late EMG response is compared between the different groups of patients, it is apparent that the amplitude varied over the time after stroke. Figure 4a gives the group means (±SEM) obtained at a displacement velocity of 300°/s. It may be seen that in the 1st month of spasticity after stroke only a low "level" of late biceps brachii EMG activity was evoked. Between the first and third month after stroke, however, the level of late reflex response increased considerably ($p < 0.01$). One year or more after

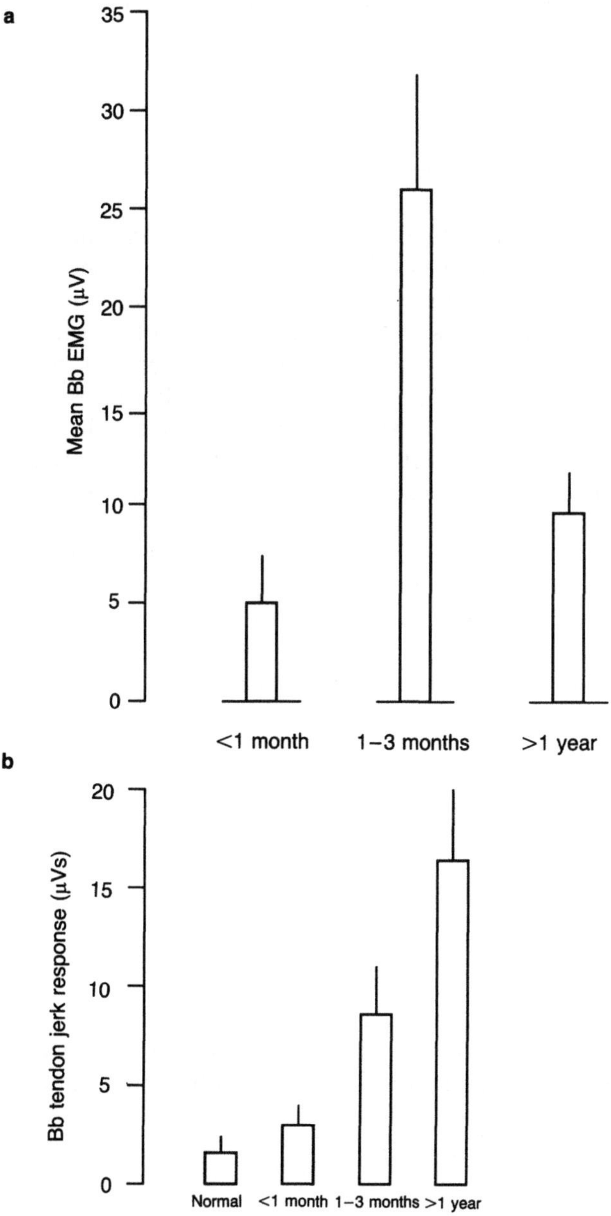

Fig. 4. **a** Group means (with standard error bar) of the average level of biceps brachii EMG activity during the "late" response evoked following an imposed extension of the relaxed arm of 30° at a velocity of 300°/s. **b** Group means of the area of the biceps brachii tendon jerk reflex EMG (with error bars) for the normal subjects and for the three groups of patients. (From Fellows et al. 1993)

stroke the late biceps brachii EMG obtained in these patient groups was considerably reduced from the level of activity seen 1–3 months after stroke ($p < 0.01$).

Tendon Jerk Responses in the Biceps Brachii Muscle

The group averages of the biceps brachii tendon jerks (which could be evoked in all patients and in all but one of the normal subjects) are shown in Fig. 4b. The tendon jerk reflex amplitude in the first month after stroke was slightly increased above normal values, but this increase was not statistically significant. One to 3 months after stroke, however, the tendon jerk reflex was markedly exaggerated ($p < 0.025$). More than 1 year after stroke this hyperreflexia increased, reaching values which were, on average eight times greater than those seen in normal subjects ($p < 0.001$). Differences between normals and patient groups existed only in the amplitude of the reflex response and not in the latency or the duration.

These results show a disparity between the behaviour of the tendon jerk reflex and the late reflex EMG activity. This means that changes in the tendon jerk reflex do not follow the same time course after stroke as those in the neurophysiological correlate of the increased muscle tone.

Summarizing the results of the elbow experiments, the following points seem important:

1. Passive extension movements of the elbow joint evoke in hemiparetic patients a reflex response in the biceps brachii muscle on the spastic side which is far more complex than the normal response.
2. Three components of this reflex response, the early component, the normal range and the late activity show different characteristics and differ over the time course after stroke.
3. The late reflex EMG is the neurophysiological correlate of the increased muscle tone and correlates linearly with displacement velocity.
4. While reflex EMG evoked by an Achilles tendon tap steadily increases over the first year after stroke, the amplitude of the late EMG activity is at maximum 1–3 months after stroke and is signficantly reduced after 1 year.

Discussion and Conclusions

Reflex responses following imposed dorsiflexion of the foot show bilateral changes after a unilateral cerebral lesion, a result which has been shown previously for the arm (Thilmann et al. 1990). The results provide evidence that the tendon jerk response of the triceps surae does not give a clear indication of the excitability of more physiological stretch reflex responses elicited by displacement at the ankle joint. Although clinically the tendon jerk is in universal use, it at no time reflected the depression of the reflexes observed on the "unaffected" side of hemiparetic patients. Additionally, it has been demonstrated that the time course of the exaggeration of the tendon jerk is different from that of movement-evoked responses, with the former becoming exaggerated more rapidly and more markedly. This can be taken as evidence that different pathways are involved in the tendon jerk and in movement-evoked reflexes. It can be postulated that after a stroke two different mechanisms may play a role for the ankle reflexes: first, a bilateral depression of reflexes which appears early and diminishes within the first year; secondly, an increase in reflexes on the contralateral side to the cerebral lesion, the time course of which is prolonged, with a later start and steady increase over the first year after stroke. It would appear that only the second effect is visible in the tendon jerk responses.

Extension movement at the elbow evokes a reflex response in the relaxed biceps brachii muscle. While only one-third of normal subjects show these movement-induced reflexes with a velocity threshold of about 150°/s, all spastic patients with increased muscle tone show a complex EMG response during the imposed movement even at stretch velocities as low as 40°/s. The arbitrary allocation into three components (early, normal range, late) seems to be useful because clear differences exist in these parts of the reflex response. The early component is observed in an increasing number of patients within the first year, but the magnitude does not differ significantly over the time course. Nevertheless, this early component is pathological and never seen in normal subjects; a similar response has been seen in other experiments on the so-called unaffected side of hemiparetic patients. Thilmann et al. (1990) discussed that the pathological changes in the early component during stretch of the biceps brachii muscles involve pathways in the spinal cord which are involved in reciprocal inhibition. The experiments reported here do not implicate any particular pathways or interneuronal groups, but the decrease in the reflex latency by up to 30 ms can only be achieved by either increased excitability of the α-motoneurones or in increased drive to the α-motoneurones by stronger afferent input. These neuronal mechanisms responsible for spastic muscle hypertonicity are discussed in Katz and Rymer (1989).

The normal-range component is not very different from the reflexes of normal subjects, but this component may be evoked at lower stretch velocities. This means that the threshold in hemispastic subjects may be lowered, as has been discussed by Powers et al. (1988).

The late activity of the stretch reflex in spastic muscles is well matched with the clinical definition of spastic muscle hypertonus and has been reported by other groups which did not separate different components of the stretch reflex response in spastics (Rondot et al. 1958; Ashby and Burke 1971). The variation of the late activity in the time course after stroke shows a maximum at 1–3 months and a significant reduction after 1 year. This reduction in the late reflex component is in contradistinction to the biceps tendon tap response which increases continuously within the first year after stroke. Again, it must be considered that different reflex pathways or spinal interneurones are involved in the tendon tap, compared to the stretch evoked reflexes. It is concluded that the tendon jerk (and its electrophysiological correlate, the H reflex) should not be taken as a measurement of increased reflexes in spasticity, although they may represent one component of that hyperreflexia and remain useful in routine clinical testing.

References

Ashby P, Burke D (1971) Stretch reflexes in upper limb of spastic man. J Neurol Neurosurg Psychiatry 34:765–771

Ashworth B (1964) Preliminary trial of Carisoprodol in multiple sclerosis. Practitioner 192: 540–542

Bathien N, Rondot P (1985) Assessment of motor functions in extrapyramidal disorders. In: Eccles J, Dimitrijevic MR (eds) Upper motor neuron functions and dysfunctions. Karger, Basel (Recent achievements in restorative neurology, vol 1), pp 211–221

Burke D, Gandevia SC, McKeon B (1984) Monosynaptic and oligosynaptic contributions to human ankle jerk and H-reflex. J Neurophysiol 52:435–448

Dietz V, Quintern J, Berger W (1981) Electrophysiological studies of gait in spasticity and rigidity: evidence that altered mechanical properties of muscle contribute to hypertonia. Brain 104:431–449

Feldmann RG, Young RR, Koella KP (1980) Spasticity: disordered motor control. Year Book Medical Publishers, Chicago

Fellows SJ, Thilmann AF (1989) The role of joint biomechanics in determining stretch reflex latency at the normal human ankle. Exp Brain Res 77:135–139

Fellows SJ, Ross HF, Thilmann AF (1993) The limitations of the tendon jerk as a marker of pathological stretch reflex activity in human spasticity. J Neurol Neurosurg Psychiatry 56:531–537

Katz RT, Rymer WZ (1989) Spastic hypertonia: mechanisms and measurement. Arch Phys Rehabil 70:144–155

Lance JW (1980) Pathophysiology of spasticity and clinical experience with baclofen. In: Feldmann RG, Young RR, Koella KP (eds) Spasticity: disordered motor control. Year Book Medical Publishers, Chicago, pp 185–203

Müller H, Zierski J, Dralle D, Hoffmann O, Michaelis G (1988) Intrathecal baclofen in spasticity. In: Müller H, Zierski J, Penn RD (eds) Local-spinal therapy of spasticity. Springer, Berlin Heidelberg New York

Powers RK, Marder-Meyer J, Rymer WZ (1988) Quantitative relations between hypertonia and stretch reflex threshold in spastic hemiparesis. Ann Neurol 23:115–124

Rondot P, Dalloz JC, Tardieu G (1958) Mesure de la force des réactions musculaires à L'étriment passiv au cours des radieurs pathologiques par lésion cérébrales. Rev Fr Etud Clin Biol 3:585–592

Thilmann AF, Fellows SJ (1991) The time-course of bilateral changes in the reflex excitability of relaxed triceps surae muscle in human hemiparetic spasticity. J Neurol 238:293–298

Thilmann AF, Fellows SJ, Garms E (1990) Pathological stretch reflexes on the "good" side of hemiparetic human subjects. J Neurol Neurosurg Psychiatry 53:208–214

Thilmann AF, Fellows SJ, Garms E (1991) The mechanism of spastic muscle hypertonia: variation in reflex gain over the time course of spasticity. Brain 114:233–244

Spastic Paresis: Reflex Activity and Muscle Tone in Elbow Muscles During Passive and Active Motor Tasks

V. Dietz[1], I.K. Ibrahim[2], M. Trippel[2], and W. Berger[2]

Introduction

Lance (1980) defined spastic hypertonia as a resistance of passive spastic muscle to stretch in a velocity-dependent manner following activation of tonic stretch reflexes. It was assumed that this definition could also be applied to the patient's spastic movement disorder, implying that therapy ought to be directed towards reduction of reflex activity. However, it was found that exaggerated reflexes during functional movements, such as locomotion, contribute little to the impaired movement (Dietz et al. 1981; Berger et al. 1984; Landau 1980). Rosenfalck and Andreassen (1980) suggested that the movement disorder is a consequence of reduced modulation of motoneurons due to the impaired function of polysynaptic reflexes and of the muscle tone. This impaired function has been suggested to be due to changes in mechanical muscle fiber properties (Berger et al. 1984, 1988; Dietz et al. 1986; Hufschmidt and Mauritz 1985; Sinkjaer et al. 1993).

The different observations of patients made during passive and active motor tasks have led to confusion and controversy. Quite often, little consideration is given to the differences in methodological approaches used. For example, one has to be aware that by focusing the approach on passive stretch of the biceps brachii without recording the stretch-induced torque, "the mechanism of spastic muscle hypertonus" (Thilmann et al. 1991) cannot be fully investigated. The conclusion drawn from the latter study was that spastic muscle hypertonus results from a pathological increase in stretch reflex gain. This is in contrast to studies on functional movements, where decreased activity of polysynaptic reflexes has been reported (Berger et al. 1984, 1988; Cody et al. 1987).

Recent investigations on spastic muscle tone including torque and EMG measurements (Powers et al. 1988, 1989) have indicated that the velocity dependence of mechanical and electrical reflex output in *active* spastic muscles was similar to that seen in normal muscles. The authors suggested

[1] Schweizer Paraplegikerzentrum, Klinik Balgrist, Forch Str. 340, CH-8008 Zürich, Schweiz
[2] Abt. Klinische Neurologie and Neurophysiologie, Universität Freiburg, Hansastr. 9, D-79104 Freiburg, Deutschland

Thilmann et al. (Eds.)
Spasticity: Mechanisms and Management

that a reduction in stretch reflex threshold is uniformly present in spastic muscles. However, no detailed analysis of elbow flexor EMG was performed, with the consequence that changes in the behavior of early and late reflexes in active and passive spastic muscles could not be specified. In addition, in most of the studies mentioned above, the EMG activity in the respective antagonistic muscles was neglected, even though they may contribute to changes in muscle tone.

The aim of the present work was to address the following questions which remain unanswered regarding the pathophysiological basis of spastic paresis and movement disorder:

1. What kind of relationship exists between reflex EMG activity and torque during passive and active motor tasks? Can the clinical assessment of spastic muscle tone be applied to a functional movement and its disorder? To study this, stretch-evoked EMG activity and torque were investigated, both in the passive limb and during a quasi-functional elbow movement.
2. How and to what extent do the early and late reflex components change during active and passive motor tasks? This was studied by quantitative analysis of reflex EMG activity during different phases of their activation following stretches of different amplitudes and velocities.
3. To what degree is the reflex modulation described for healthy subjects (Dietz et al. 1991) preserved in patients with spastic paresis? To investigate this point, displacements were induced during different modes of muscle contraction: control of position or of torque.
4. Are there basic differences in the behavior of the antagonistic muscles? Is there an influence of the respective antagonist on the stretch-evoked torque? To address this point, stretch-evoked EMG activity and torque of both the elbow flexor and extensor muscles were assessed using simultaneous recordings from the two muscle groups.

EMG Components in Passive and Active Tasks

Although recordings were made from all flexor and extensor muscles, the description of reflex behavior is concentrated on the brachioradialis EMG responses, as their modulation is most pronounced. Figure 1 shows the mean brachioradialis EMG responses obtained from unaffected (Fig. 1A) and affected (Fig. 1B) limbs of 16 patients with spastic hemiparesis following three different elbow displacements which stretched the flexor muscles by a constant amplitude of 30° at rates of 75°, 200°, and 300°/s (impulses a, b, c). Displacements were applied to the passive and active muscles. During *passive* tasks, the amplitude of the EMG responses was generally larger on the affected arm and was dependent on the displacement velocity. Its duration on the affected side approximated the duration of displacement (for impulse b from about 25 ms to 190 ms and for impulse c from about 25 ms to 150 ms).

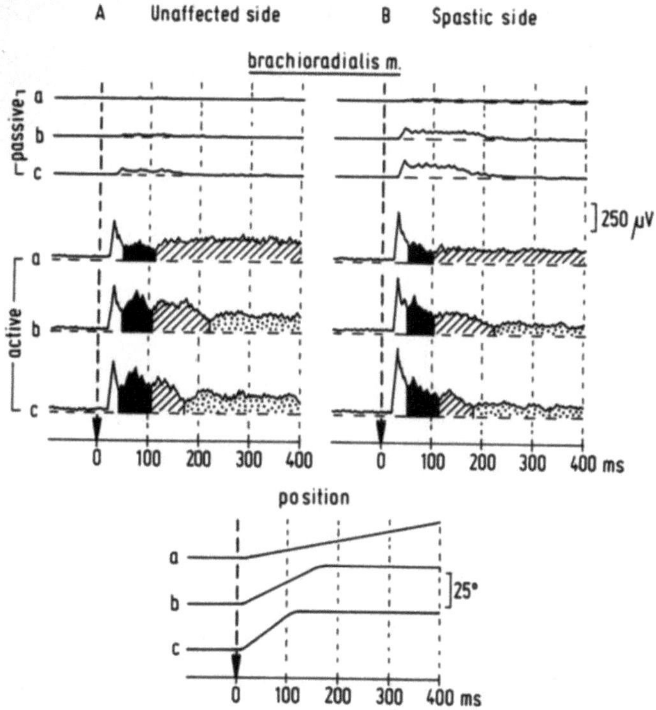

Fig. 1. Group mean values of the rectified and averaged ($n = 10$) brachioradialis EMG responses with position signals from unaffected (*A*) and affected (*B*) sides in 16 patients with spastic paresis. Displacements of *three different velocities* (75°, 200°, 300°/s *a*, *b*, and *c*) at a constant amplitude of 30° in the extending direction were induced during passive (*above*) and active (*below*) motor tasks. The different components of the EMG responses are indicated: *white*, M1; *black*, M2; *shaded*, M3; the *stippled area* indicates the basic EMG activity when the new target elbow joint angle is reached, i.e., during the "holding phase". *Arrows* indicate onset of displacement

During *active* tasks the segmentation of the EMG response (M1–M3) was roughly the same on both sides. The amplitude of all components on both sides was dependent on the displacement velocity. M1 and M2 were of constant duration, while the duration of the M3 component corresponded to the duration of the displacement. For all impulses M1 was of larger amplitude and the later components (M2 and M3) were of smaller amplitude on the affected side compared with the unaffected side. Also during the "holding phase" (maintained static stretch) the amplitude of tonic EMG activity was smaller on the affected side. The average latencies for the three components did not differ significantly between the two limbs.

Figure 2 displays the dependency of the mean values of quantified brachioradialis muscle reflex components (M1–M3) on displacement velocity for all patients on the unaffected (Fig. 2A) and affected (Fig. 2B) sides. The amplitude of all components – but more so for M2 and M3 than for M1

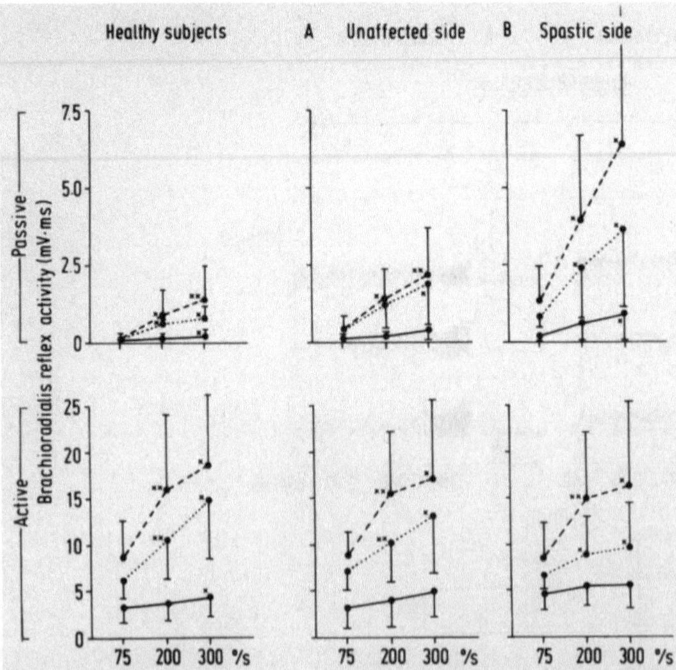

Fig. 2. Quantified group mean values (±SD) of the amplitude of the short- (20–40 ms, M1, *solid line*) and long-latency (50–100 ms, M2, *broken line*; 100–150 ms, M3, *dotted line*) reflex EMG responses in the brachioradialis muscle. Three different displacement velocities (75°, 200°, 300°/s) were applied in an extending direction on the unaffected (*A*) and affected (*B*) sides of 16 patients with spastic hemiparesis and in 15 healthy subjects. Responses during passive (*above*) and active (*below*) motor tasks are shown. Significant differences between the reflex activities are indicated using the Scheffe test (for differences between velocities): *, $p < 0.05$; **, $p < 0.01$

– from healthy subjects and both sides of the patients was velocity dependent. During *passive* tasks, the rate of rise of the slope of the three components was different between each group. The slopes were steeper on the unaffected side of patients compared with the healthy subjects ($p < 0.05$) and on the affected side compared with the unaffected side ($p < 0.01$) of patients. The M2 and M3 components increased to a greater extent on the unaffected side of patients compared with the healthy subjects. This difference was significant for M3 ($p < 0.01$) and on the affected side compared with the unaffected one. These differences have been summarized in Table 1. During *active* tasks, the main difference was a slower increase of the M3 components on the affected side compared with the unaffected side and also between the latter and healthy subjects. The slopes (done in regression format) were significantly different (*t*-test) for both ($p < 0.01$). This resulted in a smaller EMG amplitude on the affected side (see Table 2).

Table 1. Group mean values (±SD) of the amplitude of the short- and long-latency reflex EMG responses (for definition see Fig. 2) in elbow muscles following displacements in extending (biceps brachii and brachioradialis) and flexing (triceps brachii) directions during passive motor tasks; all impulses have been grouped together for calculation

Muscle	Segment	Side		
		Unaffected	Spastic	
		EMG amplitude (mV.ms)	EMG amplitude (mV.ms)	
Biceps br. m.	M1	0.11 ± 0.10	0.45 ± 0.79	$p < 0.001$
	M2–M3	1.79 ± 1.58	4.87 ± 5.33	$p < 0.001$
Brachiorad. m.	M1	0.14 ± 0.20	0.53 ± 0.52	$p < 0.01$
	M2–M3	2.35 ± 2.31	7.72 ± 7.64	$p < 0.001$
Triceps br. m.	M1	0.061 ± 0.048	0.083 ± 0.075	$p < 0.05$
	M2–M3	0.95 ± 0.69	2.14 ± 2.66	$p < 0.001$

Table 2. Values obtained during the active motor task; compare Table 1

Muscle	Segment	Side		
		Unaffected	Spastic	
		EMG amplitude (mV.ms)	EMG amplitude (mV.ms)	
Biceps br. m.	M1	2.71 ± 1.41	3.72 ± 2.99	$p < 0.001$
	M2–M3	22.21 ± 13.30	16.15 ± 8.89	$p < 0.001$
Brachiorad. m.	M1	4.71 ± 2.67	6.23 ± 4.89	$p < 0.01$
	M2–M3	32.83 ± 19.97	24.28 ± 10.21	$p < 0.001$
Triceps br. m.	M1	0.32 ± 0.31	0.48 ± 0.45	$p < 0.05$
	M2–M3	9.25 ± 4.41	9.66 ± 5.77	ns

Similar observations were made with respect to the differences between the unaffected and affected sides in the passive and active condition when different displacement amplitudes (12°, 20°, 30°) were applied with a constant rate (200°/s). During *passive* tasks, significant EMG responses were present only on the affected side. Except for the early component, no clear segmentation could be seen. The duration of the EMG responses was related to the displacement amplitude. During *active* tasks, only the duration of the M3 component corresponded to the duration of the impulse. M1 and M2 were largely independent of the amplitude of displacement. On the affected side the amplitudes of the EMG components were graded, with M1 being larger while the following components were somewhat smaller.

Although the duration of the M3 component corresponded best to the impulse duration, it was always about 60 ms shorter than the displacement.

However, when the M2 and M3 components were taken together, there was an approximately one-to-one relationship with the ramp (velocity signal) duration.

Tables 1 and 2 summarize the mean values of the EMG components (M1: 20–40 ms; M2 and M3: 50–150 ms) of unaffected and affected limbs during passive (Table 1) and active (Table 2) tasks for all patients. Under *passive* conditions the amplitudes of M1, M2, and M3 in all muscles were significantly larger on the affected side. Under *active* conditions the amplitude of M1 was larger, while those of M2 and M3 were smaller on the affected side. The M2 and M3 components were significantly different between the two sides in all but the triceps brachii muscle.

Defective Modulation of Reflex EMG Activity

Thilmann et al. (1990) suggested that the "unaffected" side of patients was itself not completely normal. During passive tasks, the stretch-evoked long-latency EMG activity (M3) was slightly stronger in the unaffected limb of the patients compared with that induced in the elbow flexors of healthy subjects. Nevertheless, in the unaffected limb, the EMG activity induced in the *passive* condition was small compared with the strong, segmented EMG responses (termed M1–M3 by Lee and Tatton 1975) recorded following displacements of *active* muscles, i.e., during the quasi-functional movements. The M1 and M2 components were of constant duration, suggesting a "coding" of both components by the acceleration signal, with a long-loop pathway serving the M2 response, explaining its longer delay (for review see Matthews 1991). The duration of M3 (together with the M2 component, which might be coded for by mixed signals) corresponded best to the duration of the ramp displacement. Therefore, this component may be coded for by the velocity signal (see Prochazka 1988). The segmentation of the EMG response seen in this study is not always seen after stretch of active muscles in other motor conditions (cf. Matthews 1991). This may be a consequence of the specific motor task used in the study. The control of elbow joint position during application of a sinusoidally changing torque signal, acting at the elbow flexor and extensors, requires appropriate firing of centrally regulated gamma motoneurons (see Dietz et al. 1991; Prochazka 1989).

In spastic limbs the three components of the EMG response were preserved. During *active* conditions there was a differential modulation of the three components compared with the unaffected side. The amplitude of the relatively small, early component was slightly increased, perhaps due to a reduced presynaptic inhibition of group Ia afferents at a segmental level (Burke and Ashby 1972; Iles and Roberts 1986). The amplitudes of the long-latency components (M2 and M3) were significantly reduced. A similar reduction of compensatory long-latency EMG responses in spastic limbs has been described in leg muscles following perturbations of stance (Berger et al. 1988), in the active elbow (Lee et al. 1987; Dietz et al. 1991), and in wrist flexor muscles (Cody et al. 1987). In contrast, during *passive* conditions, a significant "extra EMG activity" was induced in the spastic muscles. This effect was more evident in the flexors than in the extensors. The activity is due mainly to a difference in the reflex gain, which was higher in spastic than in unaffected muscles. The activity in the unaffected muscle was itself higher than that seen in muscles of normal subjects. In the passive muscle a clear segmentation of the EMG activity was not detected, although its duration corresponded to the duration of the whole EMG response in the active muscle (M1–M3), i.e., to the duration of the ramp displacement. Similar velocity-dependent extra activity in the passive biceps brachii has recently been described by Thilmann and coworkers (1991).

The wide range of responses seen in passive and active muscles argues against there being a single hypothesis explaining stretch-evoked EMG activity in spasticity. A reduction in stretch reflex threshold but preserved reflex gain (Powers et al. 1988, 1989) holds neither for the passive nor for the active conditions. The suggestion by Thilmann and coworkers (1991) that the pathological increase in stretch reflex gain is "the major factor in the genesis of spastic hypertonus" is not compatible with the observations seen here during active motor tasks. Although the velocity dependence of the long-latency EMG responses was preserved in the active flexor muscles of the spastic side, its gain was somewhat lower compared with that in the unaffected muscles.

A simple shift to a lower reflex threshold as the pathophysiological basis of spastic hypertonia (Powers et al. 1988, 1989) is also rather unlikely for the following reasons: First, such a shift could explain the observations made only during passive but not during active tasks (see Fig. 4). Second, a differential modulation of early and late EMG components takes place during active motor tasks. Reduced presynaptic inhibition of group Ia afferents may lead to the increased M1 response in both passive and active tasks (cf. Iles and Roberts 1986). A differential effect of the supraspinal lesion on Golgi tendon organ and muscle spindle afferent inputs may be responsible for the enlarged amplitude during passive and reduced amplitude during active motor tasks of the long-latency responses (cf. Houk et al. 1971). Thus long-latency reflexes in spasticity become modulated only over a small range with an impairment of the ability to "switch off" in the passive task as well as facilitation in the active condition. The degree of defective modulation obviously depends on the severity of paresis (cf. Berger et al. 1988), which was rather moderate in the experiments described here. Alternatively, the extra-activity induced during the passive condition may be due, although there is no positive indication, to a stretch-reflex origin other than that induced during the active motor task.

There was no major qualitative difference between the behavior of flexor and extensor muscles, although the differences between unaffected and spastic limb were more pronounced in the flexors. The opposite is true for the leg muscles, where the changes in the spastic limb were more obvious in the extensor muscles (Berger et al. 1988). This may be related to the antigravity function of the extensors in the leg, and of the flexors in the arm.

Influence of Motor Tasks on the Long-Latency EMG Responses

Figure 3 shows the group average of the EMG responses obtained in the brachioradialis muscle following extension applied at three velocities on the unaffected and spastic side of all subjects. The displacements were induced during sinusoidal isometric muscle contractions and the subjects had to control elbow joint position. The amplitude of both the early and the late components of EMG response on both sides was clearly dependent on displacement velocity: 300°/s (a), 200°/s (b), and 150°/s (c). On the unaffected side, the slope of the long-latency EMG response was closely correlated with the acceleration (and position) signal; that is, its ramp increase and duration correlated with the initial change in elbow joint velocity and its peak remained about constant (t = 80–95 ms) for the three displacement velocities. A similar close correlation with the acceleration signal (r = 0.82) was found in five healthy subjects. On the spastic side, amplitude of the short-latency reflex response was in general larger, that of the long-latency response smaller than on the unaffected side. In contrast to the unaffected side, the long-latency EMG response resembled a plateau and was most closely correlated with the velocity signal (see Table 1). The duration of the plateau was closely correlated with the duration of the impulse (r = 0.84). The torques evoked by the displacements were not visibly different between the unaffected and the spastic side.

Figure 4 shows the group average of the EMG responses in the brachioradialis muscle to the same displacements as in Fig. 3. The displacements were, however, induced during the sinusoidal movements when the subjects had to control elbow torque. Again, amplitudes of both the early and late EMG responses on both sides were dependent on the displacement velocity. The slope of the long-latency EMG response was more closely correlated with the velocity signal of the displacement (not shown here) on both sides. In five healthy subjects, the correlation factor was also highest for the velocity signal. Apart from the fact that the amplitude of the short-latency response was larger and the long-latency response a little smaller on the spastic side, no major differences were detected between the EMG responses of the spastic and those of the unaffected limb. On both sides, the duration of the EMG response was closely correlated with the duration of the displacement (r = 0.84 and r = 0.76, respectively).

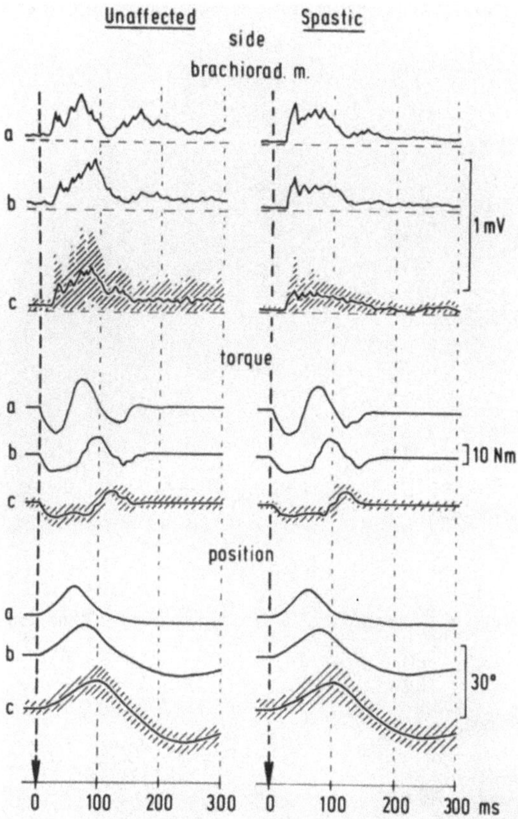

Fig. 3. Group mean values (with SD for impulse *c*) of the rectified and averaged ($n = 10$) brachioradialis EMG responses together with the torque and position signals of the unaffected and spastic side, for 15 patients with spastic hemiparesis. Displacements of three different velocities in extension direction were induced during sinusoidal isometric muscle contractions: *a*, 300°; *b*, 200°; and *c*, 150°/s. Subjects had to control elbow joint position. *Arrows* indicate onset of displacement

The results obtained from the biceps brachii recordings were similar to those from the brachioradialis muscle. This similarity was also reflected in the correlation coefficients between the long-latency responses in the biceps brachii and the biomechanical variables.

A new observation is that the long-latency reflex responses are modulated differentially during different motor tasks, and that this modulation is defective in patients with spastic paresis. In healthy subjects and, to a lesser degree, on the unaffected side of the patients the long-latency EMG responses are coded by the acceleration signal when subjects controlled elbow joint position and by the velocity signal when they controlled torque.

The difference in the generation of the EMG responses is most probably the consequence of appropriate changes in the afferent signals from second-

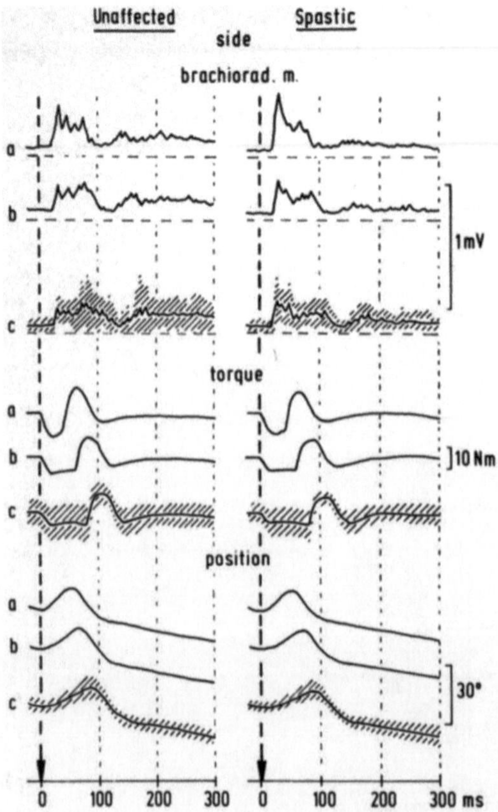

Fig. 4. EMG responses in the brachioradialis muscle to displacements in extension direction. Subjects had to control elbow joint torque. See legend to Fig. 3

ary and primary endings (for review see Matthews 1988) and in processing of this input within spinal interneuronal centers. The change in the behavior of the muscle spindles between the two motor tasks may be achieved by appropriate changes of the firing pattern of gamma motoneurons by central regulation (see Prochazka 1989). In line with this, differences in recruitment levels and firing frequencies of biceps brachii motor units were found between movement and force tasks and were attributed to a different activation of gamma motoneurons in the two tasks (Tax et al. 1990). In addition, afferent input may be different for the two tasks, regardless of whether muscle spindle sensitivity has been altered by gamma motoneuron activity .

The differential modulation of the functionally essential long-latency EMG response is impaired in the spastic limb. The ballistic movement pattern with the dynamic nature of the agonistic long-latency response was reduced or lost, while the late activation of the antagonistic muscles was preserved. This may be due to a defective integration of the afferent impulses, coding the acceleration signal on spinal interneurons (see Matthews 1988) and/or an impaired central regulation of dynamic gamma motoneuron activation. A similarly defective function of spinal interneuronal circuits has been suggested in the impaired leg muscle reflex function in patients with spastic paresis during stance (Dietz and Berger 1984) and gait (Berger et al. 1984). The dysfunction of an appropriate coding of the EMG response may explain the difficulty of these patients to perform finely coordinated finger and hand movements. A movement deficit caused by hyperactive stretch reflexes, as was found to be the case for leg movements in some patients (Corcos et al. 1986), was not seen in the present experiments. The findings are in line with the conclusion drawn from other studies that hyperactive stretch reflexes have little to do with the movement performance (Landau 1980; Levin and Hui-Chan 1993; Sahrmann and Norton 1977; Sinkjaer et al. 1993).

Torque-to-EMG Relationship

Figure 5 shows the differences in the mean torque signals between affected and unaffected sides for all arm-extending displacements during the passive

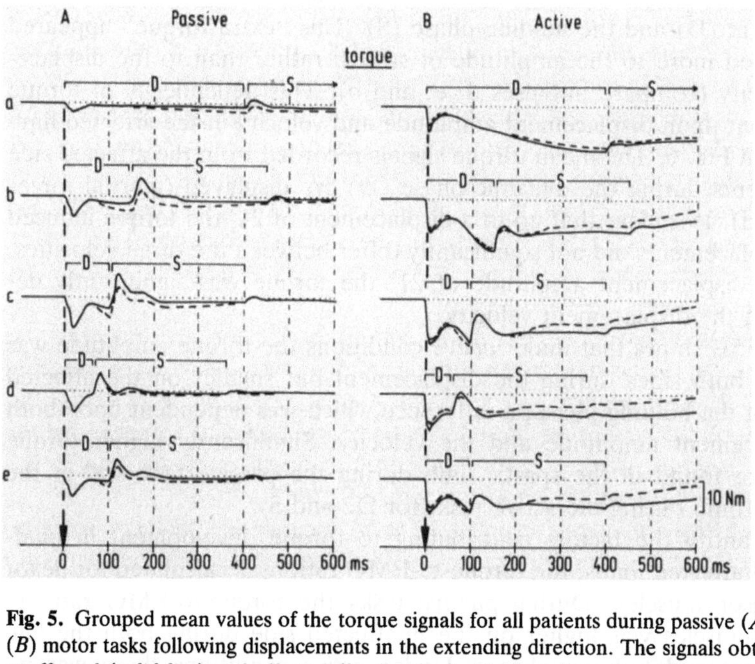

Fig. 5. Grouped mean values of the torque signals for all patients during passive (*A*) and active (*B*) motor tasks following displacements in the extending direction. The signals obtained on the unaffected (*solid line*) and affected (*broken line*) sides are shown together for every impulse. Each trace has been aligned to the torque exerted before the release of the impulse (*arrows*)

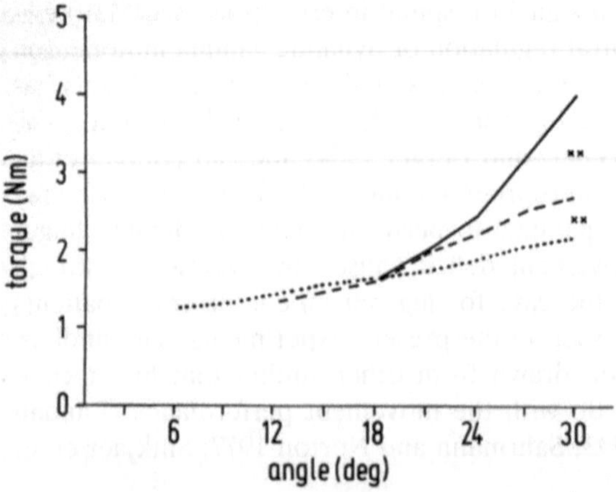

Fig. 6. Mean values of the torque signals following arm-extending displacements (phase D, without inertial forces) for three velocities: 75°/s (*dotted line*), 200°/s (*broken line*), and 300°/s (*solid line*). The average torque during the ramp displacement of all affected limbs was calculated during the passive condition for intervals of 10 ms according to the formula torque segment (Nm ms)/duration (ms) and plotted against the angle of the displacement. Significant differences between the three impulses after 20° degrees are indicated; **, $p < 0.01$

(Fig. 5A) and active (Fig. 5B) motor tasks for all patients. Under *passive* conditions, the torque was increased on the affected side during both the displacement (D) and the holding phase (S). This "extra torque" appeared to be related more to the amplitude of stretch rather than to the displacement velocity (compare impulses d, e, and b). This dependency of torque development upon displacement amplitude and velocity in the affected limb is plotted in Fig. 6. The mean torque signals recorded from the affected side of all patients during the dynamic phase (D) are displayed (inertial forces are omitted). It is clear that up to a displacement of 21° the torque induced by the displacements did not significantly differ between the three velocities. Beyond a displacement amplitude of 21° the torque was significantly dependent on the displacement velocity.

Figure 5B shows that under *active* conditions the torque amplitude was similar on both sides during the displacement but smaller on the affected side during the holding phase, a difference which was dependent upon both the displacement amplitude and the velocity. Significantly higher torque values were found in the spastic limb during the passive task and in the unaffected limb during the active task (for D2 and S).

To quantify the factors contributing to torque development in unaffected and affected limbs, the torque-to-EMG ratio was calculated for flexor and extensor muscles. During *passive* tasks the torque-to-EMG ratio in the three muscles was higher on the unaffected side during both the late displacement and holding phases. Under *active* conditions there was no significant difference between unaffected and affected sides during the two early periods. However, the torque-to-EMG ratio was significantly higher in all three muscles of the affected limb during the later part of the displacement (D2) and (except for the triceps muscle) the holding phase (S).

These observations indicate a striking increase in resistance to stretch on the spastic side during *passive* conditions. There is little doubt that this increase in muscle tone is due largely to the extra EMG activity induced under this condition. However, these results are partially at variance with the widely accepted view that the resistance of spastic muscle increases predominantly with the velocity of stretch (Lance 1980). In the present experiments, the higher resistance to stretch in the spastic flexor muscles was dependent mainly upon the amplitude of stretch. It was only with stretch amplitudes of greater than 20° that the amplitude of torque was significantly influenced by the stretch velocity. This effect may have been due to the fact that the patients used in this study were suffering only mild spastic symptoms.

During *active* conditions, the resistance to stretch of flexor muscles was different in comparison to the resistance seen during passive conditions. During the dynamic phase of stretch there was little difference in the torque between the unaffected and spastic sides. The exaggerated early reflex component contributes little to the generation of dynamic torque. During the "holding phase", the development of torque was smaller on the spastic than on the unaffected side, corresponding to a reduced late EMG activity. In this context one has to be aware that the term "spastic" may be inappropriate, as it implies muscle hypertonia (cf. Lance 1980). The influence of the respective antagonistic muscles on the torque development is suggested to be rather weak. In only a few conditions was some co-contraction present around the end of the displacement, the strength of which was similar on both sides.

For further evaluation of the factors contributing to the development of muscle tone during stretch, the torque-to-EMG ratio was calculated. Spastic muscle had a higher ratio under active conditions than under passive conditions. The increased torque-to-EMG ratio in the flexors and, to a lesser extent, in the extensors can be explained only by changes in the mechanical properties of the muscle fibers during the development of spastic tone. Because the ratio was increased only during the active task, these changes probably affect the contractile mechanism of the spastic muscle. Consequently, spastic muscles in their active state produce more torque per unit of EMG activity following stretch than unaffected muscles. Such changes in spastic paresis have already been suggested for leg extensors (Berger et al. 1988; Dietz et al. 1986) and arm flexors (Lee et al. 1987; Dietz et al. 1991). Such changes may occur to compensate for the defective modulation and reduction of EMG activity following a cerebral lesion connected with paresis (Berger et al. 1984, 1988). At least for the leg extensor muscles it was shown that the higher resistance to stretch of the tonically active leg extensors enables the patient to support his body during locomotion.

The results obtained under passive conditions will be important for the clinical evaluation of spasticity. The active condition is, however, essential for appropriate treatment of spastic movement disorder. Therefore, it is important to be aware of the differences which exist between active and passive motor tasks while prescribing any physical or drug treatment. In addition, the present study indicates that one has to recognize the limitations of the experimental approach used, as differences exist between upper and lower limbs and between flexor and extensor muscles, as well as between active and passive tasks. The limitation of the method used in this study lies mainly in the fact that only moderately affected patients were selected, who were still able to perform functional movements.

Acknowledgements. I.K. Ibrahim was on leave from the University of Alexandria (Dept. of Physical Medicine), Egypt, and held a fellowship from the Channel Program. We thank Dr. J. Gibson for scrutinizing the English text. Technical assistance was provided by Mrs. U. Römmelt. This work was supported by the DFG (SFB 325).

References

Berger W, Horstmann G, Dietz V (1984) Tension development and muscle activation in the leg during gait in spastic hemiparesis: the independence of muscle hypertonia and exaggerated stretch reflexes. J Neurol Neurosurg Psychiatry 47:1029–1033

Berger W, Horstmann GA, Dietz V (1988) Spastic paresis: impaired spinal reflexes and intact motor programs. J Neurol Neurosurg Psychiatry 51:568–571

Burke D, Ashby P (1972) Are spinal "presynaptic" inhibitory mechanisms suppressed in spasticity? J Neurol Sci 15:321–326

Cody FWJ, Richardson HC, MacDermott N, Ferguson II (1987) Stretch and vibration reflexes of wrist flexor muscles in spasticity. Brain 110:433–450

Corcos DM, Gottlieg GL, Penn RD et al. (1986) Movement deficits caused by hyperexcitable stretch reflexes in spastic humans. Brain 109:1043–1058

Dietz V, Berger W (1984) Interlimb coordination of posture in patients with spastic paresis. Impaired function of spinal reflexes. Brain 107:965–978

Dietz V, Quintern J, Berger W (1981) Electrophysiological studies of gait in spasticity and rigidity. Evidence that altered mechanical properties of muscle contribute to hypertonia. Brain 104:431–449

Dietz V, Ketelsen UP, Berger W, Quintern J (1986) Motor unit involvement in spastic paresis: relationship between leg muscle activation and histochemistry. J Neurol Sci 75:89–103

Dietz V, Trippel M, Berger W (1991) Reflex activity and muscle tone during elbow movements in patients with spastic paresis. Ann Neurol 30:767–779

Houk JC, Singer JJ, Henneman E (1971) Adequate stimulus for tendon organs with observations on mechanics of ankle joint. J Neurophysiol 34:1051–1065

Hufschmidt A, Mauritz KH (1985) Chronic transformation of muscle in spasticity: a peripheral contribution to increased tone. J Neurol Neurosurg Psychiatry 48:676–685

Iles JF, Roberts RC (1986) Presynaptic inhibition of monosynaptic reflexes in the lower limbs of subjects with upper motoneuron disease. J Neurol Neurosurg Psychiatry 49:937–944

Lance JW (1980) Pathophysiology of spasticity and clinical experience with baclofen. In: Feldmann RG, Young RR, Koella WP (eds) Spasticity: disordered motor control. Year Book Medical Publishers, Chicago, pp 185–203

Landau WM (1980) What is it? What is it not? In: Feldmann RG, Young RR, Koella WP (eds) Spasticity: disordered motor control. Year Book Medical Publishers, Chicago, pp 17–24

Lee RG, Tatton WG (1975) Motor responses to sudden limb displacements in primates with specific CNS lesions and in human patients with motor system disorders. Can J Neurol Sci 2:285–293

Lee WA, Boughton A, Rymer WZ (1987) Absence of stretch reflex gain enhancement in voluntarily activated spastic muscle. Exp Neurol 98:317–335

Levin MF, Hui-Chan C (1993) Are hand stretch reflexes in hemiparesis reproducable and correlated with spasticity? J Neurol 240:63–70

Matthews PBC (1988) Proprioceptors and their contribution to somato-sensory mapping: complex messages require complex processing. Can J Physiol Pharmacol 66:430–438

Matthews PBC (1991) The human stretch reflex and the motor cortex. TINS 14:87–91

Powers RK, Marder-Meyer J, Rymer WZ (1988) Quantitative relation between hypertonia and stretch reflex threshold in spastic hemiparesis. Ann Neurol 23:115–124

Powers RK, Campbell DL, Rymer WZ (1989) Stretch reflex dynamics in spastic elbow flexor muscles. Ann Neurol 25:32–42

Prochazka A (1989) Sensorimotor gain control: a basic strategy of motor systems? Prog Neurobiol 33:281–307

Rosenfalck A, Andreassen S (1980) Impaired regulation of force and firing pattern of single motor units in patients with spasticity. J Neurol Neurosurg Psychiatry 43:907–916

Sahrmann SA, Norton BJ (1977) The relationship of voluntary movement to spasticity in the upper motor neuron syndrome. Ann Neurol 2:460–465

Sinkjaer T, Taft E, Larsen K, Andreassen S, Hansen H (1993) Non-reflex and reflex mediated ankle joint stiffness in multiple sclerosis patients with spasticity. Muscle Nerve 16:69–76

Tax AAM, Denier van der Gon JJ, Gielen CCAM, Kleyne M (1990) Differences in central control of m. biceps brachii in movement tasks and force tasks. Exp Brain Res 79:138–142

Thilmann AF, Fellows SJ, Garms E (1990) Pathological stretch reflexes on the "good" side of hemiparetic patients. J Neurol Neurosurg Psychiatry 53:208–214

Thilmann AF, Fellows SJ, Garms E (1991) The mechanism of spastic muscle hypertonus. Brain 114:233–244

Discussion Summary

S. Miller

The paper by A.F. Thilmann reported electromyographic responses in biceps brachii to passive movement about the elbow joint in normal subjects and patients at different stages following stroke. Late velocity dependent responses were evoked as spasticity developed and were considered the neurophysiological correlate of spastic muscle hypertonia.

The discussion focused on the neurophysiological changes underlying the late responses. The possibilities considered were that the threshold of the motoneurones might be raised in addition to a change in slope (gain) of the input/output relationship. It was pointed out that the conditions under which such relationships are sought need to be standardised by taking into account the EMG/force relationship and by making the observations over the defined operating length and operating conditions of the muscle.

The basis for late responses was considered to be polysynaptic and involve spinal interneurones. Jankowska commented that the emergence of the late responses in spastic subjects might be explained by a model of spasticity that involved a dis-inhibition of group II activated spinal interneurones following damage to noradrenergic and serotonergic brainstem descending pathways (see the chapter by Jankowska). The hypothesis was considered attractive but needing proof, and other mechanisms should also be considered.

The late responses reported in biceps brachii cannot be evoked under the same stimulus parameters from triceps surae, and a distinction needs to be accepted between responses of muscles in the upper and lower limbs and those lying proximal and distal in the limb.

The study by Dietz compared the responses in muscles acting about the elbow joint to passive stretch and during an active motor task. Some initial technical questions were raised. Dietz replied that the target of muscle force was 10% of maximal power of the non-paretic limb. In the active task the subjects learned to track a force profile, and Dietz argued that the test pulse did not provoke an unwanted release of force, which might otherwise confound interpretation of the results. Further discussion established that the inertia of paretic and non-paretic limbs is not equal, and that this should

Division of Clinical Neuroscience, The Medical School, The University of Newcastle upon Tyne, Newcastle upon Tyne NE2 4HH, England, UK

Thilmann et al. (Eds.)
Spasticity: Mechanisms and Management

be taken into account when considering the relationship between EMG and torque.

The main discussion centred around the differential performance of paretic and non-paretic limbs under the two experimental conditions and the neurophysiological mechanisms underlying the changes in early and late reflex activity. Dietz argued that the passive condition represents muscles in the body which are relaxed, and the active condition defines the reorganisation of spinal reflexes underpinning a motor task. It was pointed out that the relatively small changes in early reflex responses to stretch in the two conditions and the profound differences occurring in late responses suggest involvement of different spinal mechanisms, the late responses being served by separate interneurones. The former classification of stretch reflex responses as M1, M2, M3, M4 was better described as early and late responses.

Is the active task a special case, and does it take account of lengthening contractions, shown to be common in movements of the arm (see chapter by Fellows et al.)? Dietz replied that the experimental conditions inevitably constitute a special case, although some degree of co-contraction was often present. The important feature, however, was that the late reflexes were reduced in the active versus the passive condition, and that the underlying interneuronal reorganisation is the significant emergent element. Comparisons were drawn with operant conditioning of stretch reflexes in children with cerebral palsy (Nielson and McCaughey, J Neurol Neurosurg Psychiatr 1982, 45:320–330), but Dietz considered that the disordered neural mechanisms in spastic adult subjects to be different.

The general discussion addressed the following topics.

1. Disorders of monoamine brainstem projections to the spinal cord as models for spasticity. Rymer questioned whether the bistability of α-motoneurones, which is sensitive to activity in descending serotonin and noradrenergic pathways (see the chapter by Nielsen and Hultborn), might be in conflict with the observation that locally administered monoamines depress transmission between group II muscle afferents and motoneurones (see the chapter by Jankowska). The implication of these results in the cat for the mechanisms underlying spasticity in man might suggest that on Hultborn's data transition to the higher bistable state would be mediated by increased activity in descending activity from the locus ceruleus (noradrenergic projection) and the raphe nuclei (serotonin projection). On Jankowska's model noradrenergic depression of group II transmission would require for spasticity a disfacilitation or degeneration of the locus ceruleus. Jankowska claimed there might be no conflict since the locus ceruleus and the raphe nuclei have mutual inhibitory connections. Increase in locus ceruleus activity results in less raphe activity and less efficient interneuronal excitation of motoneurones. Conversely, depression of locus ceruleus is accompanied by increased raphe activity. The consensus view was that the interrelations of the locus ceruleus and the raphe nuclei and the sensitivity of patients with spasticity to catecholamine drugs should be tested.

Hultborn also commented that changes may also take place in second messengers within motoneurones after chronic spinalisation, so that there may be intrinsic cellular changes associated with spasticity.

A further issue is the possibility that denervation hypersensitivity in interneurones and motoneurones may change in extent and type in the development of spasticity.

2. *Bistability in motoneurones.* How far does bistability in motoneurones play a role in the increased excitability of reflexes observed in human subjects with spasticity? Hultborn recalled that bistability can be evoked only in a proportion of motoneurones tested in animal preparations (see the chapter by Nielsen and Hultborn). Kernell commented that the shift with administration of monoamines to the upper level of excitability produces an increase in gain at the motoneurone membrane with a shift in the current/frequency relationship (see the chapter by Kernell). He suggested that this should be accompanied by a decrease in the after-hyperpolarisation, which Hultborn had not observed in all motoneurones tested. It became clear that further work is still required in the animal models.

Rymer commented that relatively few data exist on firing rates of motoneurones in human subjects with spasticity, although his own laboratory is investigating this issue. Reduced, not augmented, firing rates for matched force production have in fact been observed in spastic muscle. The observations were hard to interpret in view of unknown factors, such as the changes of muscle properties in spasticity, the operating temperature of the muscles, alterations in twitch contraction time, the degree of lengthening at which the muscles were tested, the structural remodelling of partially denervated motoneurones and the existence of inactive synapses from muscle afferents. More information from human subjects is clearly needed.

The role of the bistable property of motoneurones in motor performance was addressed by Hultborn, who suggested that it might come into play within a relatively short timescale in relation to muscle activation, serving to add briefly to existing depolarising activity as a kind of power amplification. One possible action might be in self-sustained firing of motoneurones involved in postural control. He noted that the monoamine control of bistability might not just result from the activation of receptors on the motoneuronal membrane but concern changes within the cell at the secondary messenger level.

Kernell developed the notion that the central nervous system must recalibrate its "hardware" following damage, and that part of the disorder of motor performance in spasticity may be due to failure of the recalibration process or faulty recalibration. It was noted that afferent input would play an important role, as some rehabilitation regimes have suggested.

3. *Reflex gain.* Rymer drew attention to the non-linear components of the stretch reflex, citing Feldman's (1973) work in the cat and Matthews' studies in cat and man. The length/tension (force) plot, which provides a measure of stiffness, begins with a flat slope, then rises and becomes linear

over a certain range and then flattens. As Kernell has argued, the changing slope provides an indication of the change of gain (see the chapter by Kernell). Stiffness thus rises with increased background force. Several speakers then drew the important conclusion that the gain of stretch reflexes should be measured under matched operating conditions, when examining the paretic and non-paretic limbs. There was the hopeful rider that the neuronal machinery driving the muscle and the contractile process should be understood as fully as possible.

A suggestion was made that where an agonist/antagonist pair of muscles (e.g., biceps and triceps brachii) are considered, the non-linearities of the stretch reflex systems of each muscle should compensate to give a more linear stiffness response. However, the consensus view was that even in co-contraction the non-linear response of one muscle would still be evident.

It was pointed out that one of the operating conditions to be standardised is obviously velocity of stretch, since it has a strong influence on the threshold of response. A high velocity stretch results in a rapidly rising excitatory post-synaptic potential in the motoneurone, and threshold is reached more rapidly. Attention should therefore be paid to the velocities at which stretches, whether ramp or sinusoidal, are applied.

Burke developed arguments that if there is no reflex gain change, and the increase in reflex excitability is purely a threshold change, then the likelihood that there is a change in afferent input to the spinal cord is much reduced. It should not be forgotten that one of the former popular theories was that the fusimotor system drives spasticity; this was its ultimate cause. He went on to cite some of the studies of Rymer, showing lack of changes in gain, which serves to undermine the old hypothesis. Furthermore, if gain changes are not evident, it becomes more difficult to invoke group II excitation as a causal factor in spasticity (see the chapter by Jankowska).

Part V
Reflex Disturbances in Spasticity: Electrophysiological Studies

Methodological Problems in the Hoffmann Reflex Study of Spasticity

N. Yanagisawa, M. Shindo, H. Morita, and S. Yanagawa

Introduction

The Hoffmann (H) reflex has been considered a monosynaptic reflex via Ia muscle afferents. Because its size represents the excitability of this reflex arc, the H reflex has been a tool to evaluate (a) the excitability of the motoneurones and (b) activities of various reflex pathways in spasticity. As the activity of the reflex arc depends on the excitability of α-motoneurones, the size of the maximal H reflex or the threshold of the H reflex are measured to assess their excitability. Since extensive studies by Magladery et al. (1950, 1951, 1952), in particular, the H reflex of calf muscles, together with a method of recovery of H reflex to double stimulation of the tibial nerve, has been used for the quantitative evaluation of spasticity. Use of the H reflex was then extended as a test monosynaptic reflex for studying the activity of spinal reflex pathways acting on the motoneurones.

Various changes in spinal reflex activities have been reported in spasticity with the H reflex method (for reviews, see Pierrot-Deseilligny and Mazières 1985; Pierrot-Deseilligny 1990). Such changes include an increase in Renshaw inhibition at rest but lack of modification by voluntary contraction (Katz and Pierrot-Deseilligny 1982), increase in excitability of α-motoneurones (Landau and Clare 1964) or γ-motoneurones (Dietrichson 1973; Tardieu et al. 1964), imbalance in reciprocal Ia inhibition (Yanagisawa et al. 1976; Yanagisawa 1980), decrease in antagonist inhibition (McLellan 1977; Knutsson 1983), decrease in presynaptic inhibition on Ia terminals (Delwaide 1969), increase in reciprocal Ia inhibition (Boorman et al. 1991), and decrease in Ib inhibition (Delwaide and Olivier 1988). However, some are speculative or nonconclusive because of indirect estimation of the activity of concerned pathways and other methodological problems.

In studies using the H reflex, one should take various methodological problems into account; some may be overcome with technical improvement, and some should be left as issues open to future studies. The problems lie in a wide range from practical matters such as the fixation of stimulating

Department of Medicine (Neurology), Shinshu University School of Medicine, Asahi 3-1-1, Matsumoto 390, Japan

Thilmann et al. (Eds.)
Spasticity: Mechanisms and Management

electrodes, determination of the motor threshold, joint angles in the extremity, subject's posture, and psychological state of the subject, ect., to more logical and physiological aspects. In the limited space of this chapter considerations are focused on the latter perspective, and merits and limitations are discussed concerning studies using the H reflex for the quantitative evaluation and elucidation of neural mechanisms of spasticity.

Assessment of Motoneuronal Excitability

The H reflex has been assumed to be the substrate of deep tendon reflex except for bypassing the activity of the muscle spindles and γ-motoneurones. Since hyperreflexia is an essential feature of spasticity, the H reflex was expected to reflect the quantity of spasticity at the α-motoneuronal level.

H/M Threshold Ratio

Increased excitability of the motoneurone pool is expected to result in lowering of the threshold of stimuli to provoke H reflexes. Since the threshold for excitation of α-motor fibers, as judged by the M wave appearance, is unaffected by the lesions in the central nervous system, the H/M threshold ratio was expected to be lowered in spasticity. Results since Magladery et al. (1952), however, have been controversial. Our results on the H/M threshold ratio in reference to clinical grades of spasticity have shown that it decreases as spasticity increases when patients are grouped into three depending on the grade of spasticity (analysis of variance, $p < 0.02$, Fig. 1, lower half). However, inter-individual variation of the H/M threshold ratio is so great that the value cannot be an indicator of spasticity in each patient.

Ratio of Maximal Values of H Reflex and M Wave

This was also expected to be an indicator of spasticity. Angel and Hoffmann (1963) reported that the H/M maximal ratio increases on the side of spastic hemiplegia due to cerebrovascular disease compared to that of the unaffected side of the same subject (see also Ongerboer de Visser et al., this volume). Our observations on spastic subjects due to cerebrovascular disease or spinal cord lesions have revealed findings similar to the H/M threshold ratio. The maximal size of the H reflex compared to that of the M wave increased as the clinical grade of spasticity increased (Fig. 1, upper half). However, inter-individual variation was great in each group (grade) of spasticity as well as in healthy subjects, in the latter of which H/M maximal values ranged from below 5% to more than 90%. The tendency of H/M

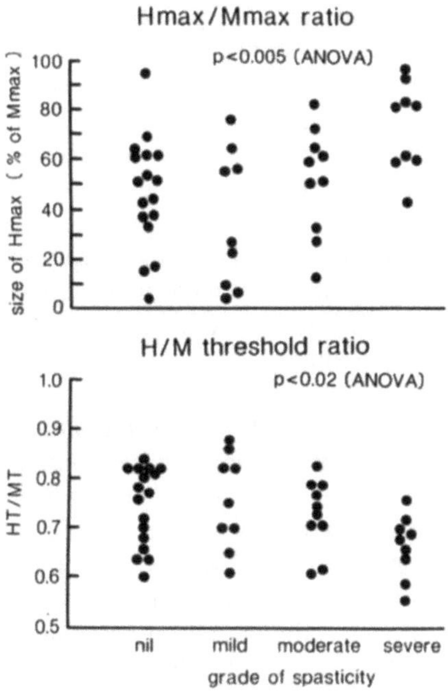

Fig. 1. The H/M amplitude ratio (*above*) and H/M threshold ratio (*below*) are plotted against the grades of spasticity. *Leftmost column*, normal subjects. The patients are classified into three groups according to the severity of spasticity. The amplitude ratio increases significantly ($p <$ 0.005, analysis of variance), and the threshold ratio decreases ($p < 0.02$) as the grade of spasticity increases as a group. However, inter-individual differences are so large that these parameters would not be suitable to evaluate the severity of spasticity in each patient. This is best recognized when we note that the H/M amplitude ratios are distributed over a wide range from below 5% to more than 90% in normal subjects

maximal values to increase with the grade of spasticity was more obvious than that of the H/M threshold ratio when they were compared among the groups composed of the same subjects.

Recovery Curve of H Reflex to Double Stimulation

Magladery et al. (1952) proposed that the course of recovery of the soleus H reflex to double stimulation of the tibial nerve reflects excitability of the motoneurones. They showed in spasticity that the minimal interstimulus interval for appearance of the H reflex to the second stimulus is shortened, and that recovery of the size of the H reflex is earlier than in normal subjects. Since then, this "recovery curve of H-reflex" has been studied extensively, and a supernormal period with interstimulus intervals of 200–300 ms, during which the size of the H reflex to the second stimulus exceeds that to the first stimulus, is a characteristic feature of spasticity. These features of the recovery of the H reflex were also observed in other muscles than the triceps surae. Courses of recovery of the H reflex to the second stimulus have been assumed to reflect the change in excitability of the motoneurones after their firing due to the first stimulus, the underlying mechanisms including absolute and relative refractoriness of the motoneurones and Renshaw inhibition.

However, these features in recovery curves of the H reflex in spastic patients may not represent increased excitability of α-motoneurones, and at least some of them can be ascribed to a change in excitability of the peripheral neuromuscular system, as discussed below.

Changes in the Nerve and Muscle Action Potentials Following Nerve Stimulation. Our recent studies on changes in excitability of the peripheral nerve and the muscle disclosed obvious changes at these "peripheral" sites, which have not previously been taken into account as mechanisms contributing to the recovery curve of H reflexes (Morita et al. 1993). The averaged time course of the effect of double stimulation of equal intensity to the nerve in six healthy subjects revealed that effects of nerve stimulation on compound nerve and muscle potentials last more than 150 ms (Fig. 2). Both potentials show a supernormal phase followed by depression, although the amount of the change differs between nerve and muscle potentials. The maximal

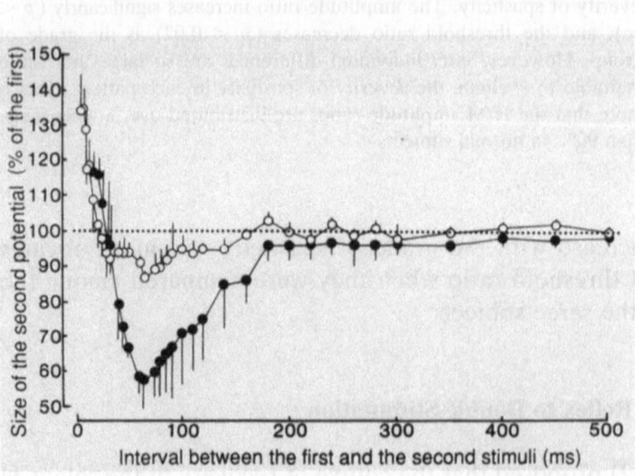

Fig. 2. Change in size of compound nerve action potentials (*open circles*) and muscle action potentials (M responses, *filled circles*) following preceding stimuli to the peripheral nerve. The data are from six subjects. Intensities of the first and second stimuli are the same, adjusted to produce 50% of the M_{max} with a single stimulus. The size of the second potential expressed as a percentage of the first is presented against the interstimulus interval. The curves show biphasic patterns, facilitation being maximally 135% (6 ms) for the nerve and 117% (16 ms) for the muscle, which is followed by inhibition to 87% (65 ms) and 58% (65 ms), respectively. Each point and bar show the means and SDs. (Quoted from Morita et al. 1993)

facilitation and depression were as much as 117% and 58% of the control in M responses, and 135% and 87% in the nerve potentials. Such changes depended on the stimulus intensity, being maximal with stimulus intensity just supraliminal for the motor threshold. No change was observed when the intensity exceeded the maximal M response. The two phases in excitability change of the muscle still remained with a similar time course when the size of the first and second nerve potentials were equalized, although the amount of the change decreased significantly. This indicates that the efficacy of neuromuscular transmission and/or excitability of the muscle itself is also influenced by the preceding stimuli.

In relation to H reflex study, the relationship of the changes between M responses and H reflexes was investigated in the soleus muscle, where the size of the two potentials was equalized, i.e., 30% of M_{max}. The M wave pattern was as described above, and the H reflex changed in a similar pattern, being nearly parallel particularly at intervals after 100 ms (Fig. 3).

Whatever mechanisms are concerned in such long-lasting effects on the nerve action potentials and muscle potentials, our study clearly demonstrates that muscle potentials show marked changes in size after activation either indirectly through changes in the excitability of the peripheral nerve or directly from those of the muscle itself. This raises a serious question: to what extent does the recovery curve of the H reflex actually represent changes in the excitability of the motoneuronal pool, as has been assumed hitherto? If one adjusts the first stimulus to the intensity subliminal to the H reflex (Táboríková and Sax 1969), factors on the neuromuscular junction or the muscle may be eliminated. However, the problem of the excitability change of Ia afferent fibers in the peripheral nerve still remains.

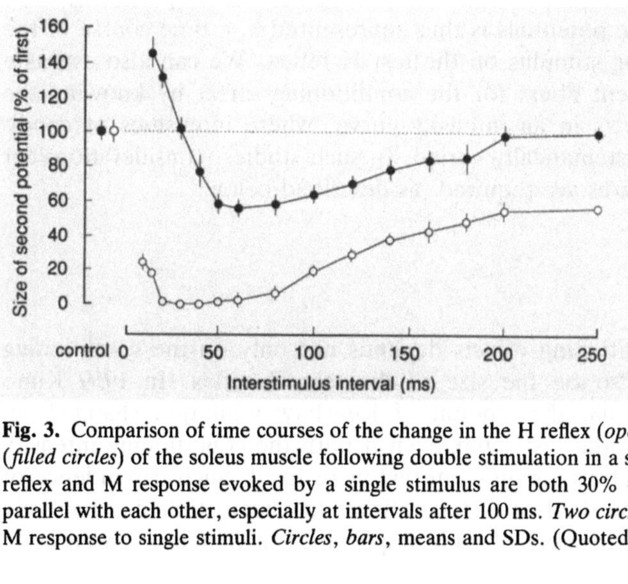

Fig. 3. Comparison of time courses of the change in the H reflex (*open circles*) and M response (*filled circles*) of the soleus muscle following double stimulation in a subject. The sizes of the H reflex and M response evoked by a single stimulus are both 30% of M_{max}. Both curves are parallel with each other, especially at intervals after 100 ms. *Two circles at left*, the H reflex and M response to single stimuli. *Circles*, *bars*, means and SDs. (Quoted from Morita et al. 1993)

Effect of Muscle Stretch on the H Reflex to the Second Stimulus. On the other hand, contraction of the muscle to the first stimulus may change Ia afferent discharge, which in turn changes the excitability of motoneurones. The supernormal phase, which is observed in spasticity at interstimulus intervals of 200–300 ms, is attributable largely to an increase in discharge of Ia afferents from the muscle spindle caused by passive lengthening of the muscle after active contraction following the first stimulus. Fixation at the ankle joints greatly reduces the supernormal phase (Hayashi et al., personal communication).

Spinal Reflex Connections Studied with the H Reflex

The major portion of the H reflex is considered a monosynaptic reflex via Ia fibers, and the size of the H reflex can be assumed to represent the excitability of the motoneurone pool and efficacy of Ia monosynaptic transmission to motoneurones. Following the methodological principle adopted by Lloyd (1943a,b, 1946a,b), who used monosynaptic reflex potentials recorded at the ventral roots as an indicator of motoneurone excitability, the H reflex has been used as a test reflex for study of various reflex connections in man with or without motor disorders.

Activity of certain reflex paths or functions of intercalated neurones which ultimately connect to motoneurones can be estimated as change in size of the test H reflex by preceding conditioning stimuli. When a conditioning stimulus evokes a postsynaptic potential in motoneurones, it alters the efficacy of motoneuronal firing to the test Ia volleys, depending on the time relationship between the conditioning and test inputs. The time course of postsynaptic potentials is thus represented in a time course of the effects of conditioning stimulus on the test H reflex. We can also estimate the responsible afferent fibers for the conditioning effect by knowing the threshold of the effects in an intensity curve, where intensities of conditioning stimuli are systematically varied. In such studies, considerations for experimental procedures are required, as described below.

Size of Test H Reflex

The amount of conditioning effects depends not only on the conditioning stimulus itself but also on the size of the test H reflex. In 1959 Kuno reported in the cat that the amount of Renshaw inhibition changed according to the size of the test reflex even if both the conditioning intensity and the conditioning test stimulus interval were fixed. Such dependence of conditioning effects on the size of test reflexes has been studied extensively and is recognized as an issue of wide importance in humans as well as in

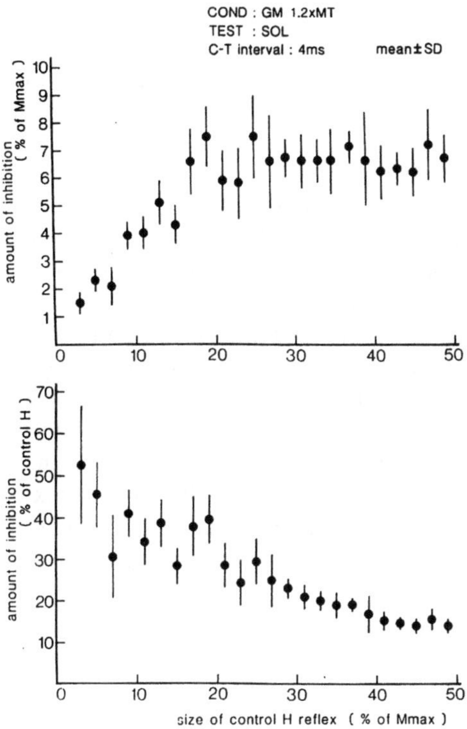

Fig. 4. Relationship between the size of the control H reflex and the amount of inhibition by conditioning stimulation to the gastrocnemius medialis nerve at rest in one subject. The amount of inhibition is the difference between the size of the conditioned reflex and that of the unconditioned one and is presented in two ways: *above*, the size of conditioned reflexes is measured as a percentage of M_{max}; *below*, it is measured as a percentage of the unconditioned control H reflex. The size of control reflex (*abscissa*) is expressed as a percentage of M_{max}. It is seen that the amount of conditioning effects is a function of the size of test H reflexes. *Circles*, *bars*, means and SDs. Both panels are from the same data, and each symbol represents 20–80 measurements. (The upper panel is adapted from Yanagawa et al. 1991)

animals (Hunt 1955; Meinck 1980; Crone et al. 1990). An example is shown in Fig. 4, where the amount of nonreciprocal group I inhibition was plotted as a function of the size of test H reflexes while the other parameters were fixed. This dependence forces us to adjust the intensity of test stimuli to obtain test reflexes of the same size when we compare the effects among subjects or among different situations, such as rest and as voluntary contraction. This is particularly important in evaluating the activity of reflex pathways in spastic patients, in whom larger and more stable H reflexes are provoked very easily than in normal subjects. The size of the reflex should thereby the normalized, usually expressing it as a percentage of the maximal M response in each subject.

Population of Test Motoneurones

On the other hand, the adjustment of the test reflex described above may raise an important question. The adjustment maneuver assumes that the motoneurones sampled in the test reflexes are identical as long as the test size is constant. If the sampled motoneurones are not necessarily the same, differences in conditioning effects may be attributable to the different populations within the motoneurone pool, not to a change of activity in reflex pathways, including interneurones. The underlying idea for the test adjustment is based on a theory of the fixed order of recruitment of motoneurones. So far, the validity of the size principle (Henneman 1957; Henneman et al. 1965a,b) has been generally granted, assuming that the population of the sampled motoneurones in the test reflex is almost identical in whatever conditions as long as the test size is equal. However, the recruitment order can change in certain situations (Kanda et al. 1977; Bawa 1981; Clamann et al. 1983), the motoneuronal population composing the H reflex in man may be different from one situation to another (Hashimoto, Shindo, Tanaka, Yanagawa, Yanagisawa, unpublished observation), and the issue can never be guaranteed especially in comparisons between normal and spastic subjects. In the latter, a different organization of neuronal connections from normal subjects may have been established during the course of their disease. This needs to be studied quantitatively and extensively.

Input-Output Relationship of the Motoneurone Pool

Quantitative comparison of conditioning effects can be made only when the input-output relationships of the motoneurone pool remain constant. If an input resulting in a certain amount of output in one situation produces a different output in another situation at the same level of excitability of the motoneurone pool, then change in the amount of conditioning effects cannot readily be ascribed to altered activity of the concerned reflex pathway. Such alteration of the input-output relationship may be understood in terms of a change in "excitability gradient" throughout the whole motoneurone pool. In fact, such change in the relationship has been proposed in human subjects as well as in animals (Hultborn and Pierrot-Deseilligny 1979; Kernell and Hultborn 1990). The input-output relationship in spasticity has been never explored in comparison to that in normal subjects, and the relationship even in the same subject may change from situation to situation depending on a motor strategy at each moment. This issue is usually ignored in daily experiments, and it needs to be clarified in more detail.

Random Alternation of Stimuli

The amount of conditioning effects is dependent on the psychological attitude of the subjects. To minimize such influence randomization of stimuli is required. In fact, randomized stimulation produces less effect than regular alternation of stimuli (Fournier et al. 1984). This is particularly true when a subject has a true insight into the significance of an examination. Computer processes make it possible to give random combination of test and conditioning stimuli, to set randomized conditioning test intervals, and to deliver randomized intensities of conditioning stimuli.

Intervals for Consecutive Tests

Cumulative depression occurs in the H reflex when it is evoked sequentially. In discussion of the methodology of the triceps surae T and H reflexes published 20 years ago (Hugon et al. 1973), it was recommended that the interstimulus interval should be set for at least 10 s. However, the depression appears even if the interstimulus interval is as long as 10 s. Furthermore, subjects anticipate the stimulus if the interstimulus interval is 5 s or longer, and this may affect the results. Considering that a large number of tests are required in recent studies on reflex connections, and that one experiment should not be longer than approximately 2 h from an ethical and a psychological point of view, an interstimulus interval of 3–4 s seems adequate in most of the studies on reflex connections. The interval should be chosen according to characteristics of a study, and it should be fixed throughout the study.

Interpretation of Results with the H Reflex

The issues pointed out above delineate difficulties in the precise interpretation of results. In the present situation, we can adjust the test stimulus intensity to obtain the test reflexes of comparable size among subjects and among situations, and we can also set reasonable intervals for test stimuli with randon alternation of conditioning and test stimuli. As for the other problems, however, no practical solution is available with the H reflex method, and it often seems impossible even to evaluate the problems. These problems remain to be elucidated in future, but we should be aware of their existence, at least of their possibility, and must be very careful in interpreting the results, especially when the amount of the effects is not remarkable.

Correlation of Motor Disturbance with Spinal Reflexes

When experiments are performed with the methodological aspects appropriately assessed, the final issue to consider is the correlation of results with clinical symptoms in spasticity. Patients with spasticity are generally characterized by exaggerated tendon reflex, increased phasic resistance to muscle stretch, and spastic gait, but the underlying mechanisms are not necessarily the same. Data on spastic patients must then be properly evaluated from the viewpoint of their functional significance in relation to various aspects of motor disturbance in these patients. As an example, changes in reciprocal Ia inhibition between ankle flexors and extensors in spasticity are discussed below from the above point of view.

Reciprocal Ia Inhibition in Spastic Conditions

In spasticity, alteration of activity in the pathway of reciprocal Ia inhibition has been controversial (Boorman et al. 1991; see also Pierrot-Deseilligny and Mazières 1985). It is easily understandable that inhibitory connections to motoneurones in a hyperexcitable state such as in spasticity would produce less inhibitory effect than in normals. However, in actual clinical conditions involving spasticity, reciprocal Ia inhibition may vary between flexors and extensors. Our previous study showed unique characteristics of reciprocal Ia inhibition in capsular hemiplegia (Yanagisawa et al. 1976).

Spastic hemiplegia caused by cerebrovascular accidents which involve the internal capsule is commonly characterized by the Wernicke-Mann posture. In this condition, the physiological extensors of the lower limb (e.g., quadriceps femoris and triceps surae) develop marked spasticity, with the muscle power relatively preserved in contrast to marked loss of strength in the physiological flexors (hamstrings and pretibial muscles). This reciprocal pattern of motor disorder, namely extensor spasticity and flexor weakness, is a direct expression of an abnormal balance of spinal reflexes affected by chronic lesion of higher centers which involve the corticospinal tract.

By interruption of descending pathways in the internal capsule, the physiological flexors of the lower leg develop an exaggeration of stretch reflex which is canceled by overwhelming spasticity of their antagonist extensors. This hyperreflexia in the flexors was confirmed by the development of clonus in the pretibial muscles after permanent block of the nerves innervating the triceps surae muscles (Yanagisawa 1980).

In subjects with capsular hemiplegia who show a sizable H reflex in the pretibial muscles, it is possible to compare the effect of Ia afferents from the extensor muscles on pretibial flexor motoneurones, and vice versa. Such reciprocal Ia inhibition from extensor nerve to flexor motoneurones is marked, but Ia inhibition from flexor nerve to extensor motoneurones is, in contrast, very weak. The amount of Ia inhibition correlates with voluntary power of the paretic muscles. Furthermore, in parallel with the temporary reduction of spasticity of the extensors by alcohol injection to their motor points, the strength of the antagonist flexors increases. These reciprocal effects, i.e., increase in muscle power of pretibial muscles by alcohol block of the nerves to the triceps surae, is, most simply and most likely, explained by reduction of Ia inhibition from the extensor nerve to pretibial motoneurones. Such an increase in the flexor power associated with reduction in extensor spasticity has been reported in different conditions and in different muscles. Local cooling of the spastic muscle depresses its stretch reflex and increases the power of its antagonists (Knutsson 1970). A decrease in spasticity by administration of the antispastic drug tizanidine produces an increase in the strength of physiological flexor hamstrings (Knutsson and Mårtensson 1984). All these phenomena may be considered to be due to the same mechanism.

Future Prospects in the Reflex Study of Spasticity

Among methodological problems in studies with an H reflex as a test reflex, the input-output relationship of the motoneurone pool and the dependence of conditioning effects on the test reflex size together with the issue of motoneuronal population are of prime importance in the interpretation of results. These problems certainly arise from the physiological characteristics of the motoneurone pool, not from those of the motoneurones themselves. The "pool" problems cannot be overcome as long as the H reflex is used a test reflex to analyze spinal circuitry.

On the other hand, methods using a single motor unit have been developed over the past two decades for studying activity of various spinal pathways and descending control of these spinal mechanisms. A poststimulus (or peristimulus) time histogram (PSTH) during weak tonic contraction of the examined muscle (Ashby and Labelle 1977; Mao et al. 1984; Fournier et al. 1986) is one such method, and it has proven to be a powerful tool for analyzing spinal circuits. The physiological significance of the parameters obtained in PSTHs has also been extensively analyzed both in animals and in human subjects (Ashby and Zilm 1982; Fetz and Gustafsson 1983; Gustafsson and McCrea 1984). This method can be applied rather easily in normal and abnormal conditions including spasticity (Mailis and Ashby 1990). The PSTH method can, in a sense, overcome the pool problems described above, and is expected to be useful in quantitative evaluation of pathological mechanisms underlying spasticity.

The PSTH, however, also has some limitations. It is available only during tonic contraction of the examined muscle, and the target motor units are restricted to those with relatively low firing thresholds. While the latter problem will be solved in near future by improving the recording electrode, a method enabling investigations in the resting state is urgently required to quantify activities of various spinal circuits in spasticity as well as in normal subjects.

Although the H reflex method has some serious problems from the strict physiological point of view, it has a great advantage in its easy applicability and in evaluating a total function of the neural connections as a "group" effect. Usefulness as well as limitations of the H reflex in studies of motor disorders including spasticity should be researched further in detail. Studies of single motor unit activities provide an method independent of those with H reflex, complementing demerits of the H reflex, and the wise combination of these methods then promises fruitful results.

References

Angel RW, Hoffmann WW (1963) The H-reflex in normal, spastic and rigid subjects. Arch Neurol 8:591–596

Ashby P, Labelle K (1977) Effects of extensor and flexor group I afferent volleys on the excitability of individual soleus motoneurones in man. J Neurol Neurosurg Psychiatry 40:910–919

Ashby P, Zilm D (1982) Relationship between EPSP shape and cross-correlation profile explored by computer simulation for studies on human motoneurons. Exp Brain Res 47:33–40

Bawa P (1981) Interaction of excitatory and inhibitory inputs to spinal motoneurones in man. Prog Clin Neurophysiol 9:212–219

Boorman G, Hulliger M, Lee RG, Tako K, Tanaka R (1991) Reciprocal Ia inhibition in patients with spinal spasticity. Neurosci Let 127:57–60

Clamann HP, Ngai AC, Kukulka CG, Goldberg SJ (1983) Motor pool organization in monosynaptic reflexes: response in three different muscles. J Neurophysiol 50:725–742

Crone C, Hultborn H, Mazières L, Morin C, Nielsen J, Pierrot-Deseilligny E (1990) Sensitivity of monosynaptic test reflexes to facilitation and inhibition as a function of the test reflex size: a study in man and the cat. Exp Brain Res 81:35–45

Delwaide PJ (1969) Approche de la physiologie de la spasticité: réflexe de Hoffmann et vibrations appliquées sur le tendon d'Achille. Rev Neurol (Paris) 121:72–74

Delwaide PJ, Olivier E (1988) Short-latency autogenic inhibition (Ib inhibition) in human spasticity. J Neurol Neurosurg Psychiatry 51:1546–1550

Dietrichson P (1973) The role of the fusimotor system in spasticity and parkinsonian rigidity. In: Desmedt JE (ed) New developments in electromyography and clinical neurophysiology, vol 3. Karger, Basel, pp 486–507

Fetz EE, Gustafsson B (1983) Relation between shapes of post-synaptic potentials and changes in firing probability of cat motoneurones. J Physiol (Lond) 341:387–410

Fournier E, Katz R, Pierrot-Deseilligny E (1984) A re-evaluation of the pattern of group I fibre projections in the human lower limb on using randomly alternated stimulations. Exp Brain Res 56:193–195

Fournier E, Meunier S, Pierrot-Deseilligny E, Shindo M (1986) Evidence for interneuronally mediated Ia excitatory effects to human quadriceps motoneurones. J Physiol (Lond) 377:143–169

Gustafsson B, McCrea D (1984) Influence of stretch-evoked synaptic potentials on firing probability of cat spinal motoneurones. J Physiol (Lond) 347:431–451
Henneman E (1957) Relation between size of neurons and their susceptibility to discharge. Science 126:1345–1346
Henneman E, Somjen G, Carpenter DO (1965a) Functional significance of cell size in spinal motoneurons. J Physiol (Lond) 28:560–580
Henneman E, Somjen G, Carpenter DO (1965b) Excitability and inhibitability of motoneurons of different sizes. J Physiol (Lond) 28:599–620
Hugon M, Delwaide P, Pierrot-Deseilligny E, Desmedt JE (1973) A discussion of the methodology of the triceps surae T- and H-reflexes. In: Desmedt JE (ed) New developments in electromyography and clinical neurophysiology, vol 3, Karger, Basel, pp 773–780
Hultborn H, Pierrot-Deseilligny E (1979) Input-output relations in the pathway of recurrent inhibition to motoneurones in the cat. J Physiol (Lond) 297:267–287
Hunt CC (1955) Monosynaptic reflex response of spinal motoneurones to graded afferent stimulation. J Gen Physiol 38:813–852
Kanda K, Burke RE, Walmsley B (1977) Differential control of fast and slow twitch motor units in the decerebrate cat. Exp Brain Res 29:57–74
Katz R, Pierrot-Deseilligny (1982) Recurrent inhibition of α motoneurons in patients with upper motor neuron lesions. Brain 105:103–124
Kernell D, Hultborn H (1990) Synaptic effects on recruitment gain: a mechanism of importance for the input-output relations of motoneurone pools? Brain Res 507:176–179
Knutsson E (1970) On effects of local cooling upon motor functions in spastic paresis. Prog Phys Ther 1:124–131
Knutsson E (1983) Analysis of gait and isokinetic movements for evaluation of antispastic drugs or physical therapies. Adv Neurol 39:1013–1034
Knutsson E, Mårtensson A (1984) Effects on maximal voluntary dynamic muscle strength in spastic patients by modulation of the inflow from reciprocal muscles. In: Conrad B, Benecke R, Bauer HJ (eds) Die klinische Wertung der Spastizität. Schattauer, Stuttgart, pp 147–155
Kuno M (1959) Excitability following antidromic activation in spinal motoneurones supplying red muscles. J Physiol (Lond) 149:374–393
Landau WM, Clare MH (1964) Fusimotor function. VI. H reflex, tendon jerks and reinforcement in hemiplegia. Arch Neurol Psychiatry 10:128–134
Lloyd DPC (1943a) Reflex action in relation to pattern and peripheral source of afferent stimulation. J Neurophysiol 6:111–120
Lloyd DPC (1943b) Neuron patterns controlling transmission of ipsilateral hind limb reflexes in cat. J Neurophysiol 6:293–315
Lloyd DPC (1946a) Facilitation and inhibition of spinal motoneurons. J Neurophysiol 9:421–438
Lloyd DPC (1946b) Integrative pattern of excitation and inhibition in two-neuron reflex arcs. J Neurophysiol 9:439–444
Magladery JW, McDougal DB Jr (1950) Electrophysiological studies of nerve and reflex activity in normal man. I. Identification of certain reflexes in the electromyogram and the conduction velocity of peripheral nerve fibers. Bull Johns Hopkins Hosp 86:265–290
Magladery JW, Porter WE, Part AM, Teasdall RD (1951) Electrophysiological studies of nerve and reflex activity in normal man. IV. The two-neurone reflex and identification of certain action potentials from spinal roots and cord. Bull Johns Hopkins Hosp 88:499–519
Magladery JW, Teasdall RD, Park AM, Languth HW (1952) Electrophysiological studies of reflex activity in patients with lesions of the nervous system. I. A comparison of spinal motoneurone excitability following afferent nerve volleys in normal persons and patients with upper motor neurone lesions. Bull John Hopkins Hosp 91:219–243
Mailis A, Ashby P (1990) Alterations in group Ia projections to motoneurons following spinal lesions in humans. J Neurophysiol 64:637–647
Mao CC, Ashby P, Wang M, McCrea D (1984) Synaptic connections from large muscle afferents to the motoneurons of various leg muscles in man. Exp Brain Res 56:341–350

McLellan DL (1977) Co-contraction and stretch reflexes in spasticity during treatment with baclofen. J Neurol Psychiatry 40:30–38

Meinck HM (1980) Facilitation and inhibition of the human H-reflex as a function of the amplitude of the control reflex. Electroencephalogr Clin Neurophysiol 48:203–211

Morita H, Shindo M, Yanagawa S, Yanagisawa N (1993) Neuromuscular response in man to repetitive nerve stimulation. Muscle Nerve 16:648–654

Pierrot-Deseilligny E (1990) Electrophysiological assessment of the spinal mechanisms underlying spasticity. EEG Suppl 41:264–273

Pierrot-Deseilligny E, Mazières L (1985) Spinal mechanisms underlying spasticity. In: Delwaide PJ, Young RR (eds) Clinical neurophysiology in spasticity. Elsevier, Amsterdam, pp 63–76

Táboríková H, Sax DS (1969) Conditioning of H-reflexes by a preceding subliminal H-reflex stimulus. Brain 92:203–212

Tardieu G, Hariga J, Tardieu C, Gagnard L, Velin J (1964) Traitement de la spasticité par infiltration d'alcool dilue au point moteur ou par injection épidurale. Rev Neurol (Paris) 110:563–565

Yanagawa S, Shindo M, Nakagawa S (1991) Increase in Ib inhibition by antagonistic voluntary contraction in man. J Physiol (Lond) 440:311–323

Yanagisawa N (1980) Reciprocal reflex connections in motor disorders in man. Prog Clin Neurophysiol 8:129–141

Yanagisawa N, Tanaka P, Ito Z (1976) Reciprocal Ia inhibition in spastic hemiplegia of man. Brain 99:555–574

Signs of the Upper Motoneuron Syndrome in Relation to Soleus Hoffmann Reflex Tests

B.W. Ongerboer de Visser, J.H.T.M. Koelman, L.J. Bour, and A.A.J. Hilgevoord

Introduction

The Hoffmann (H) reflex is evoked electrically by stimulation of afferent fibers (Ia fibers) from muscle spindles at lower intensities and has a longer latency than the direct muscle potential (M potential) evoked by motor nerve fibers (Schieppati 1987). Different tests of H reflex responses from the soleus muscle can be used for the study of central mechanisms in motor control of the leg (Delwaide 1973a).

In the upper motoneuron syndrome various alterations of these soleus H reflex tests are found. Compared to healthy controls, suppression of the H reflex response during vibration of the Achilles tendon is decreased in most patients whereas the ratio between the maximal soleus H reflex response and M potential, the so-called H/M ratio, is increased (Fig. 1; Delwaide 1973b; Taylor et al. 1984; Ongerboer de Visser et al. 1989). Furthermore, changes in the H reflex recovery curve, obtained by paired stimuli to the posterior tibial nerve, may occur (Fig. 2; Sax and Johnson 1980; Bour et al. 1991).

Clinically, in the upper motoneuron syndrome the various characteristics, such as weakness, loss of dexterity, hyperreflexia, clonus, hypertonia, and the extensor plantar response, may occur independently and in different combinations (Dohrmann and Nowack 1974). As such, a relationship between specific clinical features and changes in the H reflex test results may further contribute to elucidating underlying pathophysiological mechanisms. However, the relationship between the various clinical signs and these neurophysiological alterations is largely unknown, and in some cases evidence is conflicting (Delwaide 1973a; Taylor et al. 1984).

The purpose of this study was to investigate whether changes in soleus H reflex vibratory inhibition, H/M ratio, and homonymous recovery curve are related to specific features of the upper motoneuron syndrome.

Dept. of Clinical Neurophysiology, Academisch Medisch Centrum, Meibergdreef 9, NL-1105 AZ Amsterdam, The Netherlands

Thilmann et al. (Eds.)
Spasticity: Mechanisms and Management

Methods and Materials

For the study we selected 38 patients aged 18–75 years (mean 47) with upper motoneuron signs in the lower extremities with or without upper motoneuron signs in the arms. Of these, 9 had a cerebral lesion, 15 a myelopathy, and 14 a slowly progressive spastic paraparesis, 7 with a family history. The control group consisted of 48 healthy subjects aged 20–70 years (mean 38).

The clinical examination was performed just prior to the neurophysiological testing. The two investigations were performed independently by different investigators. Clinical examination consisted of the assessment of four elementary features of the upper motoneuron syndrome: hypertonia, muscle strength, tendon reflex activity, and the plantar response. Severity of signs were categorized in a standard fashion. Muscle tone was investigated by dorsiflexion of the ankle joint and graded according to the Ashworth scale (Ashworth 1964); a normal tone (0) up to movement of the ankle joint hardly possible (Taylor et al. 1984). Strength of triceps surae was judged according to the Medical Research Council scale (MRC 1976) and graded in four categories; normal strength (0) up to severe weakness (3). Achilles tendon reflexes were graded as diminished or absent (0) up to ankle clonus (3), that is, four reflex contractions or more produced by tendon percussion or abrupt ankle dorsiflexion. Plantar responses were graded as flexor (0), equivocal (1), and extensor (2).

During the neurophysiological examination subjects were seated in a reclining chair in a standardized position (Ongerboer de Visser et al. 1989; Hugon 1973). Square current pulses of 1 ms were delivered to the posterior tibial nerve in the popliteal fossa, and responses from the soleus muscle were digitally stored in a computer. The time interval between successive recordings was at least 30 s. An H reflex recruitment curve was constructed as a function of stimulus intensity and 12 or more H reflexes were used at different intensities (Fig. 1; Taylor et al. 1984). Simultaneously with the H reflex recruitment curve a recruitment curve of direct soleus M potentials was constructed also as a function of stimulus intensity. The H/M ratio was calculated using peak-to-peak values of the maximal H reflex response and maximal M potential. Vibration was applied to the Achilles tendon with a frequency of 100 Hz and a 1-mm amplitude. The cumulative vibratory index was used as a quantitative measure for the suppression of the H reflex response (Fig. 1; Ongerboer de Visser et al. 1989). This index is defined as the surface under the H reflex recruitment curve obtained during vibration up to a certain stimulus intensity, divided by the surface under the H reflex recruitment curve obtained without vibration up to the same stimulus intensity ×100%. The cumulative vibratory index at the stimulus intensity level yielding the maximal H reflex was used in the calculations (Bour et al. 1991). Furthermore, a test was performed to study H reflex recovery charac-

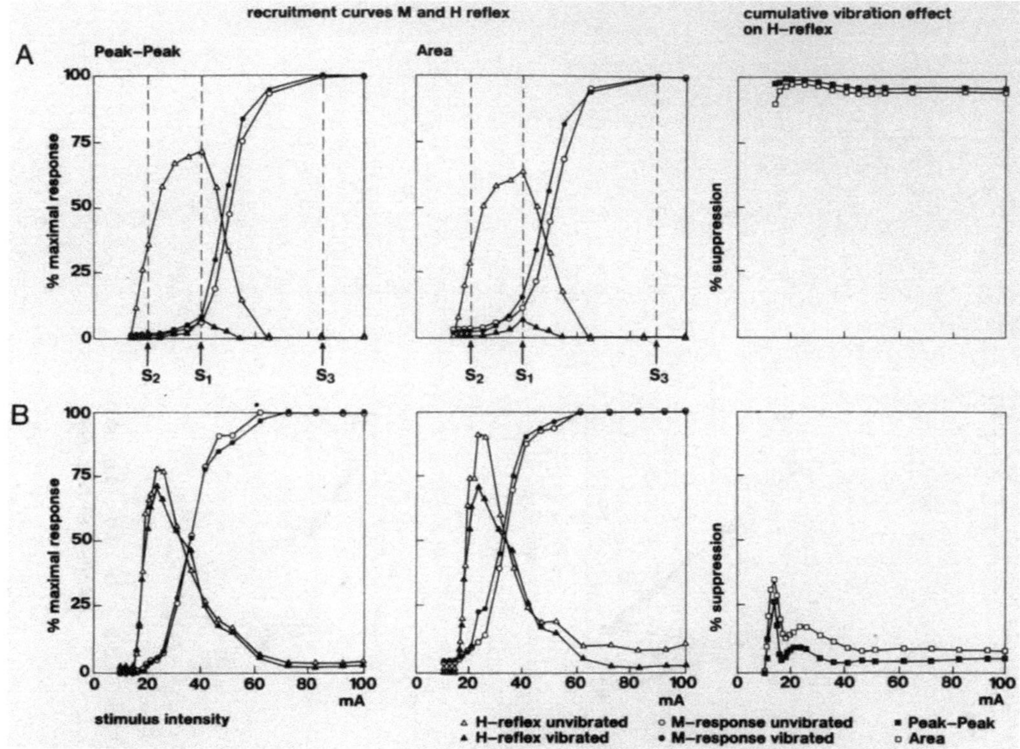

Fig. 1A,B. Example of soleus H reflex test results in a normal control subject (**A**) and in a patient with signs of an upper motoneuron syndrome (**B**). Recruitment curves of soleus H reflex response and M potential (peak-peak and area values) without Achilles tendon vibration (*open triangles*) and with tendon vibration (*solid triangles*) are shown. All values are expressed as percentages of the maximum M potential. *Right-most panels*, the cumulative vibratory effect on the peak-peak (*solid squares*) and the area (*open squares*) values of the H reflex. The amount of suppression is equal to 100% minus cumulative vibratory index

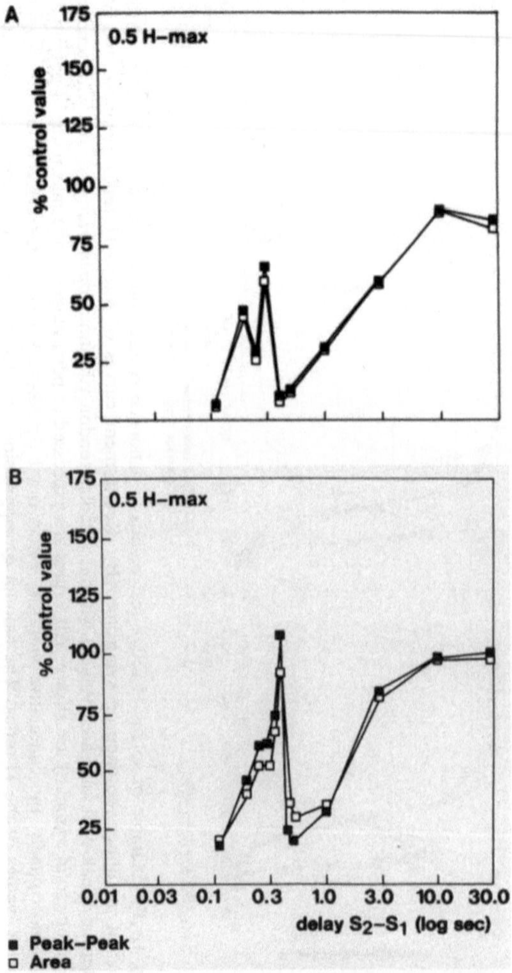

Fig. 2A,B. Recovery curve of soleus H reflex response (peak-peak and area values) at 0.5 H maximal stimulus intensity in a control subject (**A**) and in a patient with upper motoneuron syndrome signs (**B**). All values of the reflex response to the test stimulus (S_2) are presented as percentages of the preceding H reflex to the conditioning (S_1) stimulus

teristics by constructing the H reflex recovery curve (Fig. 2). Paired pulses of equal intensity (where the H reflex reached half its maximal value, 0.5 H_{max}) were applied to the tibial posterior nerve in the popliteal fossa (Bour et al. 1991). Intervals between conditioning and test stimuli increased from 100 ms to 30 s. The recovery curves were plotted as the ratio (in percentage) between area values of the test and the conditioning reflex response against the interval between the two stimuli. Two characteristic values were used for statistical analysis: (a) the local maximum of the test reflex occurring in the late facilitatory phase and (b) the local minimum response found in the late inhibitory phase of the recovery curve.

The H reflex parameters used in our study have been shown in an earlier study to discriminate most productively between patients with upper motoneuron lesions and controls (Bour et al. 1991).

The following statistical analyses were used. Analysis of variance (BMDP statistical software package 7D) was performed with use of a priori contrast tests for correlation between clinical and neurophysiological data. Differences in reflex parameters in the different patients groups were examined with pairwise *t* tests, and significance levels were adjusted for the number of pairwise comparison tests according to Bonferroni. A linear stepwise regression analysis (BMDP 1R, 2R) was performed to examine correlations among physiological parameters and between physiological parameters and age.

Results and Discussion

About 40% of the patients showed a mild clinical picture of an upper motoneuron lesion with normal strength or slight weakness combined with normal or slightly elevated muscle tone. Clonus was present in a few. In contrast, about 20% of the patients had a severe clinical picture with moderate or severe paresis combined with severely increased tone. Clonus was present in about half of them. The remainder had a clinical picture between those of the former two patient groups. The tendon reflexes were normal in one-third of the whole patient group, increased in another one-third, and clonus was present in the remaining one-third. The plantar response was extensor in the vast majority and was therefore not included in the statistical analyses.

Figure 3 shows vibratory effects in controls and patients grouped according to muscle tone. In the patient group we found a moderate but significant correlation between degree of hypertonia and vibratory effects on the H reflex expressed by the cumulative vibratory index. Patients with severe hypertonia (grade 3 or 4) showed no further increase in vibratory effects. Achilles tendon vibraton strongly activates primary spindle endings and suppresses or even abolishes the H reflex for the duration of vibration. This action is thought to be due predominantly to an autogenic axo-axonal presynaptic inhibition of Ia terminals. In contrast to previous studies (Delwaide 1973a,b), our cumulative measurement of vibratory effects integrates the vibratory inhibition at multiple stimulus intensities, by which it has some greater discriminative power (Ongerboer de Visser et al. 1989) than the ratio between maximal H reflex with and without tendon vibration. Our findings may indicate a quantitative relationship between a decrease in tonic regulation of presynaptic inhibition and an increase in muscle tone up

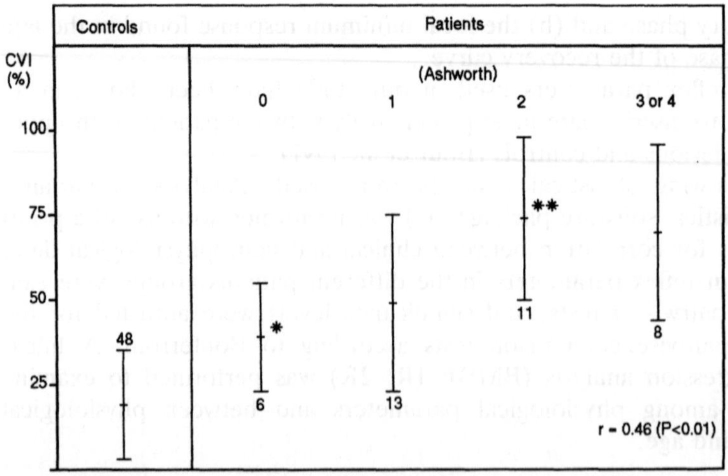

Fig. 3. Correlation (mean ± SD) between muscle tone and cumulative vibratory index (*CVI*). *, Difference in CVI between controls and patients with normal tone ($p < 0.05$); **, difference in CVI between patients with normal and patients with moderately elevated tone ($p < 0.05$, comparing six pairs of means)

to moderate levels. In addition, it may be suggested that a decrease in presynaptic inhibition results in an increased afferent input to the soleus motoneuron pool, contributing to enhanced resistance to passive stretch characteristic of hypertonia. As such, our findings would support the view that hypertonia is associated with the release of a central gain-controlling mechanism in which the ability to modulate reflex transmission is lost (Burke 1980). Furthermore, noteworthy was that in the controls vibratory effects on the H reflex were more pronounced than vibratory effects in the patients with normal tone, signifying that presynaptic inhibition may be decreased "subclinically."

Figure 4 presents H/M ratios in controls and patients grouped according to tendon reflex activity. In the patients an increase in H/M ratio with enhancement of tendon reflex activity was found. The H/M ratio also decreased with advancing age, but there was no correlation between H/M ratio and age when controlling for tendon reflex activity. The declining H/M ratio with age is therefore explained fully by lower tendon reflexes at increasing age. In all patients with ankle clonus the H/M ratio was 64% or more, up to 106%, which was significantly more than in patients with normal Achilles tendon reflexes. In patients having a normal Achilles tendon reflex the H/M ratio was similar to values found in the controls. These findings probably indicate that in clonus enhancement of motoneuron excitability is generally present. The H/M ratio correlated with a lesser degree with muscle tone. Furthermore, a small correlation was seen between H/M ratio and vibratory effects. These results could be accounted for by

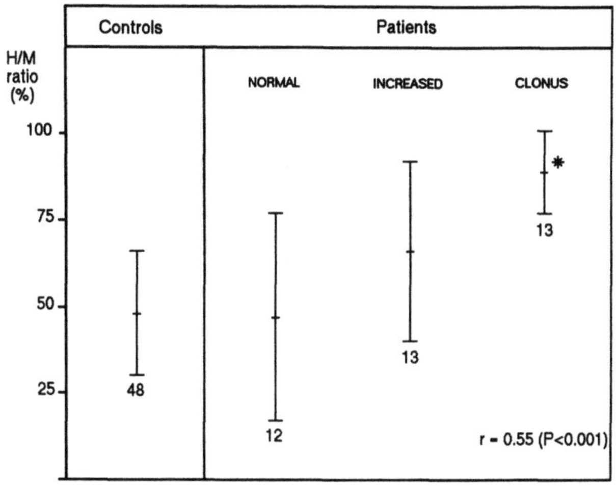

Fig. 4. Correlation (mean ± SD) between tendon reflex activity and H/M ratio. *, Difference between H/M ratio in patients with normal tendon reflexes and patients with clonus ($p < 0.01$, comparing three pairs of means)

mechanisms other than direct motoneuron excitability increase, for example, increases in transmission through the stretch reflex arc secondary to reduced presynaptic inhibition on Ia afferent terminals.

Figure 5 presents the correlations between postsynaptic late facilitation in the recovery curve and tendon reflex activity. Late facilitation in patients with normal reflexes was higher than in control subjects. In addition, in patients with clonus the late facilitation was increased compared with patients having normal tendon reflexes. Although the late facilitation of the recovery curve may be due partly to cutaneous afferents (Gassel 1970; Pierrot-Deseilligny et al. 1973; see also Yanagisawa et al., this volume), its potentiation in spasticity, Parkinsonian rigidity, and dystonia (Sax and Johnson 1980; Bour et al. 1991; Pierrot-Deseilligny et al. 1973; Yarp 1967; Zander Olsen and Diamantopoulos 1967) suggests involvement of polysynaptic pathways, either in the spinal cord or along long-loop pathways (Panizza et al. 1990). Thus in clonus interneuronal hyperactivity and increased motoneuron excitability expressed by an increase in H/M ratios may be combined. Both are consistent with conclusions based on direct recordings from Ia afferents that spinal mechanisms rather than increased fusimotor drive are involved in clonus (Hagbarth et al. 1975). Late inhibition of the recovery curve did not correlate with tendon reflex activity.

None of the reflex test results correlated with muscle weakness, and no differences in neurophysiological parameters were present among patients with a cerebral lesion, myelopathy, and spastic paraparesis.

In summary, we found that in the upper motoneuron syndrome alterations in the results of soleus H reflex tests may occur largely independently of

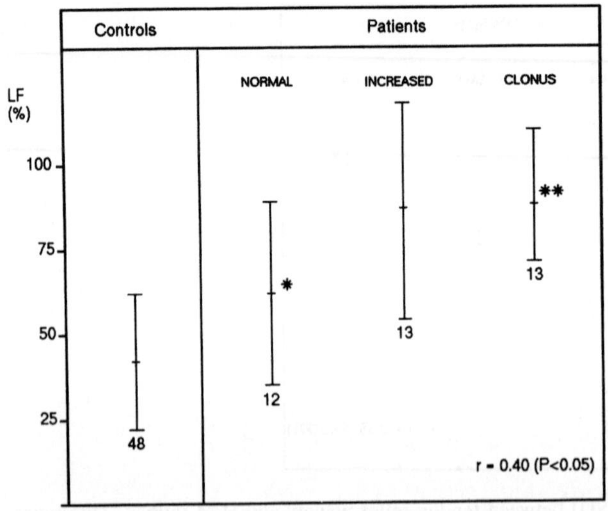

Fig. 5. Correlation (mean ± SD) between tendon reflexes and late facilitation (*LF*). *, Difference between LF in controls and patients with normal tendon reflexes ($p < 0.05$); **, difference between LF in patients with normal tendon reflexes and patients with clonus ($p < 0.05$, comparing three pairs of means)

each other, and that they relate to specific clinical features of the upper motoneuron syndrome (Koelman et al. 1993).

Acknowledgements. J.H.T.M.K. was supported by a grant from the Prinses Beatrix Fonds, and A.A.J.H. was supported by the Netherlands Organization for Scientific Research (NWO).

References

Ashworth B (1964) Preliminary trial of carispodol in multiple sclerosis. Practitioner 192:540–542

Bour LJ, Ongerboer de Visser BW, Koelman JHTM, van Bruggen GJ, Speelman JD (1991) Soleus H-reflex tests in spasticity and dystonia: a computerized analysis. J Electromyogr Kinesiol 1:9–19

Burke D (1980) A reassessment of the muscle spindle contribution to muscle tone in normal and spastic man. In: Feldman RG, Young RR, Koella WP (eds) Spasticity: disordered motor control. Symposia Specialists, Miami, pp 261–277

Delwaide PJ (1973a) Contribution of human reflex studies to the understanding and management of the pyramidal syndrome. In: Shahani BT (ed) Electromyography in CNS disorders: central EMG. Butterworth, Boston, pp 77–109

Delwaide PJ (1973b) Human monosynaptic reflexes and presynaptic inhibition; an interpretation of spastic hyperreflexia. In: Desmedt JE (ed) New developments in electromyography and clinical neurophysiology, vol 3. Karger, Basel, pp 508–522

Dohrmann GJ, Nowack WJ (1974) Relationship between various clinical signs in lesions of the descending motor system. Dis Nerv Syst 35:375–377

Gassel MM (1970) A critical review of evidence concerning long-loop reflexes excited by muscle afferents in man. J Neurol Neurosurg Psychiatry 33:358–362

Hagbarth KE, Wallin G, Löfstedt L, Aquilonius SM (1975) Muscle spindle activity in alternating tremor of parkinsonism and in clonus. J Neurol Neurosurg Psychiatry 38:636–641

Hugon M (1973) Methodology of the Hoffmann reflex in man. In: Desmedt JE (ed) New developments in electromyography and clinical neurophysiology, vol 3. Karger, Basel, pp 277–293

Koelman JHTM, Bour LJ, Hilgevoord AAJ, van Bruggen GJ, Ongerboer de Visser BW (1993) The soleus H-reflex and clinical signs of the upper motoneuron syndrome. J Neurol Neurosurg Psychiatry 56:776–781

Medical Research Council of UK (1976) Aids to the examination of the peripheral nervous system. British Crown, London, pp 6–7 (Memorandum 45)

Ongerboer de Visser BW, Bour LJ, Koelman JHTM, Speelman JD (1989) Cumulative vibratory indices and the H/M ratio of the soleus H-reflex: a quantitative study in control and spastic subjects. Electroencephalogr Clin Neurophysiol 73:162–166

Panizza M, Lelli S, Nilsson J, Hallett M (1990) H-reflex recovery curve and reciprocal inhibition of H-reflex in different kinds of dystonia. Neurology 40:824–828

Pierrot-Deseilligny E, Bussel B, Morin C (1973) Supraspinal control of the changes induced in H-reflex by cutaneous stimulation, as studied in normal and spastic man. In: Desmedt JE (ed) New developments in electromyography and clinical neurophysiology, vol 3. Karger, Basel, pp 550–555

Sax DS, Johnson TL (1980) Spinal reflex activity in man, measurement in relation to spasticity. In: Feldman RG, Young RR, Koella WP (eds) Spasticity: disordered motor control. Symposia Specialists, Miami, pp 301–313

Schieppati M (1987) The Hoffmann reflex: a means of assessing spinal reflex excitability and its descending control in man. Prog Neurobiol 28:345–376

Taylor S, Ashby P, Verrier M (1984) Neurophysiological changes following traumatic spinal lesions in man. J Neurol Neurosurg Psychiatry 47:1102–1108

Yap CB (1967) Spinal segmental and long-loop reflexes on spinal motoneurone excitability in spasticity and rigidity. Brain 90:887–896

Zander Olsen P, Diamantopoulos E (1967) Excitability of spinal motor neurones in normal subjects and patients with spasticity, Parkinsonian rigidity, and cerebellar hypotonia. J Neurol Neurosurg Psychiatry 30:325–331

Pathophysiological Mechanisms of Spasticity at the Spinal Cord Level

P.J. Delwaide

For a clinician, the term "spasticity" may have a broad meaning, sometimes synonymous with upper motor neuron syndrome. However, it is well known that disability resulting from a lesion of the descending motor pathways depends on distinct positive (hyper-reflexia, muscle hypertonia, spasms) and negative (paresis or palsy, loss of dexterity) signs which may evolve independently. For experts only, spasticity is just one of these positive signs. It was defined by Lance in 1980 as follows: "Spasticity is a motor disorder characterized by a velocity-dependent increase in tonic stretch reflexes ('muscle tone') with exaggerated tendon jerks resulting from hyperexcitability of the stretch reflex as one component of the upper motorneurone syndrome". In the 1980s, this definition enjoyed a large popularity. However, it has also been criticized on the foundation that a clinical sign is described in physiological terms. Moreover, Lance attributes spasticity only to hyperexcitability of the stretch reflex. While there is little doubt that brisk reflexes depend on that factor, the mechanisms of the tonic stretch reflex are less clear and still debated.

From a clinical point of view, spasticity designates both brisk reflexes and muscle stiffness, which increases with velocity of stretch and predominates in certain muscle groups (adductors and flexors in the upper limb; adductors and extensors in the lower). A clonus may sometimes be elicited in some muscles. Whether spasms (in flexion or in extension) belong to the spasticity definition is still open to discussion.

From a physiological point of view, the question can be raised whether dysfunctions at the spinal cord level other than hyperexcitability of the monosynaptic stretch reflex contribute or not to spasticity mechanisms. For example, if flexor spasms are accepted in the definition of spasticity, they cannot be explained solely by stretch reflex hyperexcitability.

Spasticity derives from various etiologies and may be seen in patients with complete spinal cord transection. This observation points to the fact that the responsible mechanisms are to be found at a segmental level deprived of normal descending motor control. At this level, it is likely

Départment universitaire de Neurologie, Hôpital de la Citadelle, Bd. du 12° de Ligne 1, B-4000 Liège, Belgique

Thilmann et al. (Eds.)
Spasticity: Mechanisms and Management

that peripheral mechanisms located in muscles themselves intervene in the stiffness felt during passive mobilization. Fiber types are modified in long-standing spasticity and the ratio between type-I and type-II fibers is increased (Edström 1970; Sjöstrom et al. 1980). Rheological properties of muscles are also modified (Dietz and Berger 1983; Hufschmidt and Mauritz 1985), and the number of sarcomeres is reduced (Tabary et al. 1981). However, these modifications cannot account for brisk reflexes and seem secondary to neural mechanisms. So it is worth making an inventory of the dysfunctions occurring in the spinal reflex pathways and establishing to what extent they contribute to the pathophysiology of spasticity.

It is tempting to conduct such research in experimental animals so that well-identified cells can be studied. However, though such an approach is necessary and such studies are indeed in progress (Hultborn and Malmsten 1983a; *see* also Munson et al., this volume), animal models of spasticity are often in adequate and do not mimic human spasticity exactly.

An alternative is to study patients using techniques from clinical neurophysiology. Most of them provide quantified results, so that norms with confidence intervals can be defined in control subjects and compared with the results collected from patients. Provided that differences are significant, it can be concluded that the function under study is abnormal in the patients. Since 1970, a battery of electrophysiological tests has been developed to measure motoneuron excitability (Angel and Hoffmann 1963), presynaptic inhibition (Delwaide 1971, 1973), reciprocal inhibition (Tanaka 1974), recurrent inhibition (Bussel and Pierrot-Deseilligny 1977), and the so-called autogenic Ib inhibition (Pierrot-Deseilligny et al. 1981). The last four tests are specific for one interneuron. Other tests, like the flexor reflex, give results which cannot be attributed to a single interneuron. The interested reader may refer to papers where the techniques are described along with their limitations (Delwaide et al. 1980; Delwaide 1985). Of course, all the neurons of interest have not yet been studied in man. However, the results collected so far over the past 20 years offer interesting information.

Changes in Spinal Cord Functions

Just after a stroke or a spinal cord injury many tests may be altered, but the results may vary as time passes. I will limit myself to discussing data obtained from chronic spastic patients who have reached a stable level of spasticity. From a theoretical point of view, derived from knowledge of spinal cord physiology in animals, many spinal mechanisms could explain spasticity; they include:

- Alpha motoneuron hyperactivity
- Gamma motoneuron hyperactivity

- Excitatory Ia interneuron hyperexcitability
- Reduction of presynaptic inhibition
- Reduction of recurrent inhibition (Renshaw)
- Reduction of autogenic inhibition (Ib and Ia)
- Reduction of reciprocal inhibition

Many of them can be studied in man.

Alpha Motoneuron Excitability

If the motoneurons (MNs) were in a chronic state of increased depolarization, for example because of a change in their biophysical properties, they would be nearer to threshold of discharge and respond to weaker afferent stimuli, giving an explanation for brisk reflexes.

Animal Studies. Many studies have been conducted to evaluate possible changes in motoneuron biophysical properties in chronic spinal cats. The results are essentially negative and do not support the suggestion of increased responsiveness of MNs.

Human Studies. Indirect means of evaluating excitability of the MN pool have been proposed, but they do not allow discrimination between an alteration in biophysical properties of MNs and depolarization resulting from changes in synaptic activities, be they an increase in excitatory ones or a decrease in inhibitory ones.

The F wave is a late wave recorded in a muscle after supramaximal stimulation of its motor nerve. It is due to backfiring of some MNs activated antidromically: the higher their excitability, the greater the number of discharging MNs. F-wave amplitude has been shown to be significantly increased in spastic patients and on the spastic side of hemiplegics (Schiller and Stålberg 1978; Fisher 1983). Persistence of the F waves is also greater (Eisen and Odusote 1979). However, the F-wave technique has been criticized; few MN are explored, representing only a small proportion of the MN pool.

A more global technique is offered by measuring the H_{max}/M_{max} ratio, i.e., the ratio between the maximal amplitudes of the electrically induced reflex (H or Hoffmann reflex) and the direct motor response (M max). This ratio reflects the number of MNs in the pool which can be activated by the Ia reflex pathway; if excitability is increased, a larger number of MNs will be brought to discharge. Despite large interindividual variability, due in part to age, the H_{max}/M_{max} ratio is higher in spastic patients (Angel and Hoffmann 1963; Delwaide 1971; Somerville and Ashby 1978). It is of interest to note that an H reflex can be observed in spasticity in muscles in which it is not elicited under normal conditions (Schiller and Stålberg 1978).

Taken together, the results obtained with F-wave and H_{max}/M_{max} ratio indicate that alpha motoneuron excitability is increased in spasticity. Taking into account the data obtained from experimental spasticity in the cat, however, it is likely that this change reflects chiefly modifications in synaptic input (however, see Kernell, this volume; Hultborn et al., this volume).

Gamma Motoneuron Excitability

Once the indirect route of alpha motoneuron facilitation through the gamma system had been described, hypotheses were proposed to interpret exaggerated tendon jerks and increased responses to stretch by selective hyperactivity of this system. The gamma system hyperexcitability hypothesis was popular in the 1960s and 1970s but is presently under strong criticism.

Animal Studies. It has been demonstrated that gamma system hyperactivity contributes to decerebrate rigidity (Hunt 1951; Granit and Kaada 1952; Eldred et al. 1953). However, decerebrate rigidity is not a valid model of human spasticity. When the gamma system was assessed by Ia discharge after standardized stretches in cats with spinal cord transection, the results did not argue in favor of an increased activity of the gamma system (Bailey et al. 1980).

Human Studies. Two methods have been used in the past to analyze gamma system functions in man. As the technique for eliciting the H-reflex bypasses muscle spindles, comparisons between tendon jerk and H reflex have been considered to reflect changes in muscle spindle innervation. In spastics, the increase of tendon jerk amplitude is larger than that of the H reflex (Buller and Dornhorst 1957; Dietrichson 1973). This observation has been taken as an argument in favor of a selective increase of gamma system activity.

The other technique deals with the results of procainization of motor nerves which has been supposed to paralyze the smaller fibers (including gamma fibers) selectively. As spasticity wanes after procainization, this observation has also been taken as an argument.

However, both techniques are presently under strong criticism. Burke (1983) has shown that tendon and electrically induced reflexes differ in many more respects than simply the spindle tone. The differences observed between the two responses cannot be ascribed only to changes in spindle innervation. On the other hand, it has been shown that procainization is far from selective for small fibers (Gassel and Diamantopoulos 1964). No positive evidence thus exists that the gamma system is hyperactive in spasticity.

A less controversial method of evaluating gamma activity is the direct recording in man of single Ia fiber activity (microneurography). An increase in gamma activity would result in richer discharges of Ia fibers during stretch. In spastic patients, recording Ia activity has not revealed any increased discharges (Hagbarth et al. 1973). The difficulty of quantifying data obtained by the microneurographic technique has been emphasized, however, so that no definite conclusion can be obtained before further work confirms Hagbarth's data. In addition, spindle sensitivity could be altered by mechanisms other than gamma innervation. Immobilization, for example, can change spindle response to stretch (Williams 1980). However, at present, evidence indicating a consistent change in gamma innervation is lacking.

Presynaptic Inhibition Acting on Ia Afferents

The mechanism of presynaptic inhibition has been shown to act on Ia terminals. It results in a reduced release of neurotransmitter by excitatory Ia fibers. This presynaptic inhibition, mediated through an interneuron and influenced by both peripheral and supraspinal afferents, normally attenuates the Ia effects on motoneurons. In general, the interneuron needs bursts of impulses rather than isolated spikes to be activated. Reduction of presynaptic inhibition could theoretically make the myotatic arc hyperexcitable.

Animal Studies. In cats with complete spinal transection, Naftchi et al. (1980) found a reduction in presynaptic inhibition acting on Ia afferents from the soleus. However, these results were not confirmed in cats with chronic section of the cord (Hultborn and Malmsten 1983b).

In cats it has been shown that vibration acting on the Achilles tendon, known to activate the muscle spindles very powerfully, reduces the amplitude of monosynaptic reflexes through a mechanism of presynaptic inhibition (Gillies et al. 1969).

Human Studies. In man, vibratory stimulation of a muscle normally inhibits its monosynaptic reflex. In agreement with animal experiments, it has been shown that this inhibition reflects a mechanism of presynaptic inhibition (Delwaide 1971). It is possible to express the amount of vibratory inhibition by an index: the vibratory index. All research workers who have studied vibratory inhibition in chronic spastic patients have reported a significant reduction (Delwaide 1971; Burke and Ashby 1972; Ashby and Wiens 1989; Iles and Roberts 1986; *see* Ongerboer de Visser et al., this volume). The differences with age-matched normal subjects are clear-cut, as are those between both legs in hemiplegic patients. However, no good correlation has been found between reduction of vibratory inhibition and intensity of spasticity as assessed by Ashworth's scale. The results indicate that presynaptic inhibition is clearly reduced in established spasticity.

Another way to study presynaptic inhibition in man is offered by the lumbosacral evoked potentials after stimulation of the sciatic nerve (Delwaide et al. 1985). One component of the response – the P2 wave – is thought to reflect presynaptic inhibition and is reduced in MS patients with spasticity. Reduction of the P2 area is well correlated with the degree of spasticity. These resuls corroborate those obtained with vibratory inhibition and argue for a reduction of presynaptic inhibition acting on Ia afferents. Due to this reduction, the stretch reflex pathway lacks inhibitory control and becomes more excitable.

Recurrent (Renshaw) Inhibition

During their discharge, motoneurons are submitted to a postsynaptic inhibition exerted by an interneuron which is activated by a collateral branch of the motor axon. This interneuron – called the Renshaw interneuron – is under supraspinal control and sends other inhibitory projections to gamma motoneurons and Ia inhibitory interneurons. It can be activated by antidromic activation of the motor axons. Lesions in descending pathways could reduce recurrent inhibition and, hence, explain hyperexcitability of the myotatic arc.

Animal Studies. Goldfarb (1976) provided indirect evidence for an increased activity of Renshaw cells by showing that ventral root stimulation in chronic spinal cats induced more spikes in these cells. Hultborn and Malmsten (1983b) confirmed greater Renshaw inhibition on the hemisected side of chronic spinal cats.

Human Studies. Reduction in efficacy of Renshaw inhibition has been proposed on the basis of a technique which is open to criticism (Veale et al. 1973). A more reliable method of estimating recurrent inhibition was developed later by Bussel and Pierrot-Deseilligny (1977). With this technique, recurrent inhibition is not significantly reduced at rest in hemiplegic patients (Katz and Pierrot-Deseilligny 1982). In paraplegics, the situation is less clear and some patients show a reduced or an increased inhibition. During voluntary contractions, the inhibition is no longer normally modulated. This fact indicates that voluntary contraction, but also reciprocal inhibition, is less well controlled. However, reduction of Renshaw inhibition seems to play a subsidiary role, if any, in producing hyperexcitability of the myotatic arc.

Short-latency Ia Reciprocal Inhibition

In addition to activating homonymous MNs, Ia afferents also inhibit the antagonist motor nucleus via interposed interneurons, named Ia inhibitory interneurons. On these interneurons, extensively studied by Hultborn et al. (1976), converge many other afferents, both segmental and suprasegmental. If less active, a background of postsynaptic inhibition on the MNs would be reduced and could explain an enhancement of excitability of the stretch reflex pathway.

Animal Studies. Corticospinal fibers facilitate Ia inhibitory interneurons, and their lesion could normally lead to a reduction of excitability. In chronic hemisected cats, however, reciprocal inhibition was increased and not reduced (Hultborn and Malmsten 1983b).

Human Studies. Contrary to normal subjects, in whom voluntary contraction of the tibialis anterior strongly inhibits the soleus H reflex, this inhibition is markedly reduced in spastic patients still able to dorsiflex the foot. This result suggests reduction in reciprocal inhibition. The Ia inhibitory interneuron can be tested selectively in man by conditioning a monosynaptic reflex by stimulation of Ia fibers of a nerve coming from an antagonist muscle (Tanaka 1974). This technique is applicable to both lower and upper limbs, where the inhibitory effects are more marked (Day et al. 1983). At short latencies between the conditioning and the conditioned responses, disynaptic reciprocal inhibition is first seen. It is feeble in lower limbs but clearer in upper limbs. With longer latencies, a second wave of inhibition, attributed to presynaptic inhibition, is observed. Using this technique, it has been shown that the short-latency reciprocal inhibition is reduced in spasticity from flexors to extensors (Yanagisawa et al. 1976; Delwaide 1985). This holds true also for the upper limb, where the wrist flexor H-reflex is conditioned by radial nerve stimulation (extensor to flexor).

However, using the technique of post-stimulus time histograms, Ashby and Wiens (1989) found that short-latency disynaptic inhibition from extensors (soleus) to flexors (tibialis anterior) was greater than normal. This increased inhibition could contribute to the weakness of lower limb flexor muscles. It can be mentioned that the second wave of inhibition, presumably of presynaptic origin, is clearly less marked, adding an argument in favor of a reduction in presynaptic inhibition (see above).

It can be taken for granted that at least some Ia inhibitory interneurons are less active in spasticity. However, it has not yet been proved that this reduction is due only to impaired descending facilitation of these interneurons. Segmental mechanisms could in fact intervene, such as, for example, increased activity of some Renshaw cells or reciprocal effects between Ia inhibitory interneurons of antagonist muscles.

The So-called Autogenic Ib Inhibition

The afferents originating from Golgi tendon organs (Ib fibers) project to inhibitory interneurons, which inhibit the extensor motoneurons and facilitate the flexor ones. In fact, many distinct afferents converge on the "Ib" interneurons: both from the segmental level (Ia, cutaneous, and joint afferents) and from descending pathways (cortico-, rubro-, proprio-, and reticulospinal tracts). The functional role of this interneuron may no longer be discussed only in terms of properties of Golgi tendon organs. Because of this, the concept of "nonreciprocal inhibition" tends to replace the term "autogenic Ib inhibition" (Harrison and Jankowska 1985). Reduced excitability of this inhibitory pathway could facilitate the stretch reflex by removal of a tonic inhibitory influence.

Animal Studies. Eccles and Lundberg (1959) showed, in the decerebrate cat, that inhibition (or disfacilitation) of the Ib inhibitory interneurons contributed to some extent to exaggerated tone. However, as already mentioned, decerebrate rigidity is not a good model of spasticity. To our knowledge, no specific studies of Ib interneurons have been conducted in spinal cats, but it was recently shown that activation of this interneuron by reticulospinal neurons may be responsible for hypotonia in the cat (Takakusaki et al. 1990). Conversely, its inhibition could explain muscle hypertonia.

Human Studies. Based on the assumption that afferent fibers originating from the gastrocnemius medialis have only Ib projections on soleus motoneurons, conditioning of the H-reflex by prior stimulation at low intensity of the gastrocnemius medialis nerve makes it possible to study in man the effects of Ib interneurons (Pierrot-Deseilligny et al. 1981). Applied to hemiplegic patients, this technique indicates clear-cut differences between both sides: On the affected side the normal inhibition is replaced by facilitation, which, on the mean curve, reaches 15% (Delwaide and Oliver 1988). We have confirmed these data in a larger group of spastic patients, including both hemiplegics and paraplegics.

It is interesting to note that two tone disorders – spasticity and rigidity – share a common pathophysiological disorder, i.e., a replacement of the normal Ib inhibition by a facilitation (Delwaide and Oliver 1988; Delwaide et al. 1991). Table 1 lists the major disorders which have been identified in chronic spasticity.

Other Possible Mechanisms

Despite progress in the analysis of interneuron activity in man, we are still far from being able to study selectively the functioning of all those interneurons that have been studied in the cat. However, Mailis and Ashby (1990) have provided evidence that polysynaptic pathways from Ia afferents to motoneurons are facilitated.

Many other interneurons, both facilitatory and inhibitory, might intervene to adjust the degree of excitability of the motoneurons. A gross indirect reflection of their functioning is given by the study of polysynaptic reflexes. The flexor reflex pathways are globally more active, and, in addition, the correct channeling of afferent messages is altered, leading to modified responses.

Table 1. Major pathophysiological disorders in chronic spasticity

- Increased alpha motoneuron excitability
- Facilitation of the polysynaptic pathway from Ia afferents to motoneurons
- Reduced excitability of some inhibitory interneurons
  - presynaptic inhibition
  - reciprocal Ia inhibition
  - nonreciprocal (Ib) inhibition
  - Renshaw cell (in some cases)

NB. There is no evidence for gamma system hyperactivity.

At the end of this review comparing the possible mechanisms responsible for spasticity, it appears that there are various discrepancies between animal and human data. This is not surprising, because animal models do not mimic the clinical picture and require larger cortical lesions or complete spinal hemisection to obtain increased muscle stiffness with brisk tendon jerks.

However, the contribution of animal physiology is irreplacable if we want to know how spinal circuits are organized and how supraspinal centers control them. In that respect, it may be underlined that the abnormalities revealed by clinical neurophysiology techniques fit with what is known of the supraspinal control of interneurons: Ia inhibitory interneurons (Lundberg et al. 1962), Ib interneurons (Lundberg and Voorhoeve 1962), and interneurons responsible for presynaptic inhibition (Hongo et al. 1972) are normally facilitated by the corticospinal tract. It is thus not surprising that they become less active after a lesion of this tract. The latter is necessary for the development of spasticity, even if it is not sufficient, as amply exemplified by results of its isolated lesion.

Is a Single Pathophysiological Factor Responsible for Spasticity?

As shown above, several pathophysiological changes coexist at the spinal cord level in chronic spasticity. As a rule, inhibitory mechanisms appear less active, and the myotatic arc seems uncontrolled rather than positively facilitated. The view that many factors may be involved in the explanation of spasticity contrasts with claims which prevailed years ago when a single mechanism was postulated, chiefly a selective gamma system hyperactivity.

The various pathophysiological changes could be linked and evolve in parallel, reflecting the intensity of the lesion of the descending motor tracts. To test this hypothesis, we looked for correlations between departures from normal values in the various tests. The correlation coefficients were found to be very poor, <0.3. It is possible to obtain in one single patient a higher H_{max}/M_{max} ratio combined with almost normal values of vibratory inhibition, while in another, with a similar degree of spasticity, normal H_{max}/M_{max} values with absent vibratory inhibition and moderately reduced reciprocal Ia inhibition are measured. All the combinations can be observed. From these results we concluded that the various disorders seem to be *independent*. The differences between patients could reflect the variable spectrum of descending influences which may be altered.

There are several pathophysiological changes at the spinal cord level, but, theoretically (Table 1), a single one could be enough to cause spasticity, the others appearing irrelevant. To test this possibility, we looked for possible correlations between changes in one single test and spasticity intensity, as assessed by Ashworth's scale. No good correlation was found between motoneuron excitability, vibratory inhibition, reciprocal Ia inhibition, and spasticity intensity. For nonreciprocal Ib inhibition, the correlation coefficient was better, reaching 0.6. From these data, we concluded that spasticity intensity resulted from a *combination in variable proportions* of the pathophysiological changes so far evidenced. In an individual patient the more determinant factor may vary according to the nature and number of descending fibers which function abnormally. However, two important points deserve to be underlined: First, the Ia pathway to motoneurons is liberated from inhibitory influences, explaining brisk reflexes. Second, with gradual stretching movements there is time for other afferents to interfere and contribute to the tonic stretch reflex. Mechanisms of the latter are not yet completely understood, and afferents from Ia, Ib, and II fibers could be involved (Matthews 1984, 1991). The Ib interneuron, which appears to have a key role in various types of hypertonia, could have a major importance which has not been emphasized so far and be responsible for the continuous muscle activity which persists when the muscle remains lengthened. Finally, though spasticity can be viewed as the end result of independent yet combined pathophysiological disorders, it is nevertheless possible to determine the major disturbances in a given patient by analyzing them through easy-to-perform techniques of clinical neurophysiology; thus the individual pathophysiological profile of the patient can be defined.

Conclusions

Many problems are still being debated, and information so far gleaned deals essentially with spasticity at rest in the lower limbs. However, some ideas can be taken for granted, namely that various spinal cord mechanisms are disturbed. To conclude, some amendments to Lance's definition could be proposed which take into account clinical aspects as well as data indicating that systems (namely the Ib system) other than the myotatic arc are involved in the genesis of spasticity: "Spasticity is a motor disorder characterized by brisk tendon jerks, sometimes accompanied by clonus, and a velocity-dependent muscle hypertonia during stretch which affects certain muscle groups preferentially. It results from hyperexcitability of the Ia pathway to motoneurons, combined with a disturbed processing at the spinal cord level of other peripheral afferent messages (tonic stretch reflex)."

References

Angel RW, Hoffmann WW (1963) The H-reflex in normal, spastic and rigid subjects. Arch Neurol 8:591–596

Ashby P (1990) Neuropysiological effects of chronic spinal lesions in animals and man. In: Benecke R, Emre M, Davidoff A (eds) The origin and treatment of spasticity. New Trends in Clinical Series, Lancs, Parthenon Publishing Group, pp 29–61

Ashby P, Wiens M (1989) Reciprocal inhibition following lesions of the spinal cord in man. J Physiol (Lond) 414:145–157

Bailey CS, Lieberman JS, Kitchell RL (1980) Response of muscle spindle primary endings to static stretch in acute and chronic spinal cats. Am J Vet Res 41:2030–2036

Buller A, Dornhorst AC (1957) The reinforcement of tendon reflexes. Lancet 2: 1260–1262

Burke D (1983) Critical examination of the case for or against fusimotor involvement in disorders of muscle tone. In: Desmedt JE (ed) Motor control mechanisms in health and disease, vol 39. Raven, New York, pp 133–150

Burke D, Ashby P (1972) Are spinal "presynaptic" inhibitory mechanisms suppressed in spasticity? J Neurol Sci 15:321–326

Bussel B, Pierrot-Deseilligny E (1977) Inhibition of human motoneurons, probably of Renshaw origin, elicited by an orthodromic motor discharge. J Physiol (Lond) 269:319–339

Day BL, Rothwell JC, Marsden CD (1983) Transmission in the spinal reciprocal Ia inhibitory pathway preceding willed movements of the human wrist. Neurosci Lett 37:245–250

Delwaide PJ (1971) Etude expérimentale de l'hyperréflexie tendineuse en clinique neurologique. Editions Arscia, Bruxelles

Delwaide PJ (1973) Human monosynaptic reflexes and presynaptic inhibition. In: Desmedt JE (ed) New developments in EMG and clinical neurophysiology. Karger, Basel, pp 508–522

Delwaide PJ (1985) Electrophysiological testing of spastic patients: its potential usefulness and limitations. In: Delwaide PJ, Young RR (eds) Clinical neurophysiology in spasticity. Elsevier Science, Amsterdam, pp 185–203

Delwaide PJ, Oliver E (1988) Pathophysiological aspects of spasticity. In: Marsden CD, Conrad B, Benecke (eds) Motor disturbances. I. Academic, New York, pp 153–167

Delwaide PJ, Martinelli P, Crenna P (1980) Clinical neurophysiological measurement of spinal reflex activity. In: Feldman RG, Young RR, Koella WP (eds) Spasticity: disordered motor control. Year Book Medical Publishers, Chicago, pp 345–371

Delwaide PJ, Schoenen J, De Pasqua V (1985) Lumbosacral spinal evoked potentials in patients with multiple sclerosis. Neurology 35:174–179

Delwaide PJ, Pepin JL, Maertens de Noordhout A (1991) Short-latency autogenic inhibition in patients with parkinsonian rigidity. Ann Neurol 30:83–89

Dietrichson P (1973) The role of the fusimotor system in spasticity and parkinsonian rigidity. In: Desmedt JE (ed) New developments in electromyography and clinical neurophysiology, vol 3. Karger, Basel, pp 496–507

Dietz V, Berger W (1983) Normal and impaired regulation of muscle stiffness in gait: a new hypothesis about muscle hypertonia. Exp Neurol 79:680–687

Eccles RM, Lundberg A (1959) Supraspinal control of interneurones mediating spinal reflexes. J Physiol (Lond) 147:565–584

Edström L (1970) Selective changes in the sizes of red and white muscle fibers in upper motor lesions and parkinsonism. J Neurol Sci 11:537–550

Eisen A, Odusote K (1979) Amplitude of the F wave: a potential means of documenting spasticity. Neurology 1306–1309

Eldred E, Granit R, Merton PA (1953) Supraspinal control of the muscle spindles and its significance. J Physiol (Lond) 122:498–523

Fisher MA (1983) F-response analysis of motor disorders of central origin. J Neurol Sci 62:13–22

Gassel MD, Diamantopoulos E (1964) The effect of procaine nerve block on neuromuscular reflex regulation in man (an apraisal of the role of the fusimotor system). Brain 87:729–742

Gillies D, Lance JW, Neilson PD, Tassinari CA (1969) Presynaptic inhibition of the monosynaptic reflex by vibration. J Physiol (Lond) 205:329–339

Goldfarb I (1976) Excitation of Renshaw cells via motor neuron collaterals in acute and chronic spinal cats. Brain Res 106:176–183

Granit R, Kaada B (1952) Influence of stimulation of central nervous structures on muscle spindles in cat. Acta Physiol Scand 27:130–160

Hagbarth KE, Wallin G, Löfstedt L (1973) Muscle spindle response to stretch in normal and spastic subjects. Scand J Rehabil Med 5:156–159

Hongo T, Jankowska E, Lundberg A (1972) The rubrospinal tract? IV. Effects on interneurones. Exp Brain Res 15:54–78

Harrison PJ, Jankowska E (1985) Sources of input to interneurones mediating group I non reciprocal inhibition of motoneurones in the cat. J Physiol (Lond) 361:379–401

Hufschmidt A, Mauritz KH (1985) Chronic transformation of muscle in spasticity: a peripheral contribution to increased tone. J Neurol Neurosurg Psychiatry 48:676–685

Hultborn H, Malmsten J (1983a) Changes in segmental reflexes following chronic spinal cord hemisection in the cat. I. Increased monosynaptic and polysynaptic ventral root discharges. Acta Physiol Scand 119:405–422

Hultborn H, Malmsten J (1983b) Changes in segmental reflexes following chronic spinal cord hemisection in the cat. II. Conditioned monosynaptic test reflexes. Acta Physiol Scand 119:423–433

Hultborn H, Illert M, Santini M (1976) Convergence on interneurones mediating the reciprocal Ia inhibition of motoneurones. I. Disynaptic inhibition of Ia inhibitory interneurones. Acta Physiol Scand 96:193–201

Hunt CC (1951) The reflex activity of mammalian small nerve fibres. J Physiol (Lond) 115:456–469

Iles JF, Roberts RC (1986) Presynaptic inhibition of monosynaptic reflexes in the lower limbs of subjects with upper motoneurone disease. J Neurol Neurosurg Psychiatry 49:937–944

Katz R, Pierrot-Deseilligny E (1982) Recurrent inhibition of alpha-motoneurones in patients with upper motor neuron lesions. Brain 105:103–124

Lance JW (1980) Symposium synopsis. In: Feldman RG, Young RR, Koella WP (eds) Spasticity: disordered motor control. Year Book Medical Publishers, Chicago, pp 485–494

Lundberg A, Voorhoeve P (1962) Effects from the pyramidal tract on spinal reflex arcs. Acta Physiol Scand 56:201–219

Lundberg A, Norrsell U, Voorhoeve P (1962) Pyramidal effects on lumbo-sacral interneurones activated by somatic afferents. Acta Physiol Scand 56:220–229

Mailis A, Ashby P (1990) Alterations in group Ia projections to motoneurons following spinal lesions in man. J Neurophysiol (in press)

Matthews PBC (1984) Evidence from use of vibration that the human long-latency stretch reflex depends upon spindle secondary afferents. J Physiol (Lond) 348:383–415

Matthews PBC (1991) The human stretch reflex and the motor cortex. Trends Neurosci 14:87–91

Naftchi NE, Schlosser W, Horst WD (1980) Correlation of changes in the GABA-ergic system with the development of spasticity in paraplegic cats. In: Mandel P, DeFeudis FV (eds) GABA – biochemistry and CNS functions. Plenum, New York, pp 431–450

Pierrot-Deseilligny E, Morin C, Bergego C, Tankov N (1981) Pattern of group I fibre projections from ankle flexor and extensor muscles in man. Exp Brain Res 42:337–350

Schiller HH, Stålberg E (1978) F response studied with single fiber EMG in normal subjects and spastic patients. J Neurol Neurosurg Psychiatry 41:45–53

Sjöstrom M, Fulg-Meyer R, Nordin G, Wahlby L (1980) Post-stroke hemiplegics: crural muscle strength and structure. Scand J Rehabil Med 85:53–67

Somerville J, Ashby P (1978) Hemiplegic spasticity: neurophysiologic studies. Arch Phys Med Rehabil 59:592–596

Tabary JC, Tardieu C, Tardieu G, Tabary C (1981) Experimental and rapid sarcomere loss with concomitant hypoextensibility. Muscle Nerve 4:198–203

Takakusaki K, Shimoda N, Mori S (1990) Spinal interneurons mediating the inhibition from the "medullary inhibitory region" to hindlimb alpha-motoneurons in decerebrate cats. Somatosens Mot Res 7:225–264

Tanaka R (1974) Reciprocal Ia inhibition during voluntary movements in man. Exp Brain Res 21:529–540

Veale JL, Rees S, Mark RF (1973) Renshaw cell activity in normal and spastic man. In: Desmedt JE (ed) New developments in electromyography and clinical neurophysiology. Karger, Basel, pp 523–537

Williams RG (1980) Sensitivity changes shown by spindle receptors in chronically immobilized skeletal muscle. J Physiol (Lond) 306:26p

Yanagisawa N, Tanaka R, Ito Z (1976) Reciprocal Ia inhibition in spastic hemiplegia of man. Brain 99:555–579

Dynamics of Reflex Excitability Following Intrathecal Baclofen Administration in Patients with Severe Spastic Syndromes

J. Dressnandt, A. Konstanzer, and B. Conrad

Introduction

The introduction of intrathecal baclofen therapy in 1985 (Penn and Kroin 1985; Ochs and Struppler 1987) brought significant progress in treating severe forms of spasticity. With this new therapy, high drug concentrations are reached at the desired site of action in the lumbar spinal cord while the concentration in the brain remains very low. Thus, side effects from actions at the cerebral level, such as tiredness and hallucinations, can be circumvented. The concentration of baclofen in the CSF is 50 times higher with intrathecal administration than with systemic administration (Kroin 1992).

Baclofen binds to GABA-B receptors, which are localized primarily at the presynaptic sensory neurons or interneurons (Price et al. 1987). Through a reduction of intracellular Ca^{++} (Green and Cottrell 1988; Lev-Tov et al. 1988) less neurotransmitter is released, and hereby excitatory and inhibitory neuronal circuits are depressed. In spastic patients this leads to a reduction of the exaggerated excitatory reflex mechanisms. It is known that, as a consequence of this depression, the monosynaptic H reflex and polysynaptic cutaneomuscular reflex (flexor reflex) are altered (Ochs and Struppler 1987; MacDonell et al. 1989; Sallerin-Caute et al. 1991). There are some case reports about the time course of the changes in H-reflex excitability, but there is little information on the precise time course of changes for the different reflexes or on interindividual variations. It was the aim of the present investigation to use various electrophysiological parameters which test motoneuronal excitability in order, on the one hand, to examine the precise time course of changes after different intrathecal baclofen bolus dosages and, on the other hand, to study the dose dependency of changes for different steady-state baclofen levels.

Neurologische Klinik der Technischen Universität München, Möhlstr. 28, D-81675 München, Deutschland

Thilmann et al. (Eds.)
Spasticity: Mechanisms and Management

Table 1. Characteristics of 11 patients studied

Patient	Sex	Age at time of implantation	Disease	Duration of illness (years)
JG	f	45	ms	18
WG	f	51	ms	7
SE	f	45	ms	30
TR	f	46	ms	10
VH	f	37	ms	17
DU	f	48	ms	20
KR	m	42	ms	20
SI	f	52	ms	20
SJ	f	38	ms	11
SH	f	50	ms	27
ER	m	37	ms	14

ms, Multiple sclerosis.

Patients and Methods

Eleven patients, two male and nine female, took part in this investigation (Table 1). The mean age was 45 ± 5.5 years. All had developed severe spasticity because of multiple sclerosis. The mean duration of illness was 17.6 ± 7 years at the time of the investigation. A programmable pump (Medtronic Synchromed 8611H) was implanted in all patients for continuous intrathecal baclofen infusion. Prior to implantation, a baclofen bolus was given through a lumbar puncture needle to test if and how much the patient might benefit from the clinical change in spasticity. After the implantation baclofen was given continuously. The dose was increased by steps of 30–50 μg every 24–48 h until a sufficient reduction of spasticity and spasms was observed.

Electrophysiological investigations were made prior to the baclofen bolus, every 10–20 min for 2–3 h after the bolus administration, and 24 h after the last change in dose of the continuous baclofen administration. The H reflex, the cutaneomuscular reflexes, and the F wave were investigated. The H reflex was recorded with surface electrodes over the soleus muscle, following stimulation of the ipsilateral tibial nerve in the popliteal fossa. The H reflex was evoked by increasing the stimulus voltage until its amplitude was maximal. With higher stimulus voltages, the amplitude decreased. A supramaximal stimulus was given to obtain the maximal M response (M_{max}). The data recorded were the maximal H-reflex amplitude (H_{max}) and the relation of H_{max} to M_{max} (H/M ratio). To evoke the cutaneomuscular reflex a stimulus of 5 pulses at 200 Hz (train duration 20 ms) was applied to the plantar sole, according to the technique described by Meinck et al. (1983). The EMG response was measured with surface electrodes over the ipsilateral tibial muscle and the antagonistic ipsilateral gastrocnemius muscle. The EMG activity was rectified. The F wave was measured with surface electrodes above the flexor hallucis muscle. A series of 20 supramaximal stimuli were applied to the distal tibial nerve at a rate of 1 Hz.

Results

Bolus Experiments

H Reflex. The maximal amplitude of the H reflex, and also the H/M ratio, showed a sigmoidal decrease after bolus injection of baclofen. A clear variability of the H-reflex amplitude was observed between the different patients. After a 50-μg intrathecal baclofen bolus a minimum of the H-reflex amplitude was reached in most patients after 90–120 min (Fig. 1); in one patient the minimum was reached after 150 min. The initial amplitude of the H reflex did not correlate with the rate of decrease after baclofen. In one patient four different bolus doses of baclofen (25 μg, 50 μg, 75 μg, 100 μg) were given on different days. The minimal H-reflex amplitude was reached 160 min after a bolus of 25 μg and 70 min after a 100 μg bolus (Fig. 2). With increasing doses of baclofen there was a dose-dependent faster depression of the H reflex, as can be seen in the diagram within Fig. 2.

Cutaneomuscular Reflexes. The cutaneomuscular reflex (flexor reflex) showed a decrease of the integral of the muscular activity with a minimum

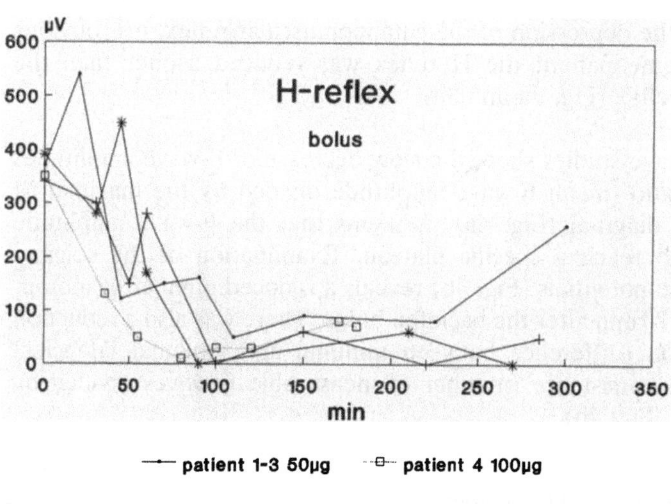

Fig. 1. Absolute amplitude of the maximal H reflex in four patients before and every 20–30 min up to 150–300 min after an intrathecal baclofen bolus. A bolus of 50 μg baclofen was given to patients 1–3 and a bolus of 100 μg to patient 4

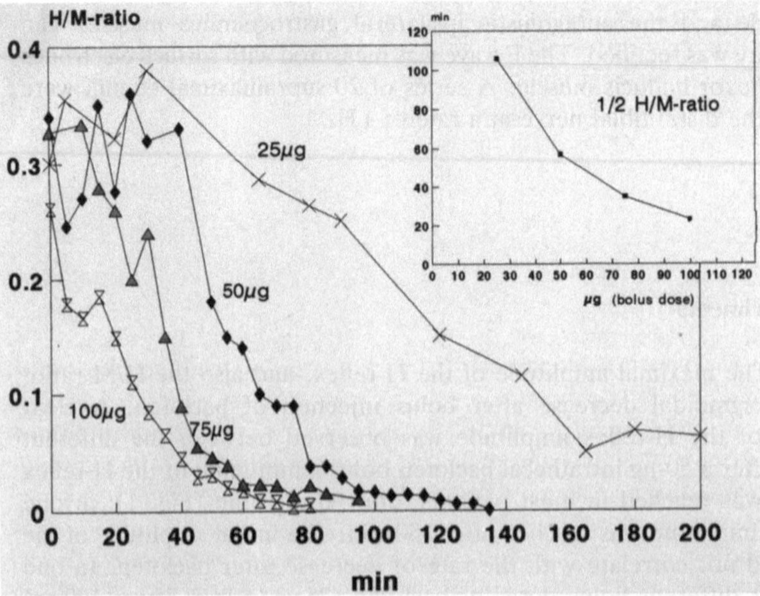

Fig. 2. H-reflex shown as H/M ratio in one patient before and after four different baclofen bolus doses on different days. *Inset* shows the course of 1/2 H/M, depending on the dose of the baclofen bolus. (*1/2 H/M* indicates when the H/M ratio reached half of the initial H/M value before baclofen application)

120 min after the bolus or later. The latency of onset of the cutaneomuscular reflex showed an increase, while the duration of cutaneomuscular reflex activity was reduced. Comparing the time course of the depression of the H reflex with the depression of the cutaneomuscular reflex, it is obvious that within the same patient the H reflex was reduced sooner than the cutaneomuscular reflex (Fig. 3a and 3b).

F-Waves. The F-wave studies showed a slow decrease of F-wave amplitudes and of the F/M ratio (mean F-wave amplitude divided by the maximal M response). In the diagram (Fig. 4a), it seems that the F-wave amplitude and the F/M ratio reach a specific plateau. Examination of the original consecutive F-wave potentials (Fig. 4b) reveals a reduced number of motoneuron discharges 120 min after the baclofen bolus. There was also a reduction in chronodispersion (difference between minimal and maximal latencies) and a decrease in persistence (number of measurable F waves divided by number of stimuli, Fig. 4b).

Steady-State (Constant-Flow) Baclofen

Baclofen infusion at a constant flow rate produced dose-dependent changes in reflex excitability. The H/M ratio decreased with higher doses, and the H

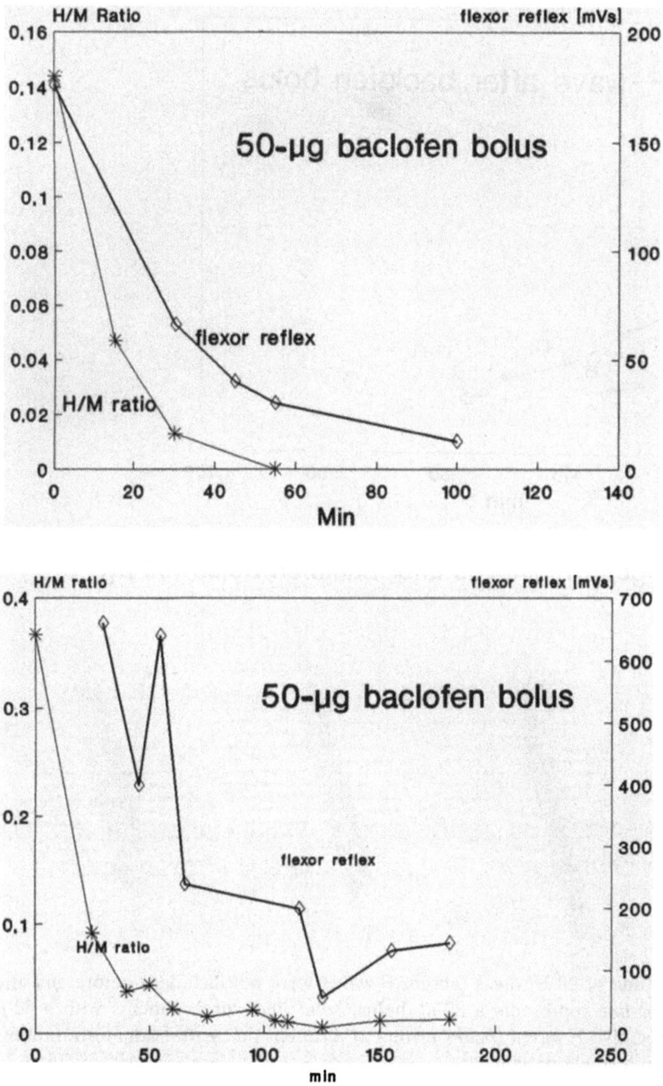

Fig. 3a,b. Changes of the H reflex and the cutaneomuscular reflex (flexor reflex) after administration of a 50 µg baclofen bolus in two different subjects

reflex was profoundly suppressed at doses between 100–220 µg/day (4.2–9.2 µg/h) baclofen (Fig. 5). The cutaneomuscular reflex was also suppressed with increasing doses of baclofen. The post-stimulus integral of muscular activity became smaller, latency increased (Fig. 6), and duration became shorter (Fig. 7). Complete suppression of the cutaneomuscular reflex was obtained at doses of 120–220 µg/day (5–9.2 µg/h) or more. Comparing the doses which led to reflex minima, it was clear that higher doses were needed to suppress the cutaneomuscular reflex than to reduce the H reflex.

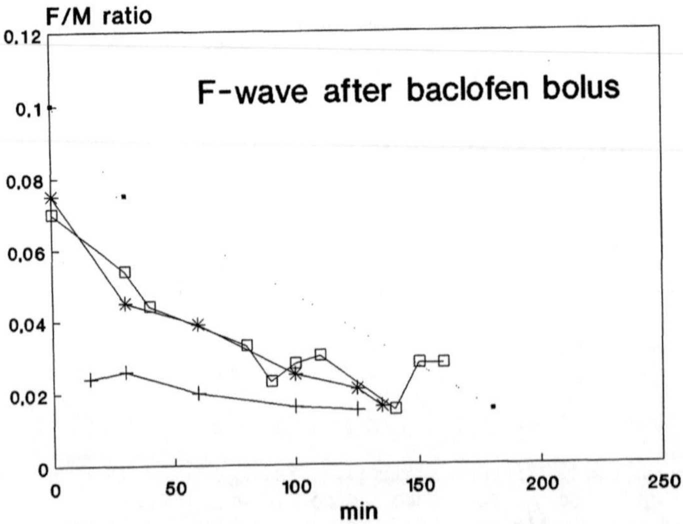

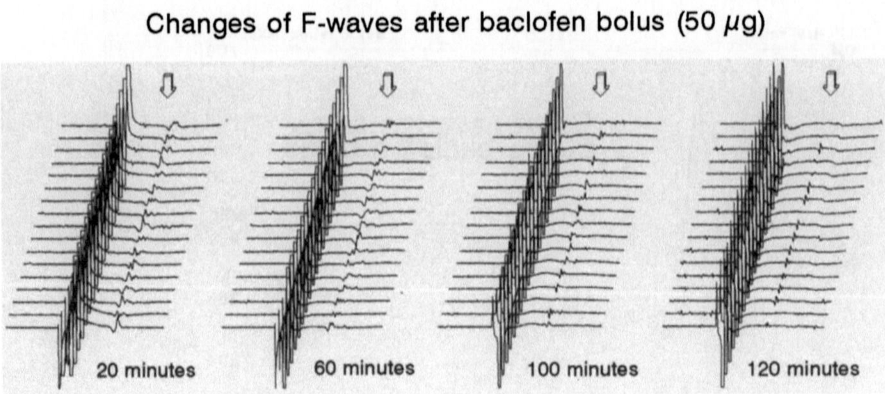

Fig. 4. **a** Mean amplitude of 20 F waves (absent F waves were not included) before and after baclofen bolus. *Dotted line* represents a 30-μg bolus, *solid lines* three patients with a 50-μg bolus. **b** Print of consecutive F waves (*below arrow*) at different times after administration of a 50-μg baclofen bolus in a single patient

The F-wave was reduced in maximal amplitude, chronodispersion, and persistence, but it was still obtainable at doses where the H reflex and the cutaneomuscular reflex could no longer be recorded.

Discussion

Intrathecal baclofen results in a dynamic process with rapid changes in motoneuronal excitability, which can be expressed in quantitative terms. In

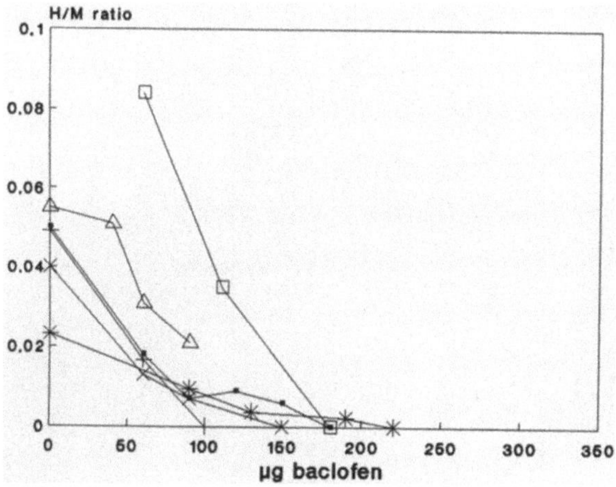

Fig. 5. H reflex with different doses of continuously infused baclofen in five patients

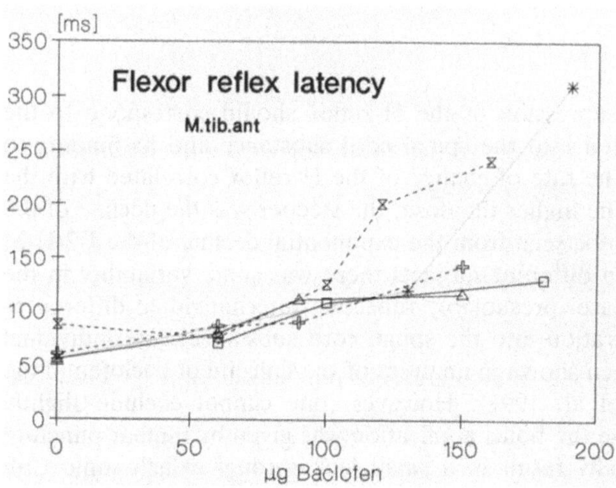

Fig. 6. Onset latency of the cutaneomuscular reflex (flexor reflex) with different doses of continuously infused baclofen

all patients examined in this study an H reflex was obtainable, and there was a decrease in H-reflex amplitude after baclofen bolus.

Doses of 100–220 µg/day baclofen by continuous infusion supressed the H-reflex completely. There was no clear correlation between the initial amplitude of the H reflex and the dose needed to suppress the reflex. These results correspond to those of others (Lazorthes et al. 1990; MacDonell et al. 1989). The reduction of the H/M ratio was caused primarily by a reduction of the H-reflex amplitude, and not by a change in the M_{max}.

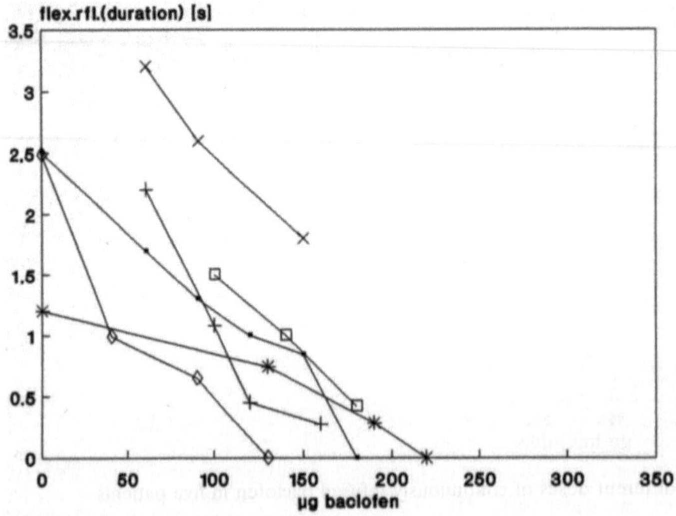

Fig. 7. Duration of the cutaneomuscular reflex (flexor reflex) with different doses of continuously infused baclofen

The dynamics of the supression of the H reflex should correspond to the diffusion of the baclofen into the spinal cord substance and its binding to GABA-B receptors. The rate of change of the H reflex correlated with the dose given by bolus. The higher the dose, the steeper was the decline of the sigmoidal curve, as can be seen from the exponential decline of the 1/2 H/M ratio (Fig. 2). Between different subjects there was some variability in the steepness of the decrease, presumably reflecting interindividual differences in diffusion and penetration into the spinal cord substance. Interindividual variations have also been shown in analyses of the half-life of baclofen in the CSF (Sallerin-Caute et al. 1991). However, one cannot exclude slightly different doses, because the bolus application was given by lumbar puncture and this can occasionally result in a small leak through which some CSF escapes into the epidural space. On the other hand, there is also a difference in the doses needed to suppress the H reflex by continuous infusion of baclofen between different subjects. This interindividual variation could also be caused by different rates of diffusion, by different receptor affinities for baclofen, or by different amounts of GABA-B receptors on the neurons. These reasons might also explain why the polysynaptic cutaneomuscular reflex took more time to be reduced in the bolus experiments and required higher doses to be suppressed in the steady-state experiments. It is also possible that some additional excitatory influence from higher spinal and supraspinal regions has more influence upon the polysynaptic cutaneomuscular reflex than upon the monosynaptic H reflex, and these descending tracts are influenced less by baclofen. Supraspinal influences cannot be excluded completely, because at least some slight voluntary movements of the toes were possible in most patients, indicating that the lesion of the spinal cord was incomplete. Monosynaptic dorsal root reflexes were abolished in animal studies (Farkas et al. 1989), while polysynaptic reflexes were only partially inhibited. Others have observed that baclofen supresses the monosynaptic Ia loop more than descending monosynaptic reflexes (Jimenez et al. 1991).

The regular occurrence of F waves can be a sign of heightened excitability of the motoneuron pool in spastic syndromes. After baclofen there was a reduction in F-wave chronodispersion and a lowering of F-wave amplitude (which was less pronounced than the reduction of the F-wave integral). This depression of the excitability of the motoneuron pool might be indirectly caused by a reduced excitatory input from other neurons because of reduced transmitter release (Lev-Tov et al. 1988). There is evidence that a subtype of the GABA-B receptor is present also post-synaptically on motoneurons and leads to a hyperpolarization through activation of a potassium channel (Wang and Dun 1990). These post-synaptically located GABA-B receptors are activated only at higher doses of baclofen than the pre-synaptically located GABA-B receptors (Wang and Dun 1990). This might explain not only the observed reduction in motoneuron excitability, but also the phenomenon that the F waves were still elicitable at a time and a dose where the H reflex and the cutaneomuscular reflex were no longer evokable. Because of side effects we did not test whether the F wave could be suppressed completely with higher intrathecal doses.

References

Farkas S, Tarnawa I, Berzsenyi P (1989) Effect of some centrally acting muscle relaxants on spinal root potentials: a comparative study. Neuropharmacology 28:161–173

Green KA, Cottrell GA (1988) Actions of baclofen on components of the Ca-current in rat and mouse DRG neurones in culture. Br J Pharmacol 94:235–245

Jimenez I, Rudomin P, Enriquez M (1991) Different effects of (−)-baclofen on Ia and descending monosynaptic EPSPs. Exp Brain Res 85:103–113

Kroin J (1992) Intrathecal drug administration: present use and future trend. Clin Pharmacokinet 22:319–336

Lazorthes Y, Sallerin-Caute B, Verdie JC, Bastide R, Carillo JP (1990) Chronic intrathecal baclofen administration for control of severe spasticity. J Neurosurg 72:393–402

Lev-Tov A, Meyers DER, Burke RE (1988) Activation of GABA-B receptors in the intact mammalian spinal cord mimics the effects of reduced presynaptic Ca++ influx. Proc Natl Acad Sci USA 85:5330–5333

MacDonell RAL, Talalla A, Swash M, Grundy D (1989) Intrathecal baclofen and the H-reflex. J Neurol Neurosurg Psychiatry 52:1110–1112

Meinck HM, Benecke R, Küster S, Conrad B (1983) Cutaneomuscular (flexor) reflex organization in normal man and in patients with motor disorders. Adv Neurol 39:787–796

Ochs G, Struppler A (1987) Treatment of spasticity by implantable drug administration devices and intrathecal catheter. In: Benecke R, Conrad B, Marsden CD (eds) Motor disturbances I. Academic, London, pp 197–208

Penn RD, Kroin JS (1985) Continuous intrathecal baclofen for severe spasticity. Lancet 2:125–127

Price GW, Kelly JS, Bowery NG (1987) The location of GABA-B receptor binding sites in mammalian spinal cord. Synapse 1:530–538

Sallerin-Caute B, Lazorthes Y, Monsarrat B, Cros J, Bastide R (1991) CSF baclofen levels after intrathecal administration in severe spasticity. Eur J Clin Pharmacol 40363–365

Wang MY, Dun NJ (1990) Phaclofen-insensitive presynaptic inhibitory action of (+/−)-baclofen in neonatal rat motoneurons in vitrol. Br J Pharmacol 99:413–421

Task-Dependent Spinal Inhibition in Spastic Hemiplegia

S. Miller[1] and R. Plant[2]

Introduction

In the spastic hemiplegia, which so frequently develops following stroke (Aho et al. 1980), voluntary control of the upper limb is impaired, leading to severe handicap in the performance of activities of daily living. The upper limb is held in a typical posture of retraction of the shoulder, medial rotation of the humerus, pronation of the forearm and flexion of the elbow, wrist and finger joints (Twitchell 1951; Bobath 1990; Carr and Shepherd 1990). In this posture increased tone is observed in several superficial muscles, particularly pectoralis major, biceps and the forearm flexors (Bobath 1990; Brunnstrom 1970; De Souza et al. 1980; Johnstone 1987). Attempts to make voluntary movements of the upper limb out of this posture are hampered by an inability to inhibit the increased tone of these muscles (e.g., Chan 1986; Musa 1986). This deficit of voluntary movement control illustrates a disorder of Sherrington's principle of spinal reciprocal inhibition, in which activation of agonist muscles is normally accompanied by centrally generated inhibition, of antagonists (Sherrington 1906). In fact, when movement is initiated voluntarily, reciprocal inhibition of antagonist muscles occurs in advance of the activation of the agonist muscles (Symoyama and Tanaka 1974).

Physical rehabilitation regimes for adults with spastic hemiplegia aim to promote relaxation of the exaggerated flexor synergy as a prerequisite for the purposeful performance of residual voluntary movement of the limb (Rood 1956; Goff 1972; Johnstone 1987; Bobath 1990). A key feature of this rehabilitation process is the relaxation of muscles acting about the shoulder, which is achieved therapeutically by appropriate positioning of the trunk and upper limb and training to relax before attempting voluntary movement.

In man the clinical syndrome of spasticity is known to involve several excitatory and inhibitory spinal mechanisms (Pierrot-Deseilligny and Maziéres 1985; Ashby and McCrea 1987), including disordered short-latency reciprocal inhibition (Tanaka 1983; Plant and Miller 1990). In a detailed

[1] Division of Clinical Neuroscience, The Medical School, The University of Newcastle upon Tyne, Newcastle upon Tyne NE2 4HH, UK
[2] Institute of Health Sciences, University of Northumbria at Newcastle, Newcastle upon Tyne NE7 7XA, UK

Thilmann et al. (Eds.)
Spasticity: Mechanisms and Management

study of inhibitory connections between muscles acting about the shoulder joint, the only muscles tested which had mutual short latency reciprocal inhibitory actions were posterior deltoid and clavicular pectoralis major (Plant and Miller 1990). These muscles were therefore chosen in the present study to determine how far positioning of the limb and the intention to perform a dynamic purposeful movement can modulate short latency reciprocal inhibition in stroke patients who have developed clinically defined spasticity. The present paper presents some preliminary results.

Methods

Two Motor Paradigms

Two experimental motor paradigms were investigated: (a) a positioning experiment, in which the influence of positioning of the arm on the magnitude of short-latency inhibition was investigated in an isometric motor task, and (b) a dynamic experiment, involving a static component (isometric contraction) and a dynamic component (the turning of a cranked wheel). The magnitudes of short-latency inhibition were compared on the two task components.

Subjects

The positioning experiment was performed on 15 normal healthy subjects (6 men and 9 women, aged 26–79 years, median 58 years) and on 19 stroke patients with spastic hemiplegia (8 men and 11 women, aged 35–85 years, median 67 years) tested 3–18 months after the stroke. The dynamic experiment was performed on 14 normal healthy subjects (6 men and 8 women, aged 35–79 years, median 62 years) and on 15 stroke patients (6 men and 9 women, aged 35–90 years, median 69 years) tested 3–18 months after the stroke. The normal subjects were recruited from university staff and the relatives of stroke patients; the patients were recruited from those attending the stroke outpatients clinic at the Royal Victoria Infirmary, Newcastle upon Tyne. The following criteria were used in selecting the patients: first-ever stroke, ability to understand the nature of the study, ability to perform the static and dynamic tasks and the absence of other pathologies which might independently impair voluntary movement control of the paretic arm. Ethical approval and informed written or witnessed consent from the subjects were obtained.

Electromyogram

The electromyogram (EMG) was recorded using surface-mounted preamplifiers (Johnson et al. 1977) placed over the clavicular fibres of pectoralis

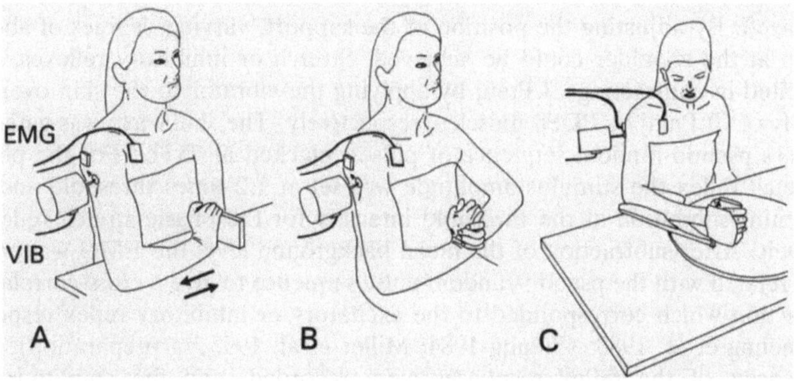

Fig. 1A–C. Experimental paradigms. **A** Static task of positioning experiment in which the subject attempted to push the hands together. **B** Static task component of dynamic experiment in which the subject attempted to push against a fixed support, which could be positioned to provide varying degrees of abduction of the arm at the shoulder. **C** Dynamic task component of dynamic experiment, in which the subject attempted to turn a cranked wheel set in a table. *EMG*, Electromyogram; *VIB*, vibrator

major (Cl.Pmaj) and over the posterior fibres of deltoid (PDelt; See Fig. 1A). The signals were filtered (sixth-order Bessel filter, −3 dB bandpass of 40–400 Hz), sampled at 400 Hz by computer and stored on magnaetic tape. For the positioning experiment and the static component of the dynamic experiment, the signals were also rectified and low-pass filtered with a cut-off of 1 Hz and displayed on an oscilloscope to the subject to provide an indication of the level of muscle contraction. In both experiments the maximum EMG level was noted in three attempts at maximal voluntary contraction.

Mechanical Stimulus

In both experiments a hand-held vibrator (Ling Altec 200), fitted with a stylus 10 mm in diameter, delivered mechanical stimuli, consisting of sine waves, 150 Hz, in the range 0.1–1.5 mm peak to peak, starting with an outward deflection from the zero crossing point of the sine wave. The experimenter applied the vibrator to the skin with a force of 4.25 N by reference to a level displayed on a meter.

Positioning Experiment

The subject sat in a chair with the shoulder braced against a fixed support, the arm in the anatomical position, elbow flexed to 90°, forearm supported in medial rotation and hand gripping a fixed support (Fig. 1A). By pushing on the support in the line of the forearm, Cl.P_{maj} contracted and PDelt relaxed. By adjusting the position of the support, varying degrees of abduction at the shoulder could be achieved. Stretch or inhibitory reflexes were elicited in contracting Cl.Pmaj by applying the vibrator to the skin over the belly of Cl.Pmaj or PDelt muscles, respectively. The sine wave was gated by a 13 s pseudo-random sequence of pulses, clocked at 75 Hz. For the phasic stretch reflex the stimulus amplitude was set at 1.2 times threshold and for eliciting inhibition at the threshold intensity for the phasic stretch reflex in PDelt. After subtraction of the mean background level the EMG was cross-correlated with the pseudo-random pulse sequence to give a cross-correlation function which corresponded to the excitatory or inhibitory reflex response (Cheung et al. 1983; Cheung 1984; Miller et al. 1993, in preparation). The latencies of the reflex responses were recorded, and the amplitudes of inhibitory reflex responses in Cl.Pmaj were compared at different degrees of abduction of the arm.

Dynamic Experiment

In this experiment the amplitudes of inhibitory and stretch reflexes were compared in a static isometric task and in a dynamic motor task with the same preamplifier electrode placement on the same test occasion. Observations in the stroke patient group were confined to the paretic arm (the arm on the hemiplegic side of the body) as the responses had to be recorded with the same placement of the preamplifier electrodes, and pilot trials showed that the subjects became too tired for prolonged testing.

For the static task the subject sat in a chair and attempted to push clasped hands together, a manoeuvre which involves bilateral contraction of Cl.Pmaj leaving PDelt of each side relaxed (Fig. 1B). A maximum voluntary contraction was obtained, and the subject attempted to maintain the level of EMG displayed on an oscilloscope set at 10% of this value. Using discrete trains of three cycles of 150 Hz sinusoidal mechanical stimuli at $1\,s^{-1}$, the maximal amplitude stretch reflex was recorded in Cl.Pmaj at full amplitude output of the stimulator (1.5 mm peak to peak). The maximal amplitude inhibitory reflex was evoked in Cl.Pmaj at 10% of maximum voluntary contraction by the same trains of stimuli with the vibrator applied to PDelt and with the amplitude set between one and two times the threshold for the phasic-stretch reflex in PDelt contracting. (This stimulus intensity was below the threshold for evoking a phasic stretch reflex in *relaxed* PDelt.) Averages were made of responses on Cl.Pmaj to 20 stimulus trains.

The dynamic task involved the turning of a cranked wheel set in a table in front of the seated subject, a task which includes elements of reaching and retrieving (Fig. 1C; Gandy et al. 1980; Hammond et al. 1982). The crank had a radius of 12.5 cm, and a frictional load of 0.45 Nm was applied. As the subject turned the wheel, the EMGs of Cl.Pmaj and PDelt were recorded, and the position of the wheel was encoded as a DC resettable voltage. A train of three cycles of 150 Hz mechanical stimuli, 0.5–1.5 mm peak to peak (as for the static task component), was delivered to either Cl.Pmaj (1.5 mm peak to peak) or PDelt (amplitude at twice threshold for the phasic-stretch reflex) once in each wheel cycle and could be triggered at any particular wheel position. The EMG in Cl.Pmaj was averaged over 20 turns of the wheel, and comparisons were made of the EMG activity with and without vibratory stimulation. The timing of the stimuli and data collection were controlled by computer. The magnitude and latency of the reflex responses were measured.

Results

Positioning Experiment

When the vibratory stimulus is applied in a pseudo-random sequence of sine waves to the tendon of Cl.Pmaj contracting at 10% of maximum voluntary contraction, cross-correlation with the EMG yields a function which approximates to the phasic stretch reflex (Cheung et al. 1983; Cheung 1984; Miller et al. 1993, in preparation). The values of the cross-correlation function are in arbitrary standard units, which have the dimension of voltage. In the upper traces of Fig. 2A the stimulus was at 1.2 times threshold, and the responses were recorded at 10% of maximal EMG, obtained in maximum voluntary contractions. When the same stimulus was applied to PDelt at the threshold for its phasic-stretch reflex, the cross-correlation function showed a downward deflection with a longer latency onset, which represents a short-latency inhibitory reflex response. The magnitude of the inhibitory responses in the non-paretic arms of stroke patients were comparable to those in normal subjects, but in the paretic arms the inhibition was of much smaller amplitude or in some subjects not recordable.

The relative latencies of the phasic-stretch reflexes and inhibitory reflexes for normal subjects and stroke patients are shown in Tables 1 and 2. Inhibitory responses were considered of short latency if the onset of the response occurred 0–7 ms later than the onset of the phasic-stretch reflex. The median difference for normal subjects was 4.5 ms and for stroke patients 3.5 ms.

In Fig. 2A the first row of inhibitory responses were obtained in normal subjects and stroke patients with the arm in 40° of abduction (See Fig. 1A). In the next row the arm was abducted to 50°. When the trials were repeated at 5° intervals over the range 35–55°, the curves of Fig. 2B were obtained for normal subjects and stroke patients, with the results from both arms of the normal subjects being pooled and the results of the paretic and non-paretic arms of the stroke patients shown as closed and open circles, respectively. In normal subjects and stroke patients the largest amplitude of

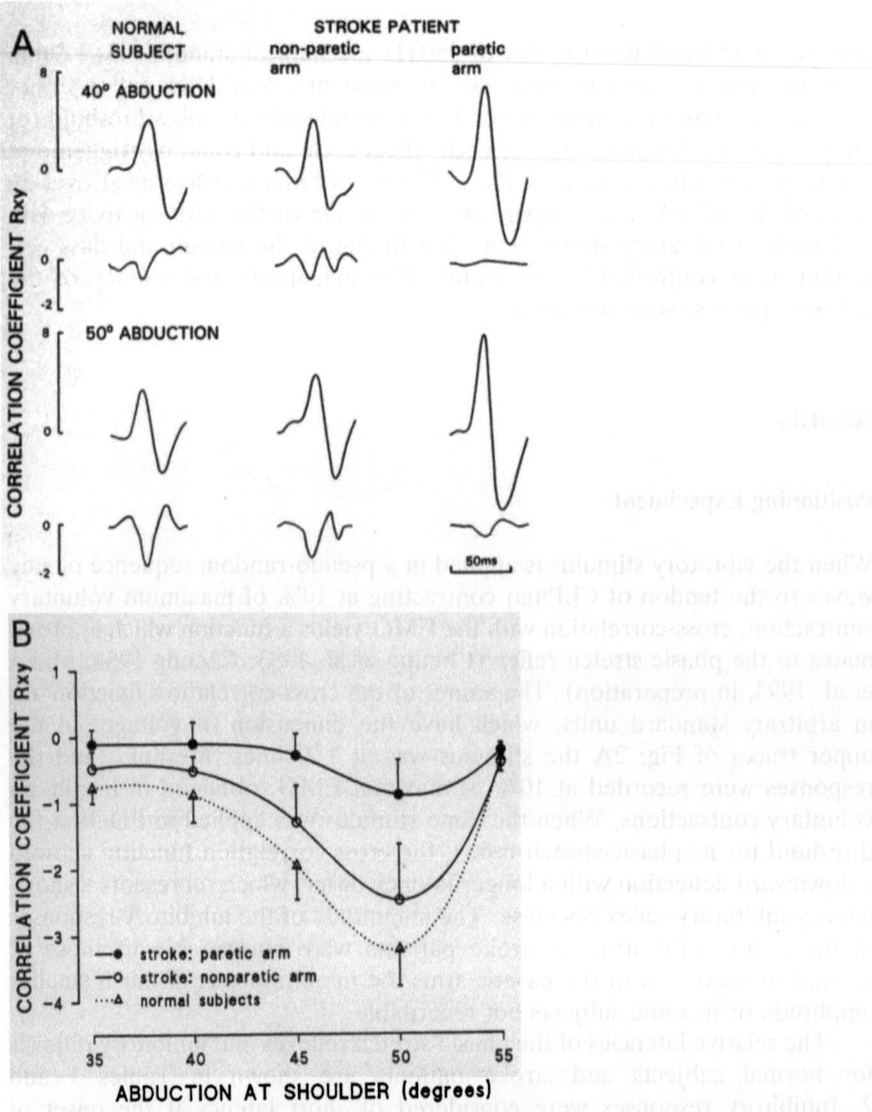

Fig. 2. A Phasic-stretch reflexes and short-latency inhibitory reflexes in contracting Cl.Pmaj at 40° and 50° abduction of the arm at the shoulder. **B** Modulation of short-latency inhibition with abduction of the arm at the shoulder. The data were pooled from all normal subjects and stroke patients, and the symbols give the mean with standard errors of the mean

inhibition occurred over a narrow range of abduction at 50°, being significantly greater than at 35, 40 or 55°. In the paretic arm of some stroke patients no inhibition was recordable at 35° but appeared at 50° abduction. Ratios of the amplitude of inhibition at 50° abduction to that at 40° abduction were calculated for each group. For the normal subject group the ratio

Table 1. Latencies of responses in postioning experiment

	Onset of phasic-stretch reflex		Onset latency difference[a]	
	Range	Median	Range	Median
Normal subjects	17–21	18	0–6	3
Stroke patients	16–21	19	0–7	4

[a] Inhibitory minus phasic stretch reflex.

Table 2. Latencies of first peaks of responses in dynamic experiment

	Stretch reflex		Onset latency difference[a]	
	Range	Median	Range	Median
Normal subjects	22–37	29	2–6	4
Stroke patients	20–35	24	2–8	6

[a] Inhibitory minus phasic stretch reflex. The onset latencies were measured in the positioning task, but the peaks of excitation and inhibition were measured in the dynamic task, where the onsets of the evoked responses were difficult to define.

was 3.7, for the non-paretic arms of stroke patients 5.2 and for the paretic arms greater than 1.0. In the latter group the ratio was difficult to calculate as some patients had no recordable inhibition at 40° abduction.

Dynamic Experiment

In this experiment the maximal amplitudes of the phasic stretch reflex and short-latency inhibitory reflexes in a static isometric task (Fig. 1B) and a dynamic task, involving the turning of a cranked wheel set in a table (Fig. 1C), were compared. For this comparison the maximal amplitudes of the phasic stretch reflexes in Cl.Pmaj obtained with optimal position of the arm were recorded following the same vibratory stimulus at maximal stimulator output and at 10% of maximum voluntary contraction. Maximal inhibitory responses in Cl.Pmaj were obtained following vibratory stimulation of PDelt at twice the threshold of its phasic-stretch reflex and at 10% maximal voluntary contraction of Cl.Pmaj. Responses comparable to those in Fig. 2A and B were obtained.

Turning of the cranked wheel results in alternate patterns of the rectified integrated EMG in PDelt and Cl.Pmaj with respect to wheel position (Fig. 3). In this dynamic task, as in the isometric task, PDelt and Cl.Pmaj have

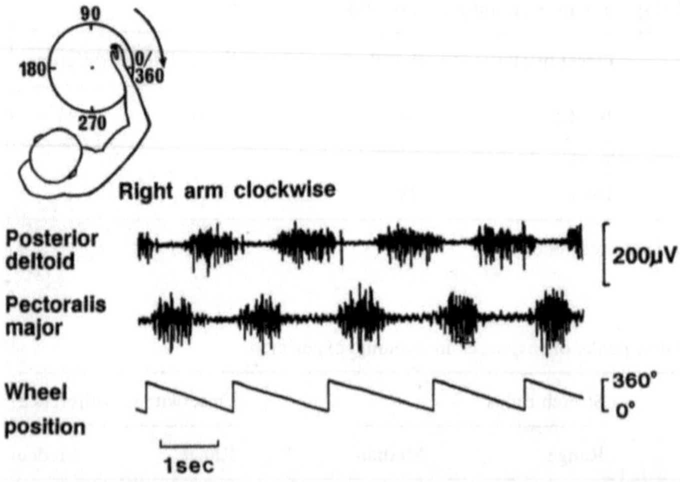

Fig. 3. EMG recordings in P.Delt and Cl.Pmaj recorded during cranking of the wheel in a right-arm clockwise direction. The position of the wheel is encoded as a DC resettable voltage

reciprocal short-latency inhibitory connections (Plant and Miller 1990). When the subject turns the cranked wheel, comparable short-latency excitatory and inhibitory responses in Cl.Pmaj can be obtained when the vibratory stimulus is applied to the tendon of Cl.Pmaj and PDelt, respectively (Fig. 4A). Table 1 gives the latencies of the first peaks of the responses. (The onsets of the excitatory and inhibitory responses were difficult to define due to the noisy background of the EMG.)

In all subject groups stretch reflexes were modulated, with peak amplitudes appearing during the first half of the EMG period of Cl.Pmaj. Inhibitory responses were also modulated with the wheel cycle, being maximal in the second half of the EMG period of Cl.Pmaj for normal subjects and the non-paretic arms of the stroke patients. However, in the paretic arms the maxima of the inhibitory response occurred in the first half of the EMG period of Cl.Pmaj.

The maximal amplitudes of the phasic-stretch reflexes and short-latency inhibitory reflexes in the static and dynamic tasks were recorded in normal subjects and in the paretic arms of the stroke patients. (The responses in the non-paretic arms were not measured; see "Methods".) In Fig. 4B the maximal amplitudes of the stretch reflex and short-latency inhibition for normal subjects and for the paretic arms of stroke patients are shown. For the phasic-stretch reflex the amplitudes are distributed about the slope drawn with a value of 1. For short-latency inhibition the values of normal subjects were distributed closely about the slope of 1, but for the paretic arms of stroke patients the maximal amplitudes of short-latency inhibition in the dynamic task were much greater than those in the static task.

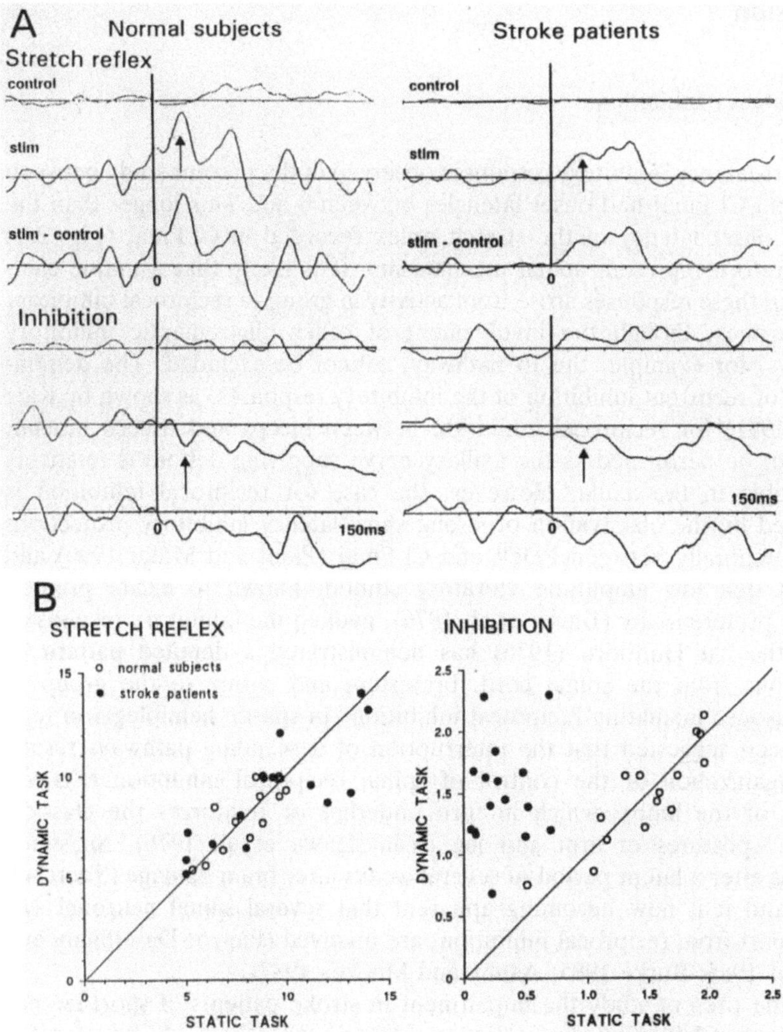

Fig. 4. A Averages of EMG recordings of Cl.Pmaj obtained during 20 cycles of the wheel. The vibratory stimulus is delivered to Cl.Pmaj during the period of its EMG to evoke a stretch reflex, and to PDelt during the period of the EMG of Cl.Pmaj to evoke a short-latency inhibitory reflex. In each set of three traces: *upper trace* (*control*), control without stimulation; *middle trace* (*stim*), with stimulation; *lowest trace* (*stim-control*), subtraction of the control from the stim trace. *Vertical line*, onset of the stimulus; *arrows*, peak of the initial responses. **B** Maximal amplitudes of stretch reflexes and short-latency inhibition obtained in the static and dynamic components of the dynamic experiment

Discussion

Short-Latency Inhibition

The short-latency inhibitory responses observed in the present study between PDelt and Cl.Pmaj had onset latencies between 0 and 7 ms longer than the shortest onset latency of the stretch reflex recorded in Cl.Pmaj (Fig. 2A) and therefore represent spinal mechanisms. It is likely that a major component of these responses arose from activity in group Ia reciprocal inhibitory interneurones, though the involvement of other oligosynaptic inhibitory pathways, for example, the Ib pathway, cannot be excluded. The demonstration of recurrent inhibition of the inhibitory responses, as shown by Katz et al. (1991) for reciprocal inhibition between biceps and triceps brachii, could not be performed as the axillary nerve supplying deltoid is relatively inaccessible in the axilla. However, the case for reciprocal inhibition is supported by the observation of strong short-latency inhibitory projections existing mutually between PDelt and Cl.Pmaj (Plant and Miller 1990) and the fact that low amplitude vibratory stimuli, known to excite primary spindles preferentially (Burke et al. 1976), evoked the inhibitory responses.

In the cat Hultborn (1976) has demonstrated a detailed pattern of projections from the spinal cord, brainstem and cortex to the group Ia interneurones mediating reciprocal inhibition. In spastic hemiplegia in man it has been suggested that the interruption of descending pathways results in disorganization of the control of spinal reciprocal inhibition between muscles of the limbs, which in turn underlies or reinforces the classical "spastic" postures of arm and leg (Yanasigawa et al. 1976). Spasticity develops after a latent period of several weeks after brain damage (Twitchell 1951), and it is now becoming apparent that several spinal neuronal systems, apart from reciprocal inhibition, are involved (Pierrot-Deseilligny and Maziéres 1985; Burke 1985; Ashby and McCrea 1987).

In the present study the impairment in stroke patients of short-latency inhibition in Cl.Pmaj during static isometric motor tasks might be expected since, of all the muscles in the upper limb, pectoralis major shows the greatest resting tone in the spastic hemiplegia following stroke (Plant and Miller 1993, in preparation). Following the arguments of Iles (1986) and Crone et al. (1987) for reciprocal inhibition in the lower limb, the tonic excitatory projection to Cl.Pmaj would also be directed to its Ia inhibitory neurones, which in turn would inhibit the Ia inhibitory interneurones in the inhibitory pathway from PDelt to Cl.Pmaj. However, the converse, i.e., augmentation of inhibition in the opposite direction from Cl.Pmaj to PDelt, has not been observed in the paretic arms of a small sample of stroke patients (Plant and Miller, unpublished observations), and the underlying mechanism must await further studies.

Task-Dependent Spinal Inhibition

Positioning of the Upper Limb. The augmentation of short-latency inhibition in Cl.Pmaj at particular degrees of abduction was unexpected, but the effect was consistent in all normal subjects and was evident in the stroke patient group, both in the paretic and non-paretic arms. In some stroke patients without demonstrable inhibition at 40°, further abduction of only 5° released the inhibition. The effect is unlikely to result from change of length in PDelt as this component of deltoid is shorter at 50° than at 35°, and the excitation of primary spindles is, if anything, reduced. Nor is the augmentation likely to be due to slight lengthening of Cl.Pmaj at 50° with increased spindle firing, as this should result in reduced inhibitory transfer from PDelt to Cl.Pmaj, due to the mutual inhibition between antagonist groups of Ia inhibitory interneurones (Hultborn et al. 1971); furthermore, any effect should increase progressively with muscle length. However, it is significant that in the cat's lower limb joint afferents can be excitatory to group Ia inhibitory interneurones (Hultborn 1976). If this were the case in the present study, it is curious that the facilitation of short-latency inhibition occurred over such a narrow range of joint position.

Increased levels of Ia reciprocal inhibition to soleus motoneurones have been reported in man during leaning backwards in standing (Rossi et al. 1988). The authors argued that the effects might be due to changes in descending vestibulo-spinal control. The present experiments differ in that no apparent control of balance is involved.

In terms of physiotherapy of patients with spastic hemiplegia, the positioning of the arm is critical for reducing unwanted muscle tone. For example, Johnstone (1980) states that positioning of the shoulder in external rotation and abduction (lengthening Cl.Pmaj) is essential for inhibiting spasm generally in the upperlimb and promoting voluntary movement.

Dynamic Versus Static Tasks for Promoting Short Latency Inhibition. The application of brief vibration to Cl.Pmaj and PDelt during the task of turning the cranked wheel evoked in normal subjects and stroke patients excitatory and inhibitory responses at latencies comparable to the phasic stretch reflex and short-latency inhibition in Cl.Pmaj. With vibration the phasic-stretch reflex of Cl.Pmaj was modulated in the wheel cycle, being maximal at the onset of the EMG. The lack of phasic stretch reflexes during the inactive period of the EMG in the wheel cycle indicates that the motoneurones receive active inhibition. In studies of locomotion in the cat Feldman and Orlovsky (1974) showed that the inactive period of EMG activity was accompanied by activity in group Ia inhibitory interneurones. In the present experiments the origin of the inhibition was not defined.

In normal subjects the inhibitory responses in Cl.Pmaj to vibration of PDelt occurred with maxima towards the end of the active EMG period of

Cl.Pmaj. Presumably this increase in inhibitory transmission helps in the termination of the EMG burst. In stroke patients the inhibitory responses were distributed throughout the active EMG period of Cl.Pmaj, which may reflect the difficulties that patients with spastic hemiplegia experience in terminating muscle activity (Bobath 1990; Gordon 1990).

In stroke patients the magnitude of inhibition in the paretic arm was greater in the dynamic task than in the static isometric task and occurred during the active period of Cl.Pmaj EMG (Fig. 4B). Unfortunately, the experimental design did not allow measurement of inhibitory reflexes during the inactive period of the EMG. The mechanisms underlying these changes of inhibitory control are unclear at present, though they are likely to reflect the influence of supraspinal control on short-latency inhibition in a purposeful task (see Tanji and Tanaguchi 1976). Iles (1986) and Crone et al. (1987) have argued that reciprocal inhibitory transmission between soleus and pretibial flexors is controlled phasically during a dynamic movement, and that descending motor pathways may control the inhibitory interneurones and α-motoneurones differentially (Iles 1986). The fact that the dynamic task facilitates short-latency inhibition in Cl.Pmaj is significant since this muscle acts about the shoulder, and most physiotherapy regimes for patients who develop spastic hemiplegia following brain damage stress the establishment of relaxation of muscles about the shoulder as a prerequisite for the recovery of purposeful voluntary movement of the arm as a whole (Brunnstrom 1970; Bobath 1990). Physiotherapists are divided as to the motor tasks which they employ to promote relaxation of unwanted muscle activity. Some advocate initially the performance of isolated components of integrated movements of the limb, for example, extension at the elbow (Johnstone 1987), while others encourage performance of a dynamic movement as a whole, for example, reaching forwards with the arm (Carr and Shepherd 1990). The present results suggest that greater inhibitory control is achieved in a dynamic task rather than a static isometric task.

Acknowledgement. Support from the Stroke Association (UK) is gratefully acknowledged.

References

Aho K, Harmsen P, Hatano S, Marquardsen J, Smirnov VE, Strasser T (1980) Cerebrovascular disease in the community: results of a WHO collaborative study. Bull WHO 58:113–130

Ashby P, McCrea DA (1987) Neurophysiology of spinal spasticity. In: Davidoff RA (ed) Handbook of the spinal cord. Dekker, New York, pp 119–143

Bobath B (1990) Adult hemiplegia: evaluation and treatment. Heinemann, London

Brunnstrom S (1970) Movement therapy in hemiplegia. Harper and Row, New York

Burke D (1985) Mechanisms underlying the tendon jerk and H-reflex. In: Delwaide PJ, Young RR (eds) Clinical neurophysiology in spasticity. Elsevier Amsterdam, pp 55–93

Burke D, Hagbarth KE, Löfstedt L, Wallin BG (1976) The responses of muscle spindle endings to vibration in non-contracting muscles. J Physiol (Lond) 261:673–693

Carr JH, Shepherd RB (1982) A motor relearning programme for stroke. Heinemann, London

Carr JH, Shepherd RB (1990) A motor learning model for rehabilitation of the movement disabled. In: Ada L, Canning C (eds) Key issues in neurological physiotherapy. Butterworth and Heinemann, Oxford, pp 1–24

Chan CWY (1986) Some techniques for the relief of spasticity and their physiological basis. Physiother Can 38:85–89

Cheung Y (1984) The examination of human stretch reflexes: an engineering approach. PhD thesis, Newcastle upon Tyne University

Cheung Y, Cozens A, Hammond GR, Miller S, Veitch ME (1983) Amplitude modulated pseudo-random excitation of stretch reflexes in arm and trunk muscles in normal subjects and stroke patients. J Physiol (Lond) 345:102P

Crone C, Hultborn H, Jespersen B, Nielsen J (1987) Reciprocal Ia inhibition between ankle flexors and extensors in man. J Physiol (Lond) 389:163–185

De Souza LH, Langton Hewer R, Miller S (1980) Assessment of recovery of arm control in hemiplegic stroke patients. I. Arm function tests. Int Rehabil Med 2:3–9

Feldman AG, Orlovsky GM (1974) Activity of interneurones mediating reciprocal Ia inhibition during locomotion. Brain Res 84:181–194

Goff B (1972) The application of recent advances in neurophysiology to Miss Rood's concept of neuromuscular facilitation. Physiotherapy 62(11):358–361

Gordon J (1990) Disorders of motor control. In: Ada L, Canning C (eds) Key issues in neurological physiotherapy. Butterworth and Heinemann, Oxford, pp 25–50

Hammond GR, Miller S, Robertson PM (1982) Central nervous control of arm movement in stroke patients. In: Sjölund B, Björklund A (eds) Brain stem control of spinal mechanisms. Elsevier North Holland, Amsterdam, pp 323–330 (Erik Fernström symposium 1)

Hultborn H (1976) Transmission in the pathway of reciprocal la inhibition to motoneurones and its control during the tonic stretch reflex. In: Homma S (ed) Understanding the stretch reflex. Prog Brain Res 44:235–255

Hultborn H, Jankowska E, Lindström S (1971) Recurrent inhibition from motor axon collaterals of transmission in the la inhibitory pathways to motoneurones J Physiol (Lond) 215:591–612

Iles JF (1986) Reciprocal inhibition during agonist and antagonist contraction. Exp Brain Res 62:212–214

Johnson SW, Lynn PA, Miller S, Reed GAL (1977) Miniature skin mounted preamplifiers for measurement of surface electromyographic potentials. Med Biol Eng Comput 15:710–711

Johnstone M (1987) Restoration of motor function in the stroke patient. Churchill Livingstone, Edinburgh

Katz R, Penicaud A, Rossi A (1991) Reciprocal Ia inhibition between elbow flexors and extensors in the human. J Physiol (Lond) 437:269–286

Musa I (1986) Recent findings on the neural control of locomotion: implications for the rehabilitation of gait. In: Banks M (ed) Stroke. Churchill Livingstone, Edinburgh, pp 79–89

Pierrot-Deseilligny E, Mazières L (1985) Spinal mechanisms underlying spasticity. In: Delwaide PJ, Young RR (eds) Clinical neurophysiology in spasticity. Elsevier, Amsterdam

Plant RD, Miller S (1990) Short latency inhibitory reflexes in shoulder and trunk muscles in spastic hemiplegia. In: Berardelli A, Benecke R, Manfredi M, Marsden CD (eds) Motor disturbances II. Academic, London, pp 367–378

Rood MS (1956) Neurophysiological mechanisms utilised in the treatment of neuromuscular dysfunction. Am J Occup Ther 4(1):220–225

Rossi A, Mazzocchio R, Scarpini C (1988) Changes in Ia reciprocal inhibition from the peroneal nerve to the soleus alpha-motoneurons with different static body positions in man. Neurosci Lett 84:283–286

Sherrington CS (1906) The integrative action of the nervous system. Yale University Press, New Haven (Silliman Memorial Lectures)

Symoyama M, Tanaka R (1974) Reciprocal Ia inhibition at the onset of voluntary movements in man. Brain Res 82:334–337

Tanaka R (1983) Reciprocal Ia inhibitory pathway in normal man and in patients with motor disorders. In: Desmedt JE (ed) Motor control mechanisms in health and disease. Raven, New York

Tanji J, Taniguchi K (1976) Presetting excitability of the spinal Ia inhibitory pathway in relation to the direction of an intended movement. Neurosci Lett 3:321–327

Twitchell TE (1951) The restoration of motor function following hemiplegia in man. Brain 74:443–480

Yanasigawa N, Tanaka R, Ito Z (1976) Reciprocal Ia inhibition in spastic hemiplegia of man. Brain 99:555–574

Late Flexion Reflex in Paraplegic Patients: Evidence for a Spinal Stepping Generator

A. Roby-Brami and B. Bussel

Introduction

Automatic spinal activity in paraplegic patients is characterized mainly by the occurrence of frequent flexor reflexes or spontaneous spasms which often participate in the disability. This has been attributed to a release of spinal networks from descending supraspinal control. Automatic spinal activity in animals has been known since the pionering work of Sherrington (1910) who described not only the myotatic reflex but also two kinds of spinal flexion reflexes. The nociceptive flexion reflex appears after a potentially noxious stimulation of the extremity by opposition to the stepping reflex, which is a distinct response related to alternating flexion-extension activity. Brown (1911) demonstrated that the rhythmic steppinglike activity was not due to a succession of reflexes but to a central generator since it persisted after a spinal cord section and deafferentation. The observation of flexion reflexes after a spinal cord section demonstrated that they are generated at the spinal level in both man and animals (Riddoch 1917; Kuhn 1950).

Animal studies showed that the flexion reflex was due to a polysynaptic activation of flexor motoneurones by group II and III afferents (Lloyd 1943). The afferents which may evoke the flexion reflex were subsequently grouped together as flexor reflex afferents (FRA) since they use common spinal interneuronal systems (Eccles and Lundberg 1959). FRA have access to alternative pathways which are mutually exclusive and differentially controlled by the higher nervous system (Holmqvist and Lundberg 1961). Particularly the injection of dopa into the acute spinal cat (which mimics the release of noradrenaline by descending pathways) depresses the usual flexion reflexes and causes the release of a long-latency flexion reflex (review in Lundberg 1979). As will be seen below, the late-flexion reflex pathways are closely linked to the spinal stepping generator (Jankowska et al. 1967; Lundberg 1981).

INSERM U215 and service de rééducation neurologique, Hôpital Raymond Poincaré, F-92380, Garches, France

Thilmann et al. (Eds.)
Spasticity: Mechanisms and Management

Electrophysiological studies in normal humans indicate that the flexion reflex is mediated by a spinal polysynaptic pathway probably driven by group III afferents since, according to animal experiments, its latency is less than 100 ms (Hagbarth 1960; Hugon 1967; Kugelberg et al. 1948; Pedersen 1954). The minimum latency of the flexion reflex in patients with a central nervous system lesion is within the same range as in normal subjects (Bathien and Bourdarias 1972; Dimitrijevic and Nathan 1968; Pedersen 1954). However, the flexion responses are usually prolonged or delayed (Rushworth 1966; Shahani and Young 1971) and have hence been considered to be disorganized (Meink et al. 1983). Our studies were undertaken to obtain a clearer picture of the flexion reflexes in patients with a complete spinal cord section and particularly to investigate the nature of the later response. The results indicate that it is similar to the late flexion reflex described by Lundberg et al. in the acute spinal cat with dopa (review Lundberg 1979) and is hence related to the spinal stepping generator. They suggest a new interpretation of the pathophysiology of the plantar reflexes (normal cutaneous plantar reflex and Babinski sign).

Flexion Reflexes in Paraplegic Patients

Methods. The effects of electrical FRA stimulation were studied in a total of 58 patients with clinically complete spinal cord sections. The examination was performed between 1 month and over 10 years after the injury, always after the spinal shock had disappeared. The FRA were stimulated by a distal electrical stimulation of the purely cutaneous sural nerve or the mixed tibialis posterior nerve. The responses were recorded by surface electromyography (EMG) of the ipsilateral flexor muscles: tibialis anterior, biceps femoris, and rectus femoris.

Presence of Two Distinct Responses. The FRA stimulation regularly induced two EMG reponses in ipsilateral flexor muscles. The early response appeared regularly on the BF muscle with a latency shorter than 100 ms and did not differ from the normal human flexion reflex. It is thus probably similar to the flexion reflex of acute spinal animals (Lloyd 1943). All the flexor muscles also showed a response with a latency longer than 130 ms. Its threshold was usually lower than that of the early reflex, and its main characteristic was that its latency gradually increased with the stimulus intensity (Fig. 1). The latency of the late response also increased with the duration of high intensity trains. The late response obviously originated from spinal reflexes in patients with a spinal section. However, the problem was to decide whether it was a true late reflex, or whether it was due to reafferences linked to a previous motor activity. Systematic studies of these

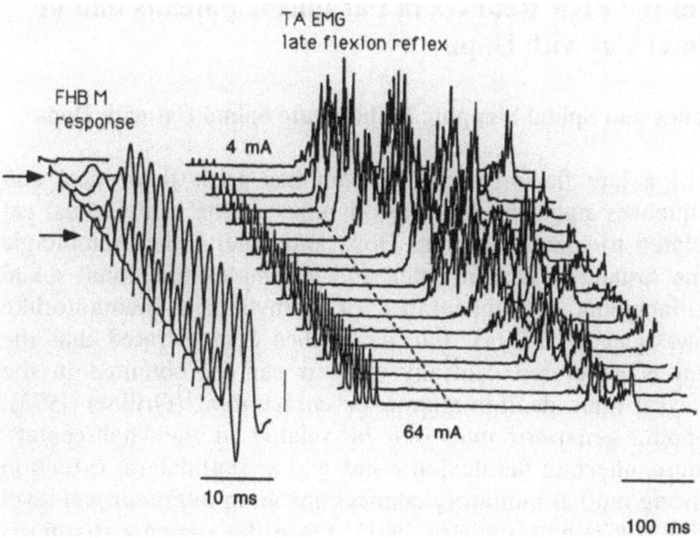

Fig. 1. Comparison of the late response in tibialis anterior (*TA*) and the simultaneous direct motor response M in flexor hallucis brevis (*FHB*) after tibial nerve stimulation in a paraplegic patient. Stimulation intensity increases from 4 to 64 mA in 4-mA steps. FHB, Each trace represents an EMG recording after the first shock of the stimulus train; TA, each trace represents the average of three individual rectified EMG recording. (From Roby-Brami and Bussel 1987)

variations showed that the late response is independent of any previous activity and is thus a genuine late reflex set up by the electrical stimulation (Roby-Brami and Bussel 1987). This late reflex is due to the recruitment of low-threshold cutaneous and probably group III cutaneous and muscular afferents. We estimate that the central delay is over 100 ms. Unmyelinated fibers are not involved since the latencies of flexion reflexes mediated by C fibers are over 1500 ms in man (Kugelberg 1948).

Nature of the Late FRA response. The prolonged central delay of the late flexion reflex indicates that it is similar to the reflex described by Lundberg and coworkers in the acute spinal cat with dopa (Andén et al. 1966a). In the acute spinal cat with dopa preparation, the latencies of the late reflex or of the related dorsal root potential (Andén et al. 1966b) increase with the stimulus intensity (Schomburg and Steffens 1988) or duration (Jankowska et al. 1967a). Lundberg (1979) and coworkers (Jankowska et al. 1967b) proposed that the increase in latency is due to opposite actions (long-latency excitation and short-latency inhibition) produced by the same FRA that converge at an interneuronal level. This schema could well explain our findings in paraplegic patients.

Comparison of the FRA Reflexes in Paraplegic Patients and in the Acute Spinal Cat with Dopa

Late Flexion Reflex and Spinal Stepping in the Acute Spinal Cat with Dopa

The presence of a late flexion reflex in man has great theoretical and practical consequences since the late flexion reflex in the acute spinal cat with dopa is related to spinal stepping. Dopa and other cathecolaminergic drugs induce the ipsilateral flexion reflex and a complex functional spinal reorganization, including the appearance of a rhythmical locomotorlike activity (Jankowska et al. 1967a). Further studies demonstrated that the stepping rhythm is generated centrally since it can be obtained in the isolated spinal cord after deafferentation or curarization (Grillner 1973). The spinal stepping generator may well be related to the "half-center" organization interconnecting the flexion reflex and a contralateral extention pathways via strong mutual inhibitory connections at an interneuronal level (Jankowska et al. 1967a,b; Lundberg 1981). Once the stepping rhythm is triggered, it may be maintained by excitation of the FRA due to the movement itself (Lundberg et al. 1987). Although the precise structure of the stepping generator remains unclear (Grillner and Wallen 1985), it is generally agreed that it is functionally related to the FRA reflexes network released by dopa (Viala et al. 1974; review in Schomburg 1990).

Spinal Reflex Organization in Paraplegic Patients

The FRA reflexes in paraplegic patients were analyzed assuming that their organization was similar to the network released by dopa in the acute spinal cat.

Effects of FRA on Extensor Muscles. Despite a careful search, a crossed extension reflex proved to be quite rare; it was found occasionally in only 3 of 16 patients (Roby-Brami and Bussel 1990). The effect of FRA on extensors was investigated using excitability curves of the monosynaptic H reflex of the soleus (Fig. 2). FRA stimulation induced two phases of facilitation of the contralateral soleus H reflex in most of the patients, indicating a subliminal activation of both early and late crossed extension reflexes. FRA stimulation also induced two corresponding inhibitory phases of the ipsilateral soleus H reflex.

Presynaptic Inhibition of Ia Fibers. Particular attention was paid to the pre- and postsynaptic nature of the H reflex inhibition since FRA-mediated bilateral presynaptic inhibition of Ia fibers has been shown in the acute spinal cat with dopa (Andén et al. 1966b; Jankowska et al. 1966); this is a characteristic of the dopa network, which has never been observed in any

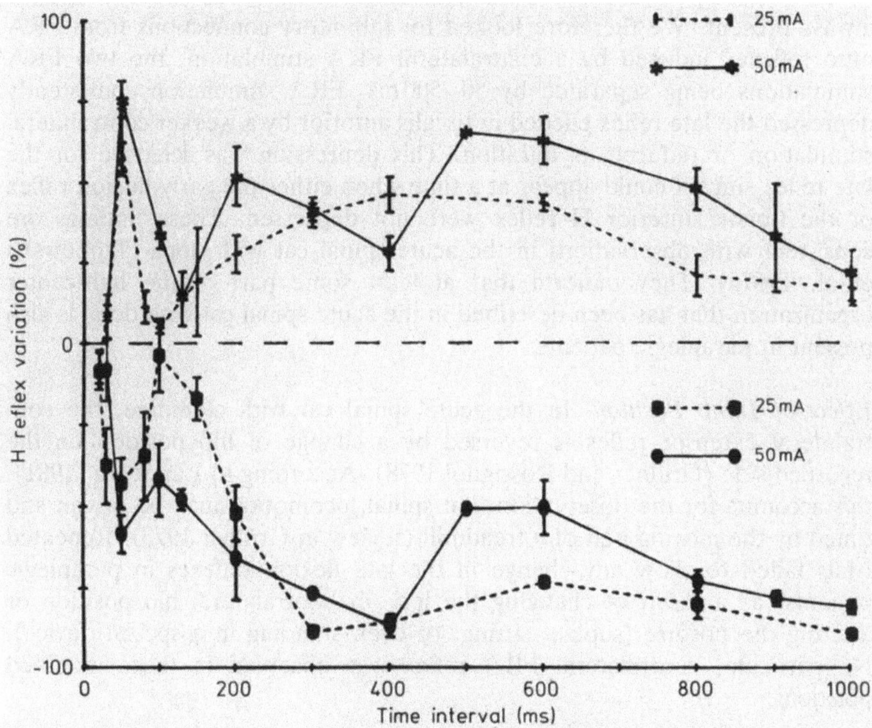

Fig. 2. Excitability curves of ipsilateral and contralateral soleus H reflexes recorded simultaneously after sural nerve stimulation. The change in the H reflex is plotted as a function of the time interval between the onset of the sural nerve stimulation and the stimulus triggering the H reflex. The contralateral H reflexes are facilitated and the ipsilateral H reflexes inhibited. Each point is the mean ± SEM of at least five values. (From Roby-Brami and Bussel 1990)

other animal preparation. We used the method described by Hultborn et al. (1987) which is based on the monosynaptic facilitation of the soleus H reflex by heteronymous Ia fibers of the crural nerve. It was not possible to analyze the strong ipsilateral H reflex inhibition (which exceeded the limits of the method), but we searched for a contralateral presynaptic Ia inhibition superimposed on the postsynaptic facilitation. FRA stimulation induced a significant reduction in the heteronymous facilitation in four paraplegic patients, indicating presynaptic inhibition (Roby-Brami and Bussel 1990). This finding was confirmed by an independant method based on vibratory inhibition. Some evidence also suggests that FRA stimulation induces bilateral presynaptic inhibition of the Ia fiber from flexor muscles (Roby-Brami and Bussel 1992).

Bilateral Inhibitory Connections. "Half-center" organization cannot be assessed directly in paraplegic patients since the crossed late reflex is not always present. We therefore looked for inhibitory connections from FRA onto reflexes induced by a contralateral FRA stimulation, the two FRA stimulations being separated by 50–500 ms. FRA stimulation consistently depressed the late reflex elicited in tibialis anterior by a weaker contralateral stimulation or reduced its duration. This depression was selective for the late reflex since it could appear at a time when either the early flexion reflex or the tibialis anterior H reflex were not depressed. These findings are consistent with observations in the acute spinal cat with dopa (Jankowska et al. 1967a). They indicate that at least some part of the half-center organization that has been described in the acute spinal cat with dopa is also present in paraplegic patients.

Effect of Limb Position. In the acute spinal cat with clonidine, the contralateral extensor reflex is reversed by a change of hip position on the recorded side (Grillner and Rossignol 1978). According to Lundberg (1981), this accounts for the observation that spinal locomotion may be driven and timed by the moving belt of a treadmill (review in Grillner 1973). Repeated trials failed to show any change in the late flexion reflexes in paraplegic patients, as a result of changing the ipsi- or contralateral hip position or altering the posture (supine, sitting, or even standing in a special device). No particular contralateral FRA reflex was observed in these modified positions.

Spinal Rhythmic Activity in Man. Iterative FRA stimulation never induces a proper rhythmic activity in paraplegic patients, unlike in animal experiments where the same preparation may display either late reflexes or locomotor activity depending on the peripheral stimulation (Viala et al. 1974). However, we had the opportunity to examine a patient with a complete cervical spinal cord section who exhibited rhythmical contractions of the trunk and lower limb extensor muscles (Bussel et al. 1988). Several arguments indicate that the rhythm is unlikely to be due to a peripheral loop but probably to an intrinsic spinal activity. The rhythmic activity (about 0.5 Hz frequency) could be induced, stopped, or modulated by FRA stimulation; iterative FRA stimulation could drive the rhythm. The sensitivity of the rhythm to peripheral stimulations suggests that it is due to the spinal stepping generator.

Plantar Reflex, Babinski Sign, and Late Flexion Reflexes

The pathophysiology of the plantar reflexes and Babinski sign remains disputed, although it is now agreed that the Babinski sign (extension of the big toe which is physiologically a flexion) is due to a lesion or a temporary dysfunction of the pyramidal tract (review in Van Gijn 1978). The spinal origin of the Babinski sign has been demonstrated by its observation in patients with a total spinal cord lesion. It has been suggested that it is due to a lowering of the threshold and a widening of the receptive field of the flexion reflex linked to a release of its spinal pathways. The normal plantar reflex induces a flexion of the toes, which is physiologically an extension and has been related either to the extensor part of a coordinated withdrawal reflex (Kugelberg et al. 1948; Grimby 1963a,b) or to a private cutaneous reflex similar to the one decribed by Engberg (1964) in the cat.

An electrophysiological study showed that a plantar scratch induced a response with a prolonged latency in both the extensor digitorum longum and the flexor hallucis brevis in spinal cord injured patients (Roby-Brami et al. 1989). We showed that the long latency is probably due to a long central conduction time and not to peripheral mechanisms (recruitment of slowly conducting afferents or afferent volley desynchronization), suggesting that

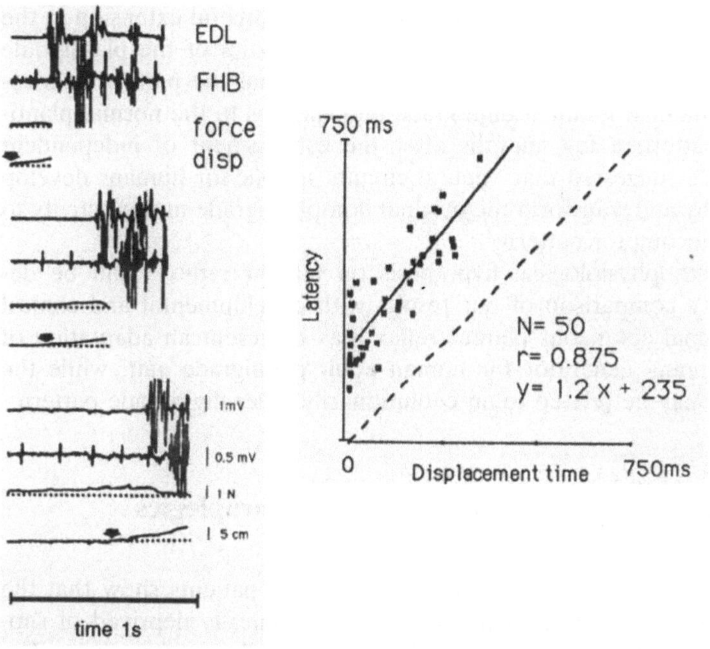

Fig. 3. Left, examples of recording obtained during a plantar scratch. *EDL*, EMG of the extensor digitorum longum; *FHB*, EMG of the flexor hallucis bervis; *Force*, force of the stimulation applied via a blunt probe and measured perpendicularly to the skin; *disp*, forward displacement of the probe during the scratch; *Arrows*, displacement of the probe by the examiner after a variable delay following initiation of the stimulation. *Right*, latency of the EDL response as a function of the delay between the stimulation initiation and the start of displacement. Each point represent one stimulation. *Continuous line*, regression line. The figure shows that the EDL response appears after a constant and prolonged delay following the start of the displacement and suggests that the latency of the late EDL response is not due to peripheral mechanisms. (From Roby-Brami et al. 1989)

the Babinski sign is in fact a late flexion reflex (Fig. 3). The same stimulation also induces late electromyographical responses of both the extensor digitorum longum and flexor hallucis brevis muscles in normal subjects. These responses are similar to, but much more variable than, the responses of paraplegic patients. Similar cocontraction EMG patterns were observed in normal subjects and in paraplegic patients by Landau and Clare (1959), who concluded that the Babinski sign is due to a stronger muscular contraction in the muscles inducing a toe dorsal flexion (the specific involvement of the extensor hallucis longus which they observed cannot be documented in our work with surface EMG recordings). Such reponses are unlikely to be due to repetitive firing in a short-latency cutaneous pathway, but we cannot exclude the contribution of supraspinal structures in normal subjects.

However, there is functional support for the hypothesis that the normal cutaneous plantar reflex is related to the spinal stepping generator. A cutaneous plantar reflex can be elicited by a forward displacement of the center of pressure on the plantar surface during stance in human adult locomotion. Therefore, the plantar flexion may take part in the forceful extension at the end of stance, which is one of the main characteristics of the plantigrade adult human gait. Forssberg (1985) demonstrated that the pattern of locomotion in the human infant is digitigrade and changes to the normal plantigrade adult pattern a few months after the establisment of independent locomotion. He suggested that "neural circuits specific for humans develop late in ontogeny and transform the original non-plantigrade motor activity to a plantigrade locomotor pattern."

A new pathophysiological hypothesis on plantar reflexes can be depeloped from a comparison of our results with developmental and clinical data. The normal cutaneous plantar reflex may represent an adaptation of the spinal stepping generator for human adult plantigrade gait, while the Babinski sign may be related to an evolutionarily older digitigrade pattern.

Significance of the Late Flexion Reflexes in Paraplegics

The neurophysiological data obtained in paraplegic patients show that the neural organization of the human spinal cord chronically deprived of supraspinal control is similar to that of the acute spinal cat after dopa. The absence of spinal locomotor activity and other related differences (absence of contralateral late reflex, insensitivity to limb position) could be attributed to species differences. However, the explanation proposed in acute animal preparations that the late flexion reflex and the stepping activity are released simultaneously by dopa from a tonic inhibition due to the early reflex pathways (Lundberg 1979) is not sufficient in paraplegic patients, since the late flexion reflex is observed without any drug injection, and the early and late FRA reflexes are present simultaneously. The functional relationships between the spinal stepping and late reflexes are clearly more complicated in the case of chronic spinal cord section. Cats whose spinal cords are sectioned as kittens spontaneously recover an almost normal locomotor activity (Forssberg et al. 1980) and display late flexion reflexes (Grillner 1973), but cats whose cords are sectioned in adulthood do not recover locomotion, unless they are rapidly and regularly trained on a treadmill (Barbeau et al. 1987). Noradrenergic drugs modulate the stepping activity of the chronic trained spinal cat, but do not induce late flexion reflexes (Barbeau et al. 1987). The locomotor capacity of monkeys with spinal lesions has been recently reexamined. Locomotion is possible in chronically trained primates after severe spinal lesions, and there is "limited evidence that rhythmic hind limb movements are possible in primates after complete spinal cord transsection" (Vilenski et al. 1992).

Further studies are obviously needed to identify the mechanism by which training interferes with the factors (development, neural plasticity, and neurochemical activity) that influence the course of a chronic spinal section to induce the recovery of spinal locomotion.

Nevertheless, the presence of functional neural activities related to spinal stepping in paraplegic patients raises a hope for the rehabilitation of locomotion. Wainberg et al. (1990) obtained encouraging results in patients with incomplete spinal lesion by combining training on a treadmill and monoaminergic drugs. In the future, grafts of fetal noradrenergic neurons could also contribute to the activation of the stepping generator.

References

Andén NE, Jukes MGM, Lundberg A, Vyklicky L (1966a) The effect of dopa on the spinal cord. I. Influence on transmission from primary afferents. Acta Physiol Scand 67:373–386

Andén NE, Jukes MGM, Lundberg A, Vyklicky L (1966b) The effect of dopa on the spinal cord. III. Depolarization evoked in the central terminals of ipsilateral Ia afferents by volleys in the flexor reflex afferents. Acta Physiol Scand 68:322–336

Barbeau H, Julien C, Rossignol S (1987) The effects of clonidine and yohimbine on locomotion and cutaneous reflexes in the adult chronic spinal cat. Brain Res 437:83–96

Bathien N, Bourdarias H (1972) Lower limb cutaneous reflexes in hemiplegia. Brain 95:447–456

Brown TG (1911) The intrinsic factor in the act of progression in the mammal. Proc R Soc Lond [Biol] 84:308–319

Bussel B, Roby-Brami A, Azouvi P, Biraben A, Yakovleff A, Held JP (1988) Myoclonus in a patient with spinal cord transection. Brain 111:1235–1245

Dimitrijevic MR, Nathan PW (1968) Studies of spasticity in man. III. Analysis of reflex activity evoked by noxious cutaneous stimulation. Brain 91:349–368

Eccles RM, Lundberg A (1959) Synaptic actions in motoneurones by afferents which may evoke the flexion reflex. Arch Ital Biol 97:199–221

Engberg I (1964) Reflexes to foot muscles in the cat. Acta Physiol Scand 62[Suppl 236]:1–64

Forssberg H (1985) Ontogeny of human locomotor control. I. Infant stepping, supported locomotion and transition to independent locomotion. Exp Brain Res 57:480–493

Forssberg H, Grillner S, Halbertsma J (1980) The locomotion of the low spinal cat. I. Coordination within a hindlimb. Acta Physiol Scand 108:269–281
Grillner S (1973) Locomotion in the spinal cat. In: Stein RB, Pearson KG, Smith RS, Redford JB (eds) Control of posture and locomotion. Plenum, New York, pp 515–535 (Advances in behavioral biology, vol 7)
Grillner S, Rossignol S (1978) Contralateral reflex reversal controlled by limb position in the acute spinal cat injected with clonidine iv. Brain Res 144:411–414
Grillner S, Wallen P (1985) Central patterns generators for locomotion with special reference to vertebrates. Annu Rev Neurosci 8:233–261
Grimby L (1963a) Normal plantar response: integration of flexor and extensor reflex components. J Neurol Neurosurg Psychiatry 26:39–50
Grimby L (1963b) Pathological plantar response: disturbances of the normal integration of flexor and extensor reflex components. J Neurol Neurosurg Psychiatry 26:314–321
Hagbarth KE (1960) Spinal withdrawal reflexes in the human lower limbs. J Neurol Neurosurg Psychiatry 23:222–227
Holmqvist B, Lundberg A (1961) Differential supraspinal control of synaptic actions evoked by volleys in the flexion reflex afférents in alpha motoneurones. Acta Physiol Scand 54[Suppl 186]:1–51
Hugon M (1967) Réflexes polysynaptiques et commandes volontaires. Contribution à la connaissance de l'organisation nerveuse spinale de l'animal et de l'homme. Thesis, University of Paris
Hultborn H, Meunier S, Morin C, Pierrot-Deseilligny E (1987) Assessing changes in presynaptic inhibition of Ia fibres: a study in man and the cat. J Physiol (Lond) 389:729–756
Jankowska E, Lund S, Lundberg A (1966) The effect of dopa on the spinal cord. IV. Depolarization evoked in the central terminals of contralateral Ia afferent terminals by volleys in the flexor reflex afferents. Acta Physiol Scand 68:337–341
Jankowska E, Jukes MGM, Lund S, Lundberg A (1967a) The affect of dopa on the spinal cord. VI. Half-center organization of interneurones transmitting effects from the flexor reflex afferents. Acta Physiol Scand 70:389–402
Jankowska E, Jukes MGM, Lund S, Lundberg A (1967b) The effect of dopa on the spinal cord. V. Reciprocal organization of pathways transmitting excitatory actions to alpha motoneurones of flexor and extensors. Acta Physiol Scand 70:369–388
Kugelberg E (1948) Demonstration of A and C fibre component in the Babinski plantar response and the pathological flexion reflex. Brain 71:304–319
Kuhn RA (1950) Functional capacity of the isolated human spinal cord. Brain 73:1–15
Landau WM, Clare MH (1959) The plantar reflex in man with special reference to some conditions where the extensor response is unexpectedly absent. Brain 82:321–355
Lloyd DPC (1943) Neuron patterns controling transmission of ipsilateral hindlimb reflexes in cat. J Neurophysiol 6:293–315
Lundberg A (1979) Multisensory control of spinal reflex pathways. Prog Brain Res 50:11–28
Lundberg A (1981) Half-centers revisited. In: Szentàgothai J, Palkovits M, Hamori J (eds) Regulatory functions of the CNS. Principles of motion and organization. Academiai Kiado, Budapest; Pergamon, Oxford, pp 155–167 (Advances in physiological sciences)
Lundberg A, Malmgren K, Schomburg ED (1987) Reflex pathways from group II muscle afferents. III. Secondary spindle afferents and the FRA: a new hypothesis. Exp Brain Res 65:294–306
Meink HM, Benecke R, Kuster S, Conrad B (1983) Cutaneomuscular (flexor) reflex organization in normal man and in patients with motor disorders. In: Desmedt JE (ed) Motor control mechanisms in health and disease. Raven, New York, pp 787–796
Pedersen E (1954) Studies on the central pathway of the flexion reflex in man and animal. Acta Psychiatr Neurol Scand [Suppl 88]:1–81
Riddoch G (1917) The reflex functions of the completly divided spinal cord in man, compared with those associated with less severe injury. Brain 40:264–402

Roby-Brami A, Bussel B (1987) Long-latency spinal reflex in man after flexor reflex afferent stimulation. Brain 110:705–725

Roby-Brami A, Bussel B (1990) Effects of flexor reflex afferent stimulation on the soleus H reflex in patients with a complete spinal cord lesion: evidence for presynaptic inhibition of Ia transmission. Exp Brain Res 81:593–601

Roby-Brami A, Bussel B (1992) Inhibitory effects on flexor reflexes in patients with a complete spinal cord lesion. Exp Brain Res 90:201–208

Roby-Brami A, Ghenassia JR, Bussel B (1989) Electrophysiological study of the Babinski sign in paraplegic patients. J Neurol Neurosurg Psychiatry 52:1390–1397

Rushworth G (1966) Some studies on the pathophysiology of spasticity. Paraplegia 4:130–141

Schomburg ED (1990) Spinal sensorimotor systems and their supraspinal control. Neurosci Res 7:265–340

Schomburg ED, Steffens H (1988) The effect of dopa and clonidine on reflex pathways from group II afferents to alpha motoneurons in the cat. Exp Brain Res 71:442–446

Shahani BT, Young RR (1971) Human flexor reflexes. J Neurol Neurosurg Psychiatry 34:616–627

Sherrington CS (1910) Flexion reflex of the limb, crossed extension reflex and reflex stepping and standing. J Physiol (Lond) 40:28–121

Van Gijn J (1978) The Babinski sign and the pyramidal syndrome. J Neurol Neurosurg Psychiatry 41:865–873

Viala D, Valin A, Buser P (1974) Relationship between the "late reflex discharge" and locomotor movements in acute spinal cats and rabbits treated with dopa. Arch Ital Biol 112:299–306

Vilenski JA, Eidelberg E, Moore AM, Walden JG (1992) Recovery of locomotion in monkeys with spinal cord lesion. J Motor Behav 24:288–296

Wainberg M, Barbeau H, Gauthier S (1990) The effects of cyproheptadine on locomotion and on spasticity in patients with spinal cord injuries. J Neurol Neurosurg Psychiatry 53:754–763

Kinematics of the Cat Hindlimb After Spinal Cord Hemisection

J.P. Kuhtz-Buschbeck, A. Boczek-Funcke, C. Weinhardt, and M. Illert

Hemisection of the spinal cord in the adult cat has been used repeatedly as an experimental model to investigate the mechanisms underlying spasticity. In many of these experiments low thoracic spinal cord lesions were performed, which spared the dorsal funiculi (McCouch et al. 1957; Murray and Goldberger 1974; Hultborn and Malmsten 1983). A recent study used hemisection of the lower lumbar spinal cord (Carter et al. 1991) to study the contributions of interneuronal relays located in these segments (Edgley et al. 1988).

The investigations have concentrated either on electrophysiological parameters such as transmission through spinal and supraspinal reflex pathways (Hultborn and Malmsten 1983; Carter et al. 1991) or on qualitative neurological tests such as tendon reflexes, clasp-knife phenomenon, or hopping reflexes (Murray and Goldberger 1974). There is general agreement that the animals develop signs of weakness and hypotonia shortly after the hemisection (McCouch et al. 1957; Teasdall et al. 1958; Hultborn and Malmsten 1983; Carter et al. 1991). Results have been conflicting, however, with regard to the development of spasticity (see Munson et al. this volume). Some reports state that typical neurological signs of spasticity are missing and conclude that spinal cord hemisection is no model for spasticity (Hultborn and Malmsten 1983); others report the long-term presence of symptoms consistent with spasticity (McCouch et al. 1957; Fujimori et al. 1966; Murray and Goldberger 1974).

The effects of spinal cord hemisections on a natural behavior like locomotion have not been investigated in the same detail, and the observations give a more qualitative than systematic account. There is, e.g., no information available as to which phases of the step cycle are affected by the thoracic spinal cord hemisections or which components show a long-term functional recovery.

We have investigated the short-and long-term effects of a low thoracic spinal cord hemisection (T-11, right side, sparing the dorsal funiculus) on locomotion. The footfall patterns and the kinematics of the hindlimb joints were examined with video and pulsed X-ray techniques during treadmill locomotion (walk, velocity range 0.3–0.9 m/s).

Abt. Physiologie, Universität Kiel, Olshausen Str. 40, D-24118 Kiel, Deutschland

Thilmann et al. (Eds.)
Spasticity: Mechanisms and Management

During the first few days after hemisection the cats showed severe locomotor deficits (flaccid monoplegia of the ipsilateral hindlimb) which improved rapidly. After 8 days they were able to walk a few steps on the treadmill. Three major deficits were found which impeded a coordinated walk: lack of coordination between the limbs, an inability of the distal joints to induce extension, and a strong circumduction during the swing phase. These symptoms improved within the next weeks, and after 2–3 months a stable condition was reached.

Gait Pattern

The low thoracic hemisection disturbed the gait pattern. It reduced the step cycle duration from 763 ± 52 ms (n = 23 steps) in the control (treadmill velocity 0.56 m/s) to 665 ± 82 ms (n = 24) 8 days after the hemisection. Sixty days after the hemisection the step cycle duration had recovered to 732 ± 38 ms (n = 24). Similar differences were found in the other treadmill velocities.

The stance phase durations of the four limbs did not differ significantly in the control animal (compare Hildebrand 1966; Halbertsma 1983). Consequently, the support lengths were about equal. Thus at a treadmill velocity of 0.56 m/s the control animal had an average support length of 28.2 ± 2.0 cm and a stance phase duration of 504 ± 36 ms (n = 23 steps). In the hemisected animals, however, these parameters had changed. They were significantly smaller in the hindlimbs than in the forelimbs (t-test, $p < 0.05$). Between the two hindlimbs they were smaller in the lesioned side (LS) than in the unlesioned side (US) limb. On the eighth postoperative day the support length of one hemisected animal was 25.3 ± 2.2 cm (n = 24) in the forelimbs, 24.2 ± 1.3 cm (n = 24) in the US hindlimb, and 21.2 ± 2.9 cm (n = 24) in the LS hindlimb. The relative stance phase durations (percentage of step cycle duration) were 68 ± 6% in the forelimbs, 65 ± 3% in the US hindlimb, and 57 ± 8% in the LS hindlimb. In the other hemisected animal the corresponding values were similar. After 60 days the support length had increased in parallel to the recovery of the step cycle duration, but the described time differences between the different limbs remained. They are still present in one of the animals (observation period longer than 200 days).

The interlimb coordination was also changed. Whereas the forelimbs moved in the alternating, symmetrical gait pattern, an asymmetrical footfall pattern was observed in the hindlimbs. Both the hindlimb/hindlimb and the hindlimb/forelimb coordination were disturbed. Figure 1 illustrates the footfall pattern of the control and of a hemisected animal, 8 and 60 days after the operation (treadmill velocity 0.56 m/s). Regarding hindlimb/hindlimb coordination in the control, the stance phase of one hindlimb starts at around half the step cycle time of the other hindlimb (see Gray 1968). Eight

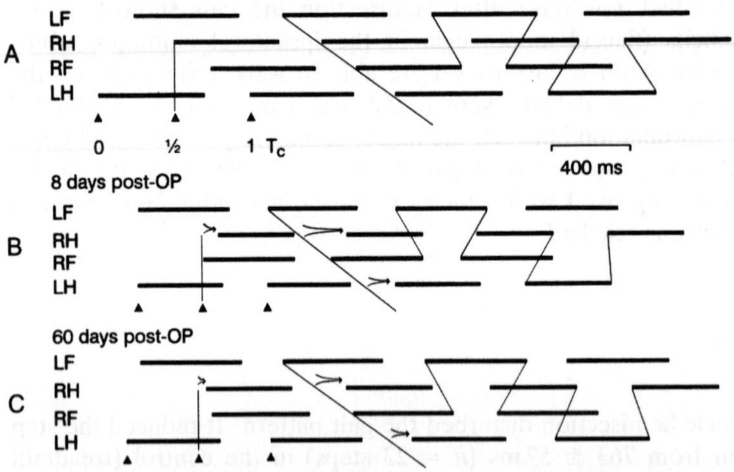

Fig. 1A–C. Gait patterns in a control animal (**A**) and in a hemisected animal 8 days (**B**) and 60 days (**C**) postoperatively. Averaged data from 23 (**A**) and 24 (**B, C**) steps. The stance phase is indicated by *horizontal bars*. *Arrows* in **B** and **C** indicate the phase delay of the lesioned side (RH) and unlesioned side (LH) hindlimb. *LF, RF*, left, right forelimb; *LH, RH*, left, right hindlimb; T_c, step cycle duration. Treadmill speed 0.56 m/s

days after the hemisection the placing of the LS hindlimb was delayed by 85 ± 31 ms (marked by an arrow in Fig. 1B), 60 days after the hemisection by 37 ± 32 ms. The hindlimb/forelimb coordination was disturbed, since placing of the LS hindlimb was delayed with respect to the placing of both forelimbs by 241 ± 56 ms 8 days, and 151 ± 41 ms 60 days after the hemisection. A similar, but smaller delay was seen in the placing of the US hindlimb (8 days: 156 ± 49 ms; 60 days: 114 ± 45 ms).

As a consequence, the forelimb/hindlimb coordination resembled a pacing gait since the ipsilateral fore- and hindlimbs struck the ground at about the same time. This gait pattern was not observed in the control. In summary, the disturbance of the gait pattern consisted of a reduced support length and a phase lag involving the LS hindlimb and, to a minor degree, the US hindlimb.

Angular Excursions of the Hindlimb Joints – Short-Term Effects

The angular excursions were established with X-ray cinematography during treadmill locomotion. Pulsed X-ray shots were used (time intervals 20 ms). The following hindlimb joints were investigated: proximal interphalangeal joint (PIP), metatarsophalangeal joint (MTP), ankle, knee, and hip. The angles were calculated off-line from X-ray video images in the parasagittal

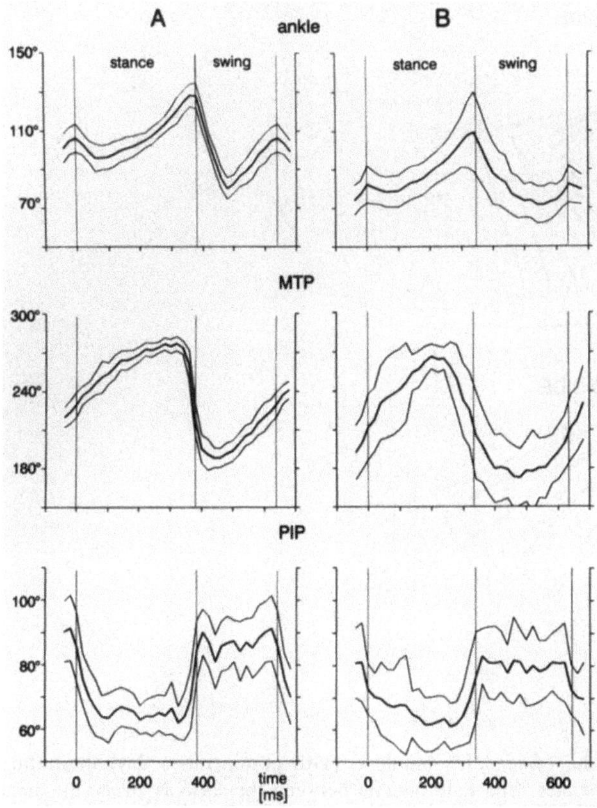

Fig. 2A,B. Angular excursions of the distal joints of the US (**A**) and LS (**B**) hindlimb (17th postoperative day). Ankle joint, metatarsophalangeal joint (*MTP*), proximal interphalangeal joint (*PIP*). Averaged data from 25 steps (*lines* indicate mean ± SD) at a treadmill velocity of 0.65 m/s

plane after the position of the joint points had been established interactively in a computer system (Nath 1990). Figure 2 illustrates from the 17th postoperative day the angular excursions of the distal joints in the LS and US hindlimbs (treadmill velocity 0.65 m/s, average of 25 steps).

The comparison shows distinct differences. Generally, the standard de-ʼiations are larger in the LS limb than in the US limb, which indicates a ;reater variability of angular excursions and timing of the single steps in the ɔrmer. At both the ankle and the MTP, the flexion leading the swing phase slowed. At the ankle the E1-phase extension is delayed and weak, the E3-hase extension is not very pronounced, and the flexion during the E2 phase he yield) is small. At the MTP, the joint angle is below 180° during the ʼing phase (see also Fig. 3). This insufficient extension (compare A) evented in many steps the clearing of the limb from the ground, with the nsequence that the paw was dragged over the treadmill belt during the

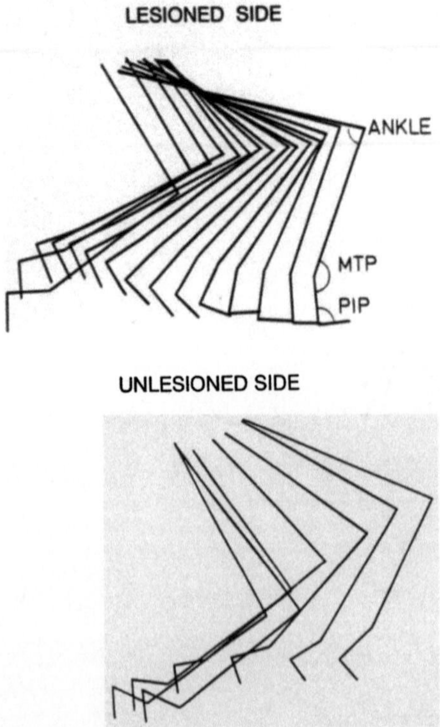

Fig. 3. Stick figures of the LS and US hindlimb (17th postoperative day) illustrating the transition from swing to stance. The time interval between the sticks is 20 ms, the animal is walking from right to left. *MTP*, Metatarsophalangeal; *PIP*, proximal interphalangeal joint

early part of the swing phase. As at the ankle, the timing and amplitude of the E1 extension are affected. At the PIP joint the extension at the transition from stance to swing is reduced, as is the flexion at the transition from swing to stance. Figure 3 shows a single step which illustrates the movement deficits during the swing phase. The paw was cleared from the ground during the last part of the swing phase (except for the first two stick figures on the right side, where it touched the treadmill belt). Note the pronounced MTP flexion, which leads to delayed elevation of the paw above the ground. It becomes evident that the duration of the swing phase is prolonged in the LS limb due to a slowing of the swing.

Angular Excursions of the Hindlimb Joints – Long-Term Effects

The deficits in the angular excursions of the distal joints improved during recovery. Figure 4 compares the angles 60 days after the hemisection. The

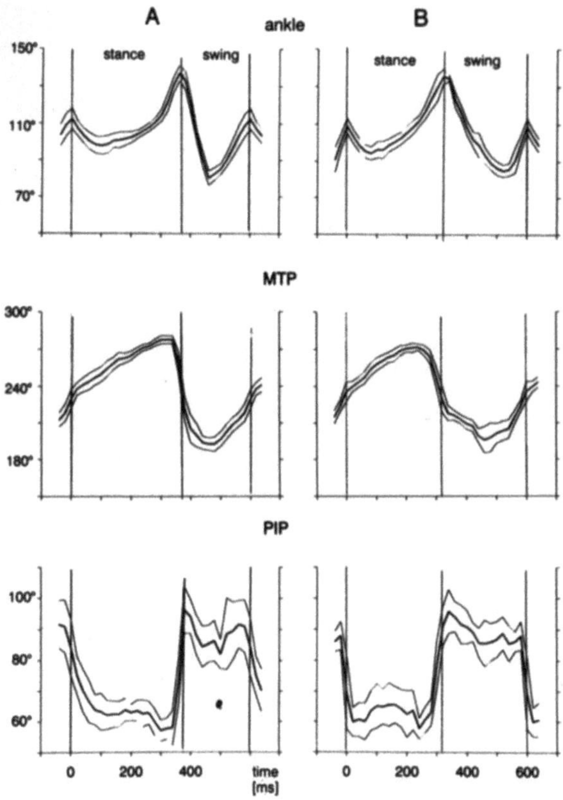

Fig. 4A,B. Angular excursions of the distal joints of the US (**A**) and LS (**B**) hindlimb (60th postoperative day). Ankle, MTP, and PIP joints. The data are averaged from 24 steps (*lines* indicate mean ± SD) at a treadmill velocity of 0.78 m/s

standard deviations in the US and LS hindlimbs are comparable, indicating a more stable and regular gait pattern than was seen immediately after the hemisection.

The LS hindlimb now reaches a degree of extension similar to that of US limb (compare Fig. 2), which indicates a general improvement of its extensor capacity. On the 60th day the angular excursions of the PIP joint are similar in the US and the LS hindlimb. Differences exist at the MTP and the ankle joints, where a slowing of the flexion leading the swing phase is present in the LS hindlimb. This is paralleled by a delay and an amplitude reduction of the E1 extension. Significant movement deficits are present at the knee and hip joints (Fig. 5). At the hip joint the monotonous extension-flexion curve is changed. Maximal extension and flexion are reduced in the LS limb by about 10°, and this leads to a decrease in the flexion-extension excursion from 75° (US limb) to around 55°. This reduction explains the decrease in the support length present in the LS limb after hemisection. The

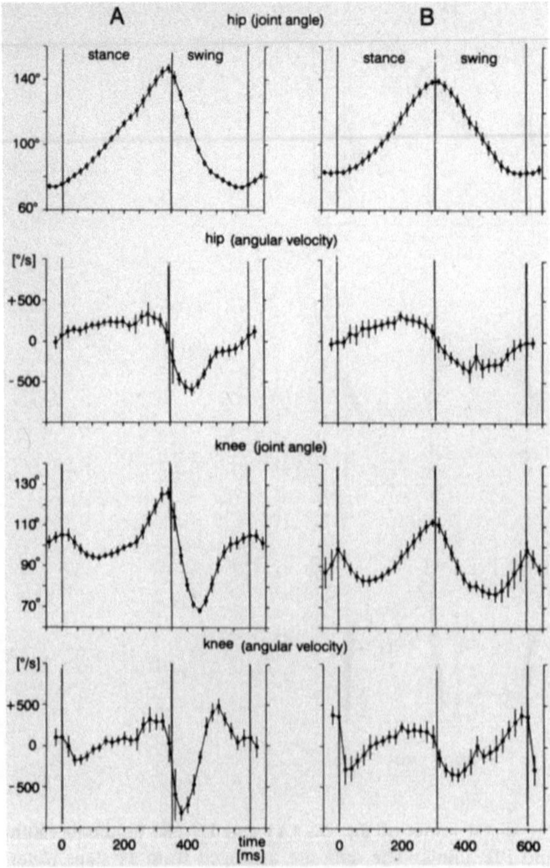

Fig. 5A,B. Hip- and knee-joint angular excursions and angular velocities of the US (**A**) and LS (**B**) hindlimb 60 days after the hemisection at a treadmill speed of 0.78 m/s. Data are averaged from 18 steps; standard deviations are indicated by *vertical bars*

angular velocities of the hip-joint movements are changed only during the swing phase. There is a slowing of hip flexion with a decrease from 600° to 200°/s. At the knee joint the deficits resemble the short-term effects present at the ankle joint (compare Fig. 2). The overall degree of knee extension is reduced in the LS limb, with a significant decrease of E1 and E3 extension. The flexion leading the swing phase is slowed from around 800° to 300°/s, and the velocity profile of the consecutive knee extension is altered. In the US limb the joint rapidly extends and decelerates after having reached a maximal velocity around 100 ms before the stance phase. In the LS limb the angular velocity increases slowly and reaches maximal values just before the stance phase. This results in a very abrupt placing of the paw and a marked, unbraked yield, both in amplitude and velocity, as can be seen in the respective velocity curves of Fig. 5.

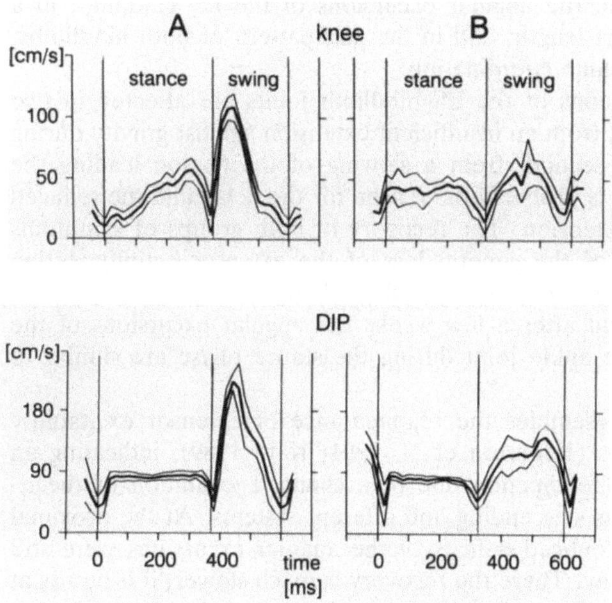

Fig. 6A,B. Velocity over ground of the knee and *DIP* (distal interphalangeal joint) joint centers, unlesioned (**A**) and lesioned (**B**) side, on 60th postoperative day; treadmill velocity 0.78 m/s. Averaged data from 20 steps; *lines* indicate mean ± SD

The velocity signals of the angular excursions of the LS hip and knee joints (Fig. 5) show a short but distinct reduction just before the middle of the swing phase, indicating a time lag during swing. This is evident on inspection of the walk of a hemisected animal on the treadmill and was substantiated by measuring the velocity over ground of the centers of the knee and of the distal interphalangeal (DIP) joint. In the US limb the velocity profile during the swing is bell-shaped and skewed to the early part of this phase (Fig. 6). In the LS limb this profile is changed. The velocity over ground is reduced in amplitude and characterized by a deceleration in the middle of the swing phase. The limb accelerates at the beginning of the swing phase, decelerates, and accelerates again in the second half of swing. This leads to a retardation of paw movement during the swing which is characteristic for the gait of the hemisected animal. This deficit is seen early after the hemisection and has a poor recovery.

Functional Considerations

Contrary to many reports in the literature, this study shows that long-term effects are present in locomotion after low thoracic spinal cord hemisections.

Deficits are observed in the angular excursions of the LS hindlimb, in a reduction of the support length, and in the gait pattern of both hindlimbs, including hindlimb/forelimb coordination.

The angular excursions of the LS hindlimb joints are affected in two ways. They suffer, first, from an insufficient extension against gravity during the stance phase and, second, from a slowing of the flexion leading the swing phase. The latter is probably the reason for the delay and the reduced amplitude of the E1 extension. The recovery of both groups of symptoms is different. Recovery of the general loss of the extensor activity during the stance phase follows a distal-proximal gradient. At the distal joints it recovers rather fast, and after a few weeks the angular excursions of the PIP, the MTP, and the ankle joint during the stance phase are similar to those of the US limb.

This time course resembles the reappearance of extensor excitability after spinal transections (Edgerton et al. 1991; Kato 1989), indicating an unspecific facilitation phenomenon due to a changed excitation of the α- and γ-motoneurons from descending and afferent systems. At the proximal joints (knee, hip) pronounced deficits of the angular excursions were still present after the 60th day. There the recovery is much slower, if it occurs at all. The group of symptoms during the swing phase, i.e., the slowed flexion at the transition from stance to swing and the delayed and reduced E1 extension, show a very poor recovery.

Early after hemisection a pronounced flexion of the MTP joint was present during the swing phase (Fig. 3). It could be due either to a contraction of the long palmar flexors or to a contraction deficit in the paw extensors. We cannot differentiate between these two possibilities since we did not perform EMG recordings. The visual inspection suggests an active palmar flexion, since paw flexion seemed stronger than one would expect from a passive behavior due to a loss in extensor strength. In this respect it resembled the palmar flexion (paw curling) observed by Miller and van der Meché (1975) in the forelimbs of decerebrate cats during swimming and air stepping. The strong paw flexion showed a fast recovery and thus behaves differently from the other locomotor deficits present during the swing phase, which all show a poor recovery.

The reduction in stride and support length is a consequence of the deficits in the angular excursions, especially at the hip and knee. It is therefore not surprising that in the LS limb the support length is still reduced after the 60th day, since at that time the amplitude of the flexion/extension of the LS hip is smaller than that in the US hindlimb. This reduction in support length is aggravated by the reduced flexion and decreased E1 extension of the knee, which both lead to a reduction of the forward movement of the paw during the swing phase.

The most striking deficit seen in our cats was the change in the gait pattern. There are no previous reports in the literature which indicate such a phenomenon. There is a lack of coordination between the two hindlimbs, with the LS hindlimb always being delayed in placing with respect to the US hindlimb. Furthermore, the gait patterns of the four limbs differ. Whereas the forelimbs move in the regular alternating pattern, the hindlimbs show a nonalternating, limping gait and hindlimb/forelimb coordination resembles a pacing gait. It is interesting that both hindlimbs are affected. Either the US limb adapts to the disturbed gait of the LS limb, or the lack of coordination between the US hindlimb and the forelimbs is due to the interruption of systems coordinating the cervical pattern generator and the generator of the US limb (e.g., Miller and Schomburg 1985). In any case, the deficits in generating a regular walking gait of the hindlimbs are persistent and severe. They indicate coordination problems within the spinal networks generating the locomotor pattern.

This is also suggested by the changes of the over-ground velocity profiles (Fig. 6) during the swing phase with the stop of the limb in mid-swing, and by the slowing of flexion in the swing phase and the delay of the E1 extension. These coordination deficits show a poor recovery. It seems that the hemisection destroyed some of the systems required to synchronize the different parts of the locomotor pattern and that, at least in the investigated time period, there is no effective restitution of this function.

Acknowledgements. The support of Prof. Dr. A.C. Nacimiento (Neurosurgical Research Laboratory of the University of Saarland), who performed the spinal cord lesions, is gratefully acknowledged. We thank Prof. Dr. H.H. Kramer, who generously allowed us to use the X-ray facilities of the Department of Pediatric Cardiology of the Christian Albrechts University at Kiel, and Privatdozent Dr. D. Onnasch and Mrs. S. Scholz, who supported us.

References

Carter RL, Ritz LA, Shank CP, Scott EW, Sypert GW (1991) Correlative electrophysiological and behavioral evaluation following L-5 lesions in the cat: a model of spasticity. Exp Neurol 114:206–215

Edgerton VR, Roy RR, Hodgson JA, Gregor, RJ, de Guzman, CP (1991) Recovery of full weight-supporting locomotion of the hindlimbs after complete thoracic spinalisation of adult and neonatal cats. In: Wernig A (ed) Plasticity of motoneuronal connections. Elsevier, Amsterdam, pp 405–418

Edgley SA, Jankowska E, Shefchyk S (1988) Evidence that mid-lumbar neurones in reflex pathways from group-II afferents are involved in locomotion in the cat. J Physiol (Lond) 403:57–71

Fujimori B, Kato M, Matsushima S, Mori S, Shimamura M (1966) Studies on the mechanism of spasticity following spinal hemisection in the cat. In: Granit R (ed) Nobel symposium on muscular afferents and motor control. Almqvist and Wiksell, Stockholm, pp 397–413

Gray J (1968) Animal locomotion. Weidenfeld and Nicholson, London, pp 265–282

Halbertsma JM (1983) The stride cycle of the cat. The modelling of locomotion by computerized analysis of automatic recordings. Acta Physiol Scand [Suppl]521:1–75

Hildebrand M (1966) Analysis of symmetrical gaits of tetrapods. Fol Bioth 6:9–22

Hultborn H, Malmsten J (1983) Changes in segmental reflexes following chronic spinal cord hemisection in the cat. I. Increased monosynaptic and polysynaptic ventral root discharges. Acta Physiol Scand 119:405–422

Kato M (1989) Chronically isolated lumbar half spinal cord, produced by hemisection and loingitudinal myelotomy, generates locomotor activities of the ipsilateral hindlimb of the cat. Neurosci Lett 98:149–153

McCouch GP, Austin GM, Liu CN, Liu CY (1957) Sprouting as a cause of spasticity. J Neurophysiol 21:205–216

Miller S, Schomburg ED (1985) Locomotor coordination in the cat. In: Bush BMH, Clarac F (eds) Co-ordination of motor behaviour, Society for Experimental Biology, Seminar 24. University Press, Cambridge, pp 201–220

Miller S, van der Meché FGA (1975) Movements of the forelimbs of the cat during stepping on a treadmill. Brain Res 91:255–269

Murray M, Goldberger ME (1974) Restitution of function and collateral sprouting in the cat spinal cord: the partially hemisected animal. J Comp Neurol 158:19–36

Nath D (1990) Räumliche Rekonstruktion von Ziel- und Greifbewegungen trainierter Katzen aus Röntgenaufnahmen. Thesis, Christian-Albrechts-Universität, Kiel

Teasdall RD, Magladery JW, Ramey EH (1958) Changes in reflex patterns following spinal cord hemisection in cats. Bull Johns Hopkins Hosp 103:223–235

Discussion Summary

E. Pierrot-Deseilligny

During the discussion of these papers, a major point was raised by Hans Hultborn and discussed by many participants: to what extent does presynaptic inhibition of Ia fibers and homosynaptic depression participate in vibratory inhibition, and what is the relative contribution of these two mechanisms in the exaggeration of the stretch reflex characterizing spasticity?

In man the administration of vibration to the Achilles tendon results in a strong discharge in Ia fibers (Burke et al. 1976) and produces a depression of the soleus H reflex (DeGail et al. 1966; Hagbarth and Eklund 1966; Delwaide, 1971, 1973). In most spastic patients, vibration fails to inhibit the H reflex to the same extent as in normal subjects (Delwaide 1971, 1973). Since the vibratory-induced depression is seen along with a motor discharge (the tonic vibration reflex), reflecting an increased excitability at the motoneuronal level, there is no doubt that this depression reflects a presynaptic mechanism which is malfunctioning in spastics. The question then arises of whether this mechanism is presynaptic inhibition of the Ecclestype and/or homosynaptic depression.

Presynaptic inhibition resulting from depolarization of primary afferent terminals was postulated (Delwaide 1971) since, in the cat, vibration of a flexor or extensor muscle causes an inhibition of the monosynaptic reflex in extensors, and this is accompanied by both dorsal root potential and primary afferent depolarization (Gillies et al. 1969; Barnes and Pompeiano 1970). A pharmacological argument favoring this view was the finding that diazepam, which is believed to facilitate transmission in the interneuronal pathways of presynaptic inhibition (Stratten and Barnes 1971), increases vibratory inhibition (Delwaide 1971). (Note, however, that little has been done about the pharmacology of homosynaptic depression.) However, it is well known that the size of the monosynaptic reflex in the cat decreases during repetitive stimulation (Eccles and Rall 1951; Lloyd and Wilson 1957), and that this frequency-related depression, or *homosynaptic depression*, has been related to "transmitter depletion" (Curtis and Eccles 1960). In man it has been shown that a subthreshold electrical stimulation, a subthreshold tendon tap, a previous voluntary contraction, and a slow passive stretch (Taborikova

Rééducation, Hôpital de la Salpètriére, 47, Bd. de l'Hôpital, F-75651 Paris Cedex 13, France

Thilmann et al. (Eds.)
Spasticity: Mechanisms and Management

and Sax 1969; Katz et al. 1977; Crone and Nielsen 1989) are able to evoke a deep depression of the reflex which lasts for several seconds. In the cat spinal cord, presynaptic inhibition resulting from depolarization of afferent terminals has been shown to last for no more than 300 ms (Schmidt 1971). The long-lasting depression is therefore more likely caused by a mechanism such as the homosynaptic depression described in the cat spinal cord (Curtis and Eccles 1960). Interestingly, this homosynaptic depression is significantly smaller in spastic patients (Nielsen, personal communication) and could contribute to stretch reflex exaggeration. It is therefore clear that homosynaptic depression may contribute to vibratory inhibition and to its reduction in spastics. Thus in experiments using long-lasting vibration of the Achilles tendon it is impossible to distinguish the contribution by presynaptic inhibition from that of homosynaptic depression.

Since most studies dealing with changes in presynaptic inhibition in spasticity are based upon changes in vibratory inhibition, the question arises of whether there is evidence for decreased presynaptic inhibition of Ia terminals in spasticity. Vibratory inhibition of the H reflex is not contaminated by a homosynaptic depression if the vibration is short (10 ms) and administered to a heteronymous tendon (tibialis anterior). The resulting depression of the H reflex (soleus and quadriceps) has been shown to last 200–300 ms (Hultborn et al. 1987), much as has been described for presynaptic inhibition in animal experiments (see Schmidt 1971). Presynaptic inhibition of Ia afferents so assessed has been shown to be decreased in patients with amyotrophic lateral sclerosis (Morin and Pierrot-Deseilligny 1988). A second method has been developed to study presynaptic inhibition in man: measurement of the amount of heteronymous monosynaptic Ia facilitation (from quadriceps to soleus) to provide an assessment of the amount of ongoing presynaptic inhibition exerted on the Ia fibers responsible for the facilitation (Hultborn et al. 1987). Current experiments using this method (Faist, personal communication) provide evidence for decreased presynaptic inhibition in spastics.

References

Barnes CD, Pompeiano O (1970) Presynaptic and postsynaptic effects in the monosynaptic pathway to extensor motoneurones following vibrations of synergic muscles. Arch Ital Biol 108:259–294

Burke D, Hagbarth KE, Löfstedt L, Wallin BG (1976) The response of human muscle spindle endings to vibration of non-contracting muscles. J Physiol (Lond) 261:673–693

Crone C, Nielsen J (1989) Methodological implications of the post activation depression of the soleus H-reflex in man. Exp Brain Res 78:28–32

Curtis DR, Eccles JC (1960) Synaptic action during and after repetitive stimulation. J Physiol 150:374–398

DeGail P, Lance JW, Neilson PD (1966) Differential effects of tonic and phasic reflex mechanisms produced by vibration of muscles in man. J Neurol Neurosurg Psychiatry 29:1–11

Delwaide PJ (1971) Etude expérimentale de l'hyperréflexie tendineuse en clinique neurologique. Arscia, Brussels

Delwaide PJ (1973) Human monosynaptic reflexes and presynaptic inhibition. An interpretation of spastic hyperreflexia. In: Desmedt JE (ed) New developments in electromyography and clinical neurophysiology, vol 3. Karger, Basel, pp 508–522

Eccles JC, Rall W (1951) Effects induced in a monosynaptic reflex path by its activation. J Neurophysiol 14:353–376

Gillies JD, Lance JW, Neilson PD, Tassinari CA (1969) Presynaptic inhibition of the monosynaptic reflex by vibration. J Physiol 205:329–339

Hagbarth KE, Eklund G (1966) Motor effects of vibratory muscle stimuli in man. In: Granit R (ed) Muscular afferents and motor control. Almqvist and Wiksell, Stockholm, pp 177–186

Hultborn H, Meunier S, Morin C, Pierrot-Deseilligny E (1987) Assessing changes in presynaptic inhibition of Ia fibres: a study in man and the cat. J Physiol (Lond) 389:729–756

Katz R, Morin C, Pierrot-Deseilligny E, Hibino R (1977) Conditioning of H-reflex by a preceding subthreshold tendon reflex stimulus. J Neurol Neurosurg Psychiatry 40:575–580

Lloyd DP, Wilson VJ (1957) Reflex depression in rythmically actived monosynaptic reflex pathways. J Gen Physiol 40:409–426

Morin C, Pierrot-Deseilligny E (1988) Mécanisme spinal de l'action antispastique de la TRH chez les malades atteints de sclérose latérale amyotrophique. Rev Neurol (Paris) 144:701–703

Schmidt RF (1971) Presynaptic inhibition in the vertebrate central nervous system. Ergeb Physiol 63:20–101

Stratten WP, Barnes CD (1971) Diazepam and presynaptic inhibition. Neuropharmacology 10:685–696

Taborikova H, Sax DS (1969) Conditioning H-reflex by a preceding subthreshold H-reflex stimulus. Brain 92:203–212

Discussion Summary

G.L. Gottlieb

The general discussion in Session V, lead by Dr. David Burke, focused on the H reflex as a tool for understanding spasticity. The question, posed in its starkest form, was whether H reflex abnormalities reveal important "pathophysiological mechanisms" or merely an "epiphenomenon."

The first point raised was that the notion that the H reflex reveals the workings of a monosynaptic pathway is, at best, an oversimplification. Electrical stimulation of Ia excitatory afferents also (and always) excites Ib inhibitory afferents as well. Jankowska reminded us that, in light of all the gaps in our knowledge about the mechanisms and pathways that create the spastic condition, we should not discard one of the few hard pieces of knowledge we have, namely, "if there is anything we can be sure about in neurophysiology, that is that there are direct connections between Ia afferents and motoneurons."

Nevertheless, Burke's point, with which most agreed, was that when a suprathreshold reflex is evoked, and we consider treatments or processes that produce changes in the amplitude of an H reflex, we are studying factors which affect the motoneurons on the fringe of firing or not firing, as opposed to the motoneurons first recruited at threshold. In these circumstances, it is not possible to consider only the state of the Ia-motoneuron synapse but also that of the Ib inhibitory interneuron. Increases in the size of an H reflex response may imply increased excitability of the Ia synapse, but this conclusion must be uncertain unless the state of Ib inhibition can be controlled or evaluated by other methods.

Ongerboer de Visser and others noted that a covariation of H reflexes with some of the symptoms of spasticity have been widely shown under a variety of conditions. However, although the correlations are often of high statistical significance, the variance accounted for does not exceed 20%. Both presenters such as Yanagisawa and discussants agreed that when studying processes or interventions that change reflex excitability in patients, it is essential to map out or at least know where on the sigmoidal, input-output relationship the measurements have been made. Even then, however,

Dept. of Neurological Sciences, Rush Medical Center, 2242 W. Harrison St., Chicago, IL 60612, USA

Thilmann et al. (Eds.)
Spasticity: Mechanisms and Management

Rymer emphasized that it is not possible to map changes in the amplitude of the H reflex with changes in hyperexcitability.

There was general agreement that H reflex techniques to map out neural circuitry, such as described by Pierrot-Deseilligny, are of indisputable value. Beyond that, however, use of the H reflex requires carefully controlled techniques and cautious interpretation. It does not currently provide a clear window into the mechanisms underlying spasticity.

Part VI
Treatment and Rehabilitation

Spasticity of the Lower Urinary Tract: Therapeutic Strategies

E. Möbius

Introduction

Continence during bladder filling is maintained by sympathetic activity in the hypogastric nerves (path 4, Fig. 1), which prevents involuntary detrusor contractions, and by tonic innervation of slow-twitch muscle fibers of the so-called rhabdosphincter – the intramural part of the striated urinary sphincter. Micturition is initiated by reflex activation of parasympathetic efferents which originate in the intermediolateral column of the sacral cord and are contained in the pelvic nerves. In addition, secondary reflexes mediated via urine flow in the urethra reinforce the micturition reflex and ensure complete voiding while the urinary sphincter is relaxed (de Groat 1990). An intact sacral micturition center consisting of parasympathetic and pudendal motoneurones was previously thought to provide for sufficient emptying of the bladder (Denny-Brown and Robertson 1933; Lapides 1970). Under this scheme, the result of a suprasacral transverse section of the spinal cord should be a pure reflex or automatic bladder in which involuntary detrusor contractions occur due to increasing bladder distention, and where sufficient relaxation of sphincter muscles is provided by recurrent inhibition of pudendal motoneurones. However, in the majority of patients with suprasacral spinal cord lesions voiding remains incomplete. Numerous clinical and experimental studies (Bradley et al. 1975; Blaivas 1982; Holstege et al. 1986; de Groat et al. 1990) have shown that micturition depends on a spinobulbospinal reflex, and that directly projecting descending pathways between the pontine and the sacral micturition center are essential for adequate urine storage and coordinated bladder emptying (path 2, Fig. 1). Their damage leads to spasticity of the lower urinary tract, characterized by overactivity of both the detrusor and the sphincter. This is correctly termed detrusor hyperreflexia–sphincter dyssynergia (DSD), according to the nomenclature of the International Continence Society (Bates et al. 1981).

Neurologische Klinik mit Klinischer Neurophysiologie des Alfried Krupp von Bohlen und Halbach Krankenhauses Essen, Alfried Krupp-Str. 21, D-45131 Essen, Deutschland

Thilmann et al. (Eds.)
Spasticity: Mechanisms and Management

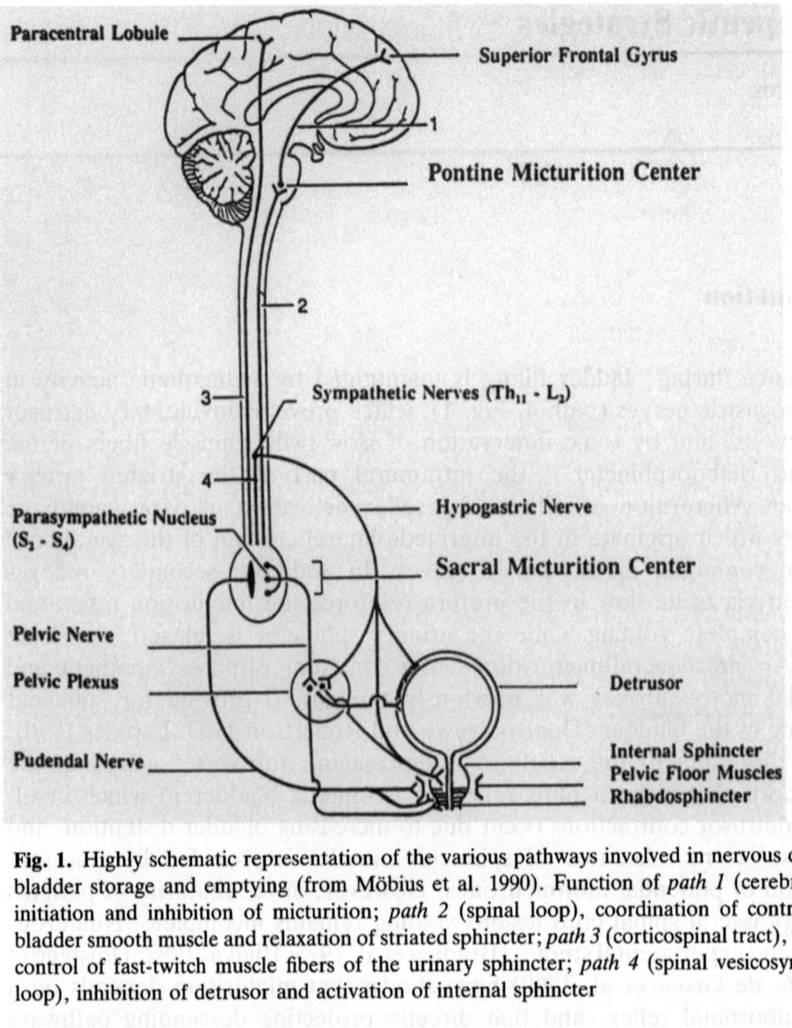

Fig. 1. Highly schematic representation of the various pathways involved in nervous control of bladder storage and emptying (from Möbius et al. 1990). Function of *path 1* (cerebral loop), initiation and inhibition of micturition; *path 2* (spinal loop), coordination of contraction of bladder smooth muscle and relaxation of striated sphincter; *path 3* (corticospinal tract), voluntary control of fast-twitch muscle fibers of the urinary sphincter; *path 4* (spinal vesicosympathetic loop), inhibition of detrusor and activation of internal sphincter

Clinical Aspects of DSD

An example of DSD is given in Fig. 2, which shows the computer-assisted combined registration of pressure, flow, and electromyogram (EMG) in a patient with a midthoracic myelitis. The recording of the detrusor pressure (the third trace) shows uninhibited detrusor contractions with a small filling volume. The lowest trace represents the integral of the rectified sphincter EMG and reveals simultaneous onset of pelvic floor muscle activity with the detrusor contractions. A weak urine flow occurs, but the bladder in this patient is incompletely emptied because of functional infravesical obstruction due to spasticity of the pelvic floor.

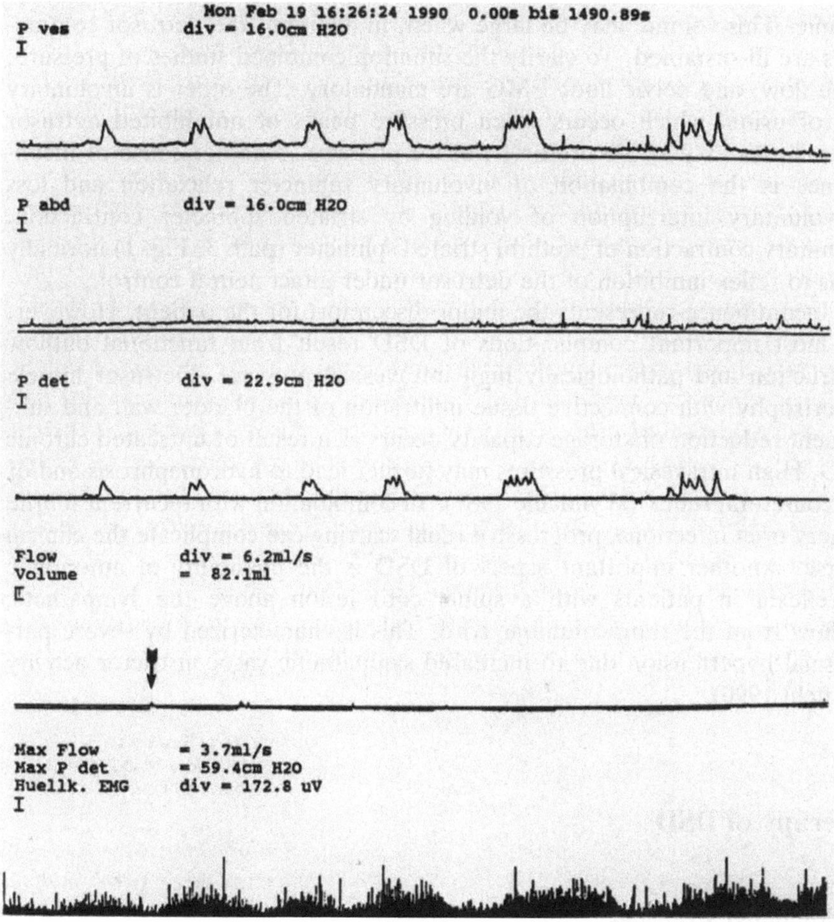

Fig. 2. Computer-assisted combined registration of pressure, flow, and EMG in a 32-year-old man with a midthoracic myelitis. *First trace*, intravesical pressure; (P_{ves}); *second trace*, abdominal pressure (P_{abd}); *third trace*, electronic subtraction of p_{abd} from p_{ves}, that is, the pressure generated intrinsically by the detrusor muscle (detrusor pressure, P_{det}; the recording of detrusor pressure reveals uninhibited detrusor contractions with a small filling volume); *fourth trace*, urine flow; *fifth trace*, the integral of the rectified sphincter EMG and reveals dyssynergic sphincter EMG. A week urine flow (*arrow*) occurs in the declining phase of involuntary detrusor contractions

The most common disorders which may cause DSD in cases of involvement of the suprasacral spinal cord include advanced multiple sclerosis, spinal cord injury, and rarer disorders such as transverse myelitis, spinal arteriovenous malformations (angioma, distal fistula, myelodysplasia, tumor) and heredoataxia.

There are two clinical characteristics of DSD. The first and most important is an impaired voiding due to sphincter spasticity, resulting in residual volume. This volume may be large when, in addition, the detrusor contractions are ill-sustained. To clarify the situation combined studies of pressure, urine flow, and pelvic floor EMG are mandatory. The other is involuntary loss of urine, which occurs when pressure peaks of uninhibited detrusor contractions exceed the urethral closure pressure. Another cause of incontinence is the combination of involuntary sphincter relaxation and loss of voluntary interruption of voiding by striated sphincter contraction. Voluntary contraction of urethral striated sphincter (path 3, Fig. 1) normally leads to reflex inhibition of the detrusor under intact neural control.

Incontinence represents the major discomfort for the patient. However, the most important complications of DSD result from functional outflow obstruction and pathologically high intravesical pressure. Detrusor muscle hypertrophy with connective tissue infiltration of the bladder wall and subsequent reduction of storage capacity occurs as a result of untreated chronic DSD. High intravesical pressures may further lead to hydronephrosis and/or vesicoureteral reflux (Wyndaele 1987). In combination with recurrent febrile urinary tract infections, progressive renal scarring can complicate the clinical course. Another important aspect of DSD is the possibility of autonomic dysreflexia in patients with a spinal cord lesion above the sympathetic outflow from the thoracolumbar cord. This is characterized by severe paroxysmal hypertension due to increased sympathetic vasoconstrictor activity (Naftchi 1990).

Therapy of DSD

Conventional Strategies

Bearing in mind the potentially dangerous complications of DSD, the therapeutic strategy in such cases should be to eliminate the functional obstruction created by the spastic sphincter. Unfortunately, there are no drugs available which effectively alter the increased urethral resistance occurring in DSD. Baclofen, for example, may provide some tonic inhibition of the micturition reflex by a GABAergic mechanism (Mallory et al. 1991), but clinically the net effect on sphincter hyperreflexia is irrelevant. Favorable results reported with intrathecal administration of baclofen (Kums and Delhaas 1991) are in contrast to our experience (Möbius et al. 1991), as demonstrated with the following case report. Figure 3 presents a combined pressure-flow-EMG registration obtained from a patient with a tentorial dural fistula with drainage into perimedullary veins and subsequent progressive cervical myelopathy. Because of severe spastic tetraplegia he underwent implantation of an Infusaid pump. An excellent effect on the spasticity of the lower extremities was observed under continuous intrathecal

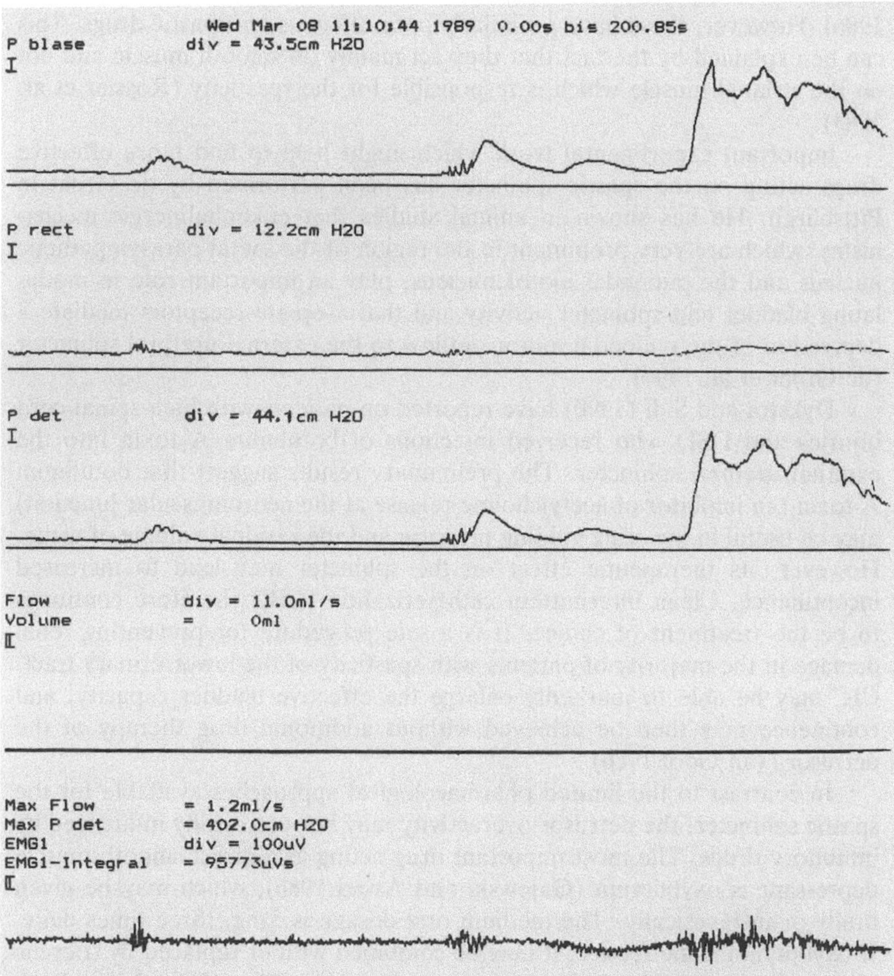

Fig. 3. Combined pressure-flow-EMG study in a 56-year-old man with a dural arteriovenous fistula localized in the tentorium with drainage into spinal and paraspinal veins. The representation of pressure and flow is the same as in Fig. 2. *Lowest trace*, simultaneous registration of sphincter and EMG, recorded with concentric needle electrodes. The study reveals a severe DSD. Detrusor pressure raises to 400 cm H_2O (*third trace*) at a maximum bladder capacity of about 70 ml. No urine flow occurs because of sphincter spasticity

administration of baclofen, and flexor spasms were completely resolved. The patient's severe DSD, however, was not influenced. Indeed, when a maximum bladder capacity of about 70 ml was reached, detrusor pressure increased to 400 cmH_2O, and typical clinical symptoms and signs of autonomic dysreflexia, such as headache, sweating, and hypertonia were observed.

α-Blocking agents, for example, phenoxybenzamine, where previously the most commonly used drugs for treating voiding dysfunction (Wein

1986). However, these have a similarly poor effect as antispastic drugs. This can be explained by the fact that they act mainly on smooth muscle and not on the striated muscle which is responsible for the spasticity (Rossier et al. 1983).

Important experimental work which might help to find more effective drugs acting on the spastic sphincter has been performed by de Groat in Pittsburgh. He has shown in animal studies that enkephalinergic mechanisms, which are very prominent in the region of the sacral parasympathetic nucleus and the pudendal motor nucleus, play an important role in modulating bladder and sphincter activity and that κ-opiate receptors mediate a depression of the pudendal motor outflow to the external urethral sphincter (de Groat et al. 1990).

Dykstra and Sidi (1990) have reported on patients with high spinal cord injuries and DSD who received injections of botulinum A toxin into the external urethral sphincter. The preliminary results suggest that botulinum A toxin (an inhibitor of acetylcholine release at the neuromuscular junction) may be useful in lowering voiding pressure and the residual volume of urine. However, its therapeutic effect on the sphincter may lead to increased incontinence. Clean intermittent catheterization (CIC) therefore continues to be the treatment of choice. It is a safe procedure for preventing renal damage in the majority of patients with spasticity of the lower urinary tract. CIC may be able to markedly enlarge the effective bladder capacity, and continence may then be achieved without additional drug therapy of the detrusor (van Gool 1986).

In contrast to the limited pharmacological approaches available for the spastic sphincter, the detrusor overactivity may be successfully influenced by inhibitory drugs. The most important drug acting as a direct smooth muscle depressant is oxybutynin (Gajewski and Awad 1986), which may be given orally or intravesically. The medium oral dosage is 5 mg, three times daily. If oxybutynin is ineffective, it may be combined with or replaced by tricyclic antidepressants such as imipramine or desipramine (25–150 mg daily). Both agents have a strong, direct, inhibitory effect on bladder smooth muscle. This action is not achieved by an anticholinergic or adrenergic mechanism but is probably due to a decrease in calcium movement (Akah 1986). When detrusor hypo- or inactivity has been induced pharmacologically, intermittent catheterization must be instituted. The spastic sphincter then provides for continence between the catheterizations.

If spasticity of the sphincter is not severe, such as in women suffering from multiple sclerosis, balanced voiding (residual urine less than 20% of bladder capacity) can be achieved by means of suprapubic tapping or perineal tactile stimulation because the skin and muscle afferents make contact with spinal circuits that elicit bladder contractions. The same is true in up to 30% of patients with transverse spinal lesions, when the disinhibition of somatovisceral pathways may convey detrusor contractions (Wyndaele

1987). Incontinence occurring between triggered micturitions can be managed by application of external collecting devices such as condom catheters or by garments and pads. However, if severe incontinence continues or secondary structural changes in the kidneys are established, it is mandatory to convert detrusor hyperreflexia into hypo- or areflexia by adequate pharmacotherapy and to perform CIC. If CIC is not acceptable to the patient or cannot be established, or if bladder capacity has been dramatically reduced due to long-standing untreated DSD, continuous catheterization, preferably via suprapublic drainage, is an alternative procedure. However, with CIC the risk of serious urinary tract infections is far lower than with indwelling catheters (Hill and Davies 1988; Timoney and Shaw 1990).

Functional Electrostimulation

Functional electrostimulation was introduced into the treatment of neurogenic voiding dysfunction by Brindley in the United Kingdom and Tanagho in the United States (Brindley et al. 1986; Tanagho et al. 1989). The procedure is as follows. To abolish detrusor overactivity and reduce sphincter spasticity, first dorsal rhizotomy of the sacral roots S_2–S_4 is performed. After simultaneous implantation of a ventral sacral root stimulator, either by an extra- or intradural approach, S_3 is stimulated bilaterally in bursts and voiding is achieved in the poststimulus phase when transitory relaxation of the sphincter occurs. Up until 1990, Brindley had implanted a total of 300 stimulators. In a long-term follow-up of the first 50 patients, who received their stimulators between 5 and 11 years earlier, 41 used their bladder pacemaker regularly for micturition, and 37 were always or usually continent (Brindley and Rushton 1990). The ideal patient for the time-consuming operation, requiring skilled surgeons and sophisticated techniques, is a woman with permanent damage of the spinal cord and a spastic bladder with incontinence (Madersbacher 1990). In male patients, loss of reflex erections must be accepted as a side effect of dorsal rhizotomy.

Summary and Conclusions

The uropharmacological approach to spasticity of the lower urinary tract is still limited to detrusor hyperreflexia. The treatment of choice in disturbed bladder emptying continues to be clean intermittent (self-)catheterization for the majority of patients. Functional electrostimulation seems to be a promising method for patients with a permanent neurological deficit and persistent severe urge or reflex incontinence in combination with a disturbed voiding. For the future the development of more effective drugs acting on both the spastic bladder and sphincter is desirable.

References

Akah PA (1986) Tricyclic antidepressant inhibition of the electrical evoked responses of the rat urinary bladder strip-effect of variation in extra cellular Ca^{2+} concentration. Arch Int Pharmacodyn Ther 284:231–238

Bates CP, Bradley WE, Glen ES, Melchior H, Rowan D, Sterling AM, Sundin T. In: Thomas D, Torres M, Turner-Warwick R, Zinner NR, Hald T (1981) Fourth report on the standardisation of terminology of lower urinary tract function. Terminology related to neuromuscular dysfunction of the lower urinary tract. Br J Urol 53:333–335

Blaivas J (1982) The neurophysiology of micturition: a clinical study of 550 patients. J Urol 127:958–963

Bradley WE, Timm GW, Scott FB (1975) Cystometry. Central nervous system organization of detrusor reflex. Urology 5:578–580

Brindley GS, Rushton DN (1990) Long-term follow-up of patients with sacral anterior root stimulator implants. Paraplegia 28:469–475

Brindley GS, Polkey CC, Rushton DN, Cardozo L (1986) Sacral anterior root stimulators for bladder control in paraplegia: the first 50 cases. J Neurol Neuro-Surg Psychiatry 49:1104–1114

De Groat WC (1990) Central neural control of the lower urinary tract. Ciba Found Symp 151:27–44

De Groat WC, Kawatani M, Hisamatsu T, Cheng CL, Ma CP, Thor K, Sleers W, Roppolo JR (1990) Mechanisms underlying the recovery of urinary bladder function following spinal cord injury. J Auton Nerv Syst 30[Suppl]:71–77

Denny-Brown D, Robertson EG (1933) On the physiology of micturition. Brain 56:149–189

Dykstra DD, Sidi AA (1990) Treatment of detrusor-sphincter dyssynergia with botulinum A toxin: a double-blind study. Arch Phys Med Rehabil 71:24–26

Gajewski JB, Awad SA (1986) Oxybutynin versus propantheline in patients with multiple sclerosis and detrusor hyperreflexia. J Urol 135:966–968

Hill VB, Davies WE (1988) A swing to intermittent clean self-catheterization as a preferred mode of management of the neuropathic bladder for the dextrous spinal cord patient. Paraplegia 26:405–412

Holstege G, Griffiths D, de Wall H, Dalm E (1986) Anatomical and physiological observations on supraspinal control of the bladder and urethral sphincter muscles in the cat. J Comp Neurol 250:449–461

Kums JJM, Delhaas EM (1991) Intrathecal baclofen infusion in patients with spasticity and neurogenic bladder disease. Preliminary results. World J Urol 9:99–104

Lapides J (1970) Neuromuscular vesical and ureteral dysfunction. In: Campbell MF, Harrison JH (eds) Urology, vol 2, 3rd edn. Saunders, Philadelphia, p 1343

Madersbacher H (1990) The various types of neurogenic bladder dysfunction: an update of current therapeutic concepts. Paraplegia 28:217–229

Mallory BS, Roppolo JR, de Groat WC (1991) Pharmacological modulation of the pontine micturition center. Brain Res 546:310–320

Möbius E, Berg-Dammer E, Thilmann AF (1990) Diagnostik und Therapie neurogener Blasenfunktionsstörungen. Nervenheilkunde 9:168–181

Möbius E, Berg-Dammer E, Nahser HC, Kühne D (1991) Autonomic dysreflexia in primary cerebral arterio-venous malformations with progressive myelopathy. 21st Annual Meeting of the International Continence Society, pp 140–141A

Naftchi NE (1990) Mechanism of autonomic dysreflexia. Contributions of catecholamine and peptide neurotransmitters. Ann NY Acad Sci 579:133–148

Rossier A, Fam B, Lee I, Sarkarati M, Evans D (1983) The respective contribution of smooth and striated components in the urethral pressure profile of the spinal cord injured. A neuropharmacological and urodynamic study. Paraplegia 21:114–116

Tanagho EA, Schmidt RA, Orvis BR (1989) Neural stimulation for control of voiding dysfunction: a preliminary report in 22 patients with serious neuropathic voiding disorders. J Urol 142:340–345

Timoney AG, Shaw PJ (1990) Urological outcome in female patients with spinal cord injury: the effectiveness of intermittent catheterisation. Paraplegia 28:556–563

Van Gool JD (1986) Spina bifida and neurogenic bladder dysfunction. A urodynamic study. Impress, Utrecht

Wein AJ (1986) Pharmacology of the bladder and urethra. In: Standon SL, Tanagho EA (eds) Surgery of female incontinence. Springer, Berlin Heidelberg New York, pp 229–250

Wyndaele JJ (1987) Urethral sphincter dyssynergia in spinal cord injury patients. Paraplegia 25:10–15

New Developments in the Medical Treatment of Spasticity

M. Emre

Introduction

Rationale and Aim of Medical Treatment of Spasticity

The medical treatment of established spasticity is principally of symptomatic nature, since spasticity is generally a consequence of irreversible central nervous system damage. This damage is not amenable to causative treatment. As spasticity presents with polymorphic clinical pictures resulting in various types and degrees of discomfort and dysfunction, the aim of the symptomatic treatment also varies. In general, it includes relief of subjective discomfort, facilitation of patient care, improvement of functional capacity, and protection of the long-term consequences of unevenly distributed excessive muscle tone. Theoretically, the main aim of an antispastic agent is to reduce increased muscle tone, given the definition of spasticity in strictly physiologic terms as a "velocity-dependent increase in tonic stretch reflexes (muscle tone)" (Lance 1980). However, spasticity is associated with other symptoms including brisk tendon reflexes, clonus, flexor, extensor, or adductor spasms which can be painful, loss of dexterity, and paresis, all of which constitute the upper motor neuron or spastic syndrome. Indeed, spasticity is sometimes used as a synonym for the upper motor neuron syndrome (Young and Delwaide 1981). The other positive signs of this syndrome, such as clonus and release phenomena (flexor reflex, spasms), are also amenable to medical treatment, whereas negative signs such as loss of dexterity and paresis are not. The latter, which are a major cause of disability, may show improvement under antispastic treatment to the extent that they represent a consequence of the positive phenomena. Given this general profile of action, antispastic drugs can be expected to be more efficacious in relieving subjective discomfort due to positive symptoms, avoiding secondary complications due to increased muscle tone such as contractures, and facilitating nursing care and physiotherapy, whereas the

Beth Israel hospital, Dept. of Neurology, Behavioral Neurology Unit, 330 Brookline Avenue, Boston, MA 02215, USA. *Present Address*: Clinical Research, CNS Dept., Sandoz Pharma Ltd., CH-4002 Basle, Switzerland

Thilmann et al. (Eds.)
Spasticity: Mechanisms and Management

other strategic aim of medical treatment, i.e., improvement of functional status, can be achieved only in a limited number of patients. In a subgroup of patients antispastic treatment may be counterproductive; if, for example, a hemiplegic patient is using the spastic leg as a crutch for ambulation, reduction of muscle tone may render the patient less functional.

Historical Perspective on the Development of Antispastic Drugs

After the introduction of mephanesin as an agent that blocks spinal interneurons (Berger and Bradley 1946), which represented the beginning of modern pharmalogical research for antispastic agents, a large number of substances from a broad range of pharmacological classes were tested in spasticity. With regard to mechanism of action, these substances can be broadly classified as those which mimic the effects of neurotransmitters utilized in descending, regulatory fiber systems (such as the noradrenergic system) or those which are used as neurotransmitters in local circuits (such as GABA) (Table 1). With regard to site of action, these substances can be divided into peripherally acting and centrally acting, the latter further into

Table 1. Major neurotransmitter systems in the spinal cord, their receptors, and drugs acting at these receptors

Transmitter	Receptor	Drug
I. Inhibitory		
GABA	GABA-A	Diazepam
	GABA-B	Baclofen
Glycine	Strychinine-sensitive spinal receptor	–
		–
II. Excitatory		
Glutamate/aspartate	NMDA	CPP, MK-801
	Non-NMDA	–
	– Kainate	
	– AMPA	
III. Modulatory		
Noradrenaline	alpha-2	Tizanidine
	alpha-1	Adrenergic blockers, phenothiazines
	beta-1	Beta blockers
	beta-2	
Serotonin	$5HT_{1A}$	No known antispastic drugs specifically acting at these receptors
	$5HT_{1B}$	
	$5HT_{1C}$	
	$5HT_{1D}$	
	$5HT_{1E}$	
	$5HT_2$	
	$5HT_3$	

those acting predominantly at the spinal cord or those acting predominantly at the supraspinal level.

There are four substances which successfully emerged from clinical trials and became established major antispastic drugs (Table 2). These include baclofen, dantrolene, diazepam, and the most recently introduced antispastic agent, tizanidine (Fig. 1). Baclofen (Lioresal®, *β*-[*p*-chlorophenyl]-GABA) is a GABA-B receptor agonist and was one of the first specific antispastic agents. Baclofen is thought to be a neuronal Ca^{++} antagonist and it decreases the release of excitatory neurotransmitters from primary afferent terminals. Baclofen is believed to exert its antispastic action directly at the spinal level by depressing monosynaptic and polysynaptic transmission in the spinal cord. In clinical trials it appeared to be effective in spinal and cerebral spasticity due to various pathological conditions, and some 75% of patients reported a decrease of varying magnitude in the muscle tone (Hattab 1980). Its usefulness in cerebral spasticity is, however, considered to be substantially less than that in spinal spasticity (Young and Delwaide 1981). The frequent adverse effects are tiredness, muscle weakness, hypotonia, and sedation, whereas nausea, vomiting, dizziness, euphoria, depression, and confusion are less frequent side effects. Dantrolene (Dantrium®) acts directly at muscle fibers. It decreases the amount of calcium released from the endoplastic reticulum, thereby reducing the effectiveness of the excitation-contraction coupling and thus reducing the contraction of skeletal muscles. Although it does depress the CNS, this does not seem to be the mechanism of its antispastic action. If effectively reduces increased muscle tone and spasms both in spinal and cerebral spasticity. Generalized weakness, diarrhea, occasional liver toxicity, which can be fatal, and to a lesser extent CNS side effects (drowsiness, dizziness) limit its use. Diazepam (Valium®), a GABA-A receptor agonist exerts a number of pharmacodynamic effects, one of which is muscle relaxation. The mechanism of antispastic action is to enhance the efficiency of GABA-ergic transmission by facilitating or potentiating the postsynaptic effects of GABA, which results in an increased chloride conductance and membrane hyperpolarization, thereby increasing the presynaptic inhibition of Ia afferents in the spinal cord. Somnolence, lightheadedness, fatigue, muscular weakness, dizziness, vertigo, ataxia, habituation, addiction, and physical dependence are among its drawbacks. Paradoxical reactions such as insomnia, anxiety, or hallucinations can occur with long-term use, especially in the elderly. Tizanidine, as the most recently introduced antispastic agent will be referred to in the next section.

In addition to the major antispastic drugs, there is a second group of agents including those substances which exert a certain antispastic action but are not primarily used for this purpose (Table 2). These include phenothiazines (such as chlorpromazine, probably by alpha-adrenergic blockade), adrenergic blocking agents (thymoxamine, propranolol), the alpha-2-adrenergic agonist clonidine, anesthetics, some anticholinergics (such as memantin), steroids, phenytoin, tetrahydrocannabinol, and morphine.

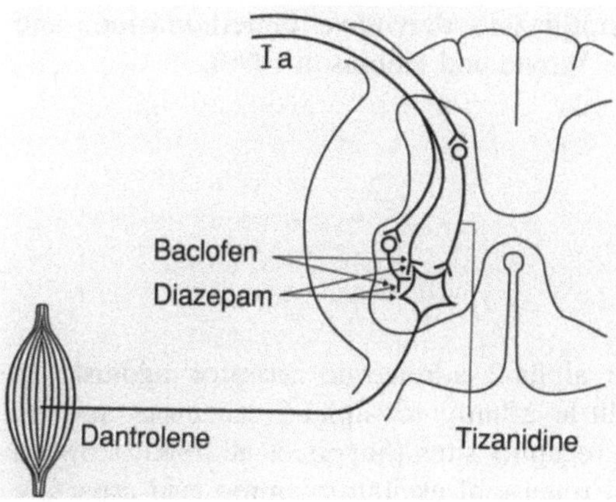

Fig. 1. Site of action of major antispastic drugs

Table 2. Drugs with antispastic action

Major antispastic drugs
Baclofen
Dantrolene
Diazepam
Tizanidine
Drugs with some antispastic action
Adrenergic blockers (thymoxamine)
Alpha-2 adrenergic agonists (clonidine)
Anticholinergics (memantin)
Anticonvulsives (phenytoin)
Corticosteroids
Opioids (morphine)
Phenotiazines (chlorpromazine)
Tetrahydrocannabinol
Experimental or discontinued drugs
Dimethothiazine (phenothiazine derivative)
Glycine
L-Threonine (glycine precursor)
Piracetam (GABA derivative)
Progabide (GABA agonist)

In fact, all drugs which have a centrally mediated analgesic effect probably also have an antispastic action.

In the third group are included substances which were tried and given up because of inefficacy, intolerance, or toxicity, or because their efficacy was not substantiated with a sufficient number of clinical trials. These include topical use of the anesthetic benzocaine, the amino acid glycine, the GABA derivative piracetam, the GABA agonist progabide, the glycine precursor L-threonine, the phenothiazine derivative dimethotiazine, and cyclobenzoprine (for a review see Whyte and Robinson 1990).

Recent Developments

Tizanidine

Tizanidine is a centrally active alpha-2 adrenergic receptor agonist. In in vitro studies it shows very little affinity to alpha-1 receptors and no appreciable affinity to the other receptor sites (Sayers et al. 1980; Coward 1981). It was found to inhibit the release of excitatory amino acid aspartate in spinal cord slices (Davies et al. 1983), to cause a post-synaptic reduction in the effectiveness of excitatory transmitters (Curtis et al. 1983) and to markedly decrease the firing rate of adrenergic locus ceruleus neurons in the rat, which was reversed by the alpha-2 antagonist yohimbine (Palmeri and Wiesendanger 1990). It is assumed that tizanidine exerts its antispastic action through one or both of these mechanisms, i.e., by reducing the excitatory transmission in local spinal circuits and by decreasing the tonic facilitatory activity of noradrenergic neurons of locus ceruleus on the spinal cord. Both of these effects are mediated via its alpha-2 agonistic action. Tizanidine also has an antinociceptive effect which is brought about by inhibition of the synaptic transmission of nociceptive stimuli in the spinal pathways through an alpha-2 adrenergic receptor-mediated action (Davies et al. 1984).

In in vivo experiments tizanidine shows a potent and selective myorelaxant action. It drastically reduces the tonic, polysynaptic component of the stretch reflex, whereas the phasic, monosynaptic component remains virtually unchanged (Coward 1981; Davies 1982; Fig. 2). In morphine-induced Straub-tail and rotarod performance tests tizanidine was found to have the highest separation index between the induction of depressant side effects and reduction of muscle tone (Novak 1982). This latter characteristic was thought to be the basis of relatively less undue muscle weakness with tizanidine seen in clinical trials (Emre 1989).

A number of clinical studies were conducted with tizanidine including patients with spasticity due to multiple sclerosis, stroke, cerebral palsy, and various other pathological conditions affecting the spinal cord or cerebral hemispheres. These studies are reviewed by Emre (1989). The overall results revealed that tizanidine was efficacious in treatment of spasticity of various origins at an average dose of 18 mg (range 12–36 mg). Increased muscle tone, spasms, and clonus were improved in 60%–80% of the patients, and in the overall efficacy analysis roughly 75% of patients reported having had some benefit from treatment, the magnitude of the effect varying from slight to very good (Table 3). In the few placebo-controlled studies, tizanidine exerted a statistically significant decrease in spastic muscle tone as measured

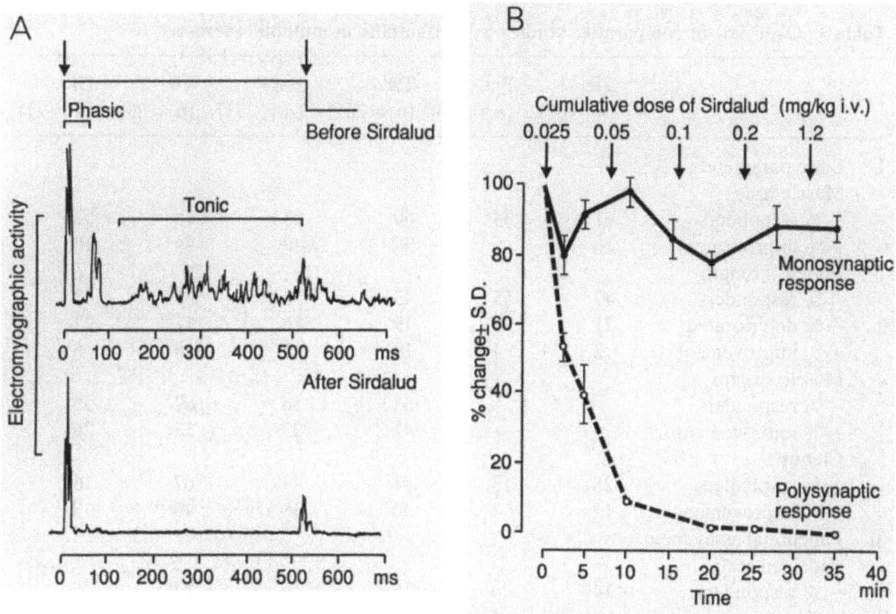

Fig. 2A,B. Experimental results showing the preferential action of tizanidine (Sirdalud) on spinal reflex activity: **A** Preferential inhibition of exaggerated tonic but not phasic component of the triceps stretch reflex in decerebrated cat after tizanidine 0.05 mg/kg i.v. (Coward 1981); **B** dose-dependent inhibition of polysynaptic but not monosynaptic reflex activity in the spinal cord of the anesthetized cat by i.v. tizanidine (Davies 1982)

with the Ashworth scale. The antispastic effect of tizanidine was by and large comparable to baclofen in most parameters, and somewhat better results were obtained with tizanidine for muscle strength. Tizanidine was found to be superior to diazepam in all parameters as well as in the overall assessment. The improvement in functional parameters was less pronounced with all compounds. The antispastic efficacy of tizanidine was maintained in long-term use and the there was no sign of habituation or chronic toxicity. Dantrolene was not included as a comparative drug in these studies.

Tizanidine was better tolerated than both baclofen and diazepam in almost all studies and discontinuations due to adverse events were fewer with tizanidine. The most frequent adverse events of tizanidine were tiredness, drowsiness, and dry mouth. It induced less undue muscle weakness than baclofen and fewer sedative effects than diazepam (Emre 1989).

In a recent double-blind, placebo-controlled, cross-over study the correlation between dose, plasma levels, and antispastic action of tizanidine was investigated in 16 patients with extensor spasticity of the legs due to multiple sclerosis (Emre et al., in preparation). The patients were given 2-, 4-, and 8-mg single doses of tizanidine and one dose of placebo on 4 separate days. An objective, quantitative method (electrogoniometer) was

Table 3. Overview of comparative studies with tizanidine in multiple sclerosis

	Tiz. (n = 65)	Pla. (n = 65)	Tiz. (n = 261)	Bac. (n = 273)	Tiz. (n = 29)	Dia. (n = 21)
I. Basic parameters						
Muscle tone						
– % responders	61	34	66	64	68	52
– % improvement	16	5	35	40	44	16
Muscle strength						
– % responders	47	55	43	33	56	33
– % deteriorated	21	29	19	26	11	22
– % improvement	2	4	19	5	18	6
Muscle spasms						
– % responders	–	–	63	56	82	56
– % improvement	–	–	41	33	37	26
Clonus						
– % responders	28	15	51	49	67	16
– % improvement	11	4	30	38	40	8
II. Functional assessment						
Pedersen scale						
– % responders	14	6	15	6	37	24
– % deteriorated	3	6	5	5	0	15
– % improvement	5	0	8	1	–	–
III. Overall assessments						
Overall efficacy (%)						
– marked (good, v.	20	5	40	37	42	34
good)	36	17	37	39	30	26
– moderate	44	78	23	24	28	40
– poor						
Patients' self-assessment	55	19	38	44	65	50
– % improved	0	10	2	0	0	0
– % deteriorated						
IV. Tolerability						
Overall tolerability (%)						
– good, excellent	61	88	71	65	66	38
– moderate	34	9	21	23	24	27
– poor	5	3	8	12	10	35
Discontinuation due to side effects (%)	5	2	6	11	11	35

Tiz, tizanidine; Pla., placebo; Bac., baclofen; Dia., diazepam.

used to assess muscle tone at knee extensors at spaced intervals on each study day, applying Wartenberg's pendulum test. Blood samples were obtained and clinical assessment of muscle tone using the Ashworth scale and muscle strength using the BMRC scale was done concomitantly. The confirmatory analysis to test the linearity of the dose-response relationship was performed using the change in the relaxation index (R2 value) 1.5 after each dose of tizanidine and placebo. This analysis revealed a statistically significant linear relationship between single doses and the antispastic action of tizanidine. Further analysis revealed a strong intraindividual linear correlation between plasma levels and antispastic action at 4- and 8-mg dose levels but a weak interindividual correlation. The analysis of dose-plasma concentration relationship revealed results compatible with linear pharmacokinetics. Thus there was a linear correlation between single doses, plasma levels, and antispastic action of tizanidine. Because there was a strong intraindividual but weak interindividual correlation between plasma levels and antispastic action, it was concluded that the effective doses of tizanidine should be determined individually. The unit doses and dose intervals should be chosen so as to maintain an effective plasma level.

Recently, a modified-release form of tizanidine was introduced which enables administration once or twice a day. This form was reported to be efficacious and well tolerated (Chantraine and Van Ouwenaller 1988).

Intrathecal Use of Baclofen

In the past few years, intrathecal use of baclofen has been extensively tested and found to be effective. The rationale for the intrathecal use of baclofen was to avoid the systemic side effects, which can be dose limiting, through direct administration of the drug in the vicinity of the spinal cord, the site at which baclofen exerts its antispastic action. In a randomized, double-blind, cross-over study in 20 patients with spasticity due to multiple sclerosis or spinal cord injury, the effect of intrathecal baclofen on abnormal muscle tone and spasms was investigated (Penn et al. 1989). The patients who had had no response to oral baclofen received an intrathecal infusion of baclofen or saline for 3 days by means of a programmable pump implanted in the lumbar subarachnoid space. Muscle tone, as assessed by the Ashworth scale, decreased in all 20 patients and spasms decreased in 18 of 19 patients thus affected. All patients were then put on long-term treatment with continuous infusion of intrathecal baclofen. During a mean follow-up period of 19 months the efficacy of baclofen on abnormal muscle tone and spasms was maintained. During the trial there were no serious adverse events, no infections were observed, and there were few technical failures. Similar experiences with intrathecal baclofen were reported by other groups (Lazorthes 1988; Müller et al. 1988b). These are summarized in a recent account which extensively covers the methods of intrathecal treatment of spasticity (Müller et al. 1988a). The efficacy of intrathecal baclofen in severe cerebral spasticity was questioned; this report represented only a single case, however (Saltuari et al. 1989).

Alternative techniques to programmable implanted pumps and site of administration were also described. Given intrathecally with a manually operated implantable pump in nine patients, baclofen proved to be effective, but significant complications occurred, including meningitis in four patients and baclofen overdosage (Patterson and Byrnes 1988). Baclofen was also found to be effective when given with an epidural catheter (Jones et al. 1988).

Intrathecal administration of morphine and midazolam was also reported to be efficacious in reducing spasticity; the systemic adverse events, however, were more frequent and more severe than with intrathecal baclofen (Müller and Zierski 1988).

Taken together, the current experience with intrathecal baclofen suggests that the method is efficacious in reducing spasticity in patients who are resistent to oral treatment. The complications are infrequent and include infection, baclofen overdose leading to sedation and reversible coma, and pump malfunction. The method is limited because of the highly technical requirements and because it is available in only a small number of centers.

Use of Botulinum Toxin

In the past decade botulinum toxin, which reduces neuromuscular transmission, has become an established treatment for various hyperkinetic-dystonic motor-disturbances. The toxin decreases the force of muscle contraction by blocking the neuromuscular transmission.

Botulinum toxin was used in a limited number of patients with multiple sclerosis who had adductor spasticity severe enough to create nursing problems. The injection of the toxin in the affected muscles resulted in a reduction of the spasticity score and also in improvement in nursing care (Snow et al. 1990). Similarly, improvement in spastic muscle tone, in the range of joint movements as well as in the overall functional status, was reported after injection of the toxin into the biceps and common flexor muscles of the forearm in patients with hemiplegia following stroke (Das et al. 1989). Long-term results are not available, however, and it remains to be demonstrated if this treatment results in sustained improvement.

Other Experimental Drugs

Oxcarbazepine, a carbamazepine derivative under investigation as an anticonvulsive drug, was accidentally found to have antispastic effects in two patients with cerebral lesions who were participating in an epilepsy trial (Bittencourt and Silvado 1985). Spastic muscle tone, pain on the hemiparetic side, movement range and gait improved. The effects were maintained for up to 20 months. Similar observations were made in one patient with spasticity following transverse myelitis and two patients with multiple sclerosis (Bittencourt 1988). Controlled studies with this substance are not available.

The clolinomimetic drug L-acetylcarnitine was found to significantly increase the amount of defective recurrent inhibition in patients with spinal spasticity (Mazzocchio et al. 1990). This effect was thought to be mediated through the cholinoceptive Renshaw cells. Spasticity was not assessed clinically in this study, and controlled clinical trials are lacking.

The essential amino acid L-threonine was administered to patients with spasticity in a few open studies and in a double-blind, cross-over study (Lee and Patterson 1990). The rationale for its use was the assumption that L-threonine would serve as a precursor for glycine, one of the two major inhibitory neurotransmitters in the spinal cord. In the controlled study, L-threonine was found to have a modest antispastic action, but there were no detectable changes in the functional status of the patients. Nevertheless, work with L-threonine suggests that manipulation of glycinergic transmission in the spinal cord may be a valuable approach to pursue for novel antispastic treatments.

Progabide, a GABA agonist binding to both GABA-A and GABA-B receptors and without any appreciable interaction with other known transmitters, was tested in a double-blind, placebo-controlled, cross-over study in 32 spastic patients with multiple sclerosis (Rudick et al. 1987). Progabide effectively lowered increased muscle tone without causing undue muscle weakness and was also preferred over placebo in the overall efficacy assessments. There was, however, no treatment effect on the spasm count or on any of the functional parameters. The use of progabide was associated with relatively frequent serious adverse events (increase in liver enzymes), which led to the discontinuation of clinical programme with this substance.

Future Outlook

Future research into the medical treatment of spasticity will involve a search for new drugs with novel mechanisms of action, improved selectivity, better tolerability or better efficacy or both, as well as a search for new technologies for selective administration. The potential candidates for the former approach which are likely to be the focus of research and to yield success include manipulation of excitatory amino acid and glycinergic receptors. These areas remain largely untapped and are thus the subject of current research in the spasticity field. Excitatory amino acids (EAA) are probably the major excitatory neurotransmitters in the central nervous system. Therefore, the inhibition of their activity, either by reduction of their release or by antagonism at their receptor site, is likely to reduce activity in spinal circuits involved in the mediation of normal and pathologically increased muscle tone. The target EAA receptors include NMDA receptors, activation of which results in a rather tonic pattern of activity, and non-NMDA receptors, which demonstrate a phasic activation. NMDA receptors can be blocked by noncompetitive and competitive antagonists or modulated by glycine-site antagonists, since the activation of glycine site at the NMDA receptor complex is necessary for the activation of the receptor-channel complex.

The non-NMDA receptors include kainate and AMPA receptor subtypes and are potentially subject to therapeutic manipulation (for a review see Lodge 1988).

Glycine is one of the two main inhibitory neurotransmitters in the spinal cord. Apart from the glycinergic modulatory site at the NMDA receptor complex, there is a distinct type of glycinergic receptor in the spinal cord, activation of which results in post-synaptic inhibition. Agonists acting at this strychinine-sensitive glycine receptor may have myorelaxant effects.

The role of serotonin, the neurotransmitter utilized in another major descending modulatory pathway, and the role of its diverse receptors in the mediation of spasticity and their potential as a target of therapeutic manipulation, also remain to be explored.

As for the treatment of other diseases of the central nervous system, drug delivery systems enabling more targeted administration of drugs at their desired site of action, thereby circumventing systemic effects, have been the subject of intensive research also in the field of spasticity. Programmable pumps for the intrathecal administration of antispastic drugs are a good example of the fruit of such research. The search for more practical, less invasive methods can be expected to continue in the future.

Conclusions

Research into the medical treatment of spasticity has yielded several practically and conceptually useful results. In terms of clinical practice, we now know that increased muscle tone and spasms can effectively be reduced by a number of compounds with different mechanisms of action. Well-controlled clinical trials have revealed that the decrease in spastic muscle tone is not always translated into functional benefits in all patients. There is no question, however, that selected patients draw functional benefit from antispastic treatment. Patients who have a demonstrable reduction in spastic muscle tone but no objectively measurable functional improvement nevertheless seem to have subjective benefit from antispastic treatment with regard to discomfort and disability due to positive symptoms, nursing care, physiotherapy, and possibly long-term prevention.

The conceptually important result of research into the medical treatment of spasticity is the confirmation of the theoretical assumption that a reduction of spastic muscle tone can be achieved through manipulation of various neurotransmitter systems, either primarily modulatory or primarily segmental in their function, by using their precursors, agonists, or antagonists.

The established oral antispastic drugs baclofen and tizanidine are the mainstay of medical treatment of spasticity. Diazepam and dantrolene have a more limited use because of their side effects. For patients who are resistent to oral treatment intrathecal baclofen represents an effective alternative. This method is limited by its high technical requirements and is confined to use in specialized centers. Botulinum toxin, if shown to be effective in long-term trials, may be another agent which can be used in specialized centers in selected patients resistant to oral treatment. Antagonists of excitatory amino acid receptors and modulation of spinal glycine receptors represent two yet unexplored fields of research in the quest for alternative treatments for spasticity.

References

Berger FM, Bradley W (1946) The pharmacological properties of a β-dihyroxy-γ (2-methylphenoxy)-propone (Myanesine). Br J Pharmacol 1:265–272

Bittencourt PRM (1988) Oxcarbazepine and spasticity. Arq Neuropsiquiatr 46:382–384

Bittencourt PRM, Silvado CES (1985) Oxcarbazepine, GP 47779, and spasticity. Lancet 21:676

Chantraine A, Van Ouwenaller C (1988) Modified-release tizanidine in the treatment of spasticity. J Int Med Res 16:459–465

Coward DM (1981) Selective muscle relaxant properties of tizanidine and an examination of its mode of action. Triangle 20:151–158

Curtis DR, Leah JD, Peet MJ (1983) Spinal interneurone depression by DS 103-282. Br J Pharmacol 79:9–11

Das TK, Park DM, Collings AD (1989) Botulinum toxin in the treatment of spasticity. J Neurol Neurosurg Psychiatry 52:1209

Davies J (1982) Selective depression of synaptic transmission of spinal neurons in the cat by a new centrally acting muscle relaxant, DS 103-282. Br J Pharmacol 76:473–481

Davies, J, Johnston SE, Lovering R (1983) Inhibition by DS 103-282 of D-(^{3}H) aspartate release from spinal cord slices. Br J Pharmacol 78:2P

Davies J, Johnston SE, Hill DR, Quinlan JE (1984) Tizanidine (DS 103-282), a centrally acting muscle relaxant, selectively depresses excitation of feline dorsal horn neurones to noxious peripheral stimuli by an action at α2-adrenoceptors. Neurosci Lett 48:197–202

Emre M (1989) Review of clinical trials with tizanidine (Sirdalud) in spasticity. In: Emre M, Benecke R (eds) Spasticity: the current status of research and treatment. Parthenon, Cornforth, pp 153–184

Hattab JR (1980) Review of European clinical trials with baclofen. In: Feldman RG, Young RR, Koella WP (eds) Spasticity: disordered motor control. Year Book Medical Publishers, Chicago, pp 71–85

Jones RF, Anthony M, Torda TA, Poulos C (1988) Epidural baclofen for intractable spasticity. Lancet 1:527

Lance JW (1980) Symposium synopsis. In: Feldman RG, Young RR, Koella WP (eds) Spasticity: disordered motor control. Year Book Medical Publishers, Chicago, pp 485–494

Lazorthes J (1988) Chronic intrathecal administration of baclofen in treatment of severe spasticity. In: Müller H, Zierski J, Penn RD (eds) Local-spinal therapy of spasticity. Springer, Berlin Heidelberg New York, pp 215–222

Lee KC, Patterson V (1990) A double-blind study of L-threonine in patients with spasticity. J Neurol Neurosurg Psychiatry 53:181

Lodge D (ed) (1988) Excitatory amino acids in health and disease. Wiley, Chichester

Mazzocchio R, Schieppati M, Scarpini C, Rossi A (1990) Enhancement of recurrent inhibition by intravenous administration of L-acetylcarnitine in spastic patients. J Neurol Neurosurg Psychiatry 53:321–326

Müller H, Zierski J (1988) Clinical experience with spinal morphine, midazolam and tizanidine in spasticity. In: Müller H, Zierski J, Penn RD (eds) Local-spinal therapy of spasticity. Springer, Berlin Heidelberg New York, pp 143–150

Müller H, Zierski J, Penn RD (eds) (1988a) Local-spinal therapy of spasticity. Springer, Berlin Heidelberg New York

Müller H, Zierski J, Dralle D, Hoffmann O, Michaelis G (1988b) Intrathecal baclofen in spasticity. In: Müller H, Zierski J, Penn RD (eds) Local-spinal therapy of spasticity. Springer, Berlin Hiedelbergy New York, pp 155–214

Novack GD (1982) Studies on the efficacy and depressant potential of muscle relaxants in mice. Drug Dev Res 2:383–386

Palmeri A, Wiesendanger M (1990) Concomitant depression of locus coeruleus neurons and of flexor reflexes by an α_2-adrenergic agonist in rats: a possible mechanism for an α_2-mediated muscle relaxation. Neuroscience 34:177–187

Patterson V, Byrnes D (1988) Management of severe spasticity with intrathecal baclofen. J Neurol Neurosurg Psychiatry 51:715

Penn R, Savoy SM, Corcos D, Latash M, Gottlieb G, Parke B, Kroin JS (1989) Intrathecal baclofen for severe spinal spasticity. N Engl J Med 320:1517–1521

Rudick RA, Breton D, Krall RL (1987) The GABA-agonist progabide for spasticity in multiple sclerosis. Arch Neurol 44:1033–1036

Saltuari L, Schmutzhard E, Kofler M, Baumgartner H, Aichner F, Gerstenbrand F (1989) Intrathecal baclofen for intractable spasticity due to severe brain injury. Lancet 2:503–504

Sayers AC, Bürki HR, Eichenberger E (1980) The pharmacology of 5-chloro-4-(2-imidazolin-2-yl-amino)-2, 1,3-benzothiadiazole (DS 103-282), a novel myotonolytic agent. Arzneimittel Forschung 30:793–803

Snow B, Tsui J, Bhatt M, Hashimoto S, Varelas M, Calne D (1990) Botulinum A toxin is effective for spastic muscle contractions. Mov Disord 5[Suppl 1]:76

Whyte J, Robinson KM (1990) Pharmacologic management. In: Glenn MB, Whyte J (eds) The practical managment of spasticity in children and adults. Lea and Febiger, Philadelphia, pp 201–226

Young RR, Delwaide PJ (1981) Drug therapy of spasticity. N Engl J Med 304:28–33, 96–99

Muscle Relaxant Action of Excitatory Amino Acid Antagonists

M. Schwarz[1,2] and F. Block[1]

Introduction

Spasticity has been defined as a motor disorder characterized by velocity-dependent increase in the resistance to passive stretch of muscles (Davidoff 1985). It develops after supraspinal or spinal lesions of descending motor fibers. Subsequent to these lesions the function of different neuronal structures involved in segmental reflex activity begins to change at the level of the spinal cord. In spasticity there is indirect experimental evidence for hyperexcitability of α-motoneurons and excitatory interneurons and of reduced activity of Ia and Ib interneurons and of axo-axonal synapses on primary afferent fibers. This results in exaggerated mono- and oligosynaptic dynamic stretch reflexes, augmented polysynaptic flexor reflexes, insufficient reciprocal and autogenic postsynaptic inhibition, and decreased presynaptic inhibition, respectively (Ashby and Wiens 1989; Delwaide and Oliver 1988; Dietz 1987; Iles and Roberts 1986; Meinck et al. 1985; Thilmann et al. 1991).

Pharmacotherapy of spasticity has been directed towards a potentiation of inhibitory transmission within the spinal cord mediated by gammaaminobutyric acid (GABA; Davidoff 1985). The mediation of presynaptic inhibition has been suggested to be the most important function of GABA at the spinal cord level (Davidoff and Hackmann 1984). It is believed that activation of receptors located on the terminals of primary afferent fibers from muscle and cutaneous receptors by GABA results in a decrease in the amount of excitatory transmitter released by inpulses in these primary afferents. This results in a reduction in the amplitude of the excitatory postsynaptic potential produced in interneurons and α-motoneurons by afferent signals from primary afferents.

Benzodiazepines are widely used in the treatment of spasticity and are thought to exert their muscle relaxant action by potentiation of GABA-

[1] Neurologische Klinik und Poliklinik der Universität Essen, Hufeland Str. 55, D-45147 Essen, Deutschland

[2] Neurologische Klinik der Rheinisch-Westfälischen Technischen Universität, Pauwels Str. 30, D-52074 Aachen, Deutschland

Thilmann et al. (Eds.)
Spasticity: Mechanisms and Management

mediated presynaptic inhibition (Schmidt et al. 1967). The potentiation of presynaptic inhibition has been suggested to reduce the release of excitatory transmitters from afferent fibers and thereby to reduce the gain in stretch and flexor reflexes. Another substance in wide clinical use as an antispastic drug is baclofen, a derivative of GABA which acts at the GABA-B receptor. It has become evident from electrophysiological and neurochemical studies on the spinal cord, on brain slices, and on synaptosomes that baclofen may exert its depressant effects on neuronal activity either by a postsynaptic hyperpolarizing action (Newberry and Nicoll 1984) or by a presynaptic action leading to a decreased release of excitatory amino acids (Stirling et al. 1989; Ault and Nadler 1982; Harrison 1990; Potashner 1979). The latter result has led to the hypothesis that the diminished reflex activity seen after administration of baclofen is due to a reduction in excitatory transmitter release by afferent fibers and, presumably, by interneurons in the spinal cord.

The excitatory amino acids (EAA) L-glutamate and L-aspertate have long been thought to be released as transmitters from terminals of primary afferent fibers and interneurons in the spinal cord (Watkins 1984). They exert their physiological effects via distinct receptor subtypes. These are characterized in terms of their preferred agonists, namely *N*-methyl-D-aspartate (NMDA), kainate (KA), and α-amino-3-hydroxy-5-methyl-4-isoxazoleproprionate (AMPA) receptors (Honore 1989). AMPA and KA receptors are often termed non-NMDA receptors. Blockade of transmission of EAA by treatment with antagonists active at NMDA and non-NMDA receptors may offer a new approach to the treatment of spasticity.

This chapter summarizes experimental evidence for the muscle relaxant action of NMDA and non-NMDA antagonists obtained from two animal models: (a) The tonic activity seen in the electromyogram (EMG) of spastic mutant rats, which may represent an useful model of pathologically increased muscle tone. Although the pathophysiology of spastic paresis in these rats is not known and may differ from that in humans, previous studies have demonstrated similar efficacy profiles of most commonly used antispastic drugs in this model and in humans (Schwarz et al. 1984, 1988a,b). (b) Mono- or oligosynaptic Hoffmann (H) reflexes and polysynaptic flexor reflexes in normal rats, as the suppression of exaggerated spinal reflexes is thought to represent a basic mechanism of action of centrally acting muscle relaxant drugs (Davidoff 1985).

Materials and Methods

Tonic EMG Activity in Genetically Spastic Rats

The experiments were performed on genetically spastic rats (Han-Wist, spa/spa: Zentralinstitut für Versuchstierzucht, Hanover, Germany) of either sex, aged 10–13 weeks and 100–150 g in weight, as has been described elsewhere (Schwarz et al. 1988b). For EMG recording, pairs of Teflon-insulated wire electrodes (Cooner Wire, Chabworth, CA, USA; AS 632 SS) were inserted percutaneously into the gastrocnemius muscle of unanesthetized rats. The animals were placed separately in well-ventilated Plexiglas boxes, and their hindlimbs, which were gently secured with adhesive tape, were allowed to hang through slots in the bottom of the boxes. The procedure for EMG recording did not produce overt discomfort for the animals. The electrical signals were amplified, band-pass filtered (5–10 kHz), full-wave rectified, and integrated. The EMG was recorded continuously, and the average activity was determined over 10-min segments. The spontaneous predrug control activity in the EMG of the gastrocnemius muscle was measured in two 10-min segments at an interval of 20 min. The changes in the activity of the EMG of the gastrocnemius muscle after systemic or intrathecal injection of solvent or drugs are expressed as a percentage of the mean value calculated from the preinjection (control) measurements.

For intrathecal injections, genetically spastic rats were chronically equipped with intrathecal polyethylene catheters (PE 10) under pentobarbitone anesthesia (Nembutal, Ceva, Neully-sur-Seine, France; 40 mg/kg). The atlanto-occipital membrane was exposed and cautiously slit on the midline. The catheter was then inserted and its tip advanced to the lumbar enlargement. The catheter was attached to the skull by means of stainless steel anchor screws and rapidly setting dental acrylic. After a recovery period of 2–3 days solvent or drugs were delivered in a volume of 5 μl vehicle to clear the catheter of the drug. After performance of experiments, correct location of the catheter tip was verified by an injection of 2% Evans blue dye through the catheter.

Spinal Reflexes in Normal Rats

For spinal reflex experiments, male Wistar rats (F. Winkelmann, Borchen, Germany) 200–250 g in weight, were used. H reflexes were recorded in pentobarbitone (50 mg/kg) anaesthetized rats, while flexor reflexes were recorded under α-chloralose (Merck, Darmstadt, Germany; 80 mg/kg) and urethane (Sigma, St. Louis, MO, USA; 400 mg/kg) anesthesia. For recording of H reflexes the left hindlimb was denervated by severing all nerves except the tibial nerve, which was mounted on bipolar platinum electrodes for stimulation (single square-wave shocks, 0.2 ms duration at 1.4–1.6 nerve threshold). EMG recordings were made with a pair of skin clip surface electrodes from the plantar foot muscles. Ten consecutive EMG responses were averaged by a laboratory computer (PDP 11/23; Digital Equipment Corporation, Maynard, MA, USA) both before (control) and 10 min after systemic or intrathecal administration of drug or solvent. The magnitude of EMG potentials was evaluated by measuring the peak-to-peak amplitude. To record the flexor reflexes the left hindpaw was electrically stimulated with a pair of fine subcutaneous pin electrodes (five square-wave shocks at 500 Hz, 0.2 ms duration at 2.0–3.0 reflex threshold). EMG recordings were made with a pair of wire electrodes inserted percutaneously into the ipsilateral tibialis muscle. Five consecutive EMG responses were rectified and averaged by the computer both before and 10 min after intrathecal administration of drugs or solvent. The magnitude of flexor reflexes was evaluated by measuring the area bounded by the averaged response and the baseline. The procedure for catheter implantation was identical to that in spastic mutant rats, except that implantation was performed immediately before the start of the experiment.

Statistical Evaluation

In all EMG and reflex experiments values measured after solvent or drug application are expressed as a percentage of the corresponding preinjection values. Statistical evaluation was performed using Student's *t* test.

Results

Tonic EMG Activity in Genetically Spastic Rats

Spinal Action. Spastic mutant rats exhibit tonic EMG activity in the gastrocnemius muscle which can be easily quantified. Administration of solvent into the lumbar subarachnoidal space did not affect EMG activity (Fig. 1). However, intrathecal injection of EAA antagonists acting on both NMDA (Fig. 1) and non-NMDA receptors potently reduces tonic EMG activity in genetically spastic rats (Schwarz et al. 1988a; Turski et al. 1985, 1986, 1987, 1988, 1992). In order of decreasing relative potency the specific competitive NMDA antagonists 3-((+)-2-carboxypiperazin-4-yl)-propyl-1-phosphonic acid (CPP) and 2-amino-7-phosphonoheptanoic acid (AP7; Fig. 1) and 2-amino-5-phosphonopentanoic acid (AP5) reduces muscle tone following intrathecal administration (Schwarz et al. 1988a; Turski et al. 1986, 1987, 1988). In contrast, the EAA agonists NMDA, KA, and quisqualate (QA) produce a potent increase in muscle tone when given intrathecally (Turski et al. 1986, 1988). The effect of NMDA on the muscle tone was antagonized by AP7. The effect of QA was prevented by the preferential KA/QA antagonist γ-D-glutamylaminomethyl sulfonate.

Supraspinal Action. Muscle relaxant activity of the competitive NMDA antagonist AP7 is also observed after intraventricular injection (Turski et al. 1985). Comparison of the efficacy of the muscle relaxant action of AP7 after intraventricular and intrathecal injection reveals that this compound is about

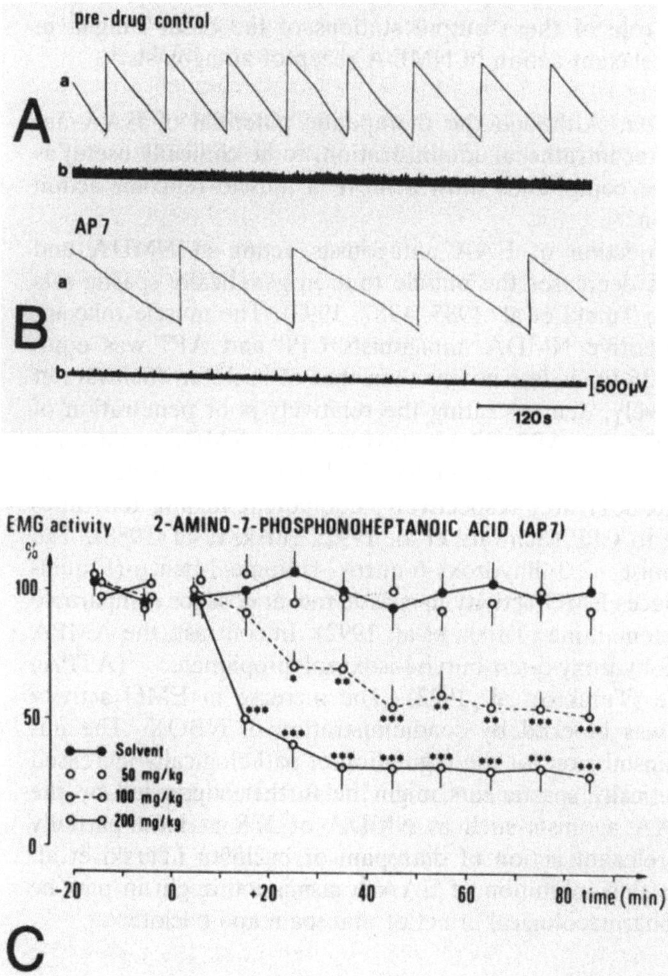

Fig. 1. Original EMG recordings 20 min before (**A**) and after (**B**) intrathecal injection of AP7. *Lower traces* (*b*), rectified EMG activity; *upper traces* (*a*), integrated EMG activity. **C** The time course of action and dose relationship of intrathecal injection of AP7 on spontaneous EMG activity of the gastrocnemius muscle in genetically spastic rats (Ha-Wist, spa/spa). *Abscissa*, time (min); *ordinate*, integrated EMG activity expressed as a percentage of the respective preinjection values. Results are presented as means + SEM. Each experimental group consisted of five to eight animals. *, $p < 0.05$; **, $p < 0.01$; ***, $p < 0.001$ versus solvent-treated rats; Student's *t* test. (From Schwarz et al. 1988a)

five times more potent after intrathecal administration. This might seem to point to the spinal cord as the main site of muscle relaxant action of NMDA receptor antagonists. However, the high efficacy of AP7 in decreasing muscle tone in spastic rats after injection into the reticular part of the substantia nigra or the entopenduncular nucleus (Turski et al. 1990b) indicates the pivotal role of these output stations of the basal ganglia in mediation of muscle relaxant action of NMDA receptor antagonists.

Systemic Administration. Although the therapeutic potential of EAA antagonists is evident after intrathecal administration, to be clinically useful as antispastic drugs these compounds must achieve a muscle relaxant action after systemic injection.

Intraperitoneal injection of EAA antagonists acting at NMDA and non-NMDA receptors decreases the muscle tone in genetically spastic rats (Schwarz et al. 1988a; Turski et al. 1985, 1987, 1992). The muscle relaxant activity of the competitive NMDA antagonists CPP and AP7 was equiponent to and about 25 times less potent than that of baclofen (Schwarz et al. 1988a,b), respectively, demonstrating the relatively poor penetration of the blood brain barrier by AP7. The noncompetitive NMDA antagonist memantine and (+)-5-methyl-10,11-dihidry-5H-dibenzo(a,d)cycloheptan-5, 10-iminemaleate (MK-801) are, respectively, equipotent to and ten times more potent relative to CPP (Schwarz et al. 1992; Turski et al. 1988). The non-NMDA anagonist 2,3-dihydroxy-6-nitro-7-sulfomoyl-benzo-(F)quinoxaline (NBQX) reduces EMG activity in spastic rats at dosages comparable to those of CPP or memantine (Turski et al. 1992). In contrast, the AMPA agonist α-amino-3-hydroxy-5-*tert*-butyl-4-isoxazolepropionate (ATPA) worsens muscle tone (Turski et al. 1992). The increase in EMG activity induced by ATPA was blocked by coadministration of NBQX. The importance of EAA transmission for the regulation of pathologically increased muscle tone in genetically spastic rats might be further supported by the observation that EAA agonists such as NMDA or KA at least partially reverse the muscle relexant action of diazepam or baclofen (Turski et al. 1988). This indicates that inhibition of EAA transmission might in part be responsible for the pharmacological effect of diazepam and baclofen.

Spinal Reflexes in Normal Rats

Spinal Action. Local application of NMDA or non-NMDA receptor antagonists in the subarachnoidal space affected spinal reflexes differently. Intrathecal injection of the competitive NMDA antagonist AP7 and of the noncompetitive NMDA antagonist memantine reduced the flexor reflex, whereas the mono- or oligosynaptic H reflex was unaffected by either drug (Figs. 2A, 3; Schwarz et al. 1992). Even the highest doses of AP7 and memantine, which reduced the flexor reflex to about 20%–40% of the preinjection control value, failed to effect the H reflex. In contrast, intrathecal injection of the non-NMDA antagonists 6,7,dinitro-quinoxaline-2,3-dione (DNQX) and 6-cyano-7-nitroquinoxaline-2,3-dione (CNQX) reduce the H reflex without affecting the M wave and do not change the magnitude of the flexor reflex (Figs. 2B, 3; Turski et al. 1990a) in normal

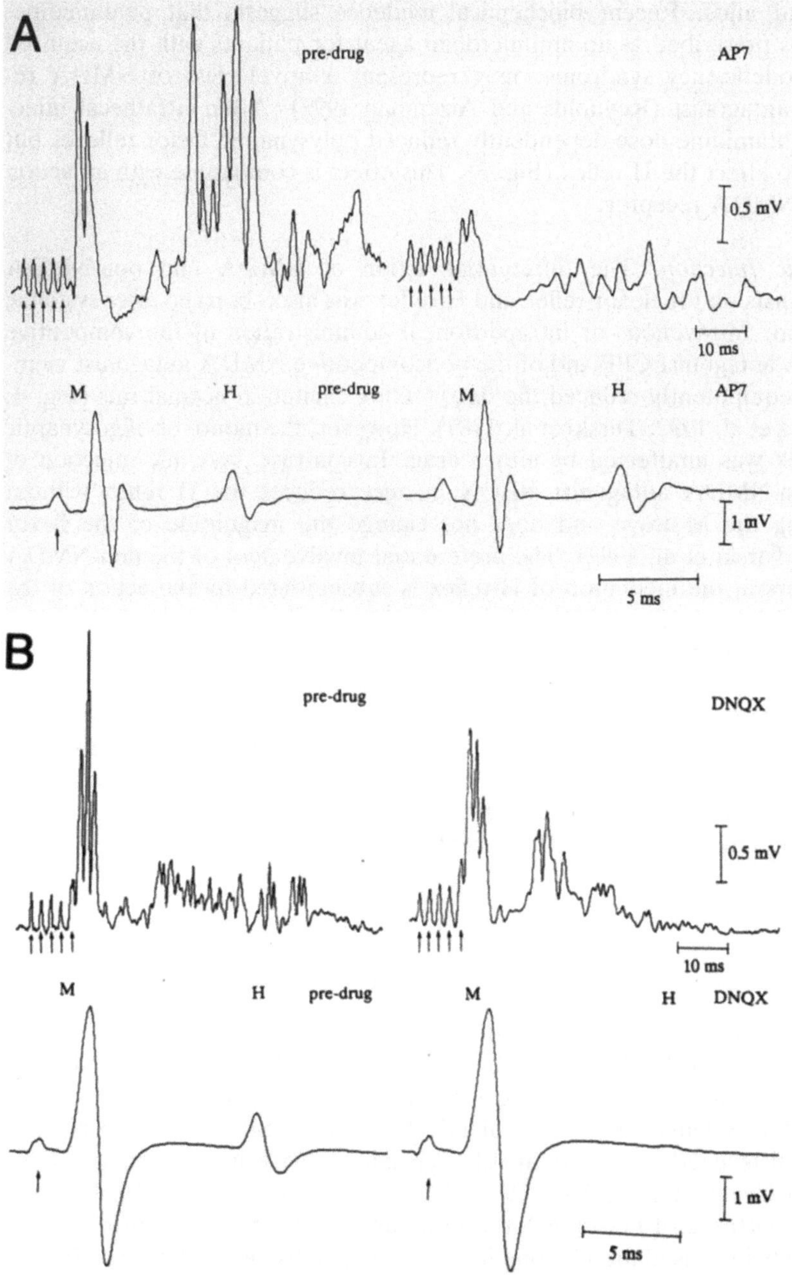

Fig. 2. Flexor reflex (*upper traces*) recorded from the tibial muscle evoked by five electrical shocks (2× reflex threshold) applied to the hindpaw and H reflex and the M wave (*lower traces*) recorded from the plantar foot muscles evoked by electrical stimulation (1.6 × nerve threshold) of the tibial nerve in normal rats before (*left*) and after (*right*) intrathecal administration of AP7 (**A**) and DNQX (**B**)

rats and mice. Recent biochemical evidence suggests that pentamidine, which is prescribed as an antimicrobial agent for patients with the acquired immunodeficiency syndrome, may represent a novel class of NMDA receptor antagonist (Reynolds and Aizenman 1992). After intrathecal injection pentamidine dose-dependently reduced polysynaptic flexor reflexes but failed to affect the H reflex (Fig. 3). This effect is compatible with an action at the NMDA receptor.

Systemic Injection. The differential action of NMDA and non-NMDA antagonists on the flexor reflex and H reflex was also observed after systemic injection. Intravenous or intraperitoneal administration of the competitive NMDA antagonist CPP and of the noncompetitive NMDA antagonist memantine equipotently reduced the flexor reflex elicited in normal rats (Fig. 4; Schwarz et al. 1992; Turski et al. 1987). However, the mono- or oligosynaptic H reflex was unaffected by either drug. In contrast, systemic injection of the non-NMDA antagonist NBQX in mice reduces the H reflex without affecting the M wave and does not change the magnitude of the flexor reflex (Turski et al. 1992). The preferential involvement of the non-NMDA receptors in the mediation of H-reflex is substantiated by the action of the non-NMDA agonist ATPA. This compound enhances the amplitude of H reflex and left flexor reflexes unaltered (Turski et al. 1992). Recent electrophysiological experiments suggest that the new 2,3-benzodiazepine GYKI 52466 might exert muscle relaxant activity via an antagonistic action at AMPA receptors (Tarnawa et al. 1989). Systemic injection of this compound reduced both monosynaptic H reflexes and polysynaptic flexor reflexes in a similar way (Fig. 4), which could argue against a specific action of this compound on non-NMDA receptors.

Discussion

These animal experiments provide evidence that EAA antagonists acting at NMDA or non-NMDA receptors may represent a new class of central muscle relaxant drugs exhibiting a distinct profile of action.

After systemic or intrathecal injection both NMDA and non-NMDA antagonists exert a potent muscle relaxant activity in genetically spastic rats (Schwarz et al. 1988a, 1992; Turski et al. 1985, 1986, 1988, 1992). The importance of both EAA receptor subtypes for the regulation of pathologically increased muscle tone in these rats is substantiated by the effect of selective receptor agonists. NMDA and the non-NMDA agonists ATPA, QA, and KA exacerbate the EMG measure of spasticity (Turski et al. 1986, 1988, 1992).

Whereas NMDA and non-NMDA antagonists depress tonic EMG activity in genetically spastic rats in a similar way, both classes of compounds

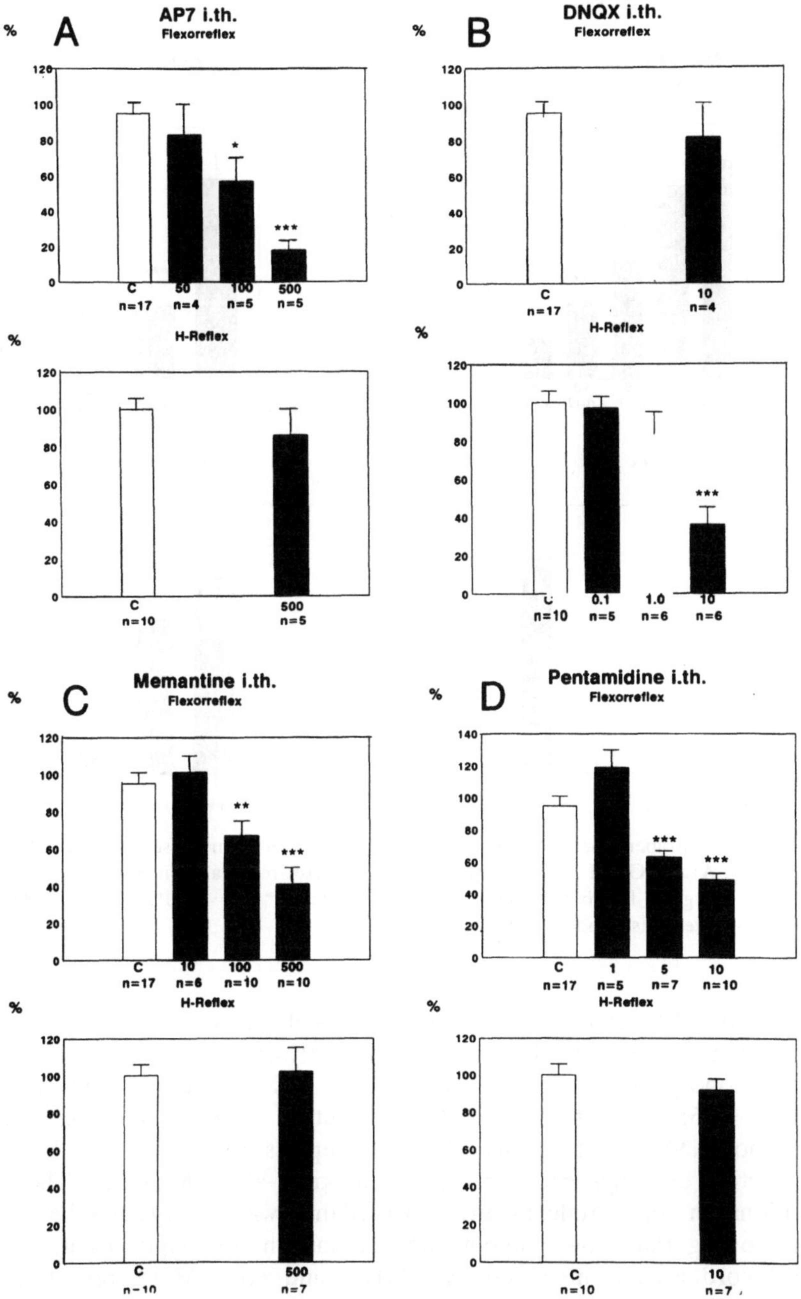

Fig. 3. Effect of intrathecal (*i. th.*) injection of solvent (*C*) or different doses of AP7 (50–500 nmol; **A**), DNQX (0.1–10 nmol; **B**), memantine (10–500 nmol; **C**), and pentamidine (1–10 nmol; **D**) on the (polysynaptic) flexor reflex and on the (monosynaptic) H reflex. Values of the flexor reflex and the H reflex are given as percentage of the respective preinjection values (mean + SEM for 4–17 rats). *, $p < 0.05$; **, $p < 0.01$; ***, $p < 0.001$ versus solvent-treated rats; Student's t test

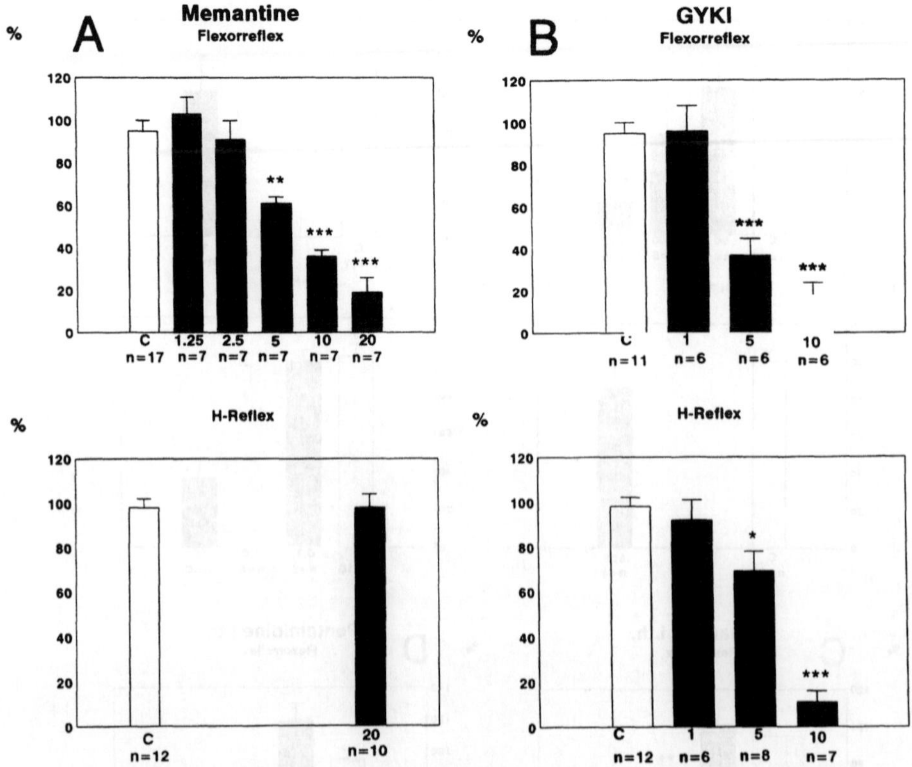

Fig. 4. Effect of intraperitoneal injection of solvent (*C*) or different doses of memantine (1.25–20 mg/kg; **A**) and GYKI 52466 (1–10 mg/kg; **B**) on flexor reflex and on H reflex. Values of the reflexes are given for six to seventeen rats. *, $p < 0.05$; **, $p < 0.01$; ***, $p < 0.001$ versus solvent-treated rats; Student's *t* test

exhibit a quite distinct profile of action on spinal reflexes in normal rats. NMDA antagonists selectively depress spinal polysynaptic reflexes without effect on monosynaptic reflexes in mice, rats, and cats (Lodge and Anis 1984; Polc 1985; Schwarz et al. 1992; Turski et al. 1987, 1990a, 1992). In contrast, non-NMDA antagonists selectively depress monosynaptic reflexes without action on polysynaptic reflexes (Turski et al. 1990, 1992). These observations on spinal reflexes are in accordance with a large number of studies showing that slow polysynaptic and fast monosynaptic excitation of spinal cord cells is mediated by NMDA and non-NMDA receptors, respectively (Davis and Watkins 1983; Fletcher et al. 1988; Watkins 1984). However, not all studies are consistent with this view. In the frog spinal cord in vitro, NMDA antagonists preferentially block the monosynaptic excitation of alpha motoneurons (Corradetti et al. 1985), whereas in the cat cord in vivo, non-NMDA antagonists fail to effect the monosynaptic response of spinal interneurons (Peet et al. 1983).

The relative importance of depressant effects on mono- or polysynaptic reflexes for optimal pharmacotherapy of spasticity is unclear (Davidoff 1985). Some arguments favor a preferential action of baclofen on monosynaptic reflexes whereas benzodiazepines are thought preferentially to effect polysynaptic reflexes (Davidoff 1985). Although the exact mechanisms of muscle relaxant action of baclofen and benzodiazepines remain unknown, some indirect evidence points to an inhibition of release of EAA, which might act on non-NMDA and NMDA receptors, respectively. This suggestion could be supported by the observation that KA at least partially blocks the muscle relaxant activity of baclofen in spastic mutant rats, whereas NMDA partially attentuates the muscle relaxant action of diazepam in this model (Turski et al. 1988). A distinct profile of action of NMDA and non-NMDA antagonist on spinal reflex activity may offer a differential approach to pharmacological treatment of spasticity. At present only the non competitive NMDA antagonist memantine is available for therapeutic use in man, and the antispastic effect of this compound has been reported in preliminary clinical studies (Mundinger and Milos 1985).

In addition to the treatment of spasticity, memantine is used clinically to treat rigidity in patients with Parkinson's disease. Since memantine interacts with noncompetetive binding sites at the NMDA receptor complex at therapeutic concentrations in Parkinson patients (Kornhuber et al. 1989), this effect is presumably achieved by blockade of pathologically enhanced EAA transmission in various nuclei of the basal ganglia, such as the subthalamic nucleus, the reticular part of substantia nigra and the internal pallidum. Both NMDA and non-NMDA receptor antagonists have been shown to reduce rigidity in animal models of parkinsonian rigidity via an action in these basal ganglia nuclei (Klockgether and Turski 1990; Klockgether et al. 1991).

References

Ashby P, Wiens M (1989) Reciprocal inhibition following lesions of the spinal cord in man. J Physiol (Lond) 414:145–157

Ault B, Nadler JV (1982) Baclofen selectively inhibits transmission at synapses made by axons of CA1 pyramidal cells in the hippocampal slice. J Pharmacol Exp Ther 223:291–297

Corradetti R, King AE, Nistri A, Rovera C, Sivilotti L (1985) Pharmacological characterization of DAPV antagonism of amino acid and synaptically evoked excitations on frog motoneurones in vitro: an intracellular study. Br J Pharmacol 86:19–25

Davidoff RA (1985) Antispasticity drugs: mechanisms of action. Ann Neurol 17:107–116

Davidoff RA, Hackmann JC (1984) Spinal inhibition. In: Davidoff RA (ed) Anatomy and physiology. Dekker, New York, pp 385–459 (Handbook of the spinal cord, vols 2, 3)

Davis J, Watkins JC (1983) Role of excitatory amino acid receptors in mono- and polysynaptic excitation in the cat spinal cord. Exp Brain Rev 4:280–290

Davis J, Watkins JC (1985) Depressant actions of alpha-D-glutamylaminomethyl sulfonate (GAMS) on amino acid induced and synaptic excitation in the cat spinal cord. Brain Res 327:113–120

Delwaide PJ, Oliver E (1988) Short-latency autogenic inhibition (Ib inhibition) in human spasticity. J Neurol Neurosurg Psychiatry 51:1546–1550
Dietz V (1987) Role of peripheral afferents and spinal reflexes in normal and impaired human locomotion. Rev Neurol (Paris) 143:241–254
Fletcher EJ, Martin D, Aram JA, Lodge D, Honore T (1988) Quinoxalinediones selectively block quisqualate and kainate receptors and synaptic events in rat neocortex and hippocampus and frog spinal cord in vitro. Br J Pharmacol 95:385–397
Harrison NL (1990) On the presynaptic action of baclofen at inhibitory synapses between cultured rat hippocampal neurones. J Physiol (Lond) 422:433–446
Honore T (1989) Excitarory amino acid receptor subtypes and specific antagonists. Med Res Rev 9:1–23
Iles JF, Roberts RC (1986) Presynaptic inhibition of monosynaptic reflexes in the lower limbs of subjects with upper motoneuron disease. J Neurol Neurosurg Psychiatry 49:937–944
Klockgether T, Turski L (1990) NMDA antagonists potentiate antiparkinsonian actions of L-dopa in monoamine-depleted rats. Ann Neurol 28:539–546
Klockgether T, Turski L, Honore T, Zhang Z, Gash DM, Kurlan R, Greenamyre JT (1991) The AMPA receptor antagonist NBQX has antiparkinsonian effects in monoamine-depleted rats and MPTP-treated monkeys. Ann Neurol 30:717–723
Kornhuber J, Bormann J, Retz W, Hubers M, Riederer P (1989) Memantine displaces (3H)MK-801 at therapeutic concentrations in postmortem human frontal cortex. Eur J Pharmacol 166:589–590
Lodge D, Anis NA (1984) Effects of ketamine and three other short acting anaesthetics on spinal reflexes and inhibition in the cat. Br J Anaesthesiol 56:1143–1151
Meinck HM, Benecke R, Conrad B (1985) Spasticity and the flexor reflex. In: Delwaide PJ, Young RR (eds) Clinical neurophysiology in spasticity. Elsevier, Amsterdam, pp 41–54
Mundinger F, Milos E (1985) Treatment of severe spastic and extrapyramidal syndromes with memantine in combination or following stereotactic surgery. Nervenarzt 56:106–109
Newberry NR, Nicoll RA (1984) Direct hyperpolarization action of baclofen on hippocampal pyramidal cells. Nature 308:450–452
Peet MJ, Leah JD, Curtis DR (1983) Antagonists of synaptic and amino acid excitation of neurones in the cat spinal cord. Brain Res 226:83–95
Polc P (1985) 2-Amino-7-phosphonoheptanoic acid depresses alpha-motoneurons and polysynaptic reflexes in the cat spinal cord. Eur J Pharmacol 117:385–389
Potashner SJ (1979) Baclofen: effects on amino acid release and metabolism in slices of guina-pig cerebral cortex. J Neurochem 32:103–109
Reynolds IJ, Aizenman E (1992) Pentamidine is a N-methyl-D-aspartate receptor antagonist and is neuroprotective in vitro. J Neurosci 12:970–975
Schmidt RF, Vogel ME, Zimmermann M (1967) Die Wirkung von Diazepam auf die präsynaptische Hemmung und andere Rückenmarksreflexe. Naunyn Schmiedebergs Arch Pharmacol 258:69–82
Schwarz M, Turski L, Sontag K-H (1984) CGS 8216, Ro 15-1788 and methyl-alpha carboline-3-carboxylate, but not EMD 41717 antagonize the muscle relaxant effects of diacepam in genetically spastic rats. Life Sci 35:1415–1421
Schwarz M, Klockgether T, Turski L, Sontag K-H (1988a) Intrathecal injection of anstispastic drugs in rats: muscle relaxant action of midazolam, baclofen, 2-amino-phosphonoheptanoid acid (AP 7) and tizanidine. In: Müller H, Zierski J, Penn RD (eds) Local spinal therapy of spasticity. Springer, Berlin Heidelberg New York, pp 65–79
Schwarz M, Klockgether T, Wüllner U, Turski L, Sontag K-H (1988b) Delta-aminovaleric acid antagonizes the pharmacological actions of baclofen in the central nervous system. Exp Brain Res 70:618–626
Schwarz M, Block F, Sontag K-H (1992) N-Methyl-D-aspartate (NMDA)-mediated muscle relaxant action of memantine in rats. Neurosci Lett 143:105–109

Stirling JM, Cross AJ, Robinson TN, Green AR (1989) The effects of GABAB receptor agonists and antagonists on potassium-stimulated (Ca^{++}) in rat brain synaptosomes. Neuropharmacology 28:699–704

Tarnawa I, Farkas S, Berzsenyi P, Pataki A, Andrasi F (1989) Electrophysiological studies with a 2,3-benzodiazepine muscle relaxant: GYKI 52466. Eur J Pharmacol 167:193–199

Thilmann AF, Fellows SJ, Garms E (1991) The mechanism of spastic muscle hypertonus: variation in reflex gain over the time course of spasticity. Brain 114:233–244

Turski L, Schwarz M, Turski WA, Klockgether T, Sontag KH, Collins JF (1985) Muscle relaxant action of excitatory amino acid antagonists. Neurosci Lett 53:321–326

Turski L, Klockgether L, Schwarz M, Turski WA, Sontag K-H (1986) Spinal cord: a site of muscle relaxant action of excitatory amiono acid antagonists. In: Roberts RJ, Storm-Mathisen J, Bradford HF (eds) Excitatory amino acids. Macmillan, London, pp 459–461

Turski L, Klockegether T, Sontag K-H, Herrling P, Watkins JC (1987) Muscle relaxant and anticonvulsant activity of 3-((+)2-carboxypiperazin-4-yl)-propyl-1-phosphonic acid, a novel N-methyl-D-aspartate antagonist in rodents. Neurosci Leff 73:143–148

Turski L, Klockgether T, Turski WA, Ikonomidou C, Schwarz M, Sontag K-H (1988) Muscle relaxant action of excitatory amino-acid antagonists: sites of action. In: Cavalheiro EA, Lehmann T, Turski L (eds) Frontiers in excitatory amino acid research. Liss, New York, pp 343–350

Turski L, Bressler K, Klockgether T, Stephens DN (1990a) Differential effects of the excitatory amino acid antagonists, 6-cyano-7-nitroquinoxaline-2,3-dione (CNQX) and 3-((+)-2-carboxypiperazin-4-yl)-propyl-1-phosphonic acid (CPP), on spinal reflex activity in mice. Neurosci Leff 113:66–71

Turski L, Klockgether T, Schwarz M, Turski WA, Sontag K-H (1990b) Substantia-nigra: a site of action of muscle relaxant drugs. Ann Neurol 28:341–348

Turski L, Jacobsen P, Honore T, Stephens DN (1992) Relief of experimental spasticity and anxiolytic/anticonvulsant actions of the alpha-amino-3-hydroxy-5-methyl-4-isoxazolepropionate antagonist 2,3-dihydroxy-6-nitro-7-sulfamoyl-benzo(F) quinoxaline. J Pharmacol Exp Ther 260:742–747

Watkins JC (1984) Excitatory amino acids and central synaptic transmission. Trends Pharmacol Sci 5:373–376

Effect of Intrathecal Baclofen on Monosynaptic Reflex in Man

B. Bussel, P. Azouvi, and A. Roby-Brami

Introduction

In patients with a spinal lesion, spasticity and the occurrence of strong flexor or extensor spasms may be responsible for significant disability. In such cases, the effects of treatments administered per os are frequently insufficient. Surgical procedures may be effective but must be reserved for very exceptional cases. Recently, Penn and Kroin (1984, 1985; Penn et al. 1989) have shown that intrathecal administration of baclofen is an effective treatment for severe spasticity. An electrophysiological study was performed in order to determine the mechanisms of action of intrathecal baclofen on spinal function: on presynaptic Ia or flexor reflex afferent (FRA) terminals or directly on alpha motoneurons. Clinical examination was focused on determining whether a lessening of spasticity could induce an improvement in voluntary movements.

Patients

Clinical evaluation was performed in 60 patients (aged 14–65) before and after they had received an intrathecal bolus of baclofen. All of them suffered from severe spasticity and disabling spasms of spinal origin.

Eighteen patients had multiple sclerosis and 35 traumatic paraplegia; this was clinically complete in 15 cases. Five patients suffered from Strümpell's familial spastic paraplegia and two from a spinal tumor.

Baclofen Administration

The doses used ranged from 50 to 300 μg: the first injection was a low dose (50 μg) and this was progressively increased (by steps of 50 μg every 3 days).

Service de Rééducation Neurologique et INSERM U215, Hôpital Raymond Poincaré, 92380 Garches, France

Thilmann et al. (Eds.)
Spasticity: Mechanisms and Management

An implanted pump was offered to the patient under three conditions. First, the mono- and polysynaptic reflexes had to be clinically decreased (decrease of tendon jerks and clonus and/or decrease of evoked spasms). Second, the bolus had to induce a clear functional benefit, appreciated by the patient himself, the examiner, and the team of nurses and physical therapists. Third, these effects had to be observed with a bolus of less than 150 μg. The implantable pump delivered baclofen via a subcutaneous intrathecal catheter. Manual pumps were implanted in two early cases and were replaced by electronically driven pumps (Medtronic DAT). The dose was adapted during a short period at the hospital and then every month on an outpatient basis; it ranged from 20 to 500 μg per day (either by continuous perfusion or by successive boluses).

To date, 19 patients have benefited from this treatment. Two have died, due to the natural evolution of multiple sclerosis. None have interrupted the treatment, and maximum follow-up has been 5 years for one patient.

Electrophysiological Investigations

Electrophysiological investigations were performed in 15 patients, aged 18–55. All gave their consent after being informed of the experimental aim of the electrophysiological study. Fourteen had a traumatic spinal cord lesion, of which six were clinically complete and eight incomplete. One had spastic paraparesis due to multiple sclerosis. The duration of the lesion was at least 6 months. At the time of the experimental session, the patients were free from pressure sores, urinary tract infection, or any other medical complications and were taking no oral medications for spasticity. During the recording sessions, the patients sat in their wheelchairs or lay on a bed with their legs in the same position throughout the experimental session. A bolus of baclofen was injected into the lumbar subarachnoid space, either by percutaneous lumbar puncture or by an implanted pump via a subcutaneous intrathecal catheter; the doses used ranged from 20 to 200 μg.

Monosynaptic Reflex

The H reflex was tested on the soleus muscle. Surface electrodes were used for both stimulation and recording. Rectangular current pulses of 1 ms duration were used for stimulation and were applied using an electrode in the popliteal fossa to stimulate the tibial nerve. The reflex responses were measured as the peak-to-peak amplitude of the evoked muscle action potential. These potentials were recorded with two nonpolarizable disk electrodes (0.9 mm diameter; interelectrode distance 2 cm) placed over the bellies of the soleus muscles (below the gastrocnemius muscles). The EMG responses were measured using an analog-digital converted linked to a microcomputer. H max value was assessed by H reflex recruitment curves obtained by 25 tibial posterior nerve stimulations with different intensities.

Polysynaptic Reflexes

Polysynaptic reflexes were obtained after an electrical FRA stimulation had been applied to the cutaneous sural nerve at the level of the ankle. As described before (Roby-Brami and Bussel 1987), FRA stimulation induced two distinct flexor reflexes: the early one, with a latency of around 100 ms, correponds to the usual flexor reflex, and the later one, with a latency of more than 150 ms, is related to more complex spinal networks (see Roby-Brami and Bussel, this volume). Three to five traces were recorded for each stimulation intensity (ranging from 5 to 50 mA in patients with sensory loss), corrected and averaged.

Presynaptic Ia Inhibition

Presynaptic inhibition of Ia terminals was assessed by two methods.

Vibratory Inhibition. Vibration (percussion of amplitude 0.7 mm, duration 2 ms, frequency 70 Hz) was applied to the Achilles tendon with a vibrator (Brüel and Kjaer model 4809) driven by monophasic rectangular pulses. H max, with and without vibration, was recorded before and every 15 min after baclofen injection in eight patients.

Heteronymous Ia Facilitation. Heteronymous Ia facilitation of the soleus H reflex was assessed according to the method described by Hultborn et al. (1987). The conditioning stimulus was obtained by supramaximal stimulation of the femoral nerve in the femoral triangle. It has been demonstrated that the first 0.5 ms of this heteronomous Ia facilitation from the quadriceps to the soleus is mediated exclusively through a monosynaptic pathway and is not contaminated by any Ib or polysynaptic Ia effect (Hultborn et al. 1987). Accordingly, the timing was adjusted in order to make the conditioning afferent volley arrive at the spinal level no more than 0.5 ms before the test soleus H reflex. The conditioning stimulus was delivered after the test stimulus, because the distance from the femoral triangle to the spinal cord is shorter than the distance from the popliteal fossa to the spinal cord. First, the shortest conditioning-test interval (using 0.1-ms steps) at which facilitation of the soleus H reflex could be elicited was determined. The interval used in the following part of the experiment was 0.2–0.3 ms longer, in order to obtain sizeable facilitation due only to the monosynaptic Ia EPSP. Then the amount of facilitation before baclofen (control situation) was compared with that obtained every 10 min after the intrathecal injection. Under these conditions, the reflex facilitation depended only on the size of the conditioning EPSP evoked by heteronomous Ia stimulation. A constant conditioning stimulus should elicit an EPSP of constant size in motoneurons, and thus a constant amount of reflex facilitation, unless the amount of presynaptic inhibition of Ia fibers changes; a decrease in the facilitation may be ascribed to an increase in presynaptic inhibition of the heteronymous Ia fibers from the quadriceps to the soleus. It has been shown (Mazières et al. 1984; Meink 1980) that the susceptibility of an H reflex to facilitation varies with the size of the unconditioned reflex. However, after baclofen injection the soleus H reflex was depressed. In order to maintain a constant size (10%–15% of M max) of the reference (unconditioned) H reflex in spite of monosynaptic reflex (MSR) inhibition, the test stimulus intensity had to be adjusted (i.e., increased). Every 10 min, a series of at least 20 unconditioned and 20 conditioned H reflexes was recorded, with stimuli being delivered every 4 s. The amount of facilitation was then calculated, the statistical analysis of the results being examined using a *t*-test. Between these series, the H max/M max ratio was measured to assess the amount of MSR inhibition due to baclofen. This experiment was performed in five patients with doses ranging from 50 to 150 μg. In two patients the experiment could not be performed because heteronymous facilitation was too small before baclofen injection. In the other patients it was not possible because femoral nerve stimulation provoked strong flexor spasms and/or a long-lasting soleus H reflex inhibition.

Results

Clinical Evaluation

The bolus dose of baclofen induced a decrease of mono- and polysynaptic reflexes in all the patients. This decrease is better appreciated with electrophysiological studies (see below). The functional benefit varied from one patient to another. In the five patients with Strümpell's familial spastic paraplegia, the decrease of tendon jerks of lower limbs was in all cases associated with an increase of lower-limb muscle weakness. This induced in all cases an increase of the subjective locomotor disability, which was clinically evident. Some spinally injured patients observed that the decrease of spasticity was not associated with a functional benefit and complained of being "too floppy" (in which case they did not receive a pump). In two paraplegics with incomplete lesions and one patient with multiple sclerosis, the intrathecal baclofen treatment induced a frank functional improvement, since they recovered some locomotor activity which was previously severely compromised. The decrease of strong quadriceps extensor spasms improved the voluntary activity in four patients. In the case illustrated in Fig. 1, a voluntary knee extension triggered a prolonged co-contraction of the quadriceps and hamstring muscles, which considerably slowed the subsequent voluntary flexion. After an intrathecal injection of baclofen, the patient was able to perform alternative movements of knee extension and flexion. Baclofen injection produced a clear improvement in the quality of life of nine paraplegics, since they could easily stay in their wheelchair. They had previously been unable to transfer from a bed to a wheelchair, to stay in a wheel chair, or to drive a car due to strongly disabling extensor or flexor spasms. In the other patients with multiple sclerosis baclofen injection decreased flexor and adductor spasms and so facilitated the sitting position and nursing.

Electrophysiological Analysis

Monosynaptic Reflex. The excitability of MSR in lower limbs was tested in all subjects. The soleus H reflex was dramatically reduced by intrathecal baclofen. However, the minimal dose necessary to obtain this inhibition varied from one patient to another. In all patients except two, we observed near-total disappearance of the reflex (H max less than 10% of the pre-treatment value) with doses of less than 150 μg. In seven patients this effect appeared with doses of less than 50 μg; in two cases higher doses (more than 150 μg) were necessary, and even with these doses inhibition was not complete (reduction of the Hmax/Mmax ratio by 50%). The onset and the rate of decrease of the H max value were also variable. In six patients the effect appeared rapidly, and when the appropriate dose was given the inhibition was complete within less than 1 h. In four cases, more than 2 h were necessary to obtain a complete inhibition of the H reflex. The minimal dose necessary to abolish the H reflex and the time to peak effect were not correlated with the Hmax/Mmax ratio before injection, or with the level of the lesion, and did not differ according to whether the spinal lesion was clinically complete or not. This interindividual variability may be due to patients' susceptibility to the drug, but modifications of CSF flow may contribute as well, at least in paraplegia of traumatic origin. For each patient there was a dose-effect relationship. This is illustrated in Fig. 2 by Hmax variation observed after different doses of baclofen (20, 50, 100 μg) in one representative patient.

Polysynaptic Reflexes. Polysynaptic reflexes were evoked on flexor muscles (tibialis anterior and biceps femoris) by electrical sural nerve stimulations in five patients. We observed that the early reflex (latency 100 ms) as well as the late (latency more than 130 ms) decreased in a parallel way (Fig. 3). The effect of baclofen on mono- and polysynaptic reflexes appeared for the same doses and seemed to be of the same magnitude.

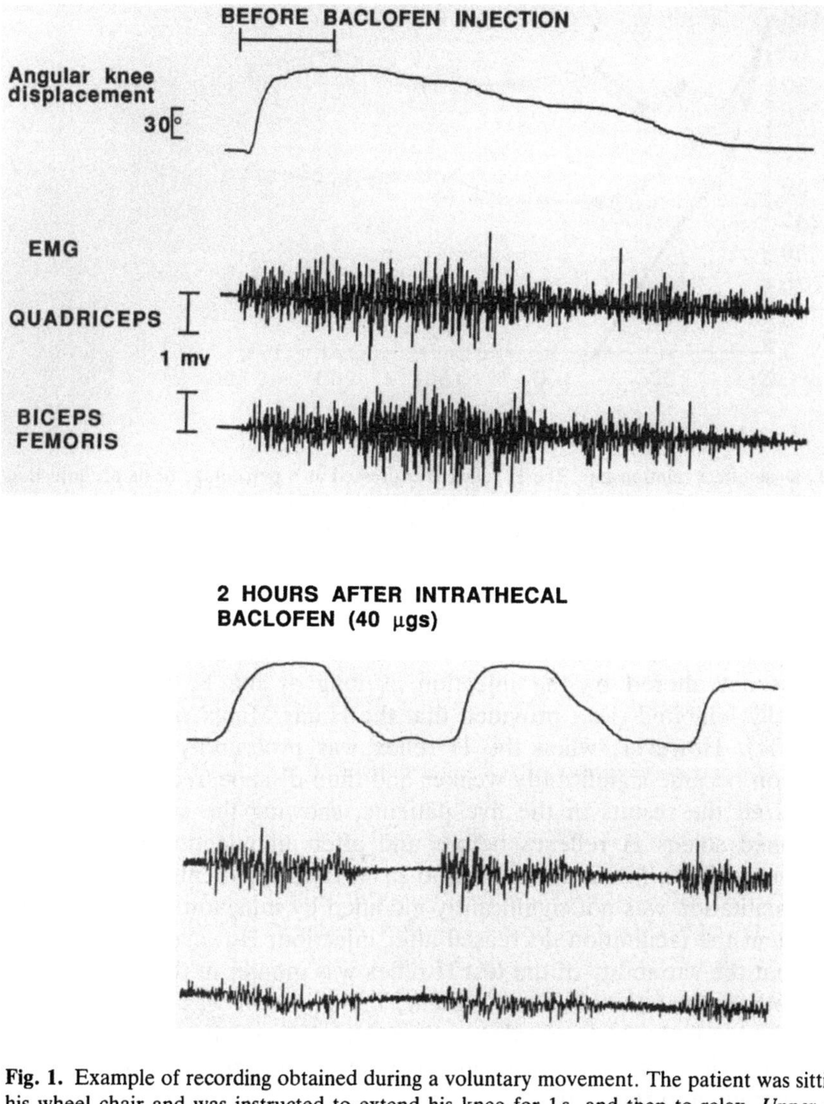

Fig. 1. Example of recording obtained during a voluntary movement. The patient was sitting in his wheel chair and was instructed to extend his knee for 1 s, and then to relax. *Upper trace*, knee angle measured with a potentiometer; *middle trace*, quadriceps EMG; *lower trace*, hamstring EMG

Effect on Soleus H Reflex Conditioned by Ia Afferent Stimulation. The effect of a vibratory stimulus was studied in eight subjects (Azouvi et al. 1993). Weak vibratory inhibition (less than 20%) was observed in only two of the eight. The injection of baclofen did not enhance vibratory inhibition in these two, nor did it make it appear in the other six patients (Fig. 4). The effect of Ia heteronymous facilitation of the soleus H reflex was studied in five

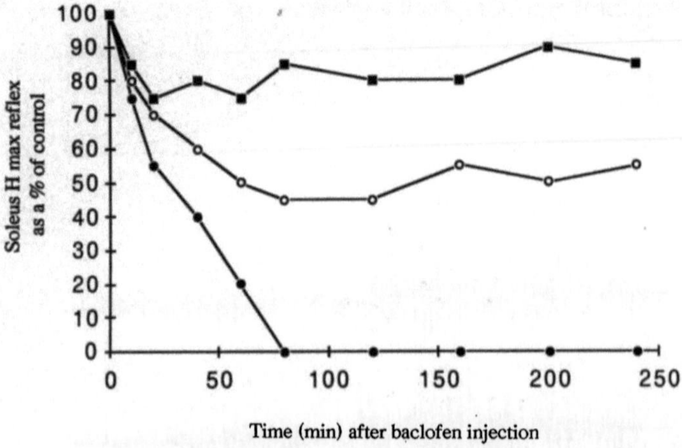

Fig. 2. Dose-effect relationship. The H reflex is expressed as a percentage of its pre-injection value. The three *curves* represent the evolution of the H max value during the 4 h following an intrathecal injection of 20 μg baclofen (*squares*), 50 μg (*open circles*), and 100 μg (*filled circles*)

patients, the test H reflex being set at 10%–15% of M max both before and after baclofen. In four patients the same findings were observed: the facilitation was not altered by the injection as long as the H reflex was not profoundly inhibited (i.e., provided that the Hmax/Mmax ratio was higher than 25%). However, when the H reflex was profoundly inhibited, the facilitation became significantly weaker and then disappeared. In Fig. 4 are summarized the results in the five patients, showing the size of test and conditioned soleus H reflexes before and after an injection producing a reduction of Hmax/Mmax ratio down to 25%. In four patients the amplitude of the facilitation was not significantly modified by injection of baclofen. In one patient the facilitation decreased after injection. However, it should be noted that the variability of the test H reflex was greater in this patient than in the other four. In this same patient, the injection was made via percutaneous lumbar puncture, while an implantable pump was used in the other four. These technical problems may have contributed to the variable results obtained in this patient.

Discussion

Clinical Effect of Baclofen

Baclofen is known to be effective in reducing muscle tone and flexor or extensor spasms in patients suffering from corticospinal tract lesions, as well as in reducing stretch reflexes in animal experiments (for review see Young and Delwaide 1981). Since this effect can be observed after spinal cord section, it is generally accepted that most antispastic activity is exerted at the spinal level. The present results document a powerful inhibitory effect of baclofen administered intrathecally, with rapid (from 35 min to 4 h), profound, and prolonged inhibition of hypertonia, clonus, and spasms in spinally injured patients. Voluntary movement may be facilitated, but the mechanism linking the depression of spasticity and the increase of voluntary movements remain controversial, as does the possible depressive action of baclofen on the voluntary activation of alpha motoneurons.

Such a dramatic effect has never been observed with oral treatment (Castaigne et al. 1973) and it confirms the previous clinical observations (Müller et al. 1988; Ochs et al. 1989; Penn and Kroin 1984, 1985; Penn et al. 1989). The mechanism linking the depression of spasticity and the increase of voluntary movements remains unclear.

Mechanism of Action of Baclofen

The precise site and mechanism of action of this drug remain unclear. Before baclofen injection, H reflex vibratory inhibition was weak or absent in our patients, all of whom had severe spasticity. This is in accordance with the findings of Delwaide (1985), who showed that vibratory inhibition was reduced in spastic patients, and that this loss of inhibition was related to the severity of hyper-reflexia. Intrathecal baclofen did not enhance or restore Ia vibratory inhibition of MSR. The same results have been reported after oral (Ashby and White 1973) or intramuscular (Delwaide 1985) administration of baclofen, but in these studies, inhibition of the MSR was much less important. As it has been shown that vibratory inhibition is probably due to Ia presynaptic inhibition brought about by Ia fibers (Delwaide 1973), this suggests that baclofen does not enhance this inhibitory mechanism. However, presynaptic inhibition on Ia fibers can be brought about by other pathways than Ia fibers themselves.

To assess more precisely the effect of baclofen on Ia afferent terminals, we used the experimental procedure described by Hultborn et al. (1987). This technique has the great advantage of allowing a study of presynaptic inhibition on Ia fibers, whatever its origin. In our study, in four of five patients in whom a reproducible Ia heteronymous facilitation of soleus motoneurons was obtained, this facilitation persisted as long as the Hmax/Mmax ratio was greater than 25%. This suggests that H reflex inhibition is not due to a presynaptic mechanism. During the Essen "Workshop on Spasticity" some alternative suggestions were made, and another interpretation of our results can be advanced. We observed a considerable decrease of the maximum H reflex (from 75% down to less than 25%), contrasting with a small (10%) and, in four cases, no significant decrease in the heteronymous Ia facilitation. This discrepancy could be explained if, due to the

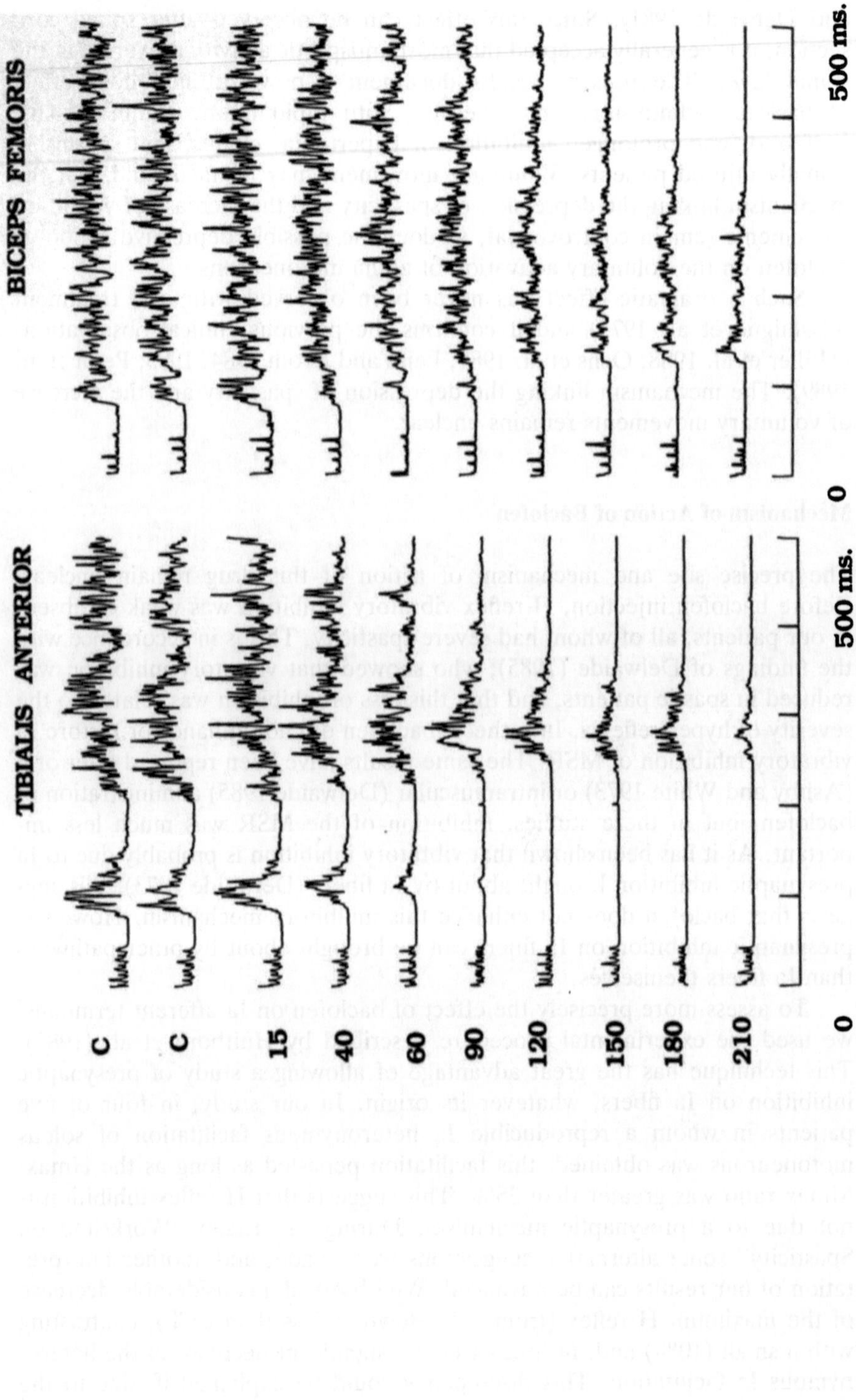

TIBIALIS ANTERIOR
BICEPS FEMORIS
C
C
15
40
60
90
120
150
180
210
0
500 ms.
0
500 ms.

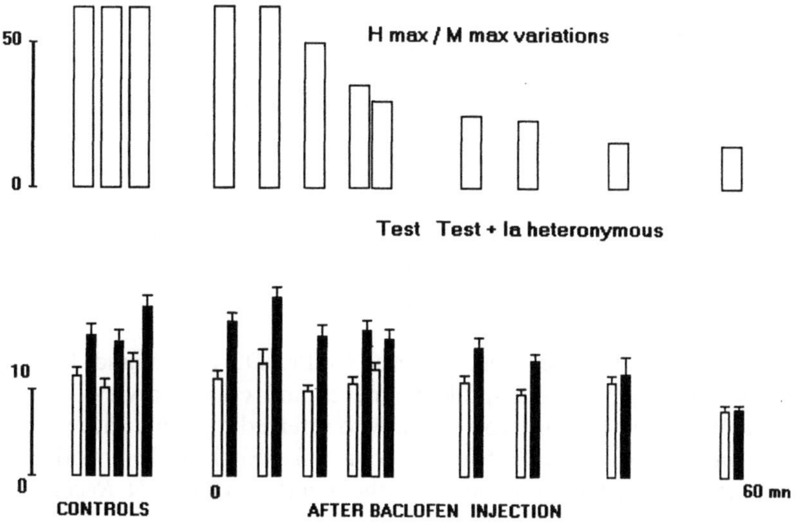

Fig. 4. Heteronymous Ia facilitation of soleus H reflex compared with Hmax/Mmax ratio before and after an intrathecal injection of baclofen 100 γ. *Top*, Hmax/Mmax variation. *Bottom*, *open bars* represent test H reflex and *solid bars* H reflex conditioned by heteronymous Ia volley. The first three measures were obtained before, the other during the 60 min following baclofen injection. Note that different scales are used for the upper and lower parts of the figure

nonlinearity of the motoneuronal activation by the Ia EPSP, a 10% decrease of the Ia EPSP induced a decrease in the size of the H max reflex from 75% to 25%. When the H reflex was more deeply inhibited, the facilitation decreased and disappeared, suggesting a presynaptic inhibition. The experimental method tests only presynaptic inhibition on heteronymous Ia afferents. It is likely that presynaptic inhibition on homonymous Ia afferents from the soleus behaves in a similar way, since Meunier and Pierrot-Deseilligny (1989) have suggested that presynaptic inhibition on these two types of Ia fibers directed to soleus alpha motoneurons could be mediated by the same interneurons.

In animal and in vitro experiments, there is no agreement about whether baclofen has a pre- or post-synaptic site of action (Zieglgansberger et al. 1988). Early studies (Pierau and Fimmermann 1973), as well as more recent ones (Edwards et al. 1989; Jimenez et al. 1991) conclude that the main site of action of baclofen is presynaptic, reducing the probability of release of

◄———

Fig. 3. Effect of intrathecal baclofen (50 μg) on the polysynaptic flexor reflexes elicited by a 25-mA stimulation of the sural nerve. Each trace represents the average of three individual recordings obtained before (*upper trace*) and after (*lower traces*) intrathecal baclofen injection. *Left*, tibialis anterior; *right*, biceps femoris neurotransmitters without any effect on the membrane time constant or EPSP decay. Nevertheless, post-synaptic effects have also been demonstrated by a number of authors, who showed that baclofen has a dose- and concentration-dependent hyperpolarizing action on some mammalian central neurons, associated with an increase in post-synaptic potassium conductance (Zieglgansberger et al. 1988). It should be noted, moreover, that most data have been obtained in experimental conditions that are quite dissimilar to human clinical conditions: first, these studies have been performed only on acute spinal animals, and second, intrathecal administration has not been studied in animal preparations.

Acknowledgements. We are most grateful to Pr. D. Burke and Pr. E. Pierrot-Deseilligny for their helpful comments and critical review of the manuscript. We thank Pr. O. Lyon-Caen, who referred to us his patients for this study, and Dr. A. Yakovleff for improving the English. This work has been supported in part by a grant from the *Fondation pour la Recherche Médicale*, received by P. Azouvi.

References

Ashby P, White DG (1973) "Presynaptic" inhibition in spasticity and the effect of β (4-chlorophenyl) GABA. J Neurol Sci 20:329–338

Azouvi P, Roby-Brami A, Biraben A, Thiebaut JB, Thurel C, Bussel B (1993) Effect of intrathecal baclofen on the monosynaptic reflex in man. Evidence for a post-synaptic action. J Neurol Neurosurg Psychiatry (in press)

Castaigne P, Held JP, Pierrot-Deseilligny E, Bussel B, Macquart-Moulin J (1973) Etude de l'effet du lioresal dans la spasticité. Rev Neurol 128:245–250

Delwaide PJ (1973) Human monosynaptic reflexes and presynaptic inhibition. An interpretation of spastic hypereflexia. In: Desmedt JE (ed) New developments in electromyography and clinical neurophysiology, vol 3. Karger, Basel, pp 508–522

Delwaide PJ (1985) Electrophysiological analysis of the mode of action of muscle relaxants in spasticity. Ann Neurol 17:90–95

Edwards FR, Harrison PJ, Jack JJ, Kullmann B (1989) Reduction by baclofen of monosynaptic EPSPs in lumbosacral motoneurones of the anaesthetized cat. J Physiol (Lond) 416:539–556

Hultborn H, Meunier S, Morin C, Pierrot-Deseilligny E (1987) Assessing changes in presynaptic inhibition of Ia fibers: a study in man and the cat. J Physiol (Lond) 389:729–756

Jimenez I, Rudomin P, Enriquez M (1991) Differential effects of (−) baclofen on Ia and descending monosynaptic EPSPs. Exp Brain Res 85:103–113

Macdonnel RAL, Talalla A, Swash M, Grundy D (1989) Intrathecal baclofen and the H reflex. J Neurol Neurosurg Psychiatry 52:1110–1112

Mazières L, Morin C, Pierrot-Deseilligny E (1984) Effet de l'amplitude du réflexe test sur le niveau de facilitation ou d'inhibition des rèponses monosynaptiques. J Physiol (Paris) 79:59A

Meinck HM (1980) Facilitation and inhibition of the human H-reflex as a function of the amplitude of the control reflex. EEG Clin Neurophysiol 48:203–211

Meunier S, Pierrot-Deseilligny E (1989) Gating of the afferent volley of the monosynaptic stretch reflex during movement in man. J Physiol (Lond) 419:753–763

Müller H, Zierski J, Dralle D, Hoffmann O, Michaelis G (1988) Intrathecal baclofen in spasticity. In: Müller H, Zierski J, Penn RD (eds) Local spinal therapy of spasticity. Springer, Berlin Heidelberg New York, pp 155–214
Ochs G, Struppler A, Meyerson BA, Linderoth B, Gybels J, Gardner BP, Teddy P, Jamous A, Weinmann P (1989) Intrathecal baclofen for long-term treatment of spasticity: a multi-centre study. J Neurol Neurosurg Psychiatry 52:933–939
Penn RD, Kroin JS (1984) Intrathecal baclofen alleviates spinal cord spasticity. Lancet 1:1078
Penn RD, Kroin JS (1985) Continuous intrathecal baclofen for severe spasticity. Lancet 2:125–127
Penn RD, Savoy SM, Corcos D, Latsh M, Gottlieb G, Parke B, Kroin JS (1989) Intrathecal baclofen for severe spinal spasticity. N Engl J Med 320:1517–1521
Pierau FK, Zimmermann P (1973) Action of a GABA derivative on post-synaptic potentials and membrane properties of cat's spinal motoneurones. Brain Res 54:376–380
Roby-Brami A, Bussel B (1987) Long-latency spinal reflex in man after FRA stimulation. Brain 110:707–725
Young RR, Delwaide PJ (1981) Drug therapy: spasticity. N Engl J Med 304:28–33, 96–99
Zieglgansberger W, Howe JR, Sutor B (1988) The neuropharmacology of baclofen. In: Müller H, Zierski J, Penn RD (eds) Local spinal therapy of spasticity. Springer, Berlin Heidelberg New York, pp 7–50

Comparison of Cyproheptadine, Clonidine and Baclofen on the Modulation of Gait Pattern in Subjects with Spinal Cord Injury

K.E. Norman and H. Barbeau

Serotonin, Noradrenaline and Gamma-aminobutyric Acid in the Spinal Cord

Underlying Rationale for Pharmacological Therapy

Serotonin and noradrenaline are present in the spinal cord as a result of projections from the brainstem: serotonin from the raphe nuclei and noradrenaline from the locus ceruleus. Both the serotonergic fibres and the noradrenergic fibres terminate in all areas of the spinal grey matter, including the ventral horns. These two neurotransmitters are known to play roles in autonomic functions, in pain, and in motor output (Marshall 1983). Gamma-aminobutyric acid (GABA), in the spinal cord is released from neurons intrinsic to the cord. Although not as common as glycine in the spinal cord, GABA plays a role in inhibition of neuronal activity (Nistri 1983).

Since these neurotransmitters play a role in regulating neuronal excitability, it is of interest to review what is known of the effects of these neurotransmitters on motor behaviour such as locomotion. In regard to serotonin and noradrenaline, there is good evidence that each has a different role in modulating locomotion. The evidence for the influence of GABA on locomotion is more indirect.

For both serotonergic drugs and noradrenergic drugs, the spinalized cat has proven to be a useful preparation in which to study the effects of such drugs on the modulation of locomotor pattern. Barbeau and Rossignol (1990) studied cats who had been trained to walk on a treadmill over a 3-month period following spinalization. Once the locomotor pattern was stable, selected drugs which act on post-synaptic receptors below the spinal cord transection were injected. Injection of the precursor or agonists of serotonin resulted in a marked increase in burst duration and amplitude of flexor and extensor electromyographic (EMG) activity, including some flexor-extensor coactivation. Furthermore, some manifestations of spasticity,

School of Physical & Occupational Therapy, McGill University, 3654 Drummond St., Montréal, Québec, Canada H3G 1Y5

Thilmann et al. (Eds.)
Spasticity: Mechanisms and Management

such as clonus and spasms, were observed during locomotion. For both the serotonergic agonists and the precursor, the effects induced by increasing the activity of serotonergic receptors were partially or completely blocked by cyproheptadine (a serotonergic antagonist), depending on the dosage. In addition, the flexion reflex to cutaneous stimulation of the dorsum of the hindpaw was increased by the serotonergic agonists, an effect that was almost completely blocked by cyproheptadine.

Clonidine is an agonist of the alpha noradrenergic receptor and has been studied extensively in spinalized cats, both acutely, after spinal cord transection, and in the chronic period. Forssberg and Grillner (1973) injected clonidine into acutely spinalized cats supported over a treadmill. No spontaneous activity was observed, but as the treadmill belt started to move the spinalized cats started to walk, supporting the hypothesis that clonidine plays an important role in initiating locomotion.

The effect of clonidine on chronic spinal cats has been described by Barbeau et al. (1987a). As described above for the serotonergic system, they reported the effects of noradrenergic drugs on the locomotor pattern of chronic spinal cats with an established stable locomotor pattern. Clonidine markedly increased the step cycle duration for the same treadmill speed, and this increase was related to larger excursions of the hip, knee and ankle. The changes in EMG, kinematic patterns and step cycle duration were reproducible with repeated administration of clonidine over time and were blocked by alpha-2 antagonists. Clonidine also reduced the flexion reflex to cutaneous stimulation, such that a current of twice the intensity was required to produce reflex amplitudes equivalent to those seen prior to clonidine.

Thus, in the spinal cat the effect of noradrenergic drugs affected primarily the duration of flexor and extensor EMG activity with minimal effect on its amplitude, whereas the serotonergic drugs affected both the amplitude and the duration.

GABAergic drugs have not been systematically studied in spinalized cats as have serotonin and noradrenaline. However, there is evidence that GABA plays a role in reducing motor output (for more information, see Bowery 1989; Bussel, this volume; Fox et al. 1978; Milanov 1992; Pedersen et al. 1974; Schwarz and Emre, this volume; Zieglgänsberger 1988). Pierau and Zimmermann (1973) reported that the action of baclofen (a $GABA_B$ agonist) on motoneuronal activity in anaesthetized cats was a depression of polysynaptic and monosynaptic excitatory post-synaptic potentials. They proposed that, in doses comparable to those given to humans with spasticity, baclofen would enhance presynaptic inhibition. Subsequent study of baclofen on spinal cord tissue has confirmed that its primary site of action in reducing motoneuronal output is presynaptic, although the exact mechanism in not clear (Davidoff and Sears 1974; Stuart and Redman 1992; Wang and Dun 1990; Wüllner and Klockgether 1990). At higher doses, baclofen has other influences on motoneuronal output, but such doses are not comparable to those given orally to humans for spasticity.

Effects on Motor Function in Spastic Paretic Subjects

The effects produced by serotonergic drugs in spinal cats were striking in their similarity to the clinical presentation of some patients with incomplete spinal cord injuries (SCI). The finding that cyproheptadine blocked these spasticity-like behaviours in animals suggests that it may be a useful medication for incomplete SCI patients with spasticity and paresis. The initial study of the effects of cyproheptadine on spastic paretic subjects focused on manifestations of spasticity (Barbeau et al. 1982). The subjects showed marked decreases in the occurrence of spontaneous spasms and in the number of beats of elicited clonus following cyproheptadine therapy. A reduction of manifestations of spasticity in similar subjects was also found by P.W. Nance (1991, Annual meeting of the NCE for neuronal regeneration and functional recovery, Montreal, unpublished abstract). Other studies of the effects of cyproheptadine on spastic paretic subjects have used EMG and kinematic and temporal data to evaluate the changes in locomotion (Fung et al. 1990; Wainberg et al. 1986, 1990). The EMG changes included a decrease in clonic discharge, an extensor burst which was generally restricted to stance phase, and a more normal pattern of reciprocal muscle activation during walking. The kinematic changes included a more normal ankle excursion pattern in swing and stance, with appropriate foot-floor contact, and more normal knee excursion in swing. An increase in comfortable treadmill speed was also reported in some cases. These studies demonstrate that cyproheptadine may be an effective means of modulating locomotion in incomplete SCI subjects.

Since clonidine has produced a very smooth and regular locomotor pattern in spinal cats it is of interest to determine whether it can modulate the locomotor pattern in humans with SCI. Clonidine has been studied by a number of workers for its potential antispastic effect (Donovan et al. 1988; Nance et al. 1985; Weingarden and Belen 1992). It has been reported to be an effective antispastic medication in a majority of the subjects studied, although adverse effects prevented it from being clinically useful in some of those subjects. Clonidine has also been studied for its effect on locomotor pattern (Fung et al. 1990; Stewart et al. 1991). Stewart et al. (1991) reported that subjects with clinically complete SCI experienced no change in locomotor ability, although a reduction in signs and symptoms of spasticity was reported for some of them. Of the three subjects studied who had incomplete SCI, one experienced a marked improvement in locomotor ability and the other two experienced relatively mild improvements. The marked improvement included more normal muscle activation patterns with decreased coactivation, as well as reduced trunk flexion, greater hip extension in late stance, more ankle dorsiflexion through early to mid stance, and a general reduction in the variability of the locomotor pattern. The subject who experienced the marked improvement was functionally the most severely affected of all the subjects with incomplete SCI.

It is interesting to note that tizanidine, also an alpha-noradrenergic agonist, is under study by others (see Emre this volume). However, tizanidine is not in general use in Canada.

Baclofen appears to be the most commonly used medication to reduce spasticity in SCI patients in Canada. It came into widespread use at a time when the alternatives, diazepam and dantrolene sodium, produced undesirable side effects. There are numerous reviews of antispastic medications which discuss baclofen (Davidoff 1985; Delwaide 1985; Little and Merritt 1988; Milanov 1992; Whyte and Robinson 1990; Young and Delwaide 1981). However, in large doses for severe spasticity, baclofen can also produce undesirable side effects. Therefore, many research groups have pursued an intrathecal approach to baclofen delivery (Armstrong et al. 1992; Broseta et al. 1989; Bussel, this volume; Hankey et al. 1986; Hugenholtz et al. 1992; Kofler et al. 1992; Latash et al. 1989, 1990; Lazorthes et al. 1990; Loubser et al. 1991; Meythaler et al. 1992; Parke et al. 1989; Penn and Kroin 1987; Penn 1991).

Baclofen is known to reduce manifestations of spasticity, such as spontaneous spasms and resistance to passive stretch, in spastic paretic subjects. However, its action on voluntary motor control is less clear. McLellan (1977) reported that oral baclofen was effective in reducing EMG responses to externally imposed passive muscle stretch but had less effect on muscle responses to stretch from voluntary shortening of its antagonist. In contrast, Latash et al. (1990) reported an increased selective control of muscle activation without antagonist coactivation following intrathecal baclofen. The conflicting results may be related to the different approaches for baclofen delivery used in the studies, or they may be related to the differences in the movements performed. There has been almost no study of locomotor pattern following baclofen therapy (Corston et al. 1981). Whether baclofen has quantifiable effects on locomotor pattern in spastic paretic subjects remains to be investigated.

Objectives of the Current Study

Cyproheptadine and clonidine have each been shown to be potentially useful medications compared with placebo in improving locomotion in incomplete SCI subjects, but they have not been systematically compared with each other. Baclofen is currently commonly used to reduce the manifestations of spasticity but its effects on gait pattern have not been systematically investigated. Furthermore, neither cyproheptadine nor clonidine has been systematically compared with baclofen. The primary objective of this study is to describe and compare the effects of cyproheptadine, clonidine and baclofen on locomotor function in subjects with spasticity and paresis due to spinal lesions. This preliminary study compares the effects of these three drugs on gait, as measured by electromyographic (EMG) and kinematic evaluation during treadmill walking.

In designing this study, it was important to consider that a subject's locomotor pattern, if modified by a medication, may not return to his/her baseline performance following withdrawal of the medication. Although such retention of effects may be clinically desirable, it demands a study design that addresses the questions thereby raised.

The Present Study

Subjects and Design

Subjects were invited to participate if they had sustained SCI between 1 and 4 years prior to entry to the study, if they were classified as having an incomplete injury with some motor function, useful or not, and if they had clinical signs of spasticity. Subjects were excluded if they had other neurological (e.g., severe brain injury) or orthopedic diagnoses, or if they were on sedative or other medications known or suspected to alter muscle tone. Each subject took each of the medications for a 3-week period, with a 2-week washout period in between (see Fig. 1); minor modifications in schedule were made when dictated by a subject's availability for laboratory evaluations. For each medication the subject started with a low dose and increased the dose every 3 days. After four such increases the maximal dosage was reached, except for clonidine, for which the maximal dosage required five increases. After at least 9 days of stable dosage, the subject was evaluated in the laboratory. The subject then tapered the medication over 6 days (8 days for clonidine) and returned to the laboratory after at least 1 week of no medication. It is intended by the end of the study that there will be an equal number of subjects following each possible medication order. Subjects were not blinded as to the identity of the medications, but possible beneficial effects and side effects for all three were presented

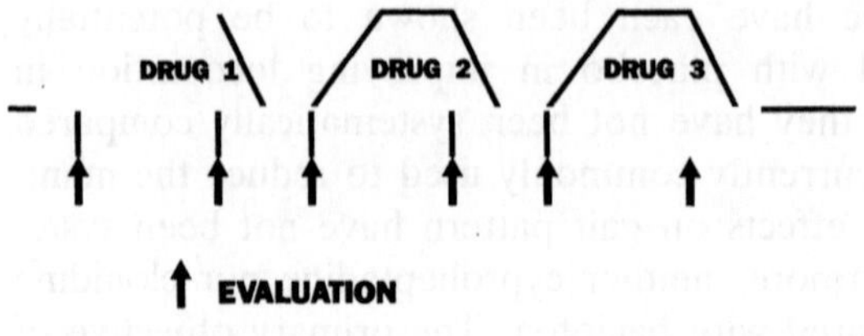

Fig. 1. Progression of each subject through the study. Each subjects is evaluated at least once at entry and then immediately begins a gradual course of increasing medication. (S)he is evaluated at the end of the period of peak dosage, usually 3 weeks after commencing the medication. The subject then tapers the dosage of the medication down to zero and is re-evaluated at the end of a period of no medication. From peak dosage to subsequent washout evaluation takes 2 weeks. The cycle is repeated for the next two medications

together. All subjects were living in their communities, on independent medication schedules, and only two of the four were even slightly acquainted with one another.

Laboratory Experimental Procedures

On each visit to the laboratory, the subject's gait was evaluated on the treadmill. When possible, overground gait was also evaluated. Preparations were made for EMG and kinematic data collection. The subjects were wheeled in their wheelchairs onto the treadmill belt. For some subjects, a modified climbing harness was worn in order to receive support from the overhead body-weight-support (BWS) system. This system has been described in more detail elsewhere (Barbeau et al. 1987b). Essentially, a modified climbing harness is attached by shoulder straps to an overhead pulley system and a force transducer. Through the pulley system the harness is raised or lowered to provide the subject's preferred level of support. The force transducer is linked to a digital reading of the amount of weight supported by the system. The actual amount of weight supported by the system varies throughout the trial. Generally, the subject "leans on" or "sits in" the harness more while trying to advance a limb, than during double limb support. This system has been used to evaluate severely disabled subjects and as a form of training. With this harness, as well as a highly adjustable treadmill, it has been possible to evaluate subjects who are too disabled to be evaluated in most gait evaluation settings and thus are not necessarily comparable to much of the literature. The harness system has also been useful in the training of subjects with spasticity and paresis (Barbeau and Blunt 1991; Visintin and Barbeau 1989).

During experiments, subjects performed a series of trials, ranging from a few steps to several minutes of walking. The first few trials served to ensure that EMG signals were relatively noise free, that video conditions were optimal, and that the harness support was suitably adjusted for those subjects who used it. Treadmill speed was adjusted according to the subject's indications of what was preferred speed and what was maximal speed.

Data Collection

Surface electrodes were placed over the tibialis anterior (TA), soleus (SOL), medial belly of gastrocnemius (GA), vastus lateralis (VL), and medial hamstrings (MH) of the leg that had more functional deficit. If both legs were equally affected, the left leg was evaluated. Surface electrodes were also placed over the contralateral TA and GA. A reference electrode was placed over a bony prominence (usually the tibial plateau or tuberosity on the ipsilateral side). EMG amplifiers provided analog filtering with a bandpass of 10–1000 Hz and a notch filter at 60 Hz. Three footswitches were taped on the sole of each shoe: one under the heel, one under the head of the fifth metatarsal and one under the big toe. EMG and footswitch signals were collected on a 14-channel FM tape recorder.

Reflective markers were placed on the side of the body with more muscles being recorded. Markers were placed over the acromion process, the greater trochanter, the lateral joint line of the knee, the lateral malleolus, the lateral aspect of the heel and the lateral aspect of the head of the fifth metatarsal. Video cameras were set at approximately 45° to the treadmill and 90° to each other.

For later synchronization of EMG and footswitch data with video recordings and subsequent kinematic measurements, the output of an SMPTE time code generator was recorded on one channel of the FM tape recorder and on the second audio track of both video tape recorders. To achieve precise synchronization of EMG and video tape data, a pulse was triggered intermittently, which simultaneously put an LED flash in both cameras as well as a square-wave pulse on one of the FM tape channels.

Data Processing

Segments of EMG and footswitch data were selected for analysis if the data on the FM tape were free of artifacts and the corresponding video tape data were optimal for viewing joint markers. The selected segments were then digitized at a sampling rate of 1000 or 1200 Hz (Data Translation: DT2821) and were processed in the DATAPAC II environment (Version 3.0, Run Technologies, Inc). Digital filtering of EMG data was performed with a bandpass filter of 20–500 Hz. For some trials the high-pass value was raised in order to minimize the artifacts created by clonus-induced vibration. Cycles were determined by the footswitch signals. Video data were digitized and processed using Peak Performance Motion Measurement system (Versions 4.2.3, 5.0). All trials were digitized at 60 Hz. The trajectories of all reflective markers were displayed as "stick figures", constructed by having segment lines drawn between certain markers: head of the fifth metatarsal to heel, head of the fifth metatarsal to lateral malleolus, heel to lateral malleolus, lateral malleolus to lateral joint line of knee, lateral joint line of knee to greater trochanter, and greater trochanter to acromion. The resultant stick figures give a visual impression of the subject's postural alignment during stance and swing phases.

Results

Despite the inclusion criteria, the participating subjects have been different from one another, both in their functional status and in their responses to the medications. Preliminary results are presented from two subjects whose initial locomotor deficits were different from one another. True randomization of medication order has not been possible. First, subjects are referred to the study with different medications. Second, ethical considerations required that a nurse in the subject's home community be available to screen for hypotension during the clonidine period, and the arrangements for this took time; thus, for these two subjects, clonidine was the last medication.

The first subject (S1) had sustained a traumatic SCI 14 months before entering the study. His neurological level of injury was C5 and his Frankel grade was C (ASIA 1982). He used a manual wheelchair independently indoors and with assistance outdoors. He was referred to the study while taking baclofen (80 mg total daily dose) and was thus evaluated. He was able to rise to stand with assistance and able to remain standing with the harness support and parallel bars. When the treadmill was started he required manual assistance to take steps. The treadmill speed averaged 0.04 m/s. His gait cycles were long and irregular (see cycle boundaries in Fig. 2A) and his EMG records show tonic or irregular activation in SOL, VL and MH, with an episode of SOL clonus (Fig. 2A). Regardless of any height adjustments made to the harness support, he required support of approximately half of his body weight in order to perform stepping. In the illustrated trial, the harness support was relatively low in height and the subject adopted a flexed posture to allow the harness to support approximately half of his body weight. The stick figures show evidence of flexion-extension in late stance related to repeated attempts to swing during stance. They illustrate that the foot is dragged forward along the treadmill belt during swing (Fig. 2B).

During the subsequent washout period, S1 required approximately the same amount of support at an average speed of 0.06 m/s. The height of the harness system was greater than in the previous experiment, but approximately the same proportion of body weight was supported. His EMG patterns are illustrated in Fig. 2C. Clonus is visible in the SOL trace in both cycles. The MH showed large bursts in early stance, coinciding approximately with the point when the contralateral foot was being unloaded in preparation for swing. The stick figures show flexion-extension in mid-stance as his knee and hip go sharply into flexion twice (Fig. 2D). As in the previous evaluation, the subject continued to require manual assistance for most cycles.

Cyproheptadine therapy (16 mg total daily dose) was associated with an improvement in locomotion. For the multiple trials on the treadmill, S1 continued to require harness support but no manual assistance. He preferred a speed of 0.06 m/s, although he was capable of keeping up with the treadmill at 0.14 m/s for several cycles. At 0.06 m/s his cycles are shorter and more regular than in previous evaluations, and his EMG patterns have more normal timing (Fig. 2E). The clonus in SOL is absent and there is a better degree of reciprocal patterning with TA. The large burst of MH is markedly

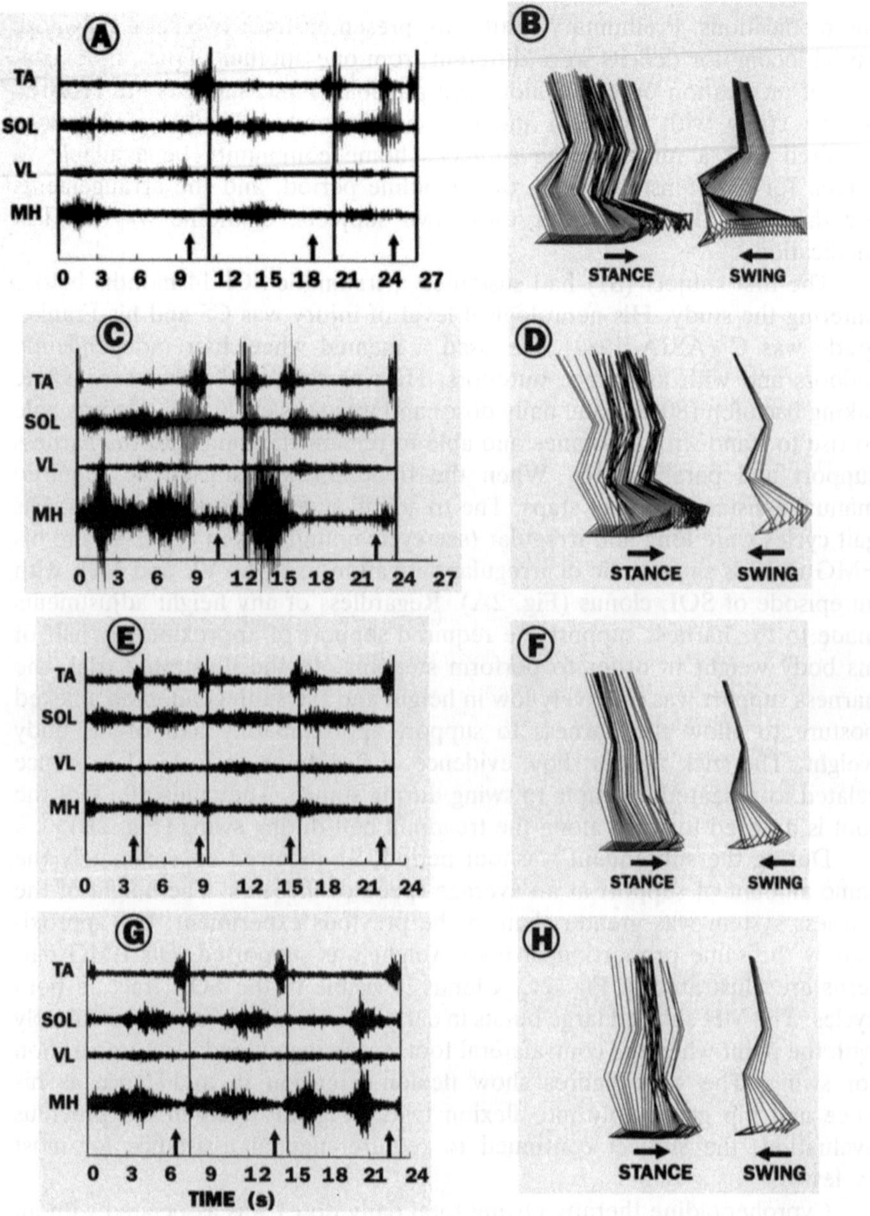
A
TA
SOL
VL
MH
0 3 6 9 12 15 18 21 24 27
B
STANCE
SWING
C
TA
SOL
VL
MH
0 3 6 9 12 15 18 21 24 27
D
STANCE
SWING
E
TA
SOL
VL
MH
0 3 6 9 12 15 18 21 24
F
STANCE
SWING
G
TA
SOL
VL
MH
0 3 6 9 12 15 18 21 24
TIME (s)
H
STANCE
SWING

reduced. The stick figures illustrate that his posture was more upright and there is less oscillation within stance (Fig. 2F). The evaluation performed following the post-cyproheptadine washout was different from the evaluation during cyproheptadine therapy, but he did not return to his baseline status (data not illustrated). He was still capable of the same treadmill speeds with a similar posture, but his EMG patterns showed a return of clonus in SOL.

Clonidine therapy (0.25 mg total daily dose) was associated with a further improvement in locomotor pattern but no change in preferred or maximal treadmill speed. The burst of SOL was generally restricted to the mid- or late-stance period, showing reciprocal patterning with TA (Fig. 2G). The subject's posture was more upright than it had been with cyproheptadine therapy (Fig. 2H).

Initially, S1 was not capable of overground walking, even with maximal assistance. Starting with the period of cyproheptadine therapy, S1 became capable of assisted overground walking with a forearm-support rolling walker for up to 30 m with supervision for safety (data not illustrated).

The second subject (S2) had sustained a traumatic SCI 3½ years prior to entering the study. His neurological level of injury was C6 and his Frankel grade was D (ASIA 1982). His primary mode of locomotion was self-propulsion in a lightweight wheelchair. In supervised situations he was able to walk with a walker for short distances indoors (approximately 15 m at less than 0.10 m/s). He wore plastic molded ankle-foot orthoses (AFOs) bilaterally. Evaluation without the AFOs was deemed unsafe. He was referred to the study while taking no antispastic medications and proceeded through the study following a medication order of first cyproheptadine, then baclofen, and finally clonidine. In general, his locomotor pattern changed little over the course of his participation in the study, and he reported no subjective benefit from any of the medications. S2 had a pattern of walking which relied heavily on parallel bars (or walking aids, when overground). During double limb support, both his feet were flat on the walking surface. To advance one leg, he pushed down hard on the parallel bars to raise himself up and his contralateral foot went into plantarflexion. The swinging

◄

Fig. 2A–H. Raw EMG traces of several cycles and stick-figure reconstruction of one cycle for subject one (S1): **A**, **B** during baclofen period; **C**, **D** during post-baclofen washout; **E**, **F** during cyproheptadine period; **G**, **H** during clonidine period. *EMG traces* are plotted as a function of time in seconds, *vertical lines* denote cycle boundaries and *upward arrows* along axes denote stance-swing transition within each cycle. *TA*, Tibialis anterior; *SOL*, soleus; *VL*, vastus lateralis; *MH*, medial hamstrings. Stick-figure reconstructions are of the left side of his body, treadmill direction is to the right. All stick figures have been additionally displaced beyond the distance actually travelled by the segments (*arrow direction*) in order to improve visual clarity. For *B*, *F* and *H*, the stance figures are a composite of images 0.167 s apart, and the swing figures are a composite of images 0.05 s apart. For *D*, because of the extraordinarily long cycles, the stance figures are a composite of images 0.20 s apart, and the swing figures are a composite of images 0.067 s apart leg showed relatively little flexion. He did not use the harness support. This pattern was essentially unchanged throughout his participation in the study. It is of interest to note small but consistent changes in his EMG patterns in response to medications, shown in Fig. 3; all data shown are from trials at a treadmill speed of 0.10 m/s. During cyproheptadine therapy (16 mg total daily dose), EMG activity in SOL was minimal at foot-floor contact (Fig. 3A). However, following the subsequent washout (Fig. 3B) and in the baclofen period (Fig. 3C), the EMG activity in SOL at foot-floor contact increased, almost as large as the mid-stance burst that occurred during the period of single-limb support. Two evaluations later, during clonidine therapy (0.25 mg total daily dose), the SOL activity at foot-floor contact was again minimal (Fig. 3D). Over the course of the seven evaluations conducted on S2, his preferred treadmill speed and his maximal treadmill speed gradually increased. In his seventh evaluation, his preferred treadmill speed was 0.16 m/s and his maximal speed for ten cycles was 0.22 m/s. His overground walking capability did not change in terms of aids required, gait speed or endurance. For S1 and S2 the adverse effects have been minimal. They reported minimal drowsiness during the periods of increasing dosage of either baclofen or cyproheptadine. S2 reported mild increase in appetite with cyproheptadine. Both reported dry mouth or dry throat from clonidine. Evaluation by a nurse during clonidine increases revealed no changes in blood pressure. In general, subjects regarded the adverse effects at these dosages as minor inconveniences.

Discussion

The preliminary results thus far are consistent with those of previous studies showing modulation of locomotor pattern with cyproheptadine and/or clonidine in human subjects (Fung et al. 1990; Stewart et al. 1991; Wainberg et al. 1990). First, cyproheptadine and clonidine have again been shown to be associated with improvements in locomotor pattern. The rationale from studies in spinalized cats for using cyproheptadine and clonidine in human beings with SCI continues to be upheld, although the effects on locomotor pattern have been less clearly specific to the individual medications. Second, of the two subjects with incomplete SCI, the one with greater manifestations of spasticity and functional deficit (S1) demonstrated more marked changes in locomotor pattern. This suggests that the potential benefit of medications, particularly cyproheptadine and clonidine for S1, may be greater for a more disabled subject – in particular, for a subject who was formerly incapable of overground gait. Third, oral cyproheptadine, clonidine and baclofen at the doses presently used produced minimal adverse effects. Our limited data thus far about baclofen suggest a minimal effect on locomotor pattern, but this needs to be further investigated.

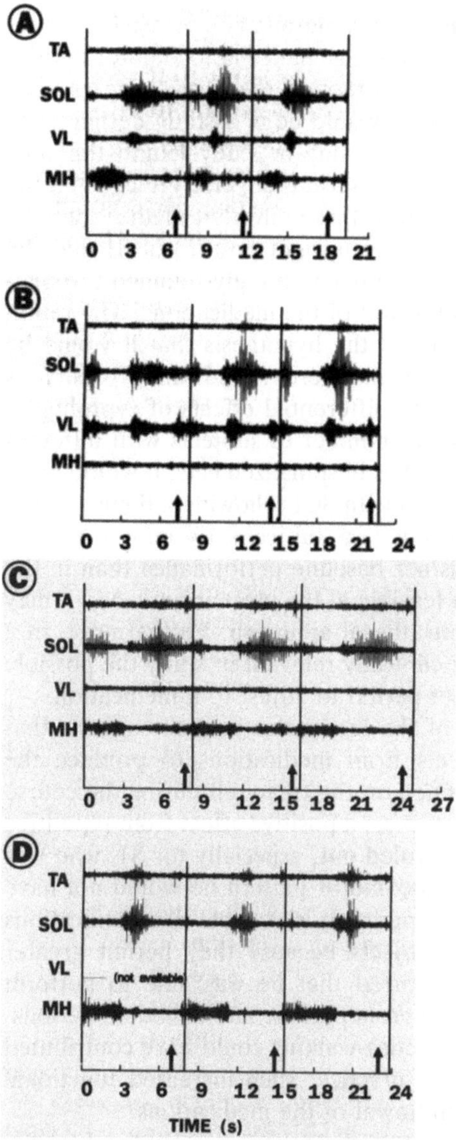

Fig. 3A–D. Raw EMG cycles of three cycles for subject two (S2): **A**, during cyproheptadine period; **B**, during post-cyproheptadine washout; **C**, during baclofen period; **D**, during clonidine period. All abbreviations and symbols are the same as in Fig. 2

The present study was designed to explore the differential effects of cyproheptadine, clonidine and baclofen. A single-subject design, whereby each subject takes all three medications, was chosen because of the heterogeneity of subjects drawn from the SCI population and the differences in their responses to the medications. For the same number of recruited subjects, there is a greater possibility of detecting a change in locomotor pattern that can be attributed to a medication. Although the design causes each subject's participation to be longer than it would be in a study design using groups, there has been no subject drop-out due to study length thus far. Presumably, this is related to the low risk of serious adverse effects, and the greater potential benefit from trying each of three different medications.

Some of the improvements, such as the upright posture in S1 and the greater treadmill speed for S1 and S2, have been partially retained throughout the washout period following withdrawal of the medications. The retention of these improvements has confirmed the hypothesis that it would be important to have different subjects follow different medication orders. A large study is in progress to compare the differential effects of cyproheptadine, clonidine and baclofen in a greater number of subjects with different medication orders. To compare with studies in spinalized cats, it would be of scientific interest to study human subjects with SCI following a bolus dose of each of the medications. In such a study, a subject's locomotor pattern would be more likely to return to his/her baseline performance than in the present study. However, this was not feasible at the present time, and it may be best accomplished only with an intrathecal approach. Furthermore, in a study of locomotor pattern it is more clinically relevant to study the possible changes once a subject has had a brief period to adjust to a medication.

The retention of some features of the locomotor pattern suggests that other factors interact with the effects from medications to produce the changes seen. S1 and S2 were not trained on the treadmill during the course of their participation in this study. However, a positive effect from repetitive evaluation in the laboratory cannot be ruled out, especially for S1, who was afforded an opportunity to practice a locomotor pattern he would not have otherwise had because of the harness support. It is possible that medications may be effective in spastic paretic subjects because they permit greater physical activity. For example, S1 reported that he was able to perform more physical activity at home in the cyproheptadine and clonidine periods. In particular, an increased ability to practice walking could have contributed to the improvements. After continued practice, such increased functional abilities were retained even upon withdrawal of the medications.

In summary, we have evidence that participation in a study of cyproheptadine, clonidine and baclofen may improve locomotor pattern, but the differential effects of the medications and the mechanism of the improvements are presently under investigation. It may be that the effectiveness of the medications is related to changes in stretch reflexes, although these changes may not necessarily correspond to improved motor control (for discussion see Armstrong 1992; Giuliani 1991; Whyte and Morrissey 1990). As more information becomes available about the quantitative effects of medications in spastic paretic subjects, we may have a greater understanding of the roles of different neurotransmitter systems in spasticity and of the possible benefits of therapies for gait rehabilitation. The Canadian multicentre trial in preparation will address these issues.

Acknowledgements. This project has been supported by grants from the Medical Research Council of Canada, and by the (Canadian) Network for Centers of Excellence (NCE) in Neuronal Regeneration and Functional Recovery. K.E. Norman is supported by a studentship from the Rick Hansen Man-in-Motion Legacy Fund. H. Barbeau is a *chercheur-boursier* of the *Fonds de recherche en santé du Québec*. Merck Sharpe & Dohme Canada are thanked for their contribution of cyproheptadine (Periactin). A. Pepin, S. Séguin, S. Carrier, M. Ladouceur, L. Glasa, K. Chitre and G. Bossé are thanked for their assistance with experiments; A. Pepin and S. Vanden Noven are thanked for their comments on earlier versions of this manuscript; Dr. R. Beaupré is thanked for referral of the subjects; and finally, the subjects are thanked for their participation.

References

Armstrong RW (1992) Intrathecal baclofen and spasticity: what do we know and what do we need to know? Dev Med Child Neurol 34:739–745

Armstrong RW, Steinbok P, Farrell K, Cochrane D, Norman MG, Kube S (1992) Continuous intrathecal baclofen treatment of severe spasms in two children with spinal-cord injury. Dev Med Child Neurol 34:731–738

ASIA (American Spinal Injury Association) (1982) Standards for neurological classification of spinal injury patients.

Barbeau H, Blunt RA (1991) A novel interactive locomotor approach using body weight support to retrain gait in spastic paretic subjects. In: Wernig A (ed) Plasticity of motoneuronal connections. Elsevier, Amsterdam, pp 461–474

Barbeau H, Rossignol S (1990) The effects of serotonergic drugs on the locomotor pattern and on cutaneous reflexes of the adult chronic spinal cat. Brain Res 514:55–67

Barbeau H, Richards CL, Bédard PJ (1982) Action of cyproheptadine in spastic paraparetic patients. J Neurol Neurosurg Psychiatry 45:923–926

Barbeau H, Julien C, Rossignol S (1987a) The effects of clonidine and yohimbine on locomotion and cutaneous reflexes in the adult chronic spinal cat. Brain Res 437:83–96

Barbeau H, Wainberg M, Finch L (1987b) Description and application of a system for locomotor rehabilitation. Med Biol Eng Comput 25:341–344

Bowery N (1989) GABAB receptors and their significance in mammalian pharmacology. Trends Pharmacol Sci 10:401–407

Broseta J, Morales F, García-March G, Sánchez-Ledesma MJ, Anaya J, Gonzalez-Darder J et al. (1989) Use of intrathecal baclofen administered by programmable infusion pumps in resistant spasticity. Acta Neurochir Suppl (Wien) 46:39–45

Corston RN, Johnson F, Godwin-Austen RB (1981) The assessment of drug treatment of spastic gait. J Neurol Neurosurg Psychiatry 44:1035–1039

Davidoff RA (1985) Antispasticity drugs: mechanisms of action. Ann Neurol 17:107–116

Davidoff RA, Sears S (1974) The effects of Lioresal on synaptic activity in the isolated spinal cord. Neurology 24:957–963

Delwaide PJ (1985) Electrophysiological analysis of the mode of action of muscle relaxants in spasticity. Ann Neurol 17:90–95

Donovan WH, Carter RE, Rossi CD, Wilkerson MA (1988) Clonidine effect of spasticity: a clinical trial. Arch Phys Med Rehabil 69:193–194
Forssberg H, Grillner S (1973) The locomotion of the acute spinal cat injected with clonidine i.v. Brain Res 50:184–186
Fox S, Krnjevic K, Morris ME, Puil E, Werman R (1978) Action of baclofen on mammalian synaptic transmission. Neuroscience 3:495–515
Fung J, Stewart JE, Barbeau H (1990) The combined effects of clonidine and cyproheptadine with interactive training on the modulation of locomotion in spinal cord injured subjects. J Neurol Sci 100:85–93
Giuliani CA (1991) Dorsal rhizotomy for children with cerebral palsy: support for concepts of motor control. Phys Ther 71:248–259
Hankey GJ, Stewart-Wynne EG, Perlman D (1986) Intrathecal baclofen for severe spasticity. Med J Aust 145:465–466
Hugenholtz H, Nelson RF, Dehoux E, Bickerton R (1992) Intrathecal baclofen for intractable spinal spasticity – a double-blind cross-over comparison with placebo in six patients. Can J Neurol Sci 19:188–195
Kofler M, Donovan WH, Loubser PG, Beric A (1992) Effects of intrathecal baclofen on lumbosacral and cortical somatosensory evoked potentials. Neurology 42:864–868
Latash ML, Penn RD, Corcos DM, Gottlieb GL (1989) Short-term effects of intrathecal baclofen in spasticity. Exp Neurol 103:165–172
Latash ML, Penn RD, Corcos DM, Gottlieb GL (1990) Effects of intrathecal baclofen on voluntary motor control in spastic paresis. J Neurosurg 72:388–392
Lazorthes Y, Sallerin-Caute B, Verdié J-C, Bastide R, Carillo J-P (1990) Chronic intrathecal baclofen administration for control of severe spasticity. J Neurosurg 72:393–402
Little JW, Merritt JL (1988) Spasticity and associated abnormalities of muscle tone. In: DeLisa JA, Currie DM, Gans BM, Gatens PF Jr, Leonard JA jr, McPhee M (eds) Rehabilitation Medicine: Principles and Practice. Lippincott, Philadelphia, pp 430–447
Loubser PG, Narayan RK, Sandin KJ, Donovan WH, Russell KD (1991) Continuous infusion of intrathecal baclofen: long-term effects on spasticity in spinal cord injury. Paraplegia 29:48–64
Marshall KC (1983) Catecholamines and their actions in the spinal cord. In: Davidoff RA (ed) Handbook of the Spinal Cord, vol 1, Pharmacology. Dekker, New York, pp 275–328
Marshall KC, Engberg I (1979) Reversal potential for noradrenaline-induced hyperpolarization of spinal motoneurons. Science 205:422–424
McLellan DL (1977) Co-contraction and stretch reflexes in spasticity during treatment with baclofen. J Neurol Neurosurg Psychiatry 40:30–38
Meythaler JM, Steers WD, Tuel SM, Cross LL, Sesco DC, Haworth CS (1992) Intrathecal baclofen in hereditary spastic paraparesis. Arch Phys Med Rehabil 73:794–797
Milanov IG (1992) Mechanisms of baclofen action on spasticity. Acta Neurol Scand 85:305–310
Nance PW, Shears AH, Nance DM (1985) Clonidine in spinal cord injury. Can Med Assoc J 133:41–42
Nistri A (1983) Spinal cord pharmacology of GABA and chemically related amino acids. In: Davidoff RA (ed) Handbook of the Spinal Cord, vol 1, Pharmacology. Dekker, New York, pp 45–104
Parke B, Penn RD, Savoy SM, Corcos DM (1989) Functional outcome after delivery of intrathecal baclofen. Arch Phys Med Rehabil 70:30–32
Pedersen E, Arlien-Soborg P, Mai J (1974) The mode of action of the GABA derivative baclofen in human spasticity. Acta Neurol Scand 50:665–680
Penn RD (1991) Intrathecal infusion of baclofen for spasticity: the RUSH and the US multicenter studies. In: Sindou M, Abbott IR, Keravel Y (eds) Neurosurgery for Spasticity. A Multidisciplinary Approach. Springer, Vienna New York, pp 103–109
Penn RD, Kroin JS (1987) Long-term intrathecal baclofen infusion for treatment of spasticity. J Neurosurg 66:181–185

Pierau F-K, Zimmermann P (1973) Action of a GABA derivative on postsynaptic potentials and membrane properties of cats' spinal motoneurones. Brain Res 54:376–380
Stewart JE, Barbeau H, Gauthier S (1991) The effects of clonidine in the modulation of the locomotor pattern and spasticity in spinal cord injured patients: a preliminary study. Can J Neurol Sci 18:321–331
Stuart GJ, Redman SJ (1992) The role of GABAA and GABAB receptors in presynaptic inhibition of la EPSPs in cat spinal motoneurones. J Physiol (Lond) 447:675–692
Visintin M, Barbeau H (1989) The effects of body weight support on the locomotor pattern of spastic paretic patients. Can J Neurol Sci 16:315–325
Wainberg M, Barbeau H, Gauthier S (1986) Quantitative assessment of the effect of cyproheptadine on spastic paretic gait: a preliminary study. J Neurol 233:311–314
Wainberg M, Barbeau H, Gauthier S (1990) The effects of cyproheptadine on locomotion and on spasticity in patients with spinal cord injuries. J Neurol Neurosurg Psychiatry 53:754–763
Wang MY, Dun NJ (1990) Phaclophen-insensitive presynaptic inhibitory action of (±)-baclofen in neonatal rat motoneurones in vitro. Br J Pharmacol 99:413–421
Weingarden SI, Belen JG (1992) Clonidine transdermal system for treatment of spasticity in spinal cord injury. Arch Phys Med Rehabil 73:876–877
Whyte J, Morrissey J (1990) Motor learning and relearning. In: Glenn MB, Whyte J (eds) The Practical Management of Spasticity in Children and Adults. Lea and Febiger, Malvern, pp 44–69
Whyte J, Robinson KM (1990) Pharmacologic management. In: Glenn MB, Whyte J (eds) The Practical Management of Spasticity in Children and Adults. Lea and Febiger, Malvern, pp 201–226
Wüllner U, Klockgether T (1990) GABAB receptors and spasticity. Is the antispastic action of baclofen mediated by GABAB receptors? Trends Pharmacol Sci 11:103
Young PR, Delwaide PJ (1981) Drug therapy. Spasticity. N Engl J Med 304:28–33, 96–99
Zieglgänsberger W (1988) Dorsal horn neuropharmacology: baclofen and morphine. Ann NY Acad Sci 531:150–156

Neurophysiological Mechanisms of Spasticity Modification by Physiotherapy

H. Hummelsheim and K.-H. Mauritz

Introduction

The term "spasticity" is used in a variety of circumstances describing impaired movement execution, enhanced muscular resistance against passive movement or abnormal limb postures. Furthermore, many clinicians equate the degree of spasticity with the excitability level of tendon reflexes.

Most pharmacological studies dealing with drug treatment of spasticity use stretch reflex excitability and/or clonus intensity as parameters to assess the efficacy of antispastic treatment. These parameters have only little functional significance as the intensity of stretch reflex responses and of clonus hardly corresponds to the quality of movement execution. Any kind of stress, higher body temperature, or raised blood levels of catecholamines enhance tendon reflex excitability but do not produce any kind of movement disorder.

Many chronic stroke patients present high degrees of spasticity especially in distal arm and hand flexors. According to clinical experience, the physiotherapeutic method of Bobath (1970; neurodevelopmental technique) is the most effective of the various physiotherapeutic techniqes (Hummelsheim und Mauritz 1993) for reducing spastic muscular hypertonus. The Bobath concept is based on the idea that active movements should be encouraged only when maximal reduction of spasticity has been achieved. Normalizing muscle tone has higher priority than promoting voluntary movements in the paretic limb.

The present paper evaluates the role of spasticity reduction in various physiotherapeutic concepts in the context of rehabilitation of motor functions. In the process of facilitating agonist activity the reduction in spastic hypertonus in the antagonist muscle is an important mechanical prerequisite to allow free range of motion in a particular joint. We argue that the excitability level in spinal α-motoneurones can be reduced by physiotherapy. This reduction in motoneuronal excitability is thought to decrease the mechanical constraints exerted by the spastic antagonist. We postulate that agonist

Klinik Berlin, Abt. für Neurologische Rehabilitation, Freie Universität Berlin, Kladower Damm 223, D-14089 Berlin, Deutschland

Thilmann et al. (Eds.)
Spasticity: Mechanisms and Management

activity itself – measured electrophysiologically – hardly depends on the degree of spasticity in the antagonist.

Sustained Muscle Stretch Reduces Motoneuronal Excitability

Spasticity (as well as raised stretch reflex excitability) can be explained at least in part by enhanced motoneurone excitability (Artieda et al. 1991; Eisen and Odusote 1979); we have therefore been interested in the effect of spasticity-reducing physiotherapeutic intervention on the characteristics of the early and late muscular response potentials to transcranial magnetic stimulation (TMS). Our hypothesis is that the physiotherapeutic method of sustained stretch of spastic muscle groups exerts an inhibitory effect on spinal α-motoneurones that should be reflected in the response characteristics to TMS.

TMS was used in a group of stroke patients with lesions in the territory of the medial cerebral artery to investigate changes in muscular responses after the administration of spasticity-reducing physiotherapy. No patient was able to perform voluntary hand or finger movements except in a coarse synergism starting in the shoulder or proximal arm musculature. Spasticity in hand and finger flexors was scored grade 4 and grade 5 on the Ashworth scale (Wade 1992):

0: No increase in muscle tone.
1: Slight increase in muscle tone, mild clasp-knife phenomenon, or minimal resistance at the end of the range of motion when the affected part(s) is moved in flexion or extension.
2: Slight increase in muscle tone, clasp-knife phenomenon, followed by minimal resistance throughout the remainder (less than half) of the range of motion.
3: More marked increase in muscle tone through most of the range of motion, but affected part(s) easily moved.
4: Considerable increase in muscle tone, passive movement difficult.
5: Affected part(s) rigid in flexion or extension.

At first, TMS was performed while the patient was seated comfortably with the affected arm in a middle position between pronation and supination and before any physiotherapeutic intervention. Electromyographic recordings were taken by means of conventional surface electrodes from the extensor and flexor carpi radialis muscles. All patients gave their informed consent to the experiments. Further methodological details are dealt with elsewhere (Hummelsheim et al. 1992).

In the relaxed state, distal arm muscles of healthy subjects regularly present a short-latency response following TMS with a relatively constant latency. In the case of voluntary preinnervation of the target muscle a second late response occurs at a latency of more than 50 ms. This is observed both in healthy subjects and in patients with central nervous system lesions (Boniface et al. 1991; Calancie et al. 1987; Mills et al. 1991).

Spasticity reduction was achieved according to the Bobath concept (1970) by stretching the spastic muscles slowly and tonically in a proximodistal sequence beginning with a prolonged outward scapula rotation. Then the arm was abducted and rotated outward in the shoulder and extended in the elbow, followed by an extension in the wrist and finger joints and an abduction of the thumb (Fig. 1). In the position in which most spastic flexor muscles were tonically stretched, muscle tone gradually decreased. After the reduction in muscle tone had reached its maximum (normally after 10–15 min) and after repositioning of the arm in the previous position response latency, response duration and stimulation threshold for the early and the late muscular responses to TMS were measured as before the physiotherapeutic intervention.

After the described physiotherapeutic intervention all parameters reflecting motoneurone excitability were reduced. The stimulation threshold for the early and the late responses were markedly increased (Fig. 2), latency was slightly longer, and the amount of the response potentials and their frequency of occurrence were clearly diminished. Surprisingly, although amplitude reduction was most pronounced in the stretched flexor muscle, the response potential in the extensor was also diminished (Fig. 2).

Besides the manual muscle stretch performed by the physiotherapist, another method for obtaining an effective reduction in the spastic muscle hypertonus is to keep the wrist and the finger joints in an extended position by means of distal splinting or plaster casting (Booth et al. 1983; Kaplan 1962; Odeen 1981). TMS after 24–48 h of casting performed in two patients elicited similar modifications in muscular response characteristics as after the physiotherapeutic stretching intervention. In comparison to physiotherapeutic spasticity reduction by sustained muscle stretch for 10–15 min distal casting for about 48 h had a clearly longer lasting effect.

Our conclusion from these observations is that sustained muscle stretch as proposed by Bobath (1970) is able to reduce enhanced motoneuronal excitability. The effect was mediated by stretch receptor adaptation to the new extended position with a relatively reduced excitatory drive from the muscle spindles when the hand was returned to a slightly flexed position. The same model of receptor adaptation or fatigue holds true for distal casting with the additional effect of a longer lasting spasticity reduction.

An inhibition of the enhanced tone of the stretched muscle by Golgi tendon organs is less probable, as these respond predominantly to actively generated tension (Houk et al. 1971). Nevertheless, Ib-inhibitory interneurones known to mediate input from Golgi tendon organs onto the α-motoneurones of homologous muscles are assumed to play an important role in transmitting inhibitory influences (Harrison and Jankowska 1985a,b; Jankowska et al. 1981) arising in even distant muscles being stretched by the

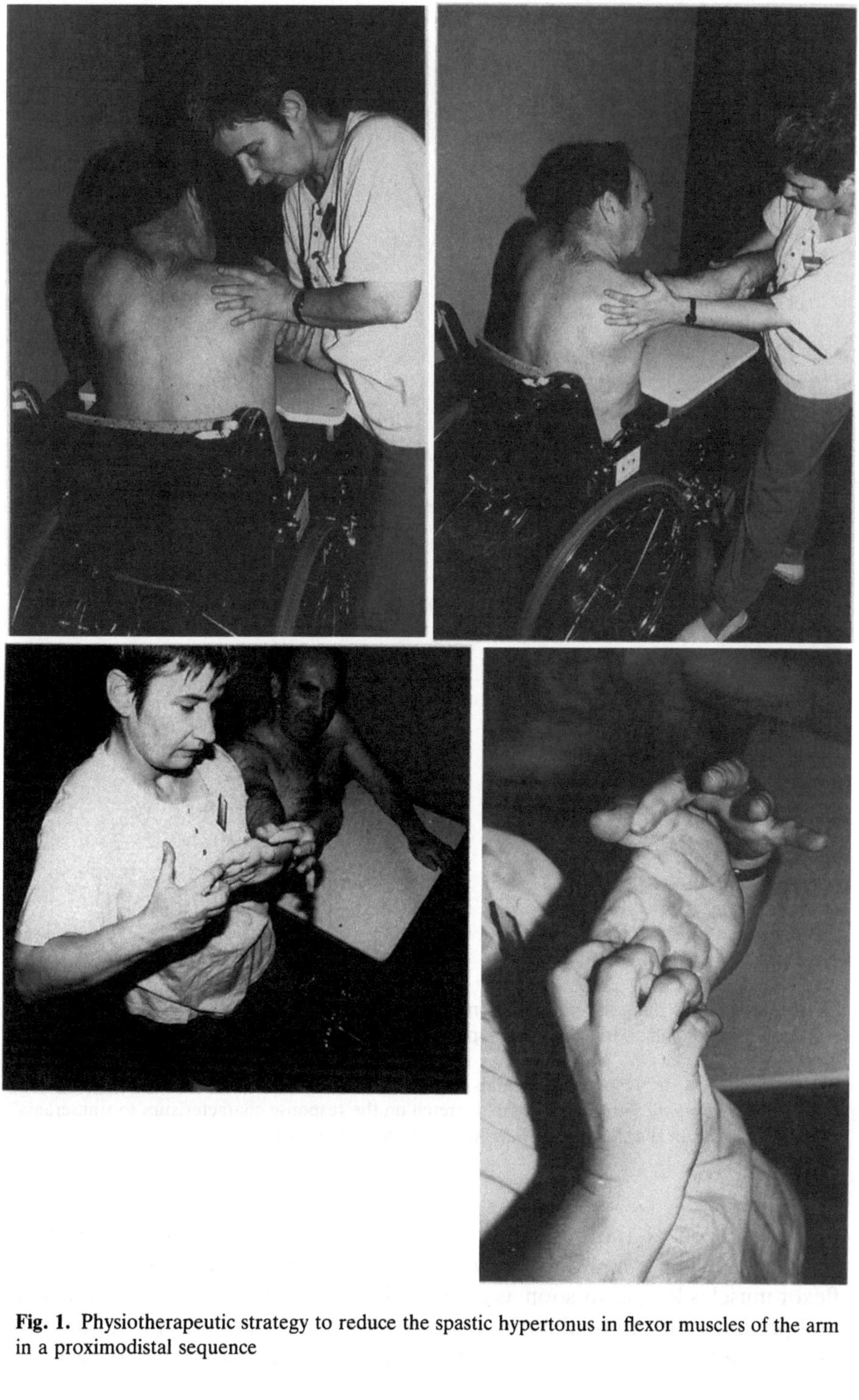

Fig. 1. Physiotherapeutic strategy to reduce the spastic hypertonus in flexor muscles of the arm in a proximodistal sequence

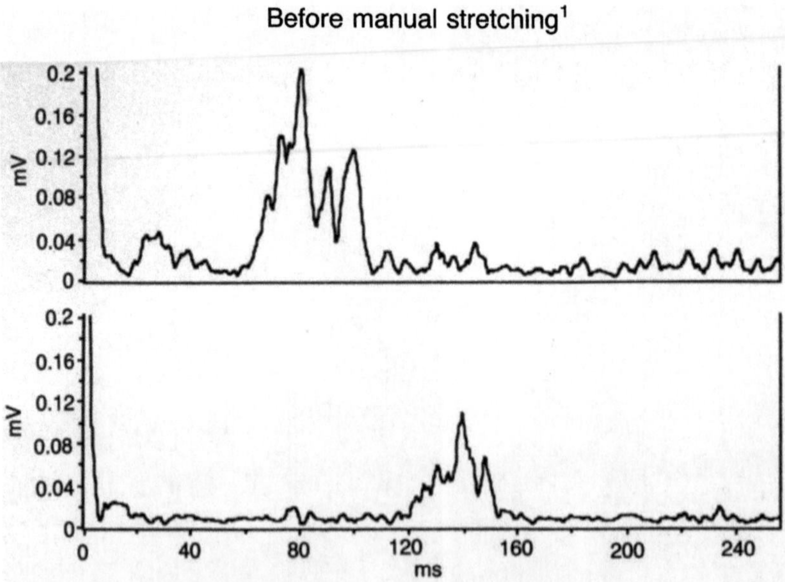

	Early Responses	Late Responses
Threshold (mean value)	78%	90%
Latency (mean value)	26 ms	79 ms
(range)	(19–41 ms)	(61–151 ms)
Duration (mean value)	25 ms	47 ms
(range)	(21–40 ms)	(28–79 ms)
Frequency	72% (in 8 patients)	62% (in 8 patients)

[1] no response in 3 patients; early response in 4 patients; late response in 4 patients; early and late responses in 4 patients

Fig. 2. *Top*, Muscular response potentials before and after sustained stretch of the hand flexors (lower trace). *Bottom*, Effect of sustained stretch on the response characteristics to transcranial brain stimulation in the flexor carpiradialis muscle of 15 patients

physiotherapist at the beginning of the described proximodistal stretching sequence. It is a common observation that the spastic hypertonus of distal flexor muscles lessens as soon as the stretching sequence starts in the shoulder and proximal arm musculature. This transmission of inhibition is thought to be mediated via Ib-inhibitory interneurones serving as an integrating relay

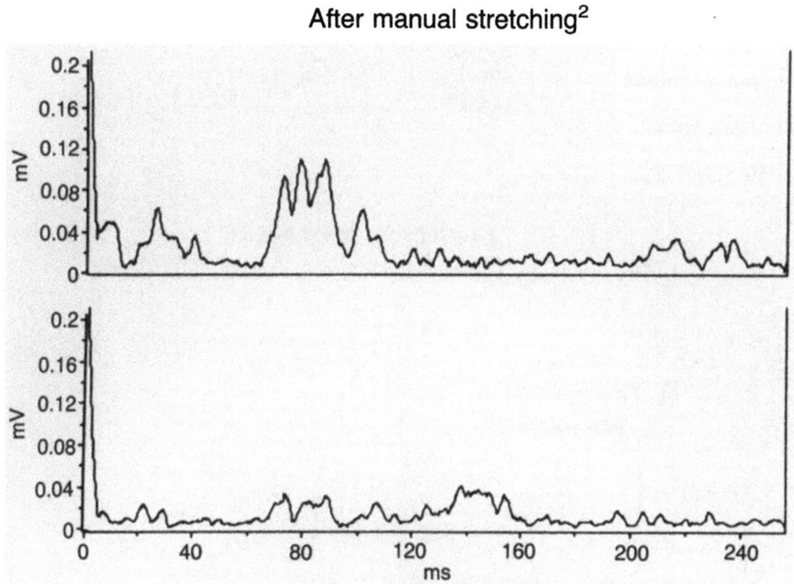

	Early Responses	Late Responses
Threshold (mean value)	93%	100%
Latency (mean value)	28 ms	110 ms
(range)	(19–44 ms)	(81–159 ms)
Duration (mean value)	22 ms	37 ms
(range)	(16–31 ms)	(35–40 ms)
Frequency	60% (in 6 patients)	25% (in 3 patients)

[2] no response in 8 patients; early response in 4 patients; late response in 1 patients; early and late responses in 2 patients

Fig. 2. (Cont.)

station of influences from muscle spindle afferents arising in more proximal muscle groups (Fig. 3). The aforementioned mild inhibitory effect on extensor motoneurones may also be explained by this mechanism.

In summary, we hypothesize that the enhanced excitability of spinal α-motoneurones and interneurones in the spastic state is reduced by sustained muscle stretch starting in the shoulder and proximal arm musculature. The inhibitory effect of this physiotherapeutic intervention can be explained by

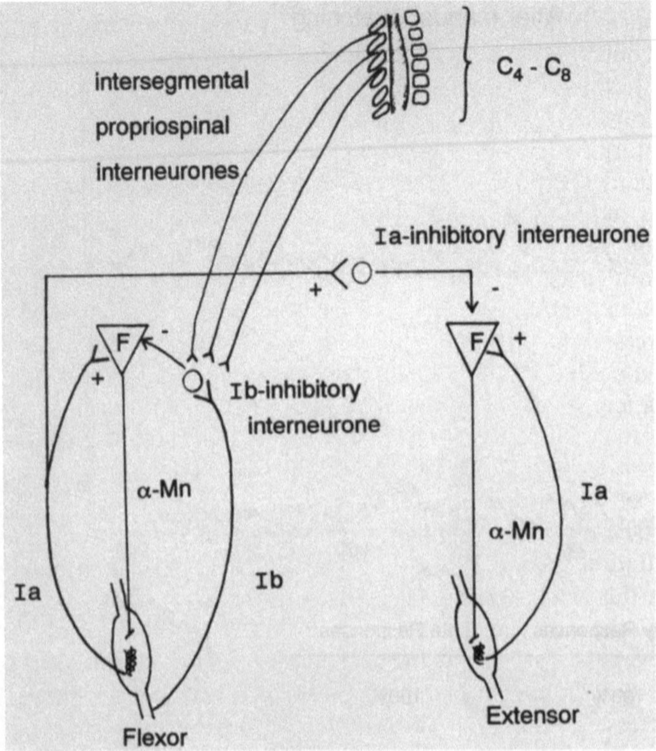

Fig. 3. Neuronal structures mediating spasticity inhibition

stretch receptor adaptation in the spastic target muscle and by an inhibition due to early stretching of more proximal muscles mediated via the integrating Ib-inhibitory interneurones projecting to the target muscle via intersegmental propriospinal interneurones (Lundberg 1979).

Spasticity Reduction and Voluntary Movements

The therapeutic strategy of Bobath (1970) always starts with the reduction in spasticity. The theory proposed by Bobath that strong recurrent Ia inhibition from the spastic muscle forms the principal basis for the weakness of its antagonist has not been confirmed. Ia inhibition is known, instead, to be less effective in hemiparetic patients (Artieda et al. 1991). Therefore, spasticity reduction by sustained stretch does not deliver a powerful disinhibition of the α-motoneurones of the weak antagonist. Probably the most important effect of sustained stretch for the release of voluntary movements is mechanical in nature. In the case of spasticity in hand and finger flexors voluntary movements of the weak hand and finger extensors could become possible if the antagonistic flexor hypertonus is reduced sufficiently to allow relatively free extension movements.

One decisive drawback of the Bobath method is the aforementioned avoidance of voluntary active movements in weak muscle groups before maximal reduction of muscle tone in spastic muscle groups has been achieved. Recent experiments in our laboratory aiming at reinforcing and training rapid voluntary hand and finger movements (Bütefisch et al. 1992a,b) in patients with a central paresis of the hand showed that the frequency of associated reactions and the spastic hypertonus of hand and finger flexors diminished (as expressed on the Ashworth scale) during the training phase of several weeks while hand and finger function improved. It can be hypothesized that the defective Ia inhibition from the weak extensors on the flexors becomes more effective parallel to the improvement in hand and finger function.

The occurrence of "associated reactions" is a frequent phenomenon in the physiotherapeutic method of Brunnstrom (1970). She emphasizes the use of central facilitation techniques, i.e., maximal innervation of corresponding muscles on the intact side of the body or innervation of less weak proximal muscles of the affected extremity to recruit more paretic distal muscle groups by overflow or irradiation mechanisms. In this context synergistic movements and the development of higher muscle tone are often observed. Nevertheless, Brunnstrom recognizes the necessity of refining the patient's movement repertoire. She therefore utilizes localized skin stimulation and tapping over the muscle belly to facilitate the activity of individual muscles or of distinct muscle groups. Her argument that facilitation of circumscribed muscle groups is appropriate to diminish spasticity of their antagonists via Ia inhibition does not take into account that Ia inhibition is defective in patients with central hemiparesis (Artieda et al. 1991). Cheney et al. (1985) described a reciprocal inhibition of corticomotoneuronal cells on antagonist muscles during voluntary innervation of the agonist in awake monkeys. Whether suppression of antagonist activity during agonist innervation at the cortical level also takes place in hemiparetic patients must be doubted (Bütefisch et al. 1992b). Nevertheless, it may be assumed that inhibition of spastic antagonists by Ia inhibition increases parallel to the improved voluntary movement capacity in the agonist. Figure 4 shows a diminished cocontraction tendency after 8 weeks of enhanced training of rapid alternating movements at the wrist.

Comparable mechanisms of spasticity reduction by reciprocal Ia inhibition are used by therapists applying proprioceptive neuromuscular facilitation (PNF) techniques (Knott and Voss 1968). Synergistic coactivation within particular multijoint movement patterns is the basic principle in the PNF technique, using irradiation effects from less paretic proximal muscles to produce and strengthen contraction in weak distal muscles. Parallel to the increase in strength of the target muscle group, Ia inhibition is thought to

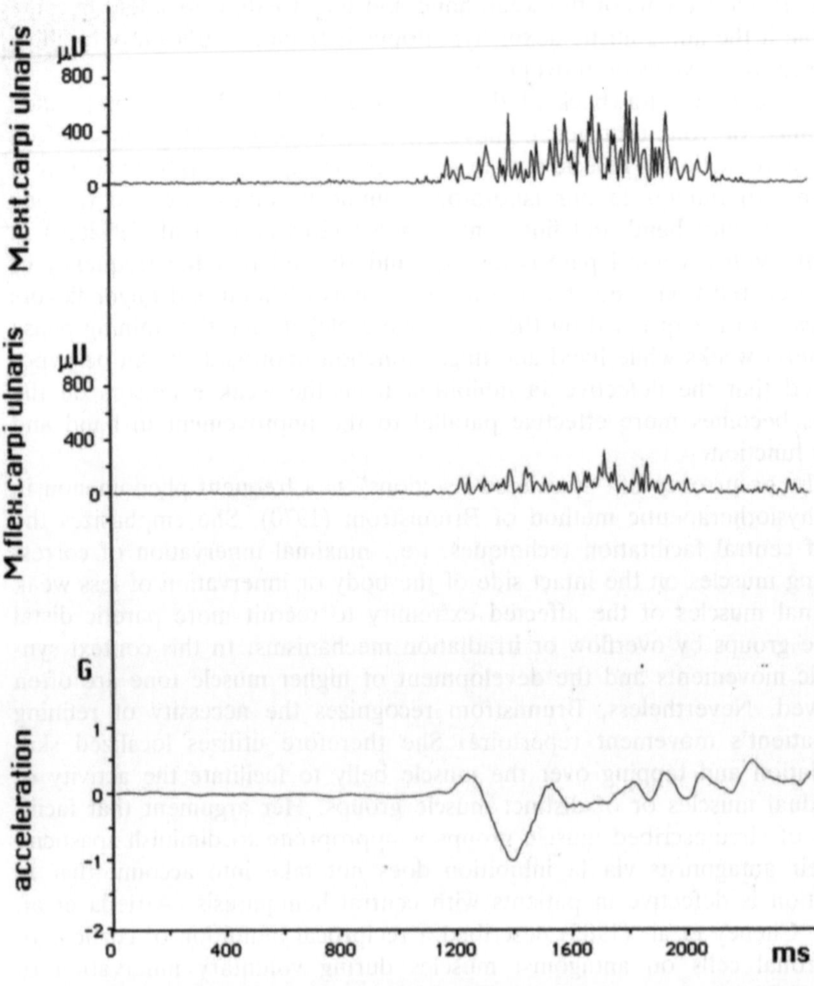

Fig. 4. Agonist/antagonist co-contraction before and after functional neuromuscular training

inhibit spasticity in the respective antagonistic muscle group. In clinical practice, however, the occurrence of associated reactions and a tendency toward undesired spastic limb postures is a common observation during PNF therapy of stroke patients with severe hemiparesis.

Mechanical Factors in Spasticity Are Often Underestimated

Apart from the neuronal mechanisms, mechanical factors concerning the viscoelastic and contractile properties of the muscle must not be under-

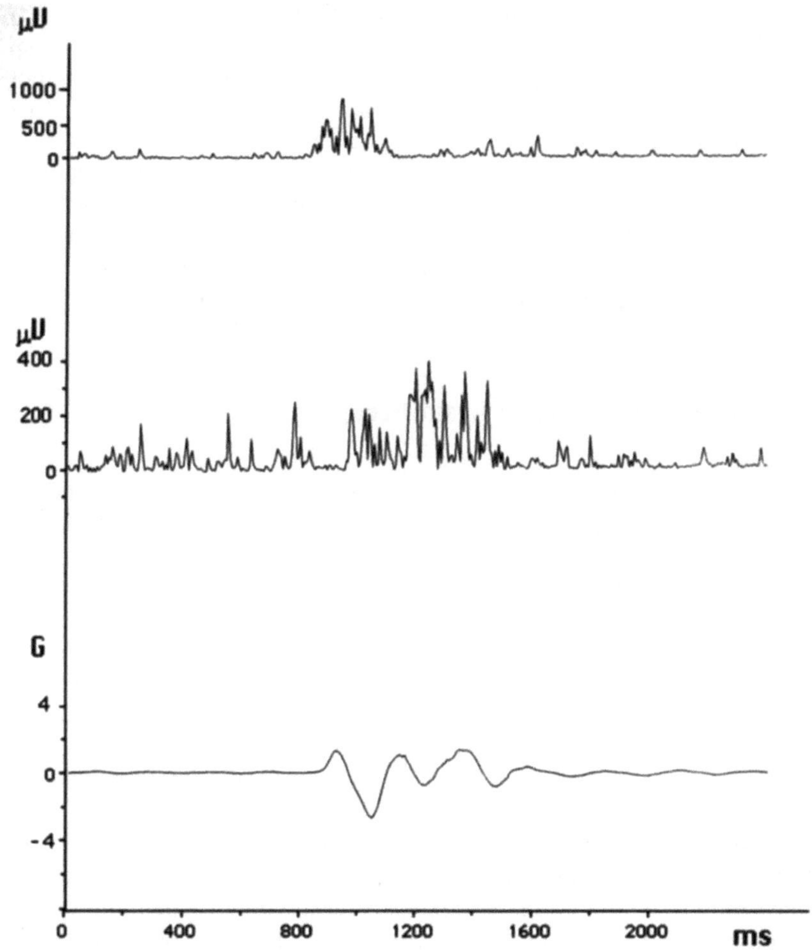

Fig. 4. (Cont.)

estimated (Dietz et al. 1981; Dietz 1990). Muscle stiffness is determined by elasticity of the contractile elements and the connective tissue as well as by the discharge frequence of the motor units (Grillner 1972; Hill 1968). Discharge frequency is influenced by sustained stretch. Nevertheless, as physiotherapeutic techniques adhere most frequently to neurophysiological principles, it is important to stress the necessity of daily passive range-of-motion exercises to prevent, at least in part, an increase in muscle stiffness and a joint and tendon tightening.

Stroke patients who spend a prolonged period in the wheel chair show a tendency to develop a marked imbalance in muscle tone in the affected lower extremity toward hip and knee flexion. This undesired flexion in hip and knee together with ankle plantarflexion may be overcome by placing the body weight on the affected joints while the patient stands with the affected joints in the extended position. The same effect, i.e., prolonged stretching of tight muscles, can be achieved by putting a wedge between the limbs that tend to approach each other. This method is most appropriate for treating leg scissoring due to hip adductor spasticity in paraplegics.

Prolonged tonic stretching of circumscribed spastic muscles or muscle groups can also be achieved by static and/or dynamic splints (Kaplan 1962; Macpherson et al. 1985; Odeen 1981), the aforementioned distal casting, or special orthoses adapted to the individual patient, for instance, ankle-foot orthoses or knee-ankle-foot orthoses (Little and Merritt 1988).

Muscle Cooling to Prepare Physiotherapy

Muscle cooling is of benefit for the spastic patient provided that it is applied long enough. This influences particularly stretch reflex activity and clonus. The initial effect of cold application is even facilitatory, as has been described by Rood (1956). Hagbarth (1952) demonstrated that phasic localized skin stimulation by "icing" induces an enhanced γ-innervation of the muscles underlying the stimulated skin. This in turn causes increased sensitivity in stretch receptors with increased stretch response and enhanced muscle tone.

The administration of prolonged and deep muscle cooling leads in many cases to a sufficient relaxation in spastic muscle groups. PNF therapists were the first to use penetrating muscle cooling in preparation for their active movement therapy to reduce muscle tone and to eliminate pain.

The almost immediate drop in stretch reflex excitability following the administration of cold cannot be due to intramuscular temperature reduction since muscle temperature requires several minutes (depending on the muscle mass) to fall significantly. The effect must therefore be mediated by a decrease in skin receptor sensitivity and its influence on α/γ-motoneurone activity (Herman and Byck 1964). Surprisingly, clonus is absent only when intramuscular temperature is markedly decreased (Miglietta 1973). It is unlikely that clonus is abolished due to transmission failures or blocks caused by cooling of the motor nerve since M and H responses show only moderate increases in latency. Unfortunately, the effect of cold on muscle spindle sensitivity has not been sufficiently clarified. Animal experiments are in part contradictory; Eldred et al. (1960) described a nearly linear decrease in muscle spindle sensitivity parallel to reduction in muscle temperature, whereas Lippold et al. (1960) found an enhanced spontaneous firing rate at muscle temperatures 2°–3°C below body temperature. Nevertheless, studies in humans describe intramuscular temperatures of 4°–5°C below body temperature so that muscle spindle discharge is considered to be reduced. Whether the proposed effect is exerted exclusively on the muscle spindle itself or is the result of a suppression of the γ-innervation (Mecomber and

Herman 1971) has not been clarified up to now. In addition to the neuronal mechanisms, the reduction in muscular hypertonus can also be caused by altered viscoelastic and/or contractile properties of the muscle. Miglietta (1973) found markedly prolonged muscle contraction times after deep muscular cooling.

Spasticity reduction by deep muscle cooling is of increasing significance in various physiotherapeutic concepts since it allows the beginning of an active functional neuromuscular training of the weak agonists during the period of cold blockage of the spastic antagonist.

References

Artieda J, Quesada P, Obeso J (1991) Reciprocal inhibition between forearm muscles in spastic hemiplegia. Neurology 41:286–289

Bobath B (1970) Adult hemiplegia: evaluation and treatment, 2nd edn. Heinemann, London

Boniface SJ, Mills KR, Schubert M (1991) Responses of single spinal motoneurons to magnetic brain stimulation in healthy subjects and patients with multiple sclerosis. Brain 114:643–662

Booth BJ, Doyle M, Montgomery J (1983) Serial casting for the management of spasticity in the head injured adult. Phys Ther 63:1960–1966

Brunnstrom S (1970) Movement therapy in hemiplegia. Harper and Row, New York

Bütefisch C, Hummelsheim H, Denzler P, Mauritz KH (1992a) Einfluß eines standardisierten Trainings auf Kraftentwicklung und schnelle Wechselbewegungen bei Patienten mit zentralen Paresen der Hand (Abstr). 4th Meeting of the Deutsche Gesellschaft für Neurologische Rehabilitation, Magdeburg

Bütefisch C, Hummelsheim H, Münch B, Mauritz KH (1992b) Kontraktions- und Bewegungsprofile der Hand bei Patienten mit zentralen Hemiparesen (Abstr). 4th Meeting of the Deutsche Gesellschaft für Neurologische Rehabilitation, Magdeburg

Calancie B, Nordin M, Wallin U, Hagbarth KE (1987) Motor-unit responses in human wrist flexor and extensor muscles to transcranial cortical stimuli. J Neurophysiol 58:1168–1185

Cheney PD, Fetz EE, Palmer SS (1985) Patterns of facilitation and suppression of antagonist forelimb muscles from motor cortex sites in the awake monkey. J Neurophysiol 53:805–820

Dietz V (1990) Spastik. Therapie der gesteigerten Reflexe oder der Bewegungsstörung? Nervenarzt 61:581–586

Dietz V, Quintern J, Berger W (1981) Electrophysiological studies of gait in spasticity and rigidity. Evidence that altered mechanical properties of muscle contribute to hypertonia. Brain 104:431–449

Eisen A, Odusote K (1979) Amplitude of the F-wave: a potential means of documenting spasticity. Neurology (Minneap) 29:1306–1309

Eldred E, Lindsley DF, Buchwald JS (1960) Effect of cooling on mammalian muscle spindles. Exp Neurol 2:144–157

Grillner S (1972) The role of muscle stiffness in meeting the changing postural and locomotor requirements for force development by the ankle extensors. Acta Physiol Scand 86:92–108

Hagbarth KE (1952) Excitatory and inhibitory skin areas for flexor and extensor motoneurons. Acta Physiol Scand 94:1–58

Harrison PJ, Jankowska E (1985a) Sources of input to interneurones mediating group I nonreciprocal inhibition of motoneurones in the cat. J Physiol (Lond) 361:379–401

Harrison PJ, Jankowska E (1985b) Organization of input to interneurones mediating group I non-reciprocal inhibition of motoneurones in the cat. J Physiol (Lond) 361:403–418

Herman R, Byck R (1964) Differential peripheral nerve blockade by procaine and local cold. Clin Pharmacol Ther 5/6(2):848–851

Hill DK (1968) Tension due to interaction between the sliding filaments in resting striated muscle: the effect of stimulation. J Physiol (Lond) 199:637–684

Houk JC, Singer JJ, Henneman E (1971) The adequate stimulus for Golgi tendon organs with observation on the mechanics of the ankle joint. J Neurophysiol 34:1051–1065

Hummelsheim H, Münch B, Bütefisch C, Hoppe S, Neumann S (1992) Transcranial magnetic stimulation in rehabilitation and physiotherapy evaluation in hemiparetic stroke patients. In: Lissens MA (ed) Clinical applications of magnetic transcranial stimulation. Peeters, Leuven, pp 290–300

Hummelsheim H, Mauritz K-H (1993) Neurophysiologische grundlagen krankengymnastischer übungsbehandlung bei Patienten mit Zentralen Hemiparesen. Fortschr Neurol Psychiat 61:208–216

Jankowska E, Johannisson T, Lipski J (1981) Common interneurones in reflex pathways from group Ia and Ib afferents of ankle extensors in the cat. J Physiol (Lond) 310:99–111

Kaplan N (1962) Effect of splinting on reflex inhibition and sensorimotor stimulation in treatment of spasticity. Arch Phys Med Rehabil 43:565–569

Knott M, Voss DE (1968) Proprioceptive neuromuscular facilitation. Harper and Row, New York

Lippold O, Nicholle J, Redfearm J (1960) A study of afferent discharge produced by cooling of mammalian muscle spindles. J Physiol (Lond) 153:218–231

Little JW, Merritt JL (1988) Spasticity and associated abnormalities of muscle tone. In: DeLisa JA (ed) Rehabilitation medicine, principles and practice. Lippincott, Philadelphia, pp 430–444

Lundberg A (1979) Multisensory control of spinal reflex pathways. Prog Brain Res 50:11–28

Macpherson JJ, Becker AH, Franszczak N (1985) Dynamic splint to reduce the passive component of hypertonicity. Arch Phys Med Rehabil 66:249–252

Mecomber SA, Herman RM (1971) Effects of local hypothermia on reflex and voluntary activity. Phys Ther 51:271–282

Miglietta O (1973) Action of cold on spasticity. Am J Phys Ther 52:198–205

Mills KR, Boniface SJ, Schubert M (1991) Origin of the secondary increase in firing probability of human motor neurons following transcranial magnetic stimulation. Brain 114:2451–2463

Odeen I (1981) Reduction of muscular hypertonus by long-term muscle stretch. Scand J Rehabil Med 13:93–99

Rood MS (1956) Neurophysiological mechanisms utilized in treatment of neuromuscular dysfunction. Am J Occup Ther 10:220–224

Wade DT (1992) Measurement in neurological rehabilitation. Oxford University Press, New York

Is Rehabilitation Effective in Spastic Syndromes?

V. Hömberg

Introduction

It is the inevitable fate of a neurologist in rehabilitation to write the closing chapter in books about spasticity or other major neurological problems. This is due to the fact that, after all, neurobiologists, neurophysiologists, and even acute neurologists, have failed to come up with a concept to fully understand the underlying crippling problem and hence fail to offer a definitive therapy. In this unfavorable situation the work of the physician in rehabilitation starts. There have been many definitions for "rehabilitation." Probably the most useful definition is to understand rehabilitation as an attempt to increase a patient's physical and psychological independence by means of a well-structured program.

There is a very simple answer to the question posed in the title of this chapter: yes! Neurological rehabilitation definitely works in spastic syndromes. In a patient with spinal cord injury today it is possible by means of a well-structured rehabilitation program to help her or him to reengage in family, vocational, and other social activities, much as before the accident. The situation is much more difficult in spastic syndromes, arising from lesions higher up the neuraxis, for example, after stroke, head injury, brain tumors, or encephalitis, because the resulting spastic disorders under these conditions are most often associated with severe problems in nonmotor domains such as cognition, perception, and memory.

A much more difficult question is to ask "Why does rehabilitation work?" or, to put it in other terms, "Is there anything specific with particular approaches in rehabilitation?" An answer to this question is difficult, and this difficulty has resulted for a long time in the fact that rehabilitation has not been looked upon as a "true" academic discipline. This is due partially to the fact that the program to be designed for a particular patient must often be extremely idiosyncratic or "custom-tailored" to fulfill all the necessary medical, pharmacological, social, vocational, recreational, and other

Direktor des NTC, Neurologisches Therapiecentrum, Heinrich-Heine-Universität, Hohensandweg 37, D-40591 Düsseldorf, Deutschland

Thilmann et al. (Eds.)
Spasticity: Mechanisms and Management

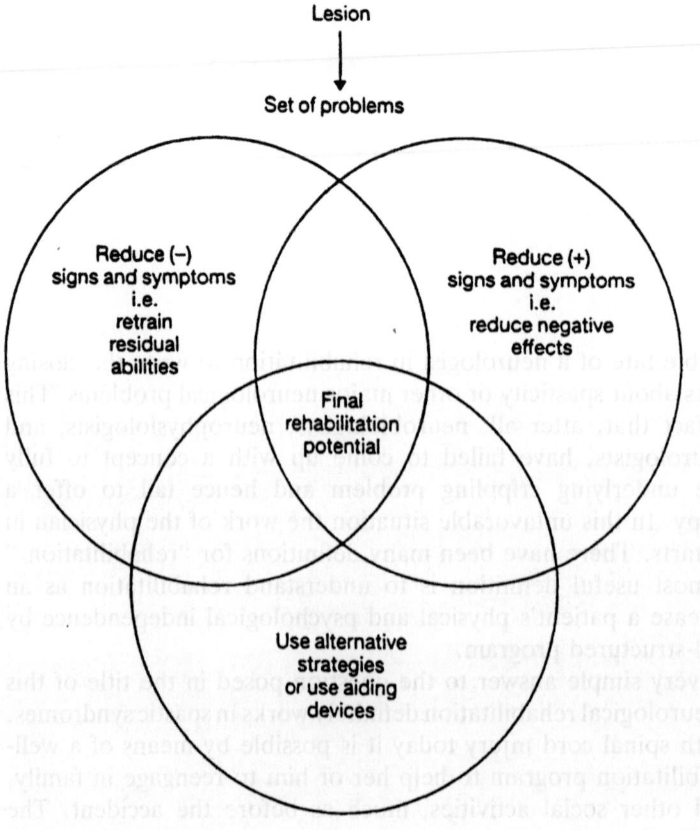

Fig. 1. Final rehabilitation potential obtainable in patients with spastic syndromes

needs of a patient in the rehabilitation process. This makes it difficult to design studies to prove the efficacy of one or the other rehabilitative technique by means of conventional group statistical designs. Fortunately over the past few years the introduction of single case statistical designs has improved this situation.

In this chapter I review some of the existing literature on nonpharmacological treatments of spasticity, including aspects of functional electrical stimulation (FES) and some aspects of more advanced neuroprosthetic technology.

As illustrated in Fig. 1, the final rehabilitation potential, obtainable in a patient with, for example, a spastic syndrome, depends on three main factors using the terminology originally introduced by Jackson, a patient with a spastic syndrome not only suffers from "plus" symptoms (e.g., spasticity, pain) but also from the typical "minus" symptoms of an upper motor neuron lesion, especially a lack of muscle strength, i.e., paresis. In the

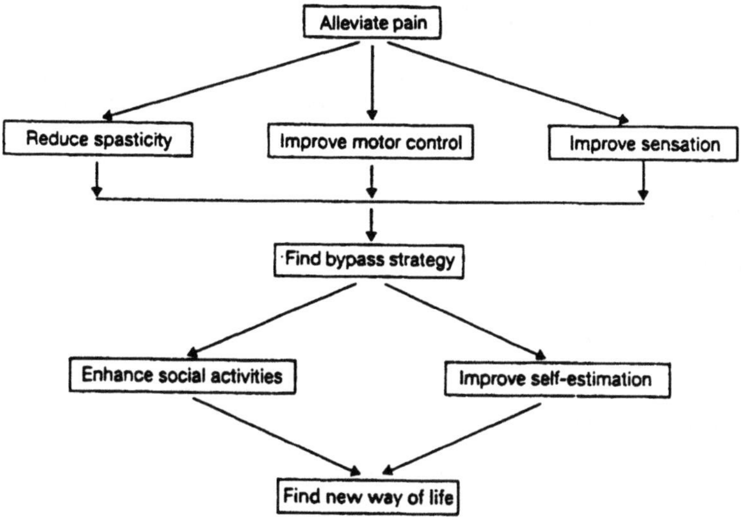

Fig. 2. Sequence of rehabilitation goals in patients with spastic syndromes

course of rehabilitation reduction in both "minus" and "plus" signs and symptoms is important. In addition, the patient learns alternative strategies to bypass the defective function or is trained to use aiding devices such as a crane or a complex neurotechnological device, for example, an environmental control system or a computerized communication tool. The final rehabilitation potential is critically dependent on the interplay of these three avenues.

Critical for the rehabilitation process is the definition of reasonable rehabilitation goals. Figure 2 gives a sequence of such rehabilitation goals for the example of a patient with a spastic hemiparesis and a painful shoulder on the affected side. The definition of such goals must always be integrated into the social context of the patient.

Table 1 gives an overview of some of the possible nonpharmacological treatment strategies in spastic syndromes, encompassing approaches to reduce undesirable "plus" symptoms, especially reduction in reflexes, and those seeking to bypass functional deficits and improve "minus" symptoms, including the cognitive relearning of motor abilities.

Cryotherapy, Vibration, and Topical Anesthesia

The local cooling of spastic limbs produces a transient decrease in muscle tension and reflex activity. The physiological basis for this effect has many facets including a decrease in nerve conduction velocity, a decrease in muscle contractility, and especially a decrease in the tuning characteristics

Table 1. Nonpharmacological ways of treating spastic syndromes

Treatment	Reduction of reflexes (+ symptoms)	Bypass function (− symptoms)	Cognitive relearning (− symptoms)
Cryotherapy/topical afferent blocking	×		
Transcutaneous or percutaneous electrical nerve stimulation	×		
Functional electrical stimulation	×	×	
EMG or kinesiological biofeedback	(×)		×
Physiotherapy	×	×	(×)
Occupational therapy (Cognitive motor therapy)	(×)	×	×

of muscle spindles. Ottoson (1965) has shown that the responsiveness of muscle spindles to stretch increases linearly with temperature in the range from 3° to about 30°C and then falls rapidly at higher temperatures. A study by Knutsson and Mattson (1969) has further shown that the decrease in sensitivity of muscle spindles to stretch is the main mediator of reduced reflex activity after cooling: the mean amplitude of tendon jerks was significantly reduced after cooling, whereas the H reflex responses, which bypass the muscle spindles, showed either an increase in amplitude immediately after cooling or insignificant variations in a later period.

The effect of cooling is usually transient, lasting 15–20 min. In our experience, the application of ice packs seems to reduce spasticity more efficiently than continuous streams of cold air given by jet stream systems, not least because the latter method may eventually elicit spasms by stimulating flexor reflex afferents, especially in patients with spinal spasticity.

Other strategies which have been employed to reduce reflex activity in spasticity include the use of vibratory muscle stimuli in man (Fisher et al. 1979; Eklund and Hagbarth 1965), which may help to decrease activity in antagonist muscles but tends to facilitate reflex activity in the stimulated muscle. Topical anesthesia or the use of percutaneous phenol blocks to motor points (Sabbahi and De Luca 1981; Garland et al. 1984) can also help to reduce spasticity in severe cases but at the expense of paralyzing the muscle for volitional use. The advantage is that it produces a long-lasting effect over weeks, months, and even years, especially after injection of phenol.

Transcutaneous or Percutaneous Electrical Nerve Stimulation

Electrical subcutaneous or transcutaneous nerve stimulation (TENS) is widely used to produce analgesic effects in patients with otherwise untreatable chronic pain states. Its precise mechanism of action is unknown, but the often wide distribution of its effect, irrespective of the site of nerve stimulation, is highly suggestive of a general, possibly humoral, mechanism, such as the release of endorphins in the central nervous system. Walker in 1982 demonstrated that subcutaneous stimulation of radial, median, and saphenous nerves helped in suppressing clonus in nine patients with spasticity. Similar to the widespread effects of TENS in pain control, he also observed a reduction in ankle clonus after contralateral leg stimulation as well as after stimulation of upper extremity nerves. Bajd et al. (1985) used TENS in six patients with spinal cord injuries and quantified the reduction of spasticity using a pendulum test. This test measures goniometrically the amount of knee motion after allowing the lower leg to fall from a horizontal position. In this study TENS was administered to nerves of the same spinal cord level as the spastic knee extensors. Noticeable decreases in spasticity were found in 50% of these patients. However, more elaborate studies using larger patient populations and looking more carefully at nonspecific effects and especially at the topical specificity of the site of TENS application are certainly needed before a definitive evaluation of this form of treatment in spasticity can be made. Our personal experience is that TENS, given at stimulation frequencies of 80–100 Hz with intensities just below pain threshold and administered either paravertebrally at the level of the spinal lesion in spinal patients or to the appropriate dermatomes associated with selected spastic muscle groups only occasionally reduces spasticity to a degree which can be measured quantitatively.

EMG Biofeedback

EMG biofeedback procedures have been developed from operant conditioning techniques. The rationale is to provide patients who have upper motoneuron lesions, and who have lost kinesiological feedback, with an auditory or visual display of EMG activity in the affected muscles to facilitate either activation or relaxation of those muscles or muscle groups. The efficacy of this form of treatment has been demonstrated in numerous studies (Brudny et al. 1974, 1976; Basmajian et al. 1977; Basmajian 1981). Studies comparing biofeedback treatment with conventional physiotherapy techniques have demonstrated the superiority of using a combination of both (Basmajian et al. 1975, 1982; Inglis et al. 1984). EMG feedback treatment seems to be most effective when used in severe cases at an early stage but can also be helpful in later stages when motor involvement is less severe. Most of these studies have used EMG recordings in only one muscle or in a group of synergistic muscles to achieve better volitional control of residual motor functions. The use of more elaborate microprocessor-assisted biofeedback systems providing improved display of the more complex signals or patterns will also make it possible to use biofeedback techniques to suppress undesired motor patterns, (e.g., co-contraction of antagonistic muscles in spinal spasticity) or to entrain more appropriate motor patterns for composite tasks in the upper extremity.

EMG feedback techniques have also been combined with electrical muscle stimulation in a way that, as soon as a certain EMG threshold is reached voluntarily by the patient, further activation is assisted by an electrical stimulator (Browman et al. 1979; Fields 1987). This technique further augments proprioception in temporal relation to the patient's attempt to move.

Physiotherapy

The use of a well-structured physiotherapy program is certainly the most effective way of retraining motor function in patients with upper motoneuron lesions and to reduce spasticity. It must be emphasized that, despite obvious rivalry between "schools" using their "specific" methods, a relative superiority of one method over another has never been demonstrated.

It is most reasonable to take a pragmatic approach, tailoring the treatment to the needs of an individual patient to achieve maximum benefit, concentrating on improvement of activities of daily living. Although many of the physiotherapy "schools" claim a neurophysiological background, there are few studies which have investigated the specific effect of physiotherapeutic techniques on, for example, reflex mechanisms (Etnyre and Abraham 1986). Only few studies have differentially assessed the effectiveness of therapeutic interventions. Wagenaar et al. (1990) compared the Brunnstrom and Bobath approaches in seven stroke patients, who received either treatment alternately with each intervention lasting 5 weeks for a total of four phases. The outcome measures taken weekly showed no significant difference in efficacy between the two methods of treatment. This study is especially remarkable as it used a single-case experimental design to cope with the inhomogeneous composition of a population of stroke patients. Several other studies have compared different physiotherapeutic approaches (Stern et al. 1970; Quin 1971; Logigian et al. 1983; Dickstein et al. 1986; Lord and Hall 1986; Jongbloed et al. 1989). The bottom line of all of these studies is that physiotherapy in general is effective in improving motor function and activities of daily living. There was, however, no significant difference between the groups treated with various approaches of treatment.

The neurophysiological framework of physiotherapeutic techniques is still generally defined in terms of inhibition or facilitation of reflexes and does not take into account the more elaborate concepts of modern motor neurophysiology. In this respect, it will in future be helpful to switch to more elaborate frameworks, exploiting also concepts of motor learning.

Functional Electrical Stimulation

The rationale for using FES in patients with upper motoneuron syndromes of cerebral or spinal origin is to try to use the normal neuromuscular system by exciting it with trains of electrical stimuli to bypass the lack of central activation of the appropriate motoneuron pools. Hence the main goal of FES is to provide a "neuroprosthetic" technique to restore motor function after disconnection from "higher motor centers." Today, FES can be used in a fairly effective manner to treat patients with central footdrop, as first proposed by Liberson in 1961, and to restore stance and sometimes even gait to patients with paraplegia, as first suggested by Kantrowitz in 1960.

In addition to its efficacy in reestablishing motor patterns (Merletti et al. 1978; Malezic et al. 1987), FES has also been shown to have effects in reducing spasticity, even when the electrical stimulation is used in a "non-functional way," for example, in a cyclic repetitive mode. Alfieri (1982) demonstrated in hemiplegic patients that FES selectively administered, not to the spastic but to the weakened antagonist muscles, is helpful in reducing spasticity in the nonstimulated spastic muscles. This effect lasted until about 60 min after the end of stimulation and was quantified by a scoring procedure (Ashworth scale). Furthermore, Vodovnik et al. (1984), administering the pendulum test, found a reduction in spasticity in five of seven patients with spinal cord injuries after using four-channel rhythmical FES of bilateral knee extensors and flexors over a period of 3 days. It is still not known whether it is more beneficial to stimulate the spastic muscle directly or to influence the spastic muscles by stimulating their antagonists, thus exploiting reciprocal inhibition.

The precise mechanism of action of FES in reducing spasticity is unknown. The stimulation of muscle spindle and tendon organ afferents is probably more likely to cause a reduction in spasticity than the stimulation of cutaneous afferents. This was corroborated by a study by Levin and Chapman (1987) in which clear inhibitory effects on the soleus H reflex after stimulation of the mixed common peroneal nerve were shown in contrast to H reflex facilitation after stimulation of the cutaneous superficial peroneal nerve both in normal subjects and in patients with spinal spasticity. It remains to be determined whether activation of agonist or antagonist afferents is more beneficial in reducing spasticity.

The routine use of FES as a neuroprosthetic tool for the restoration of stance and gait in patients with paraplegia has been precluded up to now by a wide range of clinical and technical problems, as summarized in Table 2.

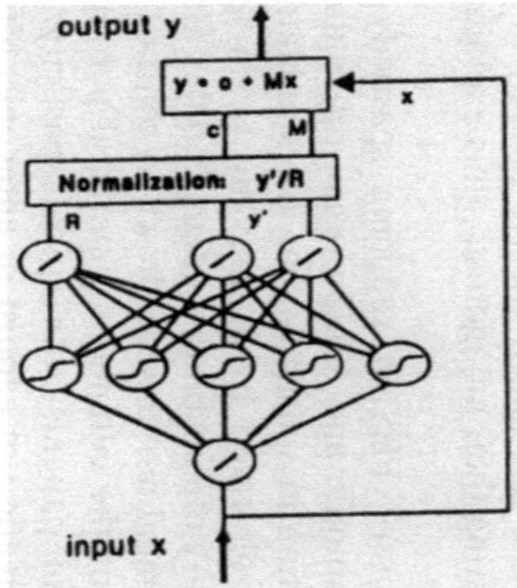

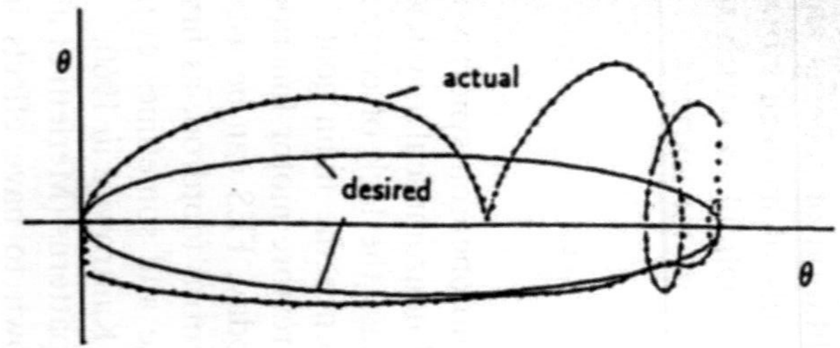

Result of P-controller application.
Sampled data are used to train a Neural Network the inverse dynamics.

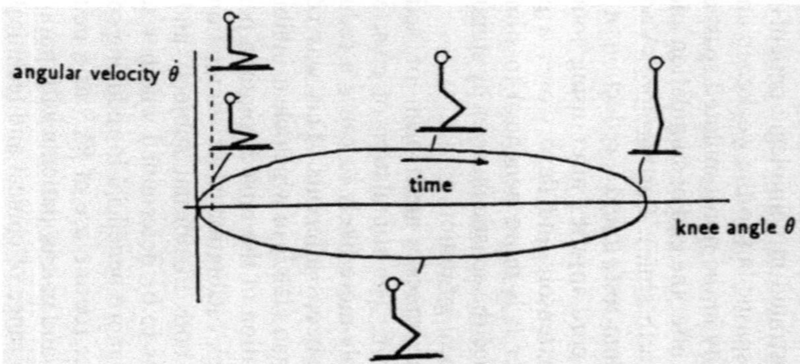

Computer simulations: Phase-space representation

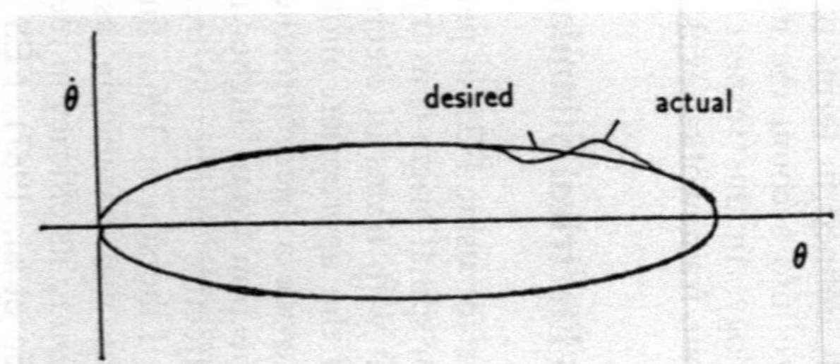

Application of the trained Neural Network.

Table 2. Problem ares in FES

Technical	*Clinical*

– Electrodes
  – Surface
  – Intramuscular
  – Epineural
  – Intramedullar
  – Intracerebellar
– Electrode-stimulator link
  – Transdermal sockets
  – Implanted with RF coupling
– Stimulator (Controller)
  – Open-loop programs
  – Shaped according to "natural" pattern
  – Closed-loop programs
– Sensors
  – Externally applicable
  – Implanted
	– Muscle wasting and fatigability
	– Contractures or other decrease in range of motion
	– Proper selection of patients
	– Time/benefit ratio
	– Motivation
	– Interaction with PT/OT
	– Long-term acceptance

The major clinical problem has remained the long-term acceptance of these neuroprotheses by the patient, generally because, unlike a cardiac pacemaker, the FES neuroprotheses are much more cumbersome to use as they most often are not fully implantable systems and require substantial time-consuming training activities on the patient's side. Another major problem so far has been that most of the FES attempts for stance and gait restauration have remained at the level of "open-loop" uncontrolled systems not being regulated by appropriate sensory feedback through a specially designed controller. For this purpose, in future the usage of artificial neural networks appears to be a promising solution for the integration of multiple sensor channels into multiple stimulation channels. These artificial neural networks are able to teach the optimal input-output relationships very rapidly (Beckmann et al. 1992). This is illustrated for a computer simulation study in Fig. 3.

Furthermore, the use of more advanced microtechnological contact probes may provide means for more effective spinal cord stimulation regulated by intercalated microelectronic networks, as illustrated in Fig. 4.

◄

Fig. 3. The use of artificial neural networks appears to be a promising solution for controllers for FES. A computer simulation of a body rising up and down is illustrated. It is evident that the desired space representation is difficult to achieve by a pure proportional controller but can easily be learned by a trained neural network

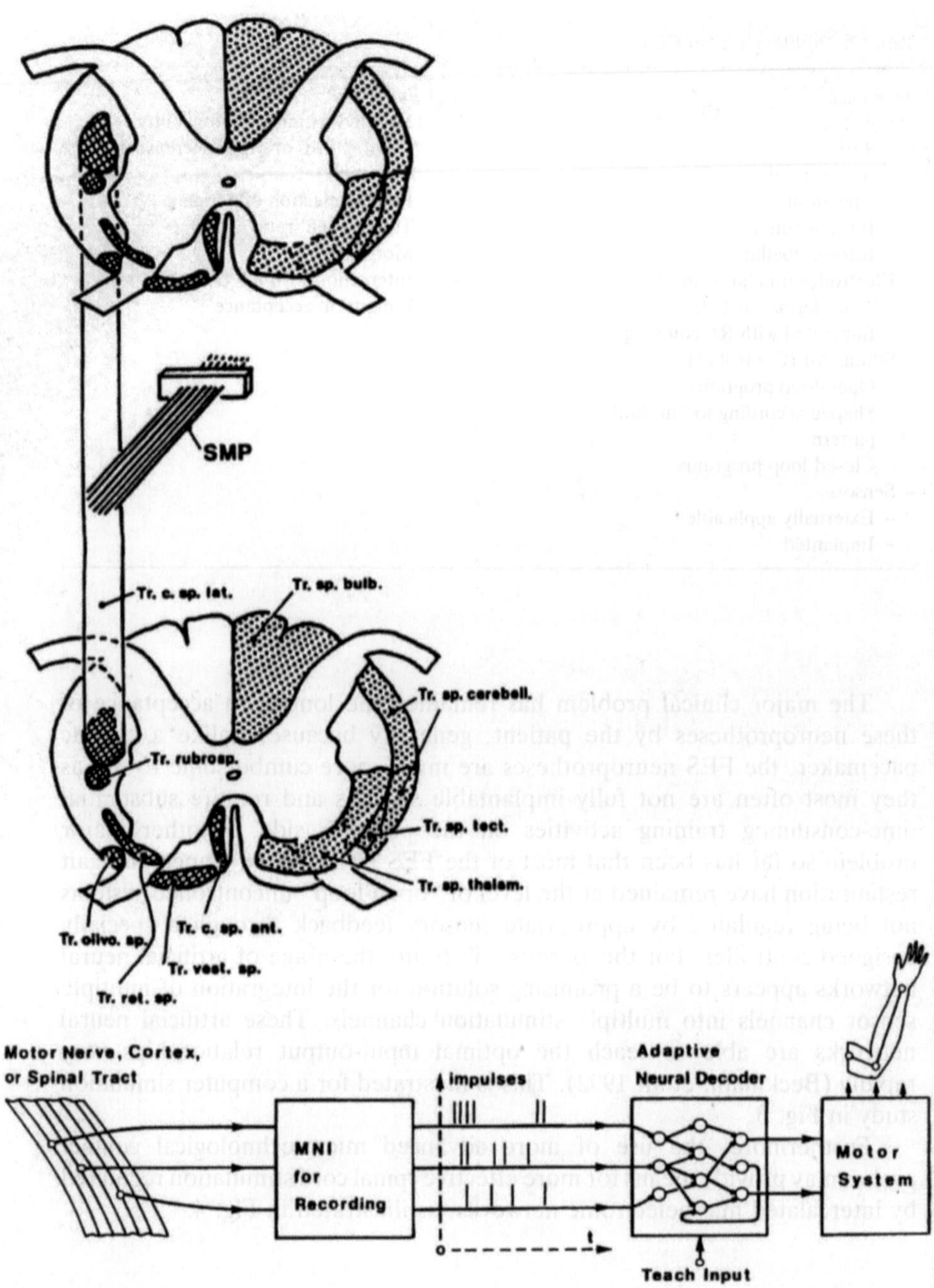

Fig. 4. Outlook for future developments in neuroprosthetic technology. By means of a surgical multicontact probe spinal tracts can be contacted directly. The signals taken up by this probe are processed in a multineuronal interface (*MNI*) and transferred into sequences of impulses which can be transformed by an adaptive neural decoder, also receiving teaching inputs into a final motor output signal, for example, by stimulation of motor nerves or muscles. (After Eckmiller 1993)

References

Alfieri V (1982) Electrical treatments of spasticity. Scand J Rehabil Med 14:177–182

Bajd T, Gregoric M, Vodovnik L, Benko H (1985) Electrical stimulation in treating spasticity resulting from spinal cord injury. Arch Phys Med Rehabil 66:515–517

Basmajian JV (1981) Biofeedback in rehabilitation: a review of principles and practices. Arch Phys Med Rehabil 62:469–475

Basmajian JV, Kukulka CG, Narayan MG, Takebe K (1975) Biofeedback treatment of footdrop after stroke compared with standard rehabilitation technique: effects on voluntary control and strength. Arch Phys Med Rehabil 56:231–236

Basmajian JV, Regenos EM, Baker MP (1977) Rehabilitating stroke patients with biofeedback. Geriatrics July I977:85–88

Basmajian JV, Gowland C, Brandstater ME, Swanson L, Trotter J (1982) EMG feedback treatment of upper limb in hemiplegic stroke patients: a pilot study. Arch Phys Med Rehabil 63:613–616

Beckmann J, Daunicht WJ, Hömberg V (1992) Control of a paraplegic patient model by neuroprothetic networks. In: Aleksander I, Taylor J (eds) Artificial neural networks 2. Elsevier, Amsterdam, pp 471–474

Browman BR, Baker LL, Waters RL (1979) Positional feedback and electrical stimulation: an automated treatment for the hemiplegic wrist. Arch Phys Med Rehabil 60:497–502

Brudny J, Korein J, Levidow L, Grynbaum BB, Lieberman A, Friedmann LW (1974) Sensory feedback therapy as a modality of treatment in central nervous system disorders of voluntary movement. Neurology (Minncap) 24:925–932

Brudny J, Korein J, Grynbaum BB, Friedmann LW, Weinstein S, Sachs-Frankel G, Belandres PV (1976) EMG feedback therapy: review of treatment of 114 patients. Arch Phys Med Rehadie 57:55–61

Dickstein R, Hocherman S, Pillar T, Schaham R (1986) Stroke rehabilitation: three exercise therapy approaches. Phys Ther 66(8):1233–1238

Eckmiller R (1993) Concerning the challenge of neurotechnology. In: Bothe HW, Samii R, Eckmiller R (eds) Neurobionics, North Holland, Amsterdam

Eklund G, Hagbarth KE (1965) Motor effects of vibratory muscle stimuli in man. Electroencephalogr Clin Neurophysiol 19:613–620

Etnyre BR, Abraham LD (1986) H-reflex changes during static stretching and two variations of proprioceptive neuromuscular facilitation techniques. Electroencephalogr Clin Neurophysiol 63:174–179

Fields RW (1987) Electromyographically triggered electric muscle stimulation for chronic hemiplegia. Arch Phys Med Rehabil 68:407–414

Fisher MA, Shahani BT, Young RR (1979) Electrophysiologic analysis of the motor system after stroke: the "suppressive" effect of vibration. Arch Phys Med Rehabil 60:11–14

Garland DE, Lilling M, Keenan MA (1984) Percutaneous phenol blocks to motor points of spastic forearm muscles in head-injured adults. Arch Phys Med Rehabil 65:243–245

Inglis J, Donald MW, Monga TN, Sproule M, Young MJ (1984) Electromyographic biofeedback and physical therapy of the hemiplegic upper limb. Arch Phys Med Rehab 65:755–759

Jongbloed L, Stacey S, Brighton C (1989) Stroke rehabilitation: sensorimotor integrative treatment versus functional treatment. Am J Occup Ther 43:391–397

Kantrowitz A (1960) Electronic physiologic aids. Report of the Maimonides Hospital. Maimonides Hospital, Brooklyn, New York, pp 4–5

Knutsson E, Mattson E (1969) Effects of local cooling on monosynaptic reflexes in man. Scand J Rehabil Med 1:126–132

Levin M, Chapman CE (1987) Inhibitory and facilitatory effects from the peroneal nerve onto the soleus H-reflex in normal and spinal man. Electroencephalogr Clin Neurophysiol 67:468–478

Logigian MK, Samuels MA, Falconer J, Zagar R (1983) Clinical exercise trial for stroke patients. Arch Phys Med Rehabil 64:364–367

Lord JP, Hall K (1986) Neuromuscular reeduction versus traditional programs for stroke rehabilitation. Arch Phys Med Rehabil 67:89–91

Malezic M, Klajic M, Acimovic-Janezic R, Gros N, Kranjnik J, Stanic U (1987) Therapeutic effects of multisite electric stimulation of gait in motor-disabled patients. Arch Phys Med Rehabil 68:553–560

Merletti R, Zelaschi F, Latella D, Galli M, Angeli S, Sessa MB (1978) A control study of muscle force recovery in hemiparetic patients during treatment with functional electrical stimulation. Scand J Rehabil Med 10:147–154

Ottoson D (1965) The effects of temperature on the isolated muscle spindle. J Physiol (Lond) 180:636–48

Quin CE (1971) Observations on the effects of proprioceptive neuromuscular facilitation techniques in the treatment of hemiplegia. Rheum Phys Med 11:186–192

Sabbahi MA, De Luca CJ (1981) Topical anesthesia: a possible treatment method for spasticity. Arch Phys Med Rehabil 62:310–314

Stern PH, McDowell F, Miller JM, Robinson M (1970) Effects of facilitation exercise techniques in stroke rehabilitation. Arch Rhys Med Rehabil 5:526–531

Vodovnik L, Bowman BR, Hufford P (1984) Effects of electrical stimulation on spinal spasticity. Scand J Rehabil Med 16:29–34

Wagenaar RC, Meijer OG, Van Wieringen PCW et al. (1990) The functional recovery of stroke: a comparison between neuro-developmental treatment and the Brunnstrom method. Scan J Rehabil Med 22:1–8

Walker JB (1982) Modulation of spasticity: prolonged suppression of a spinal reflex by electrical stimulation. Science 216:203–204

Discussion Summary

B. Conrad

The discussion of the presentation of Barbeau concentrated on a few points. He was asked why only those patients were evaluated on their antispastic drug effects who were wheelchair bound and not those who were still able to walk. Barbeau answered that so far patients who had certain locomotive capabilities did not seem to benefit much from this medication.

He was also asked why in some of his medicated patients the washout period was not the same as the baseline period, making it difficult to use these results in a cross-over study. Barbeau confirmed this observation and explained it as a problem of the study design.

Various questions centered around the new drug cyproheptadine, on its specific mechanism as well as on its potential as an antispastic agent. Barbeau pointed out that at the moment it is too early to propose any specific mechanism. The 5HT blocker cyproheptadine is nonspecific and has cholinergic effects. Cyproheptadine is likely to act on 5HT2 receptors. The drug, which can reduce spasticity and enhance locomotor function in spinal cord injury patients, is so far not available on the market.

A great deal of the discussion of the paper of Mauritz centered on how physiotherapy modifies motor behavior – whether it can modify the genuine circuitry of the central nervous mechanism or whether it means to train the patient only to use the damaged circuitry in a more appropriate way, and how to sort these two possibilities out. Mauritz stated that up to now this problem has not been solved and it seems very difficult from a methodological point of view to separate the two possibilities. These question can finally be solved only in animal experiments with a battery of new techniques.

Some discussants expressed their convictions that there might be some good reason to think that motor patterns of use will influence the way the system is organized and this is why physiotherapy should help, but so far it has not been proved decisively. In the context of these problems there was also a discussion of whether any of the current physiotherapeutic concepts, such as the Bobath method or the Brunström technique, do anything specific for the patient (like recovery of function). Up to now no final answers can be given to these questions.

Neurologische Klinik der Technischen Universität München, Möhlstr. 28, D-81675 München, Deutschland

Thilmann et al. (Eds.)
Spasticity: Mechanisms and Management

It was also asked if there has been any recent correlation between improvement through physiotherapy and concomitant changes of objective parameters (such as concentration of noradrenaline in the CSF). Mauritz knew nothing of the existence of such studies. He concluded that many of the physiotherapeutic approaches are completely empirical and thus far we do not have a good rationale and understanding of the possible underlying mechanisms of partial recovery. He expressed his belief that we are now on the verge of better implementing our fundamental knowledge and are better prepared to validate certain approaches.

Hoemberg was asked how far it is possible (at least in animals) to use afferent inputs of the organism itself. Hoemberg stated that this would make it necessary to use very delicate probes. He believes that the micromechanical techniques which might be available in the near future may make this possible. Concerning orthotic hyperdevices and the future usefulness of combining high-tech sensors with a complicated computation, Hoemberg pointed to the major problem of acceptance by the patients. After a while patients lose motivation and are not willing to have a number of orthotic devices on them.

Hoemberg was finally asked what he thinks we really gain in proposing neuronal networks to control motor behavior as long as we have not even solved simple open-loop devices. Hoemberg underlined that one has to invest on a longer run in the development of closed-loop systems because everything else will not be satisfactory. But so far it is completely "experimental medicine"; nothing is readily available, and up to now one has to warn patients that the systems might have inherent risks and make their conditions even worse.

In his final statement, Hallett discussed a few important problems. He asked what we want to treat when we are treating spastic patients. He stressed two different aspects related to the general topic of the symposium: the functional aspect of the patients on the one hand (can they walk, can they perform their daily activities?), and the aspect of increased tone that the patients have, and the treatment of that aspect on the other. There was a consensus that we clearly want to improve function. But from a simple point of view the fundamental question remains: What controls the alpha motoneuron pool that is actually controlling the muscle? How should we think of it in terms of cortical control and reflex control of this pool? They are to a certain extent inversely related. With an increasing extent of the lesion of descending pathways, reflex control seems to go up while cortical control goes down.

Maximal reflex control and minimal cortical control clearly result in the most severe forms of spasticity with painful spasms. These cases probably represent the best therapeutic model for dealing with increased tone directly, because there is little interference with function in most circumstances. The discussion of the question (do we want to treat tone directly? Is there any reason for directly trying to treat tone and attacking spasticity?) also made it clear that it will be very difficult in the majority of cases to separate "tone" and "lost functions" since they are in a sense part of the same defect and not completely different things.

It was proposed to rephrase the question: how far can therapy promote a normal special temporal pattern of inhibition and therefore mask unwanted muscle activity and promote desired muscle activity? This leads back to the problem of an inverse relation between cortical and reflex control: the less cortical control, the greater the lack of selective patterns of inhibition with unwanted muscle activity. But there are also well-known conditions where patients benefit from increased tone, mainly in locomotion and standing, even though cortical control is greatly deteriorated. Increased tone always means reduction in muscle tone modulation and decreased function.

A final aspect of this symposium was briefly discussed: whether it is worth doing tests to assess muscle tone and how to properly assess muscle tone, particularly for therapeutic trials. There was no final consensus on what to use as the best indicators, even though various functional scales and also neurophysiological techniques are clearly important.

Subject Index

Springer-Verlag and the Environment

We at Springer-Verlag firmly believe that an international science publisher has a special obligation to the environment, and our corporate policies consistently reflect this conviction.

We also expect our business partners – paper mills, printers, packaging manufacturers, etc. – to commit themselves to using environmentally friendly materials and production processes.

The paper in this book is made from low- or no-chlorine pulp and is acid free, in conformance with international standards for paper permanency.

Printed by Libri Plureos GmbH
in Hamburg, Germany

MIX
Papier aus verantwortungsvollen Quellen
Paper from responsible sources
FSC® C105338

If you have any concerns about our products,
you can contact us on
ProductSafety@springernature.com

In case Publisher is established outside the EU,
the EU authorized representative is:
Springer Nature Customer Service Center GmbH
Europaplatz 3, 69115 Heidelberg, Germany

Printed by Libri Plureos GmbH
in Hamburg, Germany